of the Elements

NONMETALS

IB	IIB	IIIA	IVA	VA	VIA	VIIA	O
						1.0079 **H** 1	4.00260 **He** 2
		10.81 **B** 5	12.011 **C** 6	14.0067 **N** 7	15.9994 **O** 8	18.9984 **F** 9	20.179 **Ne** 10
		26.9815 **Al** 13	28.086 **Si** 14	30.9738 **P** 15	32.06 **S** 16	35.453 **Cl** 17	39.948 **Ar** 18
63.546 **Cu** 29	65.38 **Zn** 30	69.72 **Ga** 31	72.59 **Ge** 32	74 9216 **As** 33	78.96 **Se** 34	79.904 **Br** 35	83.80 **Kr** 36
107.868 **Ag** 47	112.40 **Cd** 48	114.82 **In** 49	118.69 **Sn** 50	121.75 **Sb** 51	127.60 **Te** 52	126.9046 **I** 53	131.30 **Xe** 54
196.9665 **Au** 79	200.59 **Hg** 80	204.37 **Tl** 81	207.2 **Pb** 82	208.9804 **Bi** 83	(210) **Po** 84	(210) **At** 85	(222) **Rn** 86

157.25 **Gd** 64	158.9254 **Tb** 65	162.50 **Dy** 66	164.9304 **Ho** 67	167.26 **Er** 68	168.9342 **Tm** 69	173.04 **Yb** 70	174.97 **Lu** 71
(245) **Cm** 96	(245) **Bk** 97	(248) **Cf** 98	(253) **Es** 99	(254) **Fm** 100	(256) **Md** 101	(253) **No** 102	(257) **Lr** 103

Techniques and Experiments for Organic Chemistry

Techniques and Experiments for Organic Chemistry

FIFTH EDITION

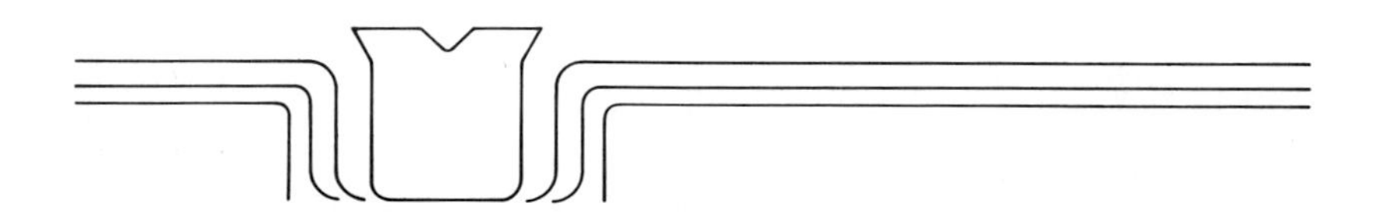

Addison Ault
Cornell College

PRENTICE HALL, Englewood Cliffs, New Jersey 07632

Editorial-Production Service: Lifland et al., Bookmakers
Cover Administrator: Linda Dickinson

Library of Congress Cataloging-in-Publication Data

Ault, Addison.
Techniques and experiments for organic chemistry.

Includes bibliographical references and indexes.
1. Chemistry, Organic—Laboratory manuals.
I. Title.
QD261.A94 1986 547'.0078 86-1152
ISBN 0-205-08752-3

A Simon & Schuster Company
Englewood Cliffs, New Jersey 07632

Printed in the United States of America

10 9 8 7 6 5 4 3 2 1

ISBN 0-205-08752-3

Prentice-Hall International (UK) Limited, *London*
Prentice-Hall of Australia Pty. Limited, *Sydney*
Prentice-Hall Canada Inc., *Toronto*
Prentice-Hall Hispanoamericana, S.A., *Mexico*
Prentice-Hall of India Private Limited, *New Delhi*
Prentice-Hall of Japan, Inc., *Tokyo*
Simon & Schuster Asia Pte. Ltd., *Singapore*
Editora Prentice-Hall do Brasil, Ltda., *Rio de Janeiro*

Contents

Preface xv

PART 1 LABORATORY OPERATIONS

Preliminary Topics 3

1 **Safety** 3
- 1.1 Fire 4
- 1.2 Explosions 5
- 1.3 Poisoning 6
- 1.4 Cuts 6
- 1.5 Spills 7
- 1.6 Chemicals in the Eye 7
- 1.7 A Short List of Hazardous Materials and Some of Their Properties 8

2 **Glassware Used in the Organic Chemistry Laboratory** 14

3 **Cleaning Up** 19
- 3.1 Care of Ground-Glass-Jointed Glassware 20
- 3.2 Separatory Funnels and Glassware with Stopcocks 21
- 3.3 Drying of Glassware 21
- 3.4 Disposal of Waste 21

4 **The Laboratory Notebook** 22

5 **The Chemical Literature** 26
- 5.1 *Chemical Abstracts* 27
- 5.2 Secondary Sources for Physical Properties of Organic Compounds 29
- 5.3 Secondary Sources for Methods of Preparation of Organic Compounds 33
- 5.4 Collections of Spectra 35
- 5.5 Miscellaneous 35

6 Tables 37
6.1 Solutions of Acids 37
6.2 Molecular Weights and Molar Volumes of Acids 37
6.3 Molecular Weights of Bases 37
6.4 Molecular Weights, Densities, and Molar Volumes of Selected Liquid Reagents 38
6.5 Solutions of Bases 38
6.6 Periodic Table of the Elements 39

Separation of Substances; Purification of Substances 41

7 Filtration 43
7.1 Gravity Filtration 44
7.2 Vacuum or Suction Filtration 44

8 Recrystallization 44
8.1 Choice of Solvent 45
8.2 Dissolving the Sample 50
8.3 Decolorizing the Solution 50
8.4 Hot Filtration 51
8.5 Cooling 52
8.6 Cold Filtration 54
8.7 Washing the Crystals 54
8.8 Drying the Crystals 55
8.9 More Techniques of Crystallization 55
Exercises 61

9 Distillation 62
9.1 Vapor Pressure 63
9.2 Distillation of a Pure Liquid 66
9.3 Miscible Pairs of Liquids 66
9.4 Fractional Distillation 71
9.5 Azeotropic Mixtures 74
9.6 Technique of Distillation 80
Exercises 85

10 Reduced-Pressure Distillation 86
10.1 Estimation of the Boiling Point at Reduced Pressure 86
10.2 Apparatus 87
10.3 Source of Vacuum 91
10.4 Pressure Measurement 91
10.5 Technique of Distillation under Reduced Pressure 92
Exercises 94

11 Distillation of Mixtures of Two Immiscible Liquids; Steam Distillation 94
11.1 Theory of Steam Distillation 95
11.2 Technique of Steam Distillation 98
Exercises 101

12 Sublimation 101
12.1 Theory of Sublimation 101
12.2 Technique of Sublimation 104
Exercises 105

13 Extraction by Solvents 105
13.1 Theory of Extraction 105
13.2 Extraction of Acids and Bases 106
13.3 Technique of Extraction 108
Exercises 115

14 Chromatography 120
14.1 Theory of Column Chromatography 120
14.2 Technique of Column Chromatography 122
Exercises 126
14.3 Theory of Thin-Layer Chromatography 127
14.4 Technique of Thin-Layer Chromatography 128
Exercises 130
14.5 Theory of Paper Chromatography 130
14.6 Technique of Paper Chromatography 131
Exercises 132
14.7 Theory of Vapor-Phase Chromatography 133
14.8 Technique of Vapor-Phase Chromatography 137
Exercises 139
14.9 High-Pressure Liquid Chromatography 141
14.10 Batchwise Adsorption; Decolorization 141

15 Removal of Water; Drying 142
15.1 Drying of Solids 142
15.2 Drying of Solutions 146
15.3 Drying of Solvents and Liquid Reagents 148
15.4 Drying of Gases 149

Determination of Physical Properties 151

16 Boiling Point 152
16.1 Experimental Determination of Boiling Point 152
16.2 Boiling Point and Molecular Structure 154
16.3 Boiling Point and the Enthalpy and Entropy of Vaporization 156

17 Melting Point 162
17.1 Experimental Determination of the Melting Point 163
17.2 The Melting Point as a Criterion of Purity 167
17.3 The Melting Point as a Means of Identification and Characterization 169
17.4 Mixture Melting Points 169
17.5 Melting Point and Molecular Structure 170

18 Density; Specific Gravity 173
18.1 Experimental Determination of the Density 173
18.2 Density and Molecular Structure 175

19 Index of Refraction 175
19.1 Experimental Determination of the Index of Refraction 175
19.2 Index of Refraction and Molecular Structure 176

20 Optical Activity 177
20.1 Experimental Determination of Optical Rotation 178
20.2 Optical Activity and Molecular Structure 180

21 Molecular Weight 182
21.1 Molecular Weight Determination by Means of Mass Spectrometry 182
21.2 Molecular Weight Determination by Other Methods 183

22 Solubility 183
22.1 Solubility of Liquids in Liquids 184
22.2 Solubility of Solids in Liquids 188
22.3 Classification of Compounds by Solubility: Relationships between Solubility and Molecular Structure 189
22.4 Techniques for Determination of Solubility 193

23 Infrared Absorption Spectrometry 195
23.1 Wavelength, Frequency, and Energy of Electromagnetic Radiation 195
23.2 Units of Light Absorption 196
23.3 Infrared Light Absorption and Molecular Structure 197
23.4 Interpretation of Infrared Spectra 198
23.5 Sample Preparation 203

24 Ultraviolet-Visible Absorption Spectrometry 211
24.1 Ultraviolet-Visible Light Absorption and Molecular Structure 211
24.2 Interpretation of UV-Visible Spectra 213
24.3 Color and Molecular Structure 214
24.4 Sample Preparation 215

25 Nuclear Magnetic Resonance Spectrometry 218
25.1 Shielding; Chemical Shift 218
25.2 Splitting 221
25.3 The Integral 228
25.4 Nuclear Magnetic Resonance and Molecular Structure 228
25.5 Interpretation of NMR Spectra 230
25.6 Sample Preparation 234

26 Mass Spectrometry 238
26.1 Theory of Mass Spectrometry 238
26.2 Interpretation of Mass Spectra 240
26.3 High-Resolution Mass Spectrometry 242

Determination of Chemical Properties; Qualitative Organic Analysis 245

27 Qualitative Tests for the Elements 246
27.1 Ignition Test; Test for Metals 246
27.2 Beilstein Test; Test for Halogens (Except Fluorine) 247
27.3 Sodium Fusion Test; Test for Nitrogen, Sulfur, and the Halogens 247

28 Qualitative Characterization Tests: Tests for the Functional Groups 251
28.1 Detection of Ammonia from Ammonium Salts, Primary Amides, and Nitriles 253
28.2 Benzenesulfonyl Chloride (Hinsberg's Test) 254
28.3 Bromine in Carbon Tetrachloride 255
28.4 Chromic Anhydride 256
28.5 2,4-Dinitrophenylhydrazine 256
28.6 Ferric Chloride Solution 257
28.7 Ferric Hydroxamate Test 257
28.8 Hydrochloric Acid/Zinc Chloride Test (Lucas's Test) 259
28.9 Iodoform Test 260
28.10 Aqueous Potassium Permanganate Solution (Baeyer's Test) 260
28.11 Alcoholic Silver Nitrate Solution 261
28.12 Sodium Hydroxide Test 262
28.13 Sodium Iodide in Acetone 263
28.14 Tollens Reagent: Silver-Ammonia Complex Ion 264

29 Characterization Through Formation of Derivatives 265
29.1 Benzoates, *p*-Nitrobenzoates, and 3,5-Dinitrobenzoates of Alcohols 271
29.2 Hydrogen 3-Nitrophthalates of Alcohols 271
29.3 Phenyl- and α-Naphthylurethans 272
29.4 Methone Derivatives of Aldehydes 272
29.5 2,4-Dinitrophenylhydrazones 273
29.6 Semicarbazones 273
29.7 Oximes 274
29.8 Carboxylic Acids by Hydrolysis of Primary Amides and Nitriles 274
29.9 9-Acylamidoxanthenes from Amides 276
29.10 Hydrolysis of N-Substituted Amides 276
29.11 Substituted Acetamides from Amines 277
29.12 Substituted Benzamides from Amines 278
29.13 *p*-Toluenesulfonamides from Amines 278
29.14 Phenylthioureas and α-Naphthylthioureas 279
29.15 Picrates 280
29.16 Quaternary Ammonium Salts: Methiodides and *p*-Toluenesulfonates 281
29.17 Carboxylic Acid Amides 281
29.18 Anilides, *p*-Toluidides, and *p*-Bromoanilides of Carboxylic Acids 282
29.19 Phenacyl and Substituted Phenacyl Esters of Carboxylic Acids 283
29.20 *p*-Nitrobenzyl Esters of Carboxylic Acids 284
29.21 N-Benzylamides from Esters 285
29.22 3,5-Dinitrobenzoates from Esters 285
29.23 Hydrolysis of Esters 286
29.24 Bromination of Aromatic Ethers 287
29.25 S-Alkylthiuronium Picrates 287
29.26 *o*-Aroylbenzoic Acids from Aromatic Hydrocarbons 288
29.27 Aromatic Acids by Oxidation by Permanganate 289
29.28 Anilides, *p*-Toluidides, and α-Naphthalides from Alkyl Halides 290
29.29 2,4,7-Trinitrofluorenone Adducts of Aromatic Hydrocarbons 290

29.30 Bromination of Phenols 291
29.31 Aryloxyacetic Acids from Phenols 291

Apparatus and Techniques for Chemical Reactions 293

30 Assembling the Apparatus 293

31 Temperature Control 294

32 Methods of Heating and Cooling 297
32.1 Heating 297
32.2 Cooling 301

33 Stirring 302

34 Addition of Reagents 304
34.1 Addition of Solids 304
34.2 Addition of Liquids and Solutions 306
34.3 Addition of Gases 306

35 Control of Evolved Gases 307

36 Concentration; Evaporation 307

37 Use of an Inert Atmosphere 310

38 Working Up the Reaction; Isolation of the Product 311

PART 2 EXPERIMENTS

Separations 321

39 Isolation of Cholesterol from Gallstones 321

40 Isolation of Lactose from Powdered Milk 323

41 Isolation of Acetylsalicylic Acid from Aspirin Tablets 325

42 Isolation of Caffeine from Tea and NoDoz 325

43 Isolation of Piperine from Black Pepper 328

44 Isolation of Clove Oil from Cloves 329

45 Isolation of Eugenol from Clove Oil 331

46 Isolation of (R)-(+)-Limonene from Grapefruit or Orange Peel 332

47 Isolation of (R)-(−)- or (S)-(+)-Carvone from Oil of Spearmint or Oil of Caraway 334

48 Resolution of (R),(S)-α-Phenylethylamine with (R),(R)-(+)-Tartaric Acid 339

Transformations 345

49 Isomerizations 345
49.1 Adamantane from *endo*-Tetrahydrodicyclopentadiene via the Thiourea Clathrate 346
49.2 *cis*-1,2-Dibenzoylethylene from *trans*-1,2-Dibenzoylethylene 348

50 Alkenes from Alcohols 351
50.1 Cyclohexene from Cyclohexanol 352
50.2 A Variation: The Dehydration of 2-Methylcyclohexanol 356

51 Cyclohexyl Bromide from Cyclohexanol 358

52 Cyclohexanol from Cyclohexene 360

53 Addition of Dichlorocarbene to Alkenes by Phase Transfer Catalysis 361
53.1 Addition of Dichlorocarbene to Cyclohexene 363
53.2 A Variation: Addition of Dichlorocarbene to Styrene 365
53.3 Another Variation: Addition of Dichlorocarbene to 1,5-Cyclooctadiene 365

54 Cyclohexanone from Cyclohexanol 366

55 Reduction of Ketones to Secondary Alcohols 369
55.1 Cyclohexanol from Cyclohexanone 369
55.2 A Variation: *cis*- and *trans*-4-*tert*-Butylcyclohexanol from 4-*tert*-Butylcyclohexanone 370
55.3 Vanillyl Alcohol from Vanillin 371

56 Alkyl Halides from Alcohols 372
56.1 *n*-Butyl Bromide from *n*-Butyl Alcohol 372
56.2 A Variation: Isoamyl Bromide from Isoamyl Alcohol 373
56.3 *tert*-Butyl Chloride from *tert*-Butyl Alcohol 373
56.4 A Variation: *tert*-Amyl Chloride from *tert*-Amyl Alcohol 375
56.5 Competitive Nucleophilic Substitution of Butyl Alcohols by Bromide and Chloride Ion 375

57 Kinetics of the Hydrolysis of *tert*-Butyl Chloride 378

58 Isoamyl Acetate: A Component of the Alarm Pheromone of the Honey Bee 384
58.1 Isoamyl Acetate from Isoamyl Alcohol and Acetic Acid; Fischer Esterification 385
58.2 Isoamyl Acetate from Isoamyl Bromide and Potassium Acetate 387

59 Preparation of Esters 388
59.1 Cholesteryl Benzoate from Cholesterol; Liquid Crystals 389
59.2 Methyl Benzoate 392
59.3 Methyl Salicylate; Oil of Wintergreen 394
59.4 Acetylsalicylic Acid; Aspirin 395
59.5 α- or β-D-Glucose Pentaacetate from D-Glucose 397
59.6 A Variation: Acetylation of Glucose in N-Methylimidazole 400

60 The Grignard Reaction: Preparation of Aliphatic Alcohols 401
60.1 Aliphatic Alcohols 402

60.2 A Variation: Synthesis of Pheromones: 4-Methyl-3-heptanol and 4-Methyl-3-heptanone 406

61 The Grignard Reaction: Preparation of Triphenylmethanol 410
61.1 Triphenylmethanol from Phenylmagnesium Bromide and Methyl Benzoate 411
61.2 A Variation: Triphenylmethanol from Phenylmagnesium Bromide and Benzophenone 413
61.3 Another Variation: Triphenylmethanol from Phenylmagnesium Bromide and Dimethyl Carbonate 413

62 Aniline from Nitrobenzene 414

63 Preparation of Amides 415
63.1 Acetanilide from Aniline 416
63.2 *p*-Ethoxyacetanilide from *p*-Phenetidine 417
63.3 N,N-Diethyl-*m*-toluamide from *m*-Toluic Acid; "Off" 419

64 Electrophilic Substitution Reactions of Benzene Derivatives 421
64.1 Methyl *m*-Nitrobenzoate from Methyl Benzoate 422
64.2 *p*-Bromoacetanilide from Acetanilide 423
64.3 2,4-Dinitrobromobenzene from Bromobenzene 424

65 Nucleophilic Aromatic Substitution Reactions of 2,4-Dinitrobromobenzene 426
65.1 2,4-Dinitroaniline 426
65.2 2,4-Dinitrophenylhydrazine 427
65.3 2,4-Dinitrodiphenylamine 427
65.4 2,4-Dinitrophenylpiperidine 428
65.5 A Variation: 4′-Substituted 2,4-Dinitrophenylanilines 428

66 Diazonium Salts of Aromatic Amines 429
66.1 Benzenediazonium Chloride from Aniline 430
66.2 *p*-Toluenediazonium Chloride from *p*-Toluidine 430
66.3 *p*-Nitrobenzenediazonium Sulfate from *p*-Nitroaniline 431

67 Replacement Reactions of Diazonium Salts 431
67.1 Chlorobenzene from Benzenediazonium Chloride 432
67.2 A Variation: *p*-Chlorotoluene from *p*-Toluenediazonium Chloride 433
67.3 Another Variation: *o*-Chlorotoluene from *o*-Toluenediazonium Chloride 434

68 Electrophilic Aromatic Substitution by Diazonium Ions; Coupling Reactions of Diazonium Salts 434
68.1 Benzenediazonium Chloride and β-Naphthol: 1-Phenylazo-2-naphthol (Sudan 1) 435
68.2 *p*-Nitrobenzenediazonium Sulfate and Phenol: *p*-(4-Nitrobenzeneazo)-phenol 435
68.3 *p*-Nitrobenzenediazonium Sulfate and β-Naphthol: 1-(*p*-Nitrophenylazo)-2-naphthol (Para Red; American Flag Red) 436
68.4 *p*-Nitrobenzenediazonium Sulfate and Dimethylaniline: *p*-(4-Nitrobenzeneazo)-dimethylaniline 436

69 The Diels-Alder Reaction 437
- 69.1 Butadiene (from 3-Sulfolene) and Maleic Anhydride 437
- 69.2 Cyclopentadiene and Maleic Anhydride 438
- 69.3 Furan and Maleic Anhydride 441

70 Aldol Condensations and Related Reactions 441
- 70.1 Self-Condensation of Propionaldehyde: 2-Methyl-2-pentenal 443
- 70.2 Mesityl Oxide and Diethyl Malonate: 5,5-Dimethyl-1,3-cyclohexanedione; Dimedon; Methone 445

71 The Wittig Reaction 447
- 71.1 The Preparation of *trans*-Stilbene 448
- 71.2 A Variation: Preparation of *trans,trans*-1,4-Diphenylbutadiene 451

72 Two Thermochromic Compounds: Dixanthylene and Dianthraquinone 452
- 72.1 Dixanthylene from Xanthone 453
- 72.2 Dianthraquinone from Anthrone via 9-Bromoanthrone 454

73 An Analgesic: *p*-Ethoxyacetanilide from *p*-Aminophenol 455
- 73.1 *p*-Acetamidophenol from *p*-Aminophenol 456
- 73.2 Phenacetin from *p*-Acetamidophenol 457

74 A Photochromic Compound: 2-(2,4-Dinitrobenzyl)pyridine 458

75 A Chemiluminescent Compound: Luminol 459

76 Thiamine-Catalyzed Formation of Benzoin from Benzaldehyde 462

77 "Coconut Aldehyde"; γ-Nonanolactone 465
- 77.1 3-Nonenoic Acid from Heptaldehyde and Malonic Acid 466
- 77.2 Coconut Aldehyde from 3-Nonenoic Acid 467

78 A Model for the Biochemical Reducing Agent NADH 469

Synthetic Sequences: Synthesis Experiments That Use a Sequence of Reactions 475

79 Steroid Transformations: Δ^4-Cholestene-3-one from Cholesterol via Cholesterol Dibromide, 5α,6β-Dibromocholestane-3-one, and Δ^5-Cholestene-3-one 476
- 79.1 Cholesterol Dibromide from Cholesterol 482
- 79.2 $5\alpha,6\beta$-Dibromocholestane-3-one from Cholesterol Dibromide 483
- 79.3 Δ^5-Cholestene-3-one from $5\alpha,6\beta$-Dibromocholestane-3-one 483
- 79.4 Δ^4-Cholestene-3-one from Δ^5-Cholestene-3-one 486

80 Tetraphenylcyclopentadienone from Benzaldehyde and Phenylacetic Acid 488
- 80.1 Benzoin from Benzaldehyde 489
- 80.2 Benzil from Benzoin 489
- 80.3 Dibenzylketone from Phenylacetic Acid 490
- 80.4 Tetraphenylcyclopentadienone from Benzil and Dibenzylketone 491

81 Sulfanilamide 492
- 81.1 *p*-Acetamidobenzenesulfonyl Chloride from Acetanilide 495

81.2 *p*-Acetamidobenzenesulfonamide from *p*-Acetamidobenzenesulfonyl Chloride 496
81.3 Sulfanilamide from *p*-Acetamidobenzenesulfonamide 496

82 A Bootstrap Synthesis: *p*-Phenetidine from *p*-Phenetidine 497
82.1 Ethyldioxyazobenzene from *p*-Phenetidine 497
82.2 Diethyldioxyazobenzene from Ethyldioxyazobenzene 498
82.3 *p*-Phenetidine from Diethyldioxyazobenzene 499

83 1-Bromo-3-chloro-5-iodobenzene 502
83.1 2-Chloro-4-bromoacetanilide from 4-Bromoacetanilide 503
83.2 2-Chloro-4-bromoaniline from 2-Chloro-4-bromoacetanilide 504
83.3 2-Chloro-4-bromo-6-iodoaniline from 2-Chloro-4-bromoaniline 506
83.4 1-Bromo-3-chloro-5-iodobenzene from 2-Chloro-4-bromo-6-iodoaniline 507

84 MOED: A Merocyanine Dye 509
84.1 1,4-Dimethylpyridinium Iodide from 4-Methylpyridine and Methyl Iodide 510
84.2 4-(*p*-Hydroxystyryl)-1-methylpyridinium Iodide from 1,4-Dimethylpyridinium Iodide and *p*-Hydroxybenzaldehyde 510
84.3 1-Methyl-4-[(oxocyclohexadienylidene)-ethylidene]-1,4-dihydropyridine (MOED) from 4-(*p*-Hydroxystyryl)-1-methylpyridinium Iodide 511

Appendix 513

Table A.1 Derivatives of Alcohols 513
Table A.2 Derivatives of Aldehydes 514
Table A.3 Derivatives of Amides 514
Table A.4 Derivatives of Primary and Secondary Amines 515
Table A.5 Derivatives of Tertiary Amines 516
Table A.6 Derivatives of Carboxylic Acids 517
Table A.7 Derivatives of Esters 518
Table A.8 Derivatives of Aromatic Ethers 519
Table A.9 Derivatives of Aliphatic Halides 520
Table A.10 Derivatives of Aromatic Halides 521
Table A.11 Derivatives of Aromatic Hydrocarbons 522
Table A.12 Derivatives of Ketones 523
Table A.13 Derivatives of Nitriles 523
Table A.14 Derivatives of Phenols 524

Chemical Substance Index 525

General Subject Index 533

Preface

This book is intended for use in the laboratory part of an introductory course in organic chemistry. The overall organization of this fifth edition is the same as that of the fourth edition. Part 1 contains general descriptions of the theory and practice of the most common laboratory techniques of organic chemistry; Part 1 also presents directions for a number of exercises that illustrate these techniques. Part 2 contains experiments that range from the purification of natural products to one-step transformations to multi-step syntheses. Again, as in the fourth edition, this book includes a number of *variations,* or alternative experiments, that call for a different starting material and that require the student to provide some of the details of the experimental procedure. By using these variations, the instructor can move away from the detailed recipe, or "cookbook," approach to laboratory work.

The first sections of Part 1 contain a discussion of laboratory safety, a description of the glassware used in the organic chemistry laboratory, advice on cleaning up, directions for writing up a laboratory notebook, and an introduction to the chemical literature. After these preliminary topics come discussions of procedures for the isolation and purification of organic substances and of techniques such as crystallization, distillation, extraction, and the chromatographic methods. These are followed by sections on physical methods for the identification and characterization of organic compounds. These methods include the determination of boiling point and melting point, the determination of properties such as density, index of refraction, and optical rotation, and the recording and interpretation of infrared and nuclear magnetic resonance spectra. Next come chemical methods of identification and characterization: qualitative tests for the elements, qualitative tests for functional groups, and procedures for the preparation of derivatives. (The physical properties of selected unknowns and the melting points of their derivatives are again listed in tables in the Appendix.) Finally, in Part 1, there are sections that describe the apparatus and techniques used in the laboratory operations of organic chemistry.

Part 2 first presents experiments that exemplify the separation and purification of substances, experiments such as the isolation of cholesterol from gallstones and of lactose from powdered milk, the recovery of (R)-(+)-limonene from grapefruit or orange peel, the isolation of the two enantiomeric forms of carvone (these enantiomers have different odors), and the resolution of α-phenylethylamine. Part 2 then continues with a variety of one-step transformations that illustrate the chemistry of the functional groups and concludes with a set of multi-step syntheses: the preparation of Δ^4-cholestene-3-one from cholesterol, the syntheses of tetraphenylcyclopentadienone, sulfanilamide, *p*-phenetidine, and 1-bromo-3-chloro-5-iodobenzene, and the preparation of a merocyanine dye.

The fifth edition retains all of the experiments of the fourth edition, and several new experiments have been added. These include the isolation of piperine from black pepper, the borohydride reduction of vanillin to vanillyl alcohol, the preparation of the analgesic *p*-ethoxyacetanilide from *p*-aminophenol, the synthesis of "coconut aldehyde," and a bootstrap synthesis—the preparation of two moles of *p*-phenetidine from one mole of *p*-phenetidine.

I believe that students find lab work more interesting when they understand why procedures work and can participate in the planning of experiments. I therefore emphasize explanations and provide for student participation in the planning of experiments by allowing students to choose one of several similar preparations of the same compound or to adapt a procedure for the preparation of one compound to the preparation of another, as in the variations.

Finally, I am happy to acknowledge the interest and other contributions of Dr. David Todd (The Worcester Polytechnic Institute, Massachusetts) and Dr. Jeffrey Keiser (Coe College, Iowa), the editorial and production assistance of Allyn and Bacon, Inc., and the continuing interest and encouragement of my wife, Janet, and of my children, Margaret, Warren, Tad, Peter, and Emily.

Techniques and Experiments for Organic Chemistry

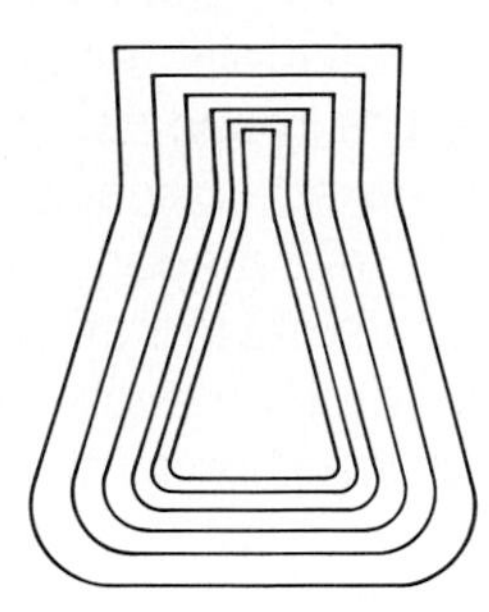

1 Laboratory Operations

- Preliminary Topics
- Separation of Substances; Purification of Substances
- Determination of Physical Properties
- Determination of Chemical Properties; Qualitative Organic Analysis
- Apparatus and Techniques for Chemical Reactions

Part 1 of this book describes many of the practical operations that one might perform in the organic chemistry laboratory. The first section in this part presents a discussion of laboratory safety, and this is followed by sections that describe the glassware used in the organic laboratory, the laboratory notebook, the literature of organic chemistry, and techniques for cleaning up. The remaining sections are presented in groups: (1) methods of separation and purification, (2) methods for the determination of physical properties, (3) methods for the determination of chemical properties, and (4) the apparatus and techniques for carrying out chemical reactions.

The sections on the theory and techniques of separation and purification include discussions of filtration, recrystallization, distillation, steam distillation, sublimation, extraction, chromatography, and removal of water.

The sections on the determination of physical properties and the dependence of physical properties on molecular structure include boiling point, melting point, density, index of refraction, optical rotation, molecular weight, and solubility characteristics, as well as the more recently developed spectrometric methods: infrared, ultraviolet-visible, nuclear magnetic resonance, and mass spectrometry.

The sections concerning the determination of chemical properties include qualitative tests for elements, qualitative tests for functional groups, and reactions for the formation of derivatives.

The sections on apparatus and techniques for chemical reactions include methods for heating and cooling, for stirring, for adding reagents, and for working up the reaction and isolating the product.

Preliminary Topics

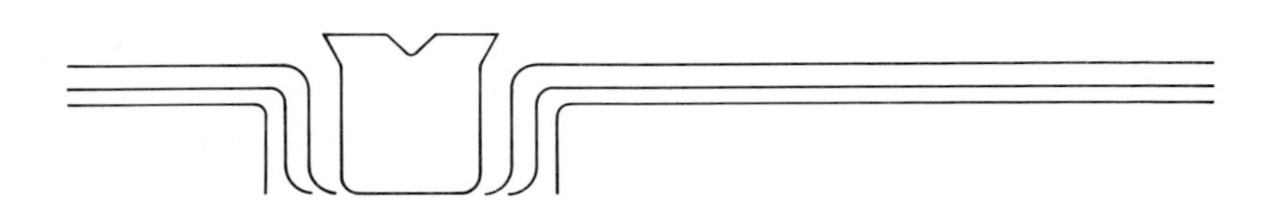

These first six sections provide information and procedures that should be of use in all experiments in the organic laboratory. In order to avoid accidents, hazards must be recognized and certain precautions must be taken; the first section describes some of these dangers and precautions. The next section describes and illustrates some of the special items of glassware that are used in the organic laboratory. The third section gives suggestions for cleaning up and disposal of waste. A fourth section describes the purpose of a laboratory notebook and recommends ways to keep such a notebook. The next section tells how to find certain kinds of information in the chemical literature, and the last section includes information about solutions of acids and bases, descriptions of several common reagents, and a periodic table of the elements with atomic weights.

1 Safety

Chemists must sometimes use hazardous materials, and therefore certain precautions must be regularly observed in order to minimize the probability and consequences of an accident.

1.1 FIRE

Know the location of the safety shower, the fire blanket, and the fire extinguishers. Never work alone in the laboratory.

The danger of fire is the most obvious hazard in the organic chemistry laboratory. This is an unavoidable result of the fact that most of the liquids used are relatively volatile and flammable (see Table 8.1). The danger and the consequences of a fire can be decreased merely by minimizing the number and size of the containers of flammable solvents that are stored in the laboratory.

The two most common reasons for a fire are

Use a condenser

1. Boiling a flammable solvent with a flame and without a condenser.

Watch your neighbor

2. Using a volatile and flammable solvent, especially during an extraction, without noticing that your neighbor's burner is on (or lighting your burner when your neighbor is using such a solvent).

Flammable liquids

If possible, then, all heating should be done with a steam bath, hot plate, heating mantle, or electric immersion heater rather than a burner. If a burner is used for heating a flammable solvent, it is absolutely essential that a condenser be used; otherwise, the vapors that escape from the neck of the flask will flow down to the flame and ignite. When working with the most volatile and flammable solvents such as carbon disulfide, ether, petroleum ether, or pentane, you must extinguish all flames on the bench, as the vapors of these solvents can travel over the desk top and be ignited by a distant burner. Solvents should never be vaporized into the atmosphere of the laboratory, but should be condensed and collected.

If the vapors from a flask do ignite, the fire can often be extinguished by

1. Turning off the burner.
2. Gently placing a notebook or clipboard over the top of the vessel that contains the burning solvent.

If solvent spilled on the desk top ignites,

Fire extinguisher

1. Move bottles and flasks of unspilled solvent away, if possible.
2. Use a carbon dioxide fire extinguisher on the fire. Discharge the extinguisher in this way:
   a. Pull out the pin. It may be necessary to break a wire or plastic strip.
   b. Aim the cone at the base of the fire from a distance of a few feet.

c. Release carbon dioxide snow by squeezing the hand grip or turning open the valve.

Do not use a fire extinguisher with wild abandon, as it is easy to make a fire worse by causing the blast from the nozzle to knock over and break bottles or flasks that contain more flammable materials.

If your clothing is burning, move under the nearest safety shower (the location of which you should know well enough that you can get there with your eyes shut) and pull the chain to turn the shower on.

If someone else's clothing has been set on fire, guide that person to the nearest safety shower and pull the chain. It is very easy for a person to panic under these conditions; the person must not be allowed to panic and run.

First aid for burns from flames, hot pieces of glass, hot iron rings, or hot plates consists *only* of the application of cold water. Never apply any kind of dressing. Call the doctor if the burn is more than a small blister.

Only water on burns

1.2 EXPLOSIONS

Always wear eye protection in the laboratory, preferably shatterproof goggles.

An explosion is an exothermic reaction that accelerates in rate until it gets out of control and shatters its container. The severity of the blast depends upon how much material is involved and how fast it all reacts.

Explosive mixtures are usually mixtures of oxidizing and reducing agents, as redox reactions are most likely to be highly exothermic. Gunpowder and many commercial explosives are mixtures of this type.

Explosive mixtures

Explosive substances are (1) compounds that can undergo internal redox reactions, such as the polynitro compounds trinitrotoluene, picric acid (trinitrophenol), and nitroglycerine, or (2) compounds that can decompose to give very stable molecules. Such compounds include acetylene, nitrogen triiodide, diazonium salts, diazo compounds, peroxides, azides, and fulminates.

Explosive compounds

If you must work with a substance or mixture that is potentially explosive or is known to be explosive, it is best to work on as small a scale as possible and behind a safety shield of shatterproof glass (hood fronts should be, but are not always, made of shatterproof glass). Just because a reaction has been run many times without incident by you or someone else does not mean that the potential danger has disappeared and that no explosion will occur the next time. New reactions should be run on a small scale, behind a safety shield, at least until their explosive potential has been estimated.

Use a safety shield

It is always a good idea to place a safety shield between yourself and the apparatus when carrying out a vacuum distillation. There is always the possibility that the flask will collapse under vacuum because of a crack or flaw in the glass. The safety shield will protect you from flying glass and the hot contents of the flask and oil bath.

Explosions are infrequent, because most people know about potentially explosive systems and try to avoid them. However, spectacular and tragic explosions continue to occur.

The three most common explosion hazards in the laboratory are

1. An exothermic reaction that gets out of control (explosion and fire).
2. Explosion of peroxide residues upon concentration of ethereal solutions to dryness (see Section 1.7).
3. Explosion upon heating, drying, distillation, or shock of unstable compounds (diazonium salts, diazo compounds, peroxides, polynitro compounds).

1.3 POISONING

Know the location of a chart that indicates first aid measures for various types of poisoning.

Almost anything is harmful in large enough doses, but it is easier to get a harmful dose of certain relatively dangerous materials in the chemical laboratory than in most other places.

There is absolutely no reason to be poisoned by mouth, as nothing should be eaten or drunk in the laboratory and pipetting should be done with a suction bulb or pipetter, not by mouth. Always wash your hands after working in the laboratory—especially if you're a fingernail biter.

Certain harmful materials can be absorbed rather quickly through intact skin. These include dimethyl sulfate, nitrobenzene, aniline, phenol, and phenylhydrazine (see Section 1.7). Fatal doses of cyanide can be acquired through a cut in the skin. Therefore, you should not work with cyanide salts or solutions if you have a cut on the hand.

When harmful or flammable gases are used (this includes almost all gases except oxygen, nitrogen, helium, neon, and argon), they should be used in the hood. The harmful properties of several gases are described in Section 1.7.

The most suitable first aid treatment for poisoning depends upon the nature of the poison. Consult a chart, which should be posted in the laboratory or the stockroom. Call the doctor.

1.4 CUTS

Most cuts occur during an attempt to force a thermometer or a glass tube into a hole in a rubber or cork stopper. The thermometer or tube breaks, and a sharp end is driven into the palm of the hand or base of the fingers. Instead of forcing, you should ream out the hole with a round file or, in the case of a rubber stopper, lubricate the hole with glycerine. The excess glycerine can be washed off with water. The use of thermometer adapters with ground-glass-jointed glassware has greatly reduced the frequency of this type of accident.

Any time glass is forced, as in trying to unfreeze a joint, protect your hands by wrapping the glass in several thicknesses of a clean cloth towel.

First aid for a cut consists first of removing any large pieces of glass and then stopping the bleeding. For venous bleeding, the cut can be pinched together or pressure can be applied with a gauze pad or clean towel. Arterial bleeding is much more dangerous and must be controlled additionally by application of hand or thumb pressure to the appropriate pressure point. For the arms and hands, this is where the pulse can be felt at the wrist or inside the upper arm just below the armpit.

Control of bleeding

Cuts should be treated by a medical doctor.

1.5 SPILLS

In general, spills on your skin or clothing should be treated by washing with plenty of water, either in the sink or under the safety shower. Section 1.7 describes how several particular spills should be treated. Refer to this section after you wash thoroughly, or while you wash have someone look up the substance you have spilled. You should reread Section 1.7 before you work with any of the substances listed there.

Safety shower

Spills on the desk or floor should be treated as recommended in Section 1.7.

Both acids and bases should be neutralized with sodium bicarbonate. Never use strong acids or bases for this purpose.

1.6 CHEMICALS IN THE EYE

Always wear eye protection, preferably shatterproof goggles. Know the location of the eyewashing fountain.

If your eyes are protected by shatterproof goggles, you would have to be very unlucky indeed to get something in your eyes.

First aid for chemicals in the eyes is to wash them thoroughly for several minutes, either by means of a special water fountain, which can direct a large but gentle flow of water into the eye, or with the eyewash dispenser. If these cannot be found immediately, use a gentle stream from a hose attached to a water faucet or a beaker of water. Contact lenses *must* be removed in order to wash the eyes. Be alert to the possibility that a person with something in his or her eyes may not be

Remove contact lenses

able to see well enough to get to the eyewash fountain and may have to be led there by someone else.

1.7 A SHORT LIST OF HAZARDOUS MATERIALS AND SOME OF THEIR PROPERTIES

A discussion of the dangerous properties of many substances can be found in *Dangerous Properties of Industrial Materials*, 6th edition, by N. I. Sax (New York: Van Nostrand Reinhold, 1984). If you plan to work with an unfamiliar substance, you should look up its dangerous properties in Section 6 of that book.

The purpose of the following list is to alert you to the greatest dangers presented by substances that are often found in the organic laboratory. The laboratory assistants and the instructor in charge should determine the dangerous properties of all substances present in the laboratory.

Abbreviations include O.S.H.A.: Occupational Safety and Health Administration; A.C.G.I.H.: American Conference of Governmental Industrial Hygienists; STEL: Short Term Exposure Limit; IDLH: Immediately Dangerous to Life or Health; ppm: parts per million (molecules per million molecules of air); mg/cubic meter (milligrams per cubic meter of air).

- *Carcinogens:* Benzidine, α-naphthylamine, β-naphthylamine, certain polynuclear hydrocarbons.
- *Chemicals that can be rapidly absorbed in fatal doses through intact skin:* Aniline, dimethyl sulfate, nitrobenzene, phenylhydrazine, phenol, 1,1,2,2-tetrachloroethane.
- *Explosives:* Polynitro compounds such as picric acid, trinitrotoluene, trinitrobenzene, 2,4-dinitrophenylhydrazine.
- *Lachrymators and vesicants:* Benzylic halides, allylic halides, α-halocarbonyl compounds (such as the phenacyl halides), dicyclohexylcarbodiimide, isocyanates.
- *Volatile and flammable solvents:* Pentane, petroleum ether, diethyl ether, and carbon disulfide present the greatest fire hazard of the common solvents (see Section 1.1). Carbon disulfide vapors can spontaneously ignite at steam bath temperatures.

These are some of the specific materials:

- *Acetic acid:* First aid for spills: dilute with large amounts of water.
- *Acetic anhydride:* Corrosive; will quickly blister the skin if not washed off. First aid for spills on the skin: wash off with water and finally with dilute ammonia solution.
- *Acetonitrile:* Poisonous on its own, but it can also produce hydrogen cyanide.

- *Acetyl chloride:* Reacts violently with water to produce HCl and acetic acid; see *Acid chlorides*.
- *Acid chlorides (and other acid halides):* Acetyl chloride, benzoyl chloride, benzenesulfonyl chloride, *p*-toluenesulfonyl chloride: corrosive, lachrymators. First aid for spills on the skin: wash with water and finally with dilute ammonia solution.
- *Acids:* Corrosive. First aid for spills: dilute with large amount of water, then wash with sodium bicarbonate solution.
- *Aluminum chloride:* Corrosive; reacts violently with water to produce hydrogen chloride.
- *Ammonia:* First aid for inhalation: fresh air; inhalation of steam. First aid for ammonia in the eyes: irrigation with water for 15 minutes.
- *Aniline:* Fatal doses can be absorbed through intact skin.
- *Benzene:* According to the *Merck Index,* chronic toxicity involves bone marrow depression and aplasia; rarely, leukemia. Under the O.S.H.A. emergency standard, effective May 21, 1977, benzene exposure should be no more than 1 part per million during an 8-hour day (a time-weighted average; 3.1 mg/cubic meter), and no more than 5 ppm during a 15-minute period. A.C.G.I.H. standards are 10 ppm, or 30 mg/cubic meter. STEL (tentative): 25 ppm; IDLH: 2000 ppm. Benzene should be dispensed in the hood so that vapors and spills will not contaminate the lab.
- *Benzoyl peroxide:* Explosive; should not be heated for recrystallization or for melting-point determination; see *Peroxides*.
- *Bromine:* Exceedingly corrosive. Should be poured wearing gloves, face shield, and laboratory apron. Should be dispensed by means of a buret with a Teflon stopcock. First aid for spills on skin: wash *instantly* with water, rinse with ethanol, and rub in glycerine. First aid for inhalation: see *Chlorine*.
- *Carbon tetrachloride:* Poisoning by carbon tetrachloride can be accomplished by inhalation, ingestion, or absorption through the skin. O.S.H.A. regulations, effective as of May, 1976, state that exposure to carbon tetrachloride vapors should be limited to 10 ppm (65 mg/cubic meter) during an 8-hour day. STEL (proposed): 20 ppm; IDLH: 300 ppm.
- *Chlorine:* Exceedingly corrosive. Should be handled in the hood. First aid for inhalation: fresh air; inhalation of vapors from a very dilute solution of ammonia. First aid for spills on skin: see *Bromine*.
- *Chloroform:* O.S.H.A. regulations, effective as of May, 1976, state that exposure to chloroform vapors should be limited to 50 ppm (240 mg/cubic meter) during an 8-hour day. STEL: 50 ppm; IDLH: 1000 ppm.
- *Chlorosulfonic acid:* Exceedingly corrosive; reacts with water with

explosive violence to form sulfuric acid and hydrogen chloride. First aid for spills on skin or clothing: wash with very large amounts of water (safety shower).

- *Dichloromethane (methylene chloride):* Dichloromethane vapors are said to be narcotic in high concentrations. O.S.H.A. regulations, effective as of May, 1976, state that exposure to methylene chloride vapors should be limited to 500 ppm (1740 mg/cubic meter) during an 8-hour day. STEL: 500 ppm; IDLH: 5000 ppm. Dichloromethane appears to be the least hazardous of the halogenated solvents.
- *Diethyl ether* (see also *Ethers*): O.S.H.A. regulations: 400 ppm (1200 mg/cubic meter); STEL (tentative): 500 ppm; IDLH: 19,000 ppm.
- *Dimethyl sulfate:* Very dangerous; fatal doses can be absorbed quickly through intact skin. First aid for spills on skin: remove by washing with dilute ammonia solution; remove contaminated clothing.

DMSO

- *Dimethyl sulfoxide:* Not particularly harmful in itself, but is rapidly absorbed through intact skin and can apparently aid the absorption of other materials through the skin.
- *Ethers:* Ethyl ether, isopropyl ether, tetrahydrofuran, dioxane. In addition to being highly flammable, ethers can also absorb and react with oxygen upon storage to form dangerously explosive peroxides. Ether that has not been stored in full, airtight, amber bottles should be routinely discarded within two months. Ether of unknown vintage should be treated with the respect due to an equal amount of an exceedingly unstable high explosive; if crystals can be observed in the ether or if it appears to contain a viscous layer, the bottle should not even be touched but should be disposed of by explosives experts. Large amounts of ether should never be concentrated unless the absence of peroxides has been experimentally verified. Ethereal solutions should never be concentrated to dryness by heating with a flame. Isopropyl ether seems to be especially treacherous with respect to peroxide formation.

Test for peroxides

Test for presence of peroxides in ethers or hydrocarbons: Add 0.5–1 mL of the material to be tested to an equal volume of glacial acetic acid to which has been added about 100 mg of sodium or potassium iodide. A yellow color indicates a low concentration of peroxides, and a brown color a high concentration. A blank determination should be run.

Removal of peroxides

Removal of peroxides from ethers: Stir or shake the ether with portions of a solution prepared by dissolving 60 grams of ferrous sulfate and 6 mL of conc. sulfuric acid in 100 mL of water.

For information concerning the formation of peroxides, detection and estimation of peroxides, inhibition of peroxide formation, and removal of peroxides, see N. V. Steere, *Handbook of Laboratory Safety,* 2nd edition (Cleveland, Ohio: The Chemical Rubber Company, 1971), pp. 190–194, and H. L. Jackson et al., *J. Chem. Educ.* **47,** A175 (1970).

- *Fuming nitric acid:* 95% nitric acid, containing oxides of nitrogen: extremely corrosive. First aid for spills: wash with large amounts of water and finally with sodium bicarbonate solution.
- *Fuming sulfuric acid:* Oleum; concentrated sulfuric acid containing dissolved sulfur trioxide: extremely corrosive. First aid for spills: wash with large amounts of water and finally with sodium bicarbonate solution.
- *Halogenated solvents:* Carbon tetrachloride, chloroform, etc. 1,1,2,2-Tetrachloroethane is the most dangerous, as it can be absorbed rapidly through the skin. Avoid breathing the vapors of these solvents.
- *Hydrazine:* Explosive; dangerous in combination with oxidizing agents.
- *Hydrides:* Lithium aluminium hydride, sodium hydride: react instantly and explosively with water, liberating and possibly igniting hydrogen. Calcium hydride is slightly less vigorous in its reaction with water. Contact with water must be scrupulously avoided. Borohydrides react less vigorously with water, but rapidly with acidic solutions, to liberate hydrogen. Further information concerning the properties and procedures recommended for the safe handling of individual hydrides can be found in Reference 5.
- *Hydriodic acid:* First aid for spills: wash with large amounts of water and finally with dilute sodium bicarbonate solution.
- *Hydrobromic acid:* First aid for spills: wash with large amounts of water and finally with dilute sodium bicarbonate solution.
- *Hydrochloric acid:* First aid for spills: wash with large amounts of water and finally with dilute sodium bicarbonate solution.
- *Hydrogen bromide gas:* First aid for inhalation: fresh air; lie down and rest.
- *Hydrogen chloride gas:* First aid for inhalation: fresh air; lie down and rest.
- *Hydrogen peroxide:* Concentrated solutions can explode; powerful oxidizing agent; see *Peroxides*.
- *Hydrogen sulfide gas:* First aid for inhalation: fresh air; artificial respiration if necessary.
- *Lithium metal:* Reacts with water to produce hydrogen gas.
- *Lithium aluminum hydride:* Reacts instantly and explosively with

water, liberating and possibly igniting hydrogen gas. Excess lithium aluminum hydride can be decomposed by dropwise addition of ethyl acetate. See *Hydrides*.

- *Methylene chloride:* see *Dichloromethane*.
- *Nitric acid:* Corrosive. First aid for spills: dilute with large amounts of water, then wash with sodium bicarbonate solution.
- *Nitrobenzene:* Fatal doses can be absorbed through intact skin.
- *Oleum:* See *Fuming sulfuric acid*.
- *Peracids:* Peracetic acid, trifluoroperacetic acid. Concentrated solutions can explode; powerful oxidizing agent; see *Peroxides*.
- *Perchloric acid:* Concentrated solutions can explode; powerful oxidizing agent; see *Peroxides*. For dangers in the use of perchloric acid and perchlorates, see N. V. Steere, *Handbook of Laboratory Safety*, 2nd edition (Cleveland, Ohio: The Chemical Rubber Company, 1971), pp. 205–216.
- *Peroxides:* All peroxides are potentially explosive, especially if heated. They are also powerful oxidizing agents, and mixtures with oxidizable materials are also potentially explosive. Ethers and other substances, such as tetralin, decalin, cumene, and certain other hydrocarbons, form peroxides when stored without exclusion of oxygen. Such liquids should not be distilled or concentrated without first determining that peroxides are not present; see *Ethers*, test for presence of peroxides.
- *Phenol:* Corrosive. Should not be handled with bare hands. Fatal doses can be absorbed through intact skin.
- *Phenylhydrazine:* Fatal doses can be absorbed through intact skin.
- *Phosphoric acid:* First aid for spills: wash with large amounts of water and finally with sodium bicarbonate solution.
- *Potassium cyanide:* Source of cyanide ion; fatal in small amounts; fatal doses can be ingested via a cut in the skin. Spills should be carefully cleaned up and disposed of immediately. Acidification of cyanide solutions will release deadly hydrogen cyanide gas.
- *Potassium hydroxide:* Caustic; corrosive solution. First aid for spills: wash with large amounts of water and finally with dilute bicarbonate solution.
- *Potassium metal:* Reacts instantly and explosively with water to form and ignite hydrogen gas. On storage without exclusion of oxygen, it forms exceedingly dangerous and explosive peroxides. If such material is possibly present, the sample should be disposed of by explosives experts. See *Sodium metal*.
- *Sodium amide:* Reacts violently with water. Aged samples have been reported to decompose explosively.
- *Sodium cyanide:* See *Potassium cyanide*.
- *Sodium hydride:* See *Hydrides*.

- *Sodium hydroxide:* Caustic; corrosive solution. First aid for spills: wash with large amounts of water and finally with dilute bicarbonate solution.
- *Sodium metal:* Reacts instantly and explosively with water to form and usually ignite hydrogen gas. Sodium and potassium metals are usually stored under xylene. If oxide-free metal is required, transfer a piece of the metal to a mortar containing xylene and cut off the oxide coating with a knife. To weigh the metal, remove a piece from the xylene, briefly blot it with a piece of filter paper and add it to a tared beaker of xylene. For further details and information concerning the handling of sodium and potassium metals, see Reference 5. Scrap sodium should be disposed of by adding little pieces to a large volume of methanol; scrap potassium can be disposed of similarly, using *tert*-butanol.
- *Sulfuric acid:* Corrosive. First aid for spills: dilute with large amounts of water, then wash with sodium bicarbonate solution.
- *1,1,2,2-Tetrachloroethane:* Fatal doses can be absorbed through intact skin.
- *Thionyl chloride:* Corrosive, volatile. First aid for spills: wash with large amounts of water and finally with dilute ammonia solution.

References

Sources from which one can obtain more information about laboratory safety in general, specific information about the storage and handling of dangerous substances, and hazardous properties of specific chemicals include

1. *Handbook of Laboratory Safety,* 2nd edition, N. V. Steere, editor, The Chemical Rubber Co., Cleveland, Ohio, 1971.
2. N. I. Sax, *Dangerous Properties of Industrial Materials,* 6th edition, Van Nostrand Reinhold, New York, 1984.
3. Wall-size chart: "Emergency Procedures for Dangerous Materials," Sargent-Welch Scientific Co., Catalog No. S-18812.
4. *Prudent Practices for Handling Hazardous Chemicals in Laboratories,* Committee on Hazardous Substances in the Laboratory, National Research Council, National Academy Press, Washington, D.C., 1981.

Information about properties and methods of handling of many compounds can be found in

5. L. F. Fieser and M. Fieser, *Reagents for Organic Synthesis,* Wiley, New York; Volume 1, 1967; Volume 2, 1969; Volume 3, 1972; Volume 4, 1974; Volume 5, 1975; Volume 6, 1977; Volume 7, 1979; Volume 8, 1980; Volume 9, 1981; Volume 10, 1982; Volume 11, 1984.

Information about the toxicity of many compounds can be found in

6. *The Merck Index,* 10th edition, Merck and Co., Inc., Rahway, New Jersey, 1983.

7. M. Sittig, *Handbook of Toxic and Hazardous Chemicals,* Noyes Data Corporation, Park Ridge, New Jersey, 1981. As the preface puts it, "This handbook presents concise chemical, health and safety information on nearly 600 toxic and hazardous chemicals, so that responsible decisions can be made by chemical manufacturers, safety equipment producers, toxicologists, industrial safety engineers, waste disposal operators, health care professionals, and the many others who may have contact with or interest in these chemicals due to their own or third party exposure." A useful book.
8. *Hazardous Chemicals Data Book,* G. Weiss, editor, Noyes Data Corporation, Park Ridge, New Jersey, 1980. "Instant information for decision making in emergency situations. . . ." A compilation of information on 1350 hazardous chemicals; raw information; no interpretation.
9. *Registry of Toxic Effects of Chemical Substances, 1979 Edition,* R. S. Lewis, Sr., and R. L. Tatken, editors, DHHS(NIOSH) Publication No. 80-111, September, 1980. It is very difficult to find or recognize what you want in this book.

Information about methods of disposal of many compounds can be found in

10. *Catalog Handbook of Fine Chemicals,* Aldrich Chemical Co., Inc., Milwaukee, Wisconsin; 1984–1985.

Questions

1. Draw a plan of your laboratory that shows
   a. the location of the safety shower, the fire blanket, and the fire extinguishers.
   b. the location of the eyewash fountain.

2. Describe where the *Poison Chart* is located.

3. Provide a definition for each word.
   a. carcinogen
   b. lachrymator
   c. vesicant
   d. caustic
   e. corrosive

4. The O.S.H.A. standard for exposure to ethyl alcohol is 1000 ppm.
   a. Express this standard in mg/cubic meter.
   b. How many mL of liquid ethyl alcohol, density = 0.79 gm/mL, would have to be vaporized in a laboratory 10 meters by 14 meters by 5 meters to provide a concentration of vapor in the air of 1000 ppm?

5. Both acids and bases can be neutralized by sodium bicarbonate. Explain. Illustrate your explanation with balanced equations.

2 Glassware Used in the Organic Chemistry Laboratory

Organic chemists routinely use a variety of special items of glassware. Flasks with a round bottom and a narrow neck (*boiling flasks;* Figure 2.1) are used to contain a reaction mixture that is to be heated or boiled.

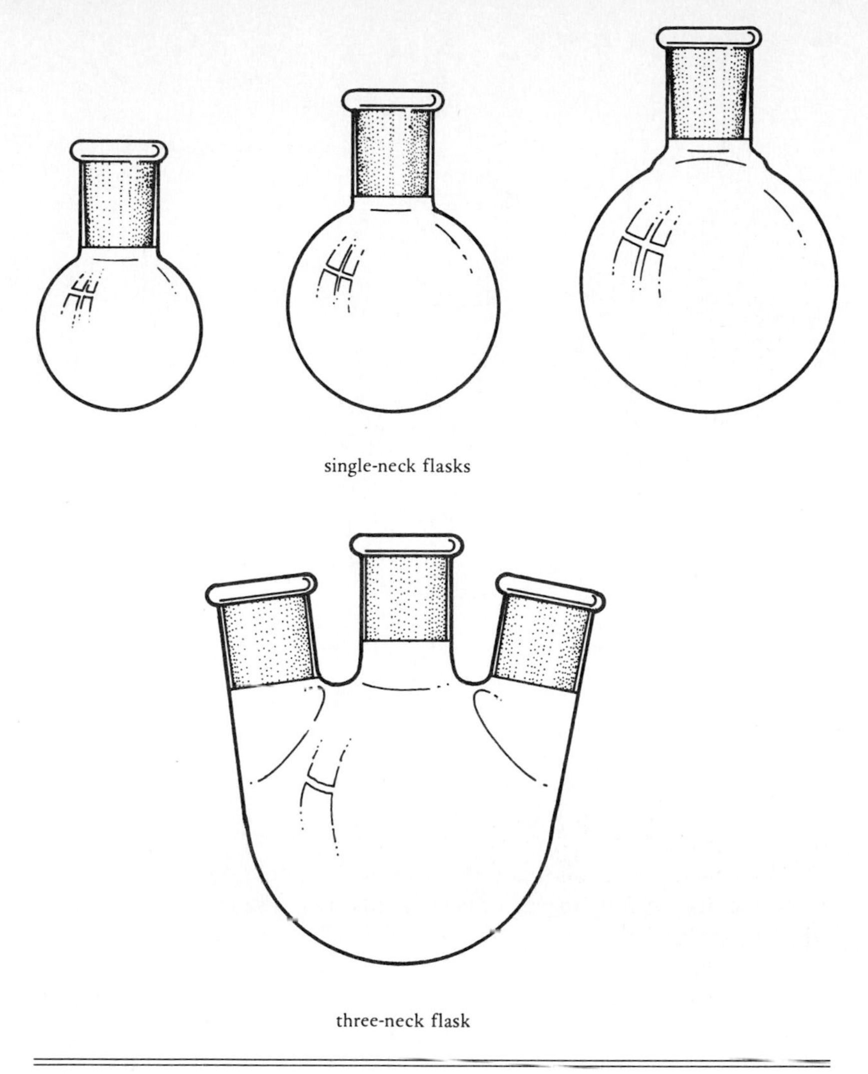

Figure 2.1. Boiling flasks.

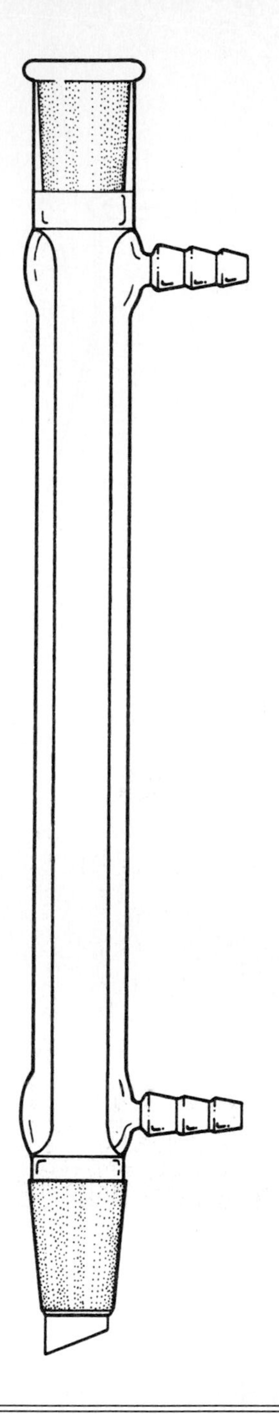

Figure 2.2. Condenser.

In order to condense and return the vapors, the flask is usually fitted with a *reflux condenser* (Figure 2.2) as described in Section 30. Boiling flasks are made with multiple necks as well (Figure 2.1).

Flasks with a flat bottom and a narrow neck (*Erlenmeyer flasks;* Figure 2.3) are used for reaction mixtures that will be allowed to stand for a period of time, or to hold a solution that is to be set aside for crystallization (Section 8). The narrow neck retards evaporation and can accept a cork, rubber, or glass stopper.

Beakers (Figure 2.4) are used only as temporary containers or when a flask with a narrow neck would be inconvenient.

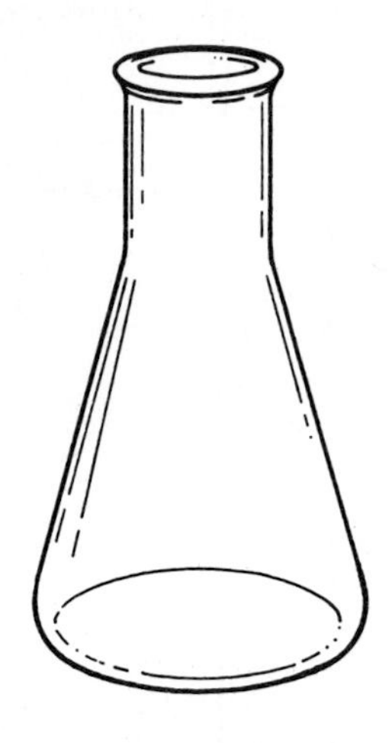

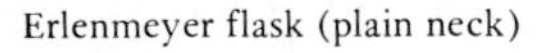

Erlenmeyer flask (plain neck)

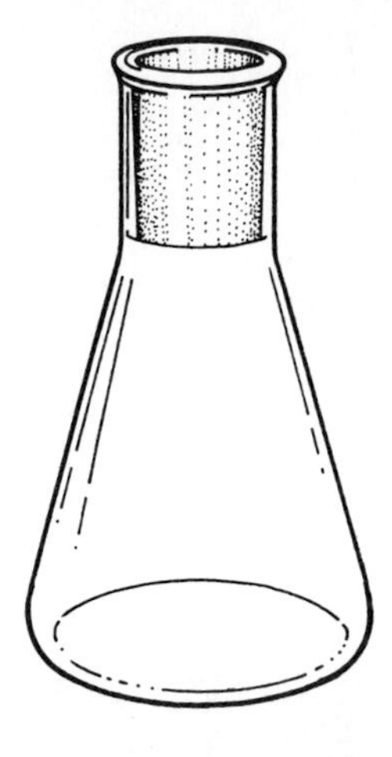

Erlenmeyer flask (ground-glass-jointed neck)

Figure 2.3. Erlenmeyer flasks.

Condensers and flasks can be connected together and fit with other items of equipment by means of various *adapters* (Figure 2.5). As examples, two common arrangements for distillation are shown in Figures 9.17 and 10.2.

A *separatory funnel* (Figure 2.6) is used to separate mixtures of immiscible liquids (Section 13.3), or it can be used as a dropping funnel to add a liquid to a reaction mixture, as shown in Figure 30.2.

Porcelain filtering funnels, either the Buchner or Hirsch style, are used with a *suction filtration flask* (Figure 2.7) in order to separate solids from liquids, as described in Section 8.

Other items of glassware commonly used by organic chemists include *graduated cylinders* for measuring liquids, *distillation flasks* (round bottom flasks each with a long neck and a side-arm), and *drying tubes* (Figure 2.8).

Figure 2.4. Beaker.

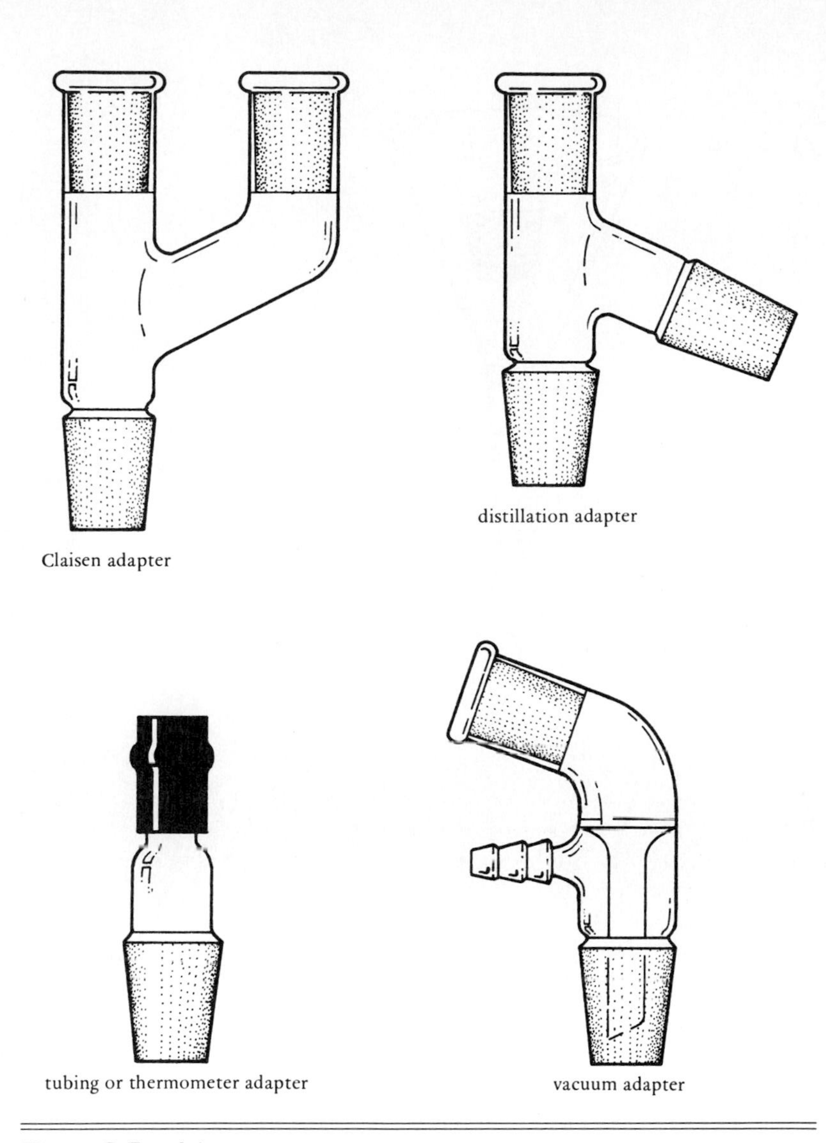

Figure 2.5. Adapters.

Figure 2.6. Separatory funnel.

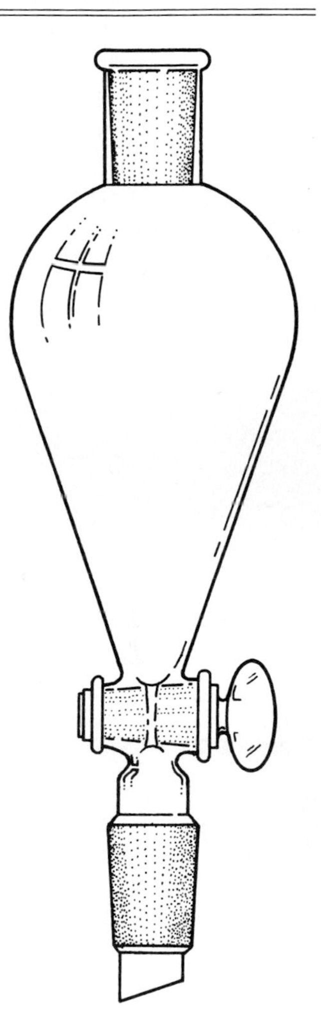

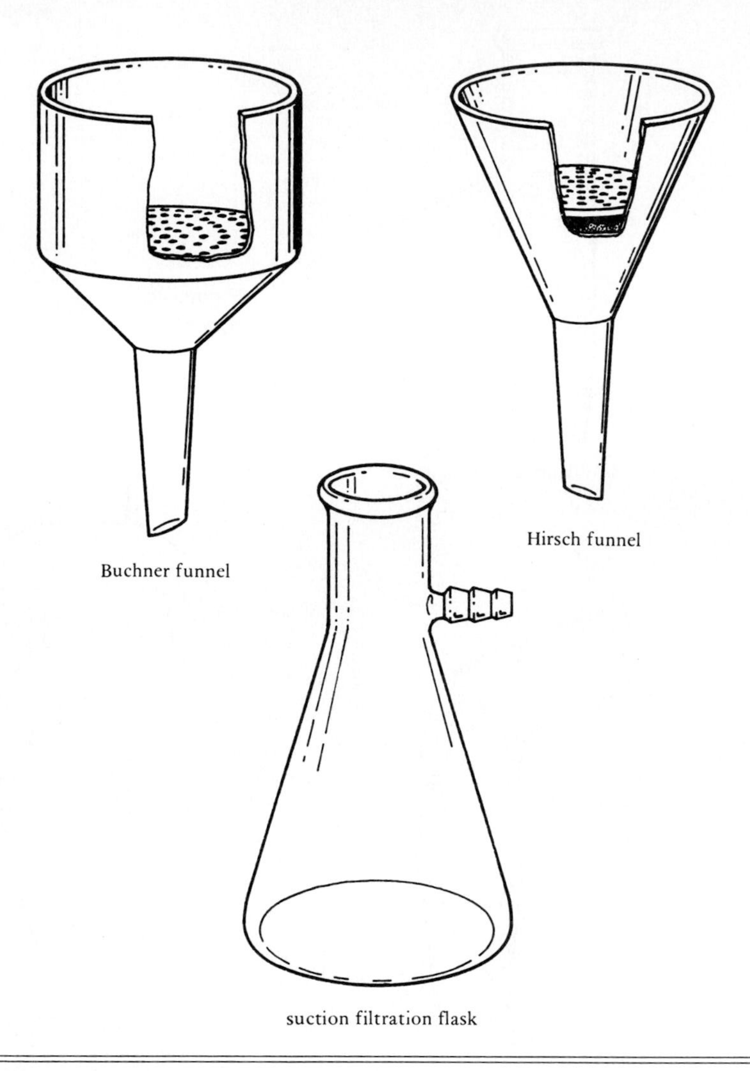

Figure 2.7. Porcelain filtering funnels and suction filtration flask.

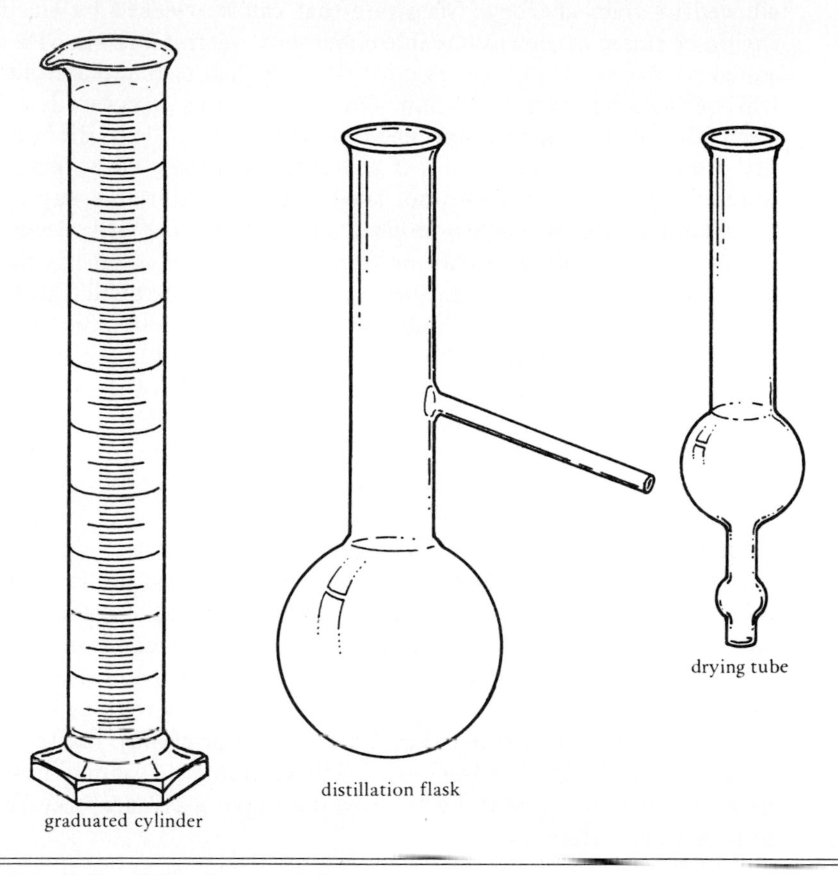

Figure 2.8. Other items of glassware.

3 Cleaning Up

In chemistry, as in many things, cleaning up is a part of the job that seems to take a lot more time than it should. Some ways by which time and effort of cleaning can be minimized are described in this section.

One way to expedite cleanup is to distinguish glassware that is merely wet with a volatile solvent, glassware that can be cleaned simply by rinsing, and glassware that is too dirty to be cleaned by rinsing. Glassware that is just wet with a volatile solvent does not need to be cleaned, but simply set aside on a towel spread out on the desk top and allowed to drain and dry. Glassware that can be cleaned by rinsing should be rinsed as soon as possible either with water if it is wet with an aqueous solution, or with an organic solvent such as technical acetone if it is wet with an organic solution. Once rinsed, this glassware is only wet with a volatile solvent and need merely be set aside to drain and dry. You can save a lot of work if you avoid scrubbing with soap and water any glassware that does not need to be cleaned in this way.

Glassware that will not come clean merely by rinsing—the flasks in which a reaction has been run or from which a distillation has been carried out are examples—can almost always be cleaned by first adding an organic solvent such as technical acetone (or a solvent appropriate to the nature of the material in the flask) and allowing the flask to stand with occasional swirling in order to dissolve the bulk of the material. After pouring the dirty solvent into the waste solvent container and repeating this procedure if necessary, remove any stubborn residue by scrubbing with technical acetone and a brush, or soapy water and a brush, or both in succession. It is rarely necessary to use anything stronger than soap, water, and acetone to clean glassware.

Just soap, water, and acetone

Much cleaning of glassware can be done while waiting for something to filter or cool or warm up or crystallize or distill or react. Using odd moments to rinse, clean, and put away will save a lot of time at the end of the day. Also, the sooner something is cleaned after using, the easier it is to clean, as you should know from your dishwashing experience. This is especially true of items that cannot be cleaned inside with a brush, such as pipets or Buchner or Hirsch funnels. It should not be necessary to point out how shortsighted it is to put glassware away dirty to be cleaned before use!

Clean up instead of watching

3.1 CARE OF GROUND-GLASS-JOINTED GLASSWARE

Ground-glass-jointed glassware is very nice and most convenient, but it is expensive (about $5.00 per joint). All glassware can be broken; ground-glass-jointed glassware, however, can also become useless if the pieces cannot be separated because the joints have become "frozen." The best way to prevent freezing is by dissassembling the apparatus immediately after use, before the apparatus cools if it is hot, and by not lubricating the joints except as described in Section 30. When a lubricant has been used, the joints should be wiped as clean of grease as possible before washing so that the grease will not be spread over the rest of the piece. Hydrocarbon-based greases can be removed by rinsing with dichloromethane, but silicone greases resist removal almost completely. Use dichloromethane or ether followed by soap and water. Glassware that has been used with silicone-based greases can easily be recognized because it is not wet by water; the water stands up in beads on the surface.

Avoid frozen joints

No grease

3.2 SEPARATORY FUNNELS AND GLASSWARE WITH STOPCOCKS

Separatory funnels should be stored with the plug removed so that it cannot become frozen. Exceptions are separatory funnels with Teflon stopcocks or those whose plugs have been removed, cleaned, and re-greased just before storage. When rinsing a separatory funnel with a glass stopcock plug, the plug should be removed and the plug and barrel wiped free of grease unless the apparatus has been used only briefly and the grease has not been leached out. The plug should be regreased (Teflon plugs need no lubricant) by applying a thin stripe of lubricant along the plug on the two sides without holes. After the plug is inserted into the barrel, the grease is spread out evenly by rotating the plug a few times in the barrel. Too much grease will cause the holes in the plug to become partially or completely stopped up. Other items with stopcocks, such as burets and distilling heads, should be treated similarly.

If a stopcock plug becomes frozen, pullers are available. These will either pull out the plug or break the piece in the attempt. While the jack screw of the puller is being tightened, the item should be wrapped in towels to contain the fragments if the piece breaks.

It is sometimes possible to loosen frozen stopcocks or joints by very judicious heating or tapping; these techniques are best learned by watching an expert.

3.3 DRYING OF GLASSWARE

Glassware that has been rinsed with either water or acetone will dry upon storage. The outsides of all items and the insides of beakers can be dried with a towel. If it is necessary to dry a piece of glassware such as a graduate or a flask quickly for immediate reuse, a stream of air can be drawn through the inside by inserting to the bottom a length of glass tubing that is connected to the aspirator by a length of hose. Acetone or other volatile solvents will be removed from the insides of most items within a minute. A pipet can be connected to the aspirator, and air drawn through the pipet by the aspirator. If the piece is wet with water, it should be rinsed with technical acetone to remove the water, and then dried as just described. Obviously, something to be used with water or with an aqueous solution will not normally need to be dried of water before use! It will often be fastest and most satisfactory to rinse the piece with a little of the solvent that is to be used next.

Use it wet

3.4 DISPOSAL OF WASTE

Make use of the different containers provided for waste paper and broken glass.

Chemicals should be disposed of according to specific instructions from the laboratory assistant. Relatively harmless solids may be discarded along with other solid waste, but others may have to be decomposed before discard or washed down the sink drain. Organic liquids should not be poured down the drain but put into the containers provided for them. Aqueous solutions or water-soluble liquids may usually be disposed of down the sink drain along with a lot of water.

Radioactive wastes should be placed in the containers reserved for them.

The *Aldrich Catalog Handbook of Fine Chemicals* suggests methods to be used to dispose of various substances.

4 The Laboratory Notebook

The lab notebook serves two purposes. First, it is a place to record and keep information that should be available in the lab while you are doing an experiment. Second, it is a place to write down and preserve both the description of the experiment as it was actually done and the results of the experiment.

The information you need to have in the lab always includes a description of the procedure you plan to follow. When you are running a reaction, you need to know the properties of the materials with which you are working—properties such as molecular weight, melting and boiling points, density, and solubility behavior. At some time, a balanced equation must be written down, and the suitability of the molar and gram amounts called for in the procedure should be verified by reference to the balanced equation and the molecular weights. All of this information should be recorded in the lab notebook so that at some later time it will be possible to discover the source of any mistakes that may have been made.

During the course of the experiment try to record all relevant information about what was done and what happened. The easiest way to do this is to state how the actual experimental conditions and results differed from the conditions and results that were anticipated. Sometimes, for example, it is difficult or unnecessary to duplicate exactly certain of the conditions called for (conditions such as time and temperature, if indeed they are precisely specified), but the actual conditions should be recorded in the notebook. Similarly, the behavior of the reaction may not be as anticipated, and the variations should be written down. An adequate notebook will enable another person to know exactly what you did and what happened, and your record should make it possible for your results to be verified. An adequate notebook is an essential part of good graduate school and industrial research.

A general lab notebook format might be as follows:

1. *Descriptive title and date.* This will serve to identify the experiment that is the subject of the notebook entry. If the notebook is the record of a research project, the different attempts to carry out a particular procedure should be distinguished by a notation such as "Run One" or "Run Two" in the title. The record of the first experiment should start on page 3 or page 5 of the notebook so as to leave room for a table of contents at the beginning of the book.

2. *Balanced equation.* If you are planning to run a reaction, the equation will show at a glance exactly what you are trying to do. It will also show the basis for the stoichiometric calculations.

3. *Molecular weights and molar, gram, and volume amounts of reagents, solvents, and products.* When an experiment is carefully planned, all of this information will be needed. It makes sense to record it in a well-organized and permanent way.

4. *Relevant physical properties of reagents, solvents, and products.* It is often necessary to know the melting point, boiling point, density, and solubility properties of the material with which you are working. In a recrystallization, for example, it is necessary to know which solvents might freeze at the temperature of the ice bath. In a distillation, you need to know at what temperature the various components of the mixture can be expected to boil. In an extraction, you need to know which solvent should contain each substance and which solution should be the more or less dense. You must apply a certain amount of judgment as to what is relevant. For example, it should seldom be necessary to know the exact melting point or boiling point of solid sodium chloride or anhydrous magnesium sulfate. Similarly, while it makes sense to measure liquids by volume, solids are usually measured by weight, and so the densities and volumes of solids are irrelevant.

   Items 3 and 4 may be combined, perhaps into a comprehensive table. The necessary information about the physical properties can usually be found in one of the handbooks or dictionaries (see Section 5.2).

5. *A description of the experimental procedure.* If you plan to follow a procedure from the lab manual, a reference to the page in the book where the procedure is given should be sufficient, since you will have the manual in the lab with you. If you are following a procedure from a handout or from the chemical literature, you can tape or rubber-cement a copy in your notebook; be sure to include a reference to the source of the procedure. It is usually not permitted to bring a library book into the lab, where it might be damaged.

6. *Departures from a planned procedure.* Here, with reference to the planned procedure given in item 5, you tell what you actually did. Perhaps a slightly different amount was used than was called for. The actual time, temperature, and concentration should be recorded if they were different from those called for in the procedure or if they were unspecified. Here is where you admit that something was spilled or boiled over. Don't forget that if you have an extraordinary loss, the amounts of materials used in the rest of the experiment must be reduced in proportion to the amount of the loss.
7. *Results, including percent yield.* The outcome of the experiment should be described. If a substance has been prepared, you should report the amount obtained, both in grams and as a percent of what could theoretically be expected on the basis of the balanced equation. For example, if you prepare methyl benzoate from benzoic acid and methanol (Section 59.2), the balanced equation and the gram and molar amounts of starting materials are

$$C_6H_5-C(=O)-O-H + CH_3OH \longrightarrow C_6H_5-C(=O)-O-CH_3 + H_2O$$

benzoic acid	methanol	methyl benzoate
12.2 g	19.7 g	
m.w. = 122 g/mole	m.w. = 32 g/mole	m.w. = 136 g/mole
0.1 mole	0.62 mole	

It should be clear that the limiting reagent in this case is benzoic acid and that the maximum possible (or theoretical) yield of methyl benzoate is 0.1 mole, which, for a molecular weight of 136 g/mole, would correspond to 13.6 grams. If 10 grams were actually obtained, the actual yield would be 10/13.6, or 74% of theoretical.

If you have determined any physical or chemical properties, such as color, melting point, boiling point, index of refraction, density, or specific rotation, these should be recorded along with theoretical or literature values. The sources of the literature values (lab manual, handbook, etc.) should also be recorded. If you determine the infrared spectrum of your product, you may wish to fasten the spectrum into your notebook. If you are working on a research project, it may be preferable to file the spectra separately. When this is done, you should make a notation in your book that a spectrum was obtained.

8. *Comments*. In case you or someone else might wish to repeat the experiment, you should point out any difficulties you encountered and write down any suggestions for changes in the procedure.

Figure 4.1. Sample page from lab notebook.

Cyclohexene from Cyclohexanol 10-27-87

In this experiment we will convert cyclohexanol to cyclohexene by heating the alcohol with 85% H_3PO_4 :

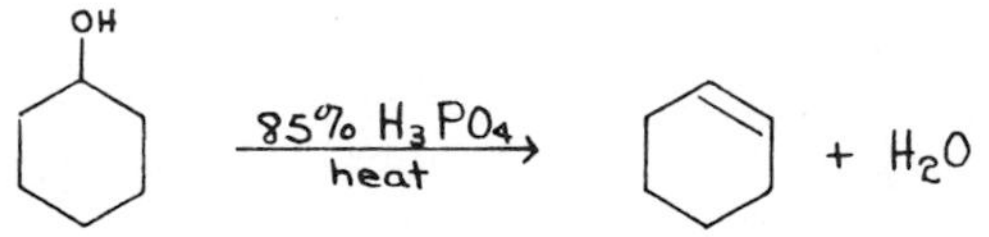

$$\text{cyclohexanol (OH)} \xrightarrow[\text{heat}]{85\%\ H_3PO_4} \text{cyclohexene} + H_2O$$

0.20 mole → 0.20 mole

20.0 grams → 16.4 grams

The product will be isolated by distillation.

Substance	Amount: moles	grams	ml	M.W.	density	m.p.	b.p.	solubility: H_2O	organics
cyclohexanol	0.20	20.0	21	100.2	0.96	25	161	no	yes
85% H_3PO_4			5	98.0		liquid	dec.	yes	no
xylene			20	106.2	mixture of isomers			no	yes
anh. $MgSO_4$		~0.5		120.4		high		yes	no
cyclohexene	0.20	16.4	20.2	82.2	0.81	-131	83	no	yes

Procedure: Ault: Techniques and Experiments for Organic Chemistry, 2nd ed., Holbrook Press, Boston (1976) p. 177

Departures from procedure: washed with 20 ml. of water instead of saturated NaCl solution.

Yield: I obtained 5.2 grams of product boiling between 82° and 84° C. Theoretical yield of cyclohexene: 16.4 g
Percent yield: (5.2/16.4)(100) = 32%

Since a good lab notebook will *show all original data* (such as the weight of the paper upon which a substance was to be weighed and the weight of the paper plus the sample) and will *indicate all calculations* (such as the difference between the two weights just mentioned), the notebook tends to get messy. Yet the notebook will be the only record of your lab work, and you should want it to be intelligible and an example of your best efforts. One way out of this dilemma is to use the left-hand pages for recording data and calculations in a preliminary form, if desired, and the right-hand pages for the final entries. Thus if the product of a reaction were being weighed as just described, the two weights and the subtraction could be recorded on the left-hand page, and the difference (the weight of the sample) would be recopied in the appropriate place on the right-hand page. Figure 4.1 illustrates a sample page from a laboratory notebook.

Questions

1. Look up in a handbook the molecular weight, boiling point, and density of acetic anhydride.
2. Look up in a handbook the molecular weight and melting point of salicylic acid.
3. Aspirin can be made by treating salicylic acid with acetic anhydride according to the procedure described in Section 59.4. Assume that the experiment calls for 10 grams of salicylic acid and 20 mL of acetic anhydride:
   a. Write a balanced equation for the reaction.
   b. Calculate the number of moles of salicylic acid and of acetic anhydride that will be put into the flask.
   c. Which of these substances will determine the maximum amount of product that will be produced? That is, which substance is the limiting reagent?
   d. What is the maximum amount (moles and grams) of aspirin that could be formed in the experiment?
   e. If all the salicylic acid were converted to aspirin, how many moles of acetic acid would be formed, and how many moles of acetic anhydride would remain unused?
   f. If 11.0 grams of aspirin is isolated after the reaction takes place, what is the percent yield of aspirin?

5 The Chemical Literature

Journals

The results of chemical research are normally recorded in notes, communications, or papers in various chemical journals. If you wanted to look up the reported physical properties of a compound, methods of analysis for a compound, or methods of preparation or purification, you could search this primary literature through the index volumes of the various journals. Since there are several dozen major journals, and thousands of journals altogether, the time and effort required would be enormous.

To cope with this problem, abstracting journals have been established. Abstracting journals publish a brief summary, or abstract, of a paper and then provide detailed indexes to the abstracts. Today, *Chemical Abstracts* is the most important abstracting journal in the field of chemistry. *Chemical Abstracts* was started in 1907 by the American Chemical Society, and in 1984—to take a single year—it published abstracts of over 443,000 papers, reports, and patents. *Chemical Abtracts* currently produces five principal indexes: Author, General Subject, Chemical Substance, Formula, and Numerical Patent. An index entry refers you to the abstract, and if the abstract indicates that the desired information is in the original article or patent, you then look up the original by means of the reference provided with the abstract. A complete search of *Chemical Abstracts* requires that you look in the various cumulative indexes and, for recent volumes, the two semiannual indexes for each year. The use of *Chemical Abstracts* is outlined briefly in Section 5.1.

Chemical Abstracts

While this is an improvement over looking things up in the indexes of individual journals, it is still a lot of work if all you want to know is the melting point of a compound or what would be a good solvent for recrystallization. For this reason, a great many specialized secondary sources have been developed. The most familiar is probably the *Handbook of Chemistry and Physics,* published by the Chemical Rubber Company, or *Lange's Handbook of Chemistry.* These and other secondary sources make certain kinds of information easy to obtain. For example, there are collections of methods of synthesis; methods of analysis; physical properties such as melting point, boiling point, solubility, vapor pressure, heat of combustion, and infrared, ultraviolet, NMR, and mass spectra; physiological properties; and hazardous properties. The key to success in quickly finding answers to specific questions about a compound is to be familiar with the secondary sources for that kind of information. Sections 5.2 and 5.3 describe the most commonly used secondary sources for information about physical properties and methods of preparation of organic compounds, and Section 5.4 lists several collections of infrared, ultraviolet, NMR, and mass spectra. A few of the other books that organic chemists find useful are listed in Section 5.5.

Secondary sources

The three articles by J. E. H. Hancock provide a more complete introduction to the literature of organic chemistry (Reference 1).

5.1 CHEMICAL ABSTRACTS

Chemical Abstracts consists of print or microfilm abstracts of notes, papers, reviews, patents, etc., that have been published to report results of research and scholarship in the field of chemistry and related areas. The abstracts have been classified and indexed in order to make it as easy as possible to locate those abstracts that contain or refer to specific information. Without the indexes, the abstracts themselves would be of very limited value. In addition to the indexes referred to above—Author, General Subject, Chemical Substance, Formula, Numerical Patent—there are the Index of Ring Systems, the Patent Concordance, and the Index Guides. An introduction to the Index Guide appears in Volume 76, pages 1I through 140I, and it provides descriptions and instructions for use of the various indexes and of the Index Guide that accompanies Volume 76 (1972). More recent Index Guides were published in 1982, 1984, and 1985. The current Index Guide replaces all former versions.

The first job in searching the literature by means of *Chemical Abstracts* is to locate the relevant entries in the Subject or Chemical Substance Index. If you wish to find information about a specific compound, the task is not particularly difficult, but it may take more than a few minutes. The initial step is to determine the name used to index the compound (the *index name*). An approach that usually works is to look for the substance under its molecular formula in the Formula Index. All references to the compound in which you are interested will be made under the particular name employed as the index name by *CA* at that time. Then you look up the index name in the corresponding Subject or Chemical Substance Index, where the entries will be much more descriptive. The 10th edition of *The Merck Index* also provides the *CA* index name for most of the compounds that it describes.

Determine the index name

Each entry in the index will refer to an abstract by means of two numbers. The first, in bold face, is the *CA* volume number. The second serves to locate the abstract within the volume. For entries from 1907 through 1933, the second number refers to the page, with a smaller superscript number as a suffix indicating the fractional distance down the page on a scale of 1–9, thus: **21**: 138^5. For entries from 1934 through 1966, the second number refers to the column, since two columns have been printed per page from 1934 to the present. In addition, for entries from 1947 to 1966, a letter takes the place of the superscript number, and the scale of distance down the column runs from a–i. In 1967, a new system was adopted wherein the abstracts of each volume are consecutively numbered and where reference is made to an abstract by volume number and, within the volume, by abstract number. The letter that follows these abstract numbers has no meaning except as a computer check character.

Find the abstract

Occasionally the abstract itself contains the desired information, perhaps a melting point, but usually one has to take the final step of looking up the original article by means of the reference cited in the abstract. The reference will include the name of the author, an abbreviated title of the journal, the volume and page numbers, and the year of publication.

Look up the reference

Table 5.1. Partial summary of indexes for *Chemical Abstracts*

Name	*Years*	*Volumes*	S	GS	CS	A	F	IG
Decennial Index	1907–1916	1–10	x			x		
Decennial Index	1917–1926	11–20	x			x		
Decennial Index	1927–1936	21–30	x			x		
Decennial Index	1937–1946	31–40	x			x		
Collective Formula Index	1920–1946	14–40					x	
Fifth Decennial Index	1947–1956	41–50	x			x	x	
Sixth Collective Index	1957–1961	51–55	x			x	x	
Seventh Collective Index	1962–1966	56–65[a]	x			x	x	
Eighth Collective Index	1967–1971	66–75[a]	x			x	x	x
Ninth Collective Index	1972–1976	76–85[a]	x	x	x	x	x	x
Tenth Collective Index	1977–1981	86–95[a]	x	x	x	x	x	x
Semiannual	1982	96, 97[a]	x	x	x	x	x	x
Semiannual	1983	98, 99[a]	x	x	x	x	x	x
Semiannual	1984	100, 101[a]	x	x	x	x	x	x
Semiannual	1985	102, 103[a]	x	x	x	x	x	x

S = Subject Index   GS = General Subject Index   CS = Chemical Substance Index
A = Author Index   F = Formula Index   IG = Index Guide

[a] Two volumes per year.

A search of all the volumes of *Chemical Abstracts*, from 1907 to the present, will require the use of several cumulative indexes and, in addition, the individual volume indexes for the most recent years. Table 5.1 presents a partial summary of the indexes available for *Chemical Abstracts*. In 1972, the Subject Index was divided into two parts: the Chemical Substance Index and the General Subject Index. The first contains references to chemical substances by name, and the second contains references to all other subjects.

5.2 SECONDARY SOURCES FOR PHYSICAL PROPERTIES OF ORGANIC COMPOUNDS

For finding such physical properties as molecular weight, melting point, density, index of refraction, color, and solubility, the following handbooks are very convenient:

- *Handbook of Chemistry and Physics*, 65th edition, R. C. Weast, editor, CRC Press, Cleveland, Ohio, 1984–1985. Contains physical properties for about 14,000 organic compounds in addition to much other information. A reference to Beilstein (see below) is given for almost every compound.

- *Lange's Handbook of Chemistry*, 12th edition, J. A. Dean, editor, McGraw-Hill, New York, 1979. Contains physical properties for about 6500 organic compounds, in addition to much other information. A reference to Beilstein is given for each compound. It is much easier to locate the entry for a compound in the *Lange's Handbook* than in the *Handbook of Chemistry and Physics*.
- *Aldrich Catalog Handbook of Fine Chemicals*, Aldrich Chemical Co., Inc., Milwaukee, Wisconsin, 1984–1985. In addition to molecular weight and selected physical properties, it gives methods for disposal, references to Beilstein and *The Merck Index*, and references to infrared and NMR spectra published in *The Aldrich Library of Infrared Spectra* and *The Aldrich Library of NMR Spectra* (see Section 5.4).
- *Handbook of Tables for Identification of Organic Compounds*, 3rd edition, The Chemical Rubber Co., Cleveland, Ohio, 1967. Contains melting point and boiling point data, and melting points of derivatives for over 8000 organic compounds. The organization is by functional group and by increasing melting point or boiling point within each functional group.

Three other useful secondary sources are

- *Dictionary of Organic Compounds* (Heilbron), 5th edition, Volumes 1–7, Chapman and Hall, New York, 1982. This dictionary is an alphabetical listing of over 25,000 organic compounds. It contains valuable information about solvents used for recrystallization, as well as some reactions, derivatives, and literature references. An annual supplement is published.
- *The Merck Index*, 10th edition, Merck and Co., Inc., Rahway, New Jersey, 1983. This index contains information about solubility, purification, and hazardous properties as well as medicinal uses for 10,000 compounds. The *Chemical Abstracts* index name is provided for most compounds.

Beilstein

- *Beilsteins Handbuch Der Organischen Chemie*, 4th edition, Springer-Verlag, Berlin, 1918–present. This German language work is the most complete secondary source for information about the properties, preparation, and reactions of organic compounds. For all information, references are given to the primary literature sources. The fourth edition, in 31 volumes (Bände), was published during 1918–1938. This principal edition (Hauptwerk; H) covers the organic chemical literature from the beginning until 1909. Since 1938, there have been published a First Supplement (Erstes Ergänzungswerk; E I) covering from 1910 to 1919 in an organization parallel to that of the Hauptwerk, and a Second Supplement

(Zweites Ergänzungswerk; E II) similarly covering the period from 1920 to 1929. The Hauptwerk and the first two supplements list every organic compound known through 1929. A Third Supplement (Drittes Ergänzungswerk; E III) covers the literature from 1930 through 1949, with many more recent references. The publication of a Fourth Supplement (Viertes Ergänzungswerk) was begun in 1972. This supplement covers the chemical literature from 1950 through 1959. Starting with Volume 17, published in 1974, the Third and Fourth Supplements are being published together in common volumes that review the literature from 1930 through 1959.

How to find it in Beilstein

There are four ways to find a compound in Beilstein. Possibly the easiest is to first look it up in either *Lange's Handbook of Chemistry* or the *Aldrich Catalog Handbook of Fine Chemicals*. For example, *p*-bromoacetanilide is found in the *Lange's Handbook* under "Bromoacetanilide (*p*)"; the Beilstein reference is given as XII-642, which means that an entry for this compound is given on page 642 of Volume 12 of the Hauptwerk. In the *Aldrich Catalog Handbook* the reference given under "*p*-Bromoacetanilide" is *Beil.* 12, 642, which provides the same information as the *Lange's Handbook*. The *Handbook of Chemistry and Physics* may also be used. Under the entry "Acetic acid, amide, N(4-bromophenyl)" the handbook gives the reference $B12^2$ 348, which means that an entry for *p*-bromoacetanilide can be found in Beilstein in Volume 12 of the Second Supplement on page 348. However, as the example indicates, it is sometimes difficult to find the entry in the *Handbook of Chemistry and Physics*.

A second method for finding a substance in Beilstein is to use the cumulative Formula Index (General-Formelregister) of the Second Supplement (Volume 29). For each entry, the names of all the isomers are given with references to the Hauptwerk, the Erstes Ergänzungswerk, and the Zweites Ergänzungswerk. For example, there are eighteen entries under C_8H_8BrNO. The fifteenth is "4-Brom-acetanilid **12** 642, I 319, II 348," which means that information about *p*-bromoacetanilide can be found in Volume 12 of the Hauptwerk on page 642, in Volume 12 of the Erstes Ergänzungswerk on page 319, and in Volume 12 of the Zweites Ergänzungswerk on page 348. This method requires a little understanding of the German nomenclature in order to recognize the name of the desired isomer.

A third way is to use the corresponding cumulative Subject Index (General Sachregister) of the Second Supplement (Volume 28). Since this requires a working knowledge of German nomenclature, it is relatively unsatisfactory for most Americans.

The fourth way requires some understanding of the organization of Beilstein. It sounds a little complicated, but after some familiarity with it, it is usually the fastest. Beilstein is divided into four major parts:

1. Acyclische Reihe (nonring compounds) Vols. 1–4
2. Isocyclische Reihe (ring compounds; only carbon atoms in ring) Vols. 5–16
3. Heterocyclische Reihe (ring compounds; atoms other than carbon in ring) Vols. 17–27
4. Natural Products Vols. 30 and 31

(The Subject and Formula Indexes are in Volumes 28 and 29.) Within each of the first two major divisions, compounds are listed in the following order of functioning classes:

1. Kohlenwasserstoffe (hydrocarbons)
2. Oxy-Verbindungen (alcohols)
3. Oxo-Verbindungen (aldehydes and ketones)
4. Carbonsäuren (carboxylic acids)
5. Sulfinsäuren (sulfinic acids)
6. Sulfonsäuren (sulfonic acids)
8. Amine
9. Hydroxylamine
10. Hydrazine
11. Azo-Verbindungen (azo compounds)

Numbers 7 and 12–28 are for other more unusual classes of compounds.

Polyfunctional compounds are found under the class that comes latest in the list: hydroxy acids are found under carboxylic acids, while amino acids are under amines (principle of latest position). Within each class, compounds appear in order of increasing unsaturation, and within these groups in order of increasing molecular weight.

Compounds that are not members of functioning classes are organized in this way:

1. Halogen, nitroso, nitro, and azido substitution products are found following the unsubstituted (or parent) compound. For example, the halo-, nitro-, and halonitrobenzenes appear after benzene.
2. Compounds that give members of the functioning classes upon hydrolysis are found under the last possible entry (principle of latest position). For example, methyl propionate is found under propionic acid, while N-methylpropionamide is found under methylamine. Propionic anhydride, propionamide, and propionitrile are found after propionic acid in that order. Also, methyl benzoate is found under benzoic acid, and phenyl acetate

under phenol, as is anisole (methyl phenyl ether). Phenyl acetate is after anisole (principle of latest position again).

Ø—C(=O)—O—Me
methyl benzoate

Ø—O—C(=O)—Me
phenyl acetate

Ø—O—Me
anisole

Once the entry is found in either the Hauptwerk or any Supplement, the corresponding entry in any other series can be found easily through a system of double numbering of pages. In each volume of each Supplement, in addition to the ordinary page numbers, a cross-reference page number appears at the top center of each page. These numbers refer to the corresponding pages in the Hauptwerk. Thus, having found a compound in the Hauptwerk (or having located the page where it would have been entered had it been known before 1909), one can find and search the corresponding pages of the same volume in each Supplement using these cross-reference page numbers. Similarly, after an entry is located in a volume of the supplement, the cross-reference page numbers can be used to search for corresponding entries in the Hauptwerk or other Supplements. Finally, direct cross-references are often given in the Second and Third Supplements. For example, at the beginning of the entry for *p*-bromoacetanilide in Volume 12, page 348, of the Second Supplement appears the notation (H 642; E I 319), which states that the corresponding entries appear (in Volume 12) on page 642 of the Hauptwerk and on page 319 of the First Supplement.

5.3 SECONDARY SOURCES FOR METHODS OF PREPARATION OF ORGANIC COMPOUNDS

Lab manuals

Some of the most available and easily used sources of detailed directions for the preparation of specific compounds are the various laboratory manuals used in undergraduate courses in organic chemistry. Since these procedures are designed for use by students, a moderately competent organic chemist can usually make them work. Because most of the compounds whose preparation is described in a laboratory manual are available commercially, this source of information is of limited value. The most useful book of this type is

- *Vogel's Textbook of Practical Organic Chemistry, Including Qualitative Organic Analysis,* 4th edition, revised by B. S. Furniss, A. J. Hannaford, V. Rogers, P. W. G. Smith, and A. R. Tatchell, Longman Inc., New York, 1978.

There are two useful collections of specific procedures for the preparation of individual compounds. The first (Shirley) is a single volume, and the second (*Organic Syntheses*) is a continuing series.

- *Preparation of Organic Intermediates,* D. A. Shirley, Wiley, New York, 1951. A collection of more than 500 preparations of compounds not available commercially at the time of publication.

- *Organic Syntheses*, Wiley, New York. A continuing series of annual volumes, initiated in 1921. Each volume presents 30 to 35 detailed and tested preparations of compounds not available commercially at the time of publication. Five Collective Volumes have been published, which include, respectively, annual volumes 1–9, 10–19, . . . , 40–49. In addition to a general index, the Collective Volumes contain indexes for type of compound, type of reaction, molecular formula, apparatus, and author.

There are four reference works that describe general methods for the preparation of different classes of compounds. The first two (Wagner and Zook, and Migrdichian) are one- and two-volume works, respectively, and the last two (*Organic Reactions* and *Newer Methods*) are continuing series.

- *Synthetic Organic Chemistry,* R. B. Wagner and H. D. Zook, Wiley, New York, 1953. Each chapter summarizes known methods for the preparation of a type of organic compound. The tables at the end of each chapter list compounds of the type discussed in the chapter and indicate for each compound the methods used for its preparation, the yield, and references to procedures in the chemical literature.

Organic Synthesis

- *Organic Synthesis,* V. Migrdichian, 2 volumes, Reinhold, New York, 1957. This work is similar to that of Wagner and Zook.

Organic Reactions

- *Organic Reactions,* Wiley, New York. A continuing, approximately biennial, series of volumes, initiated in 1943. A typical volume presents several thorough discussions of a type of reaction or the methods of preparation of a type of compound. Exhaustive tables are included, which summarize the applications of each reaction. The information given includes reaction conditions, yields, and references to the literature.
- *Newer Methods of Preparative Organic Chemistry,* W. Foerst, editor, Academic Press, New York. A series of occasional volumes. Each volume is a collection of review articles that originally appeared in *Angewandte Chemie.* Each article discusses a particular type of reaction and includes some specific preparations and many references to the literature.

A more general reference work that contains information and references concerning the preparation of a great many compounds is the revision of the multivolume treatise originally edited by E. H. Rodd:

- *Rodd's Chemistry of Carbon Compounds,* 2nd edition, S. Coffey, editor; 4 volumes in 33 parts, including supplements; Elsevier, Amsterdam and New York, 1964–1982. A systematic presentation of the preparation and properties of organic compounds. The overall organization is by structural type.

Finally, the most comprehensive and most generally useful reference work for the preparation of organic compounds, as well as their properties and reactions, is Beilstein:

- *Beilsteins Handbuch Der Organischen Chemie,* 4th edition, Springer-Verlag, Berlin, 1918–present.

The use of Beilstein was described in the preceding section.

5.4 COLLECTIONS OF SPECTRA

- *A Handy and Systematic Catalog of NMR Spectra,* Addison Ault and Margaret R. Ault, University Science Books, Mill Valley, California, 1980. A collection of 350 nuclear magnetic resonance spectra that includes, in addition to 290 proton NMR spectra, examples of fluorine and carbon NMR spectra.
- *The Aldrich Library of Infrared Spectra,* C. J. Pouchert, editor, Aldrich Chemical Co., Inc., Milwaukee, Wisconsin; 1st edition (8000 spectra), 1970; 2nd edition (10,000 spectra), 1975; 3rd edition (12,000 spectra), 1981.
- *High Resolution NMR Spectra Catalog,* Varian Associates, Palo Alto, California; Volume 1, 1962; Volume 2, 1963; 700 spectra.
- *The Aldrich Library of NMR Spectra,* C. J. Pouchert and J. R. Campbell, editors, Aldrich Chemical Co., Inc., Milwaukee, Wisconsin; ten volumes plus index; 1st edition (6000 spectra), 1974; 2nd edition (two volumes, 8500 spectra), 1983.
- *Sadtler Standard Spectra,* Sadtler Research Laboratories, Inc., Philadelphia, Pennsylvania. The Sadtler standard spectra are large and continuing collections of infrared, ultraviolet, and NMR spectra. They are available both in printed form and on microfilm. The largest collection, that of infrared spectra obtained with prism instruments, contains over 40,000 spectra. In addition, there are special collections available (pharmaceuticals, commonly abused drugs, etc.) and commercial special collections (agricultural chemicals, food additives, etc.).

5.5 MISCELLANEOUS

Other useful books include the following:

- L. F. Fieser and M. Fieser, *Reagents for Organic Synthesis,* Wiley, New York; Volume 1, 1967; Volume 2, 1969; Volume 3, 1972; Volume 4, 1974; Volume 5, 1975; Volume 6, 1977; Volume 7,

1979; Volume 8, 1980; Volume 9, 1981; Volume 10, 1982; Volume 11, 1984. These volumes contain a wealth of interesting information about the availability, preparation, properties, and use of many substances in organic synthesis. The more recent volumes update and supplement the first volume.

- D. D. Perrin, W. L. F. Armarego, and D. R. Perrin, *Purification of Laboratory Chemicals,* Pergamon Press, New York, 1966. This book first offers a brief discussion of various methods of purification and then gives procedures for purification of individual organic, inorganic, and organometallic compounds.
- J. A. Riddick and W. B. Bunger, *Organic Solvents,* Volume II in *Techniques of Chemistry,* 3rd edition, A. Weissberger, editor, Wiley-Interscience, New York, 1970. This volume contains physical properties and purification methods for 354 solvents.
- N. I. Sax, *Dangerous Properties of Industrial Materials,* 6th edition, Van Nostrand Reinhold, New York, 1984. This book provides discussions of methods of hazard control followed by specific hazard-analysis information for more than 18,000 common industrial and laboratory materials.

References

1. J. E. H. Hancock, "An Introduction to the Literature of Organic Chemistry," *J. Chem. Educ.* **45,** 193, 260, 336 (1968).
2. E. H. Huntress, *A Brief Introduction to the Use of Beilstein's Handbuch der Organischen Chemie,* 2nd edition, Wiley, New York, 1938.
3. O. Weissbach, *The Beilstein Guide.* A Manual for the Use of *Beilsteins Handbuch der Organischen Chemie,* Springer-Verlag, New York, 1976.
4. *Searching the Chemical Literature,* revised and enlarged edition, Advances in Chemistry Series #30, R. F. Gould, editor, American Chemical Society, Washington, D.C., 1961. A collection of papers on various topics concerning the chemical literature.
5. M. G. Mellon, *Chemical Publications,* 4th edition, McGraw-Hill, New York, 1965. An introduction to the nature and use of the chemical literature.

6 Tables

Table 6.1. Solutions of acids

Solution	*Density (grams/mL)*	*Concentration (moles/liter)*[a]	*To Make a Liter of Solution*
95–98% H_2SO_4	1.84	18.1	Concentrated sulfuric acid
6 *M* H_2SO_4	1.34	6.0	332 mL conc. H_2SO_4 + 729 mL H_2O
3 *M* H_2SO_4	1.18	3.0	166 mL conc. H_2SO_4 + 875 mL H_2O
10% H_2SO_4	1.07	1.09	60 mL conc. H_2SO_4 + 957 mL H_2O
1 *M* H_2SO_4	1.06	1.0	55 mL conc. H_2SO_4 + 962 mL H_2O
69–71% HNO_3	1.42	15.7	Concentrated nitric acid
6 *M* HNO_3	1.19	6.0	382 mL conc. HNO_3 + 648 mL H_2O
3 *M* HNO_3	1.10	3.0	191 mL conc. HNO_3 + 831 mL H_2O
10% HNO_3	1.06	1.67	106 mL conc. HNO_3 + 905 mL H_2O
1 *M* HNO_3	1.03	1.0	64 mL conc. HNO_3 + 943 mL H_2O
36.6–38% HCl	1.18	12.0	Concentrated hydrochloric acid
6 *M* HCl	1.10	6.0	500 mL conc. HCl + 510 mL H_2O
3 *M* HCl	1.05	3.0	240 mL conc. HCl + 756 mL H_2O
10% HCl	1.05	2.9	242 mL conc. HCl + 763 mL H_2O
1 *M* HCl	1.02	1.0	83 mL conc. HCl + 920 mL H_2O
99.7% CH_3COOH	1.05	17.5	Glacial acetic acid
6 *M* CH_3COOH	1.04	6.0	343 mL glacial acetic acid + 682 mL H_2O
3 *M* CH_3COOH	1.03	3.0	171 mL glacial acetic acid + 845 mL H_2O
10% CH_3COOH	1.01	1.69	97 mL glacial acetic acid + 912 mL H_2O
1 *M* CH_3COOH	1.01	1.0	57 mL glacial acetic acid + 949 mL H_2O
48% HBr	1.50	8.9	Constant-boiling HBr
57% HI	1.70	7.6	Constant-boiling HI
85% H_3PO_4	1.70	14.7	Syrupy phosphoric acid
70% $HClO_4$	1.67	11.7	Concentrated perchloric acid

[a] Also mmole/mL.

Table 6.2. Molecular weights and molar volumes of acids

Compound	*Molecular Weight*	*Molar Volume, mL*
H_2SO_4, 98%	98.1	55
HNO_3, 70%	63.0	64
HCl, 37%	36.5	83
CH_3COOH	60.0	57
HBr, 48%	80.9	51
HI, 57%	127.9	132
H_3PO_4, 85%	98.0	68
$HClO_4$, 70%	100.5	86

Table 6.3. Molecular weights of bases

Compound	*Molecular Weight*
NaOH	40.0
KOH	56.1
NH_3	17.0
K_2CO_3	138.2
Na_2CO_3	106.0
$NaHCO_3$	84.0
$NaC_2H_3O_2$	82.0
$NaC_2H_3O_2 \cdot 3H_2O$	136.1

Table 6.4. Molecular weights, densities, and molar volumes of selected liquid reagents

Compound	*Molecular Weight*	*Density*	*Molar Volume, mL*
acetic anhydride	102	1.08	94
ammonium hydroxide, 28%	17	0.90	67
aniline	93	1.02	91
bromine	160	3.12	51
hydrazine, 64%	32	1.04	48
phosphorus oxychloride, $POCl_3$	153	1.68	91
phosphorus trichloride, PCl_3	137	1.58	87
pyridine	79	0.98	81
sodium hydroxide, 50%	40	1.53	52
thionyl chloride, $SOCl_2$	119	1.66	72

Table 6.5. Solutions of bases

Solution	*Density (grams/mL)*	*Concentration (moles/liter)*[a]	*To Make a Liter of Solution*
50% NaOH	1.53	19.1	Concentrated NaOH solution
6 *M* NaOH	1.21	6.0	314 mL 50% NaOH + 734 mL H_2O
3 *M* NaOH	1.11	3.0	157 mL 50% NaOH + 873 mL H_2O
10% NaOH	1.11	2.77	145 mL 50% NaOH + 889 mL H_2O
1 *M* NaOH	1.04	1.0	52 mL 50% NaOH + 963 mL H_2O
45% KOH	1.45	11.7	Concentrated KOH solution
6 *M* KOH	1.26	6.0	512 mL 45% KOH + 512 mL H_2O
3 *M* KOH	1.14	3.0	256 mL 45% KOH + 764 mL H_2O
10% KOH	1.09	1.95	167 mL 45% KOH + 850 mL H_2O
1 *M* KOH	1.05	1.0	85 mL 45% KOH + 927 mL H_2O
28–30% NH_3	0.90	15.0	Concentrated ammonium hydroxide
6 *M* NH_3	0.95	6.0	400 mL conc. NH_4OH + 590 mL H_2O
10% NH_3	0.96	5.63	375 mL conc. NH_4OH + 620 mL H_2O
3 *M* NH_3	0.98	3.0	200 mL conc. NH_4OH + 797 mL H_2O
1 *M* NH_3	0.99	1.0	67 mL conc. NH_4OH + 933 mL H_2O
10% K_2CO_3	1.09	0.79	109 g anh. K_2CO_3 + 982 mL H_2O
5% K_2CO_3	1.04	0.38	52.2 g anh. K_2CO_3 + 992 mL H_2O
10% Na_2CO_3	1.10	1.04	52.5 g anh. Na_2CO_3 + 998 mL H_2O
5% Na_2CO_3	1.05	0.50	110.3 g anh. Na_2CO_3 + 993 mL H_2O
Saturated $NaHCO_3$	1.06	1.0	85 g $NaHCO_3$ + 973 mL H_2O
5% $NaHCO_3$	1.04	0.62	52.8 g $NaHCO_3$ + 983 mL H_2O

[a] Also mmole/mL.

Table 6.6. Periodic table of the elements

METALS — NONMETALS

TRANSITION METALS

PERIODS	IA	IIA	IIIB	IVB	VB	VIB	VIIB	VIII	VIII	VIII	IB	IIB	IIIA	IVA	VA	VIA	VIIA	O
1	1.0079 **H** 1																1.0079 **H** 1	4.00260 **He** 2
2	6.94 **Li** 3	9.01218 **Be** 4											10.81 **B** 5	12.011 **C** 6	14.0067 **N** 7	15.9994 **O** 8	18.9984 **F** 9	20.179 **Ne** 10
3	22.9898 **Na** 11	24.305 **Mg** 12											26.9815 **Al** 13	28.086 **Si** 14	30.9738 **P** 15	32.06 **S** 16	35.453 **Cl** 17	39.948 **Ar** 18
4	39.098 **K** 19	40.08 **Ca** 20	44.9559 **Sc** 21	47.90 **Ti** 22	50.9414 **V** 23	51.996 **Cr** 24	54.9380 **Mn** 25	55.847 **Fe** 26	58.9332 **Co** 27	58.71 **Ni** 28	63.546 **Cu** 29	65.38 **Zn** 30	69.72 **Ga** 31	72.59 **Ge** 32	74.9216 **As** 33	78.96 **Se** 34	79.904 **Br** 35	83.80 **Kr** 36
5	85.4678 **Rb** 37	87.62 **Sr** 38	88.9059 **Y** 39	91.22 **Zr** 40	92.9064 **Nb** 41	95.94 **Mo** 42	98 9062 **Te** 43	101.07 **Ru** 44	102.9055 **Rh** 45	106.4 **Pd** 46	107.868 **Ag** 47	112.40 **Cd** 48	114.82 **In** 49	118.69 **Sn** 50	121.75 **Sb** 51	127.60 **Te** 52	126.9046 **I** 53	131.30 **Xe** 54
6	132.9054 **Cs** 55	137.34 **Ba** 56	57–71 *	178.49 **Hf** 72	180.9479 **Ta** 73	183.85 **W** 74	186.2 **Re** 75	190.2 **Os** 76	192.22 **Ir** 77	195.09 **Pt** 78	196.9665 **Au** 79	200.59 **Hg** 80	204.37 **Tl** 81	207.2 **Pb** 82	208.9804 **Bi** 83	(210) **Po** 84	(210) **At** 85	(222) **Rn** 86
7	(223) **Fr** 87	(226.0254) **Ra** 88	89–103 †	104	105	106	107	108										

* LANTHANIDE SERIES	138.9055 **La** 57	140.12 **Ce** 58	140.9077 **Pr** 59	144.24 **Nd** 60	(145) **Pm** 61	150.4 **Sm** 62	151.96 **Eu** 63	157.25 **Gd** 64	158.9254 **Tb** 65	162.50 **Dy** 66	164.9304 **Ho** 67	167.26 **Er** 68	168.9342 **Tm** 69	173.04 **Yb** 70	174.97 **Lu** 71
† ACTINIDE SERIES	(227) **Ac** 89	232.0381 **Th** 90	231.0359 **Pa** 91	238.029 **U** 92	237.0482 **Np** 93	(242) **Pu** 94	(243) **Am** 95	(245) **Cm** 96	(245) **Bk** 97	(248) **Cf** 98	(253) **Es** 99	(254) **Fm** 100	(256) **Md** 101	(253) **No** 102	(257) **Lr** 103

Separation of Substances; Purification of Substances

A pure substance contains only one kind of molecule; an impure substance is a mixture of molecules. When the different molecules of a mixture each behave in a different way under the conditions of some procedure, the procedure can result in a separation of the different molecules. The theory of each of the separation procedures described in this section is presented from the point of view of the different behavior that can be expected from different molecules under the experimental conditions.

According to the definition just given for a pure substance, purification procedures must be separation procedures. The definition also implies that the ultimate experimental criterion for purity is that a substance has been shown to be inseparable by all known separation procedures. Thus the possibility always remains that a substance currently believed to be pure may someday be shown to be separable into components. This situation is appreciated especially well by biochemists, who have repeatedly had the experience of finding that a new separation procedure shows that a substance considered pure is, in fact, a mixture. Separation of many biochemical materials is very difficult because often the molecules to be separated behave in nearly the same way in almost all separation procedures.

In practice, the question of the purity of a particular substance is usually settled indirectly. That is, the properties of a sample of the substance in question are compared with the properties of a sample that is judged to be pure because it could not be separated into components. If the two samples agree in all properties, they may be judged to be of the same purity—both pure or, sometimes, both impure. Some of the properties by which substances can be characterized, such as boiling point, index of refraction, and infrared spectrum, and the ways by which they can be determined, are discussed in Part I, Determination of Physical Properties, Sections 16–26. If the comparison shows that the samples differ in some property, at least one is not pure.

As an example, suppose you isolate eugenol from oil of cloves (Section 45). After purification of the sample, you determine its boiling point, index of refraction, and infrared spectrum. Comparison of these properties with those determined for a sample of eugenol obtained from a chemical supply house shows agreement within experimental precision. The strongest statement that can be made concerning the purity of your product is that *you have not shown the composition of your material to be different* from that of the sample obtained from the commercial source. It is possible that if both samples were obtained from the same type of cloves, and purified using the same methods, both could be contaminated by the same impurities. If each sample had been prepared in a different way from different starting materials (so that the product mixtures would contain different impurites) and each had been subjected to extensive purification by powerful separation procedures, your confidence in the degree of purity of both samples would be much greater. You would have to be very unlucky to have both samples contaminated to the same degree with the same impurities.

Usually, however, the comparison with a reference sample is not a direct comparison like the one just described. Most often, you simply compare your experimental values with values given in the chemical literature. To use the example of eugenol from cloves again, you would compare your experimentally determined boiling point and index of refraction with the literature values (which may not be self-consistent) and, if possible, your infrared spectrum with a photoduplicated reference spectrum. In this kind of indirect comparison, you are usually more tolerant of discrepancies between your experimental values and the literature values, since it is unlikely that the experimental conditions were exactly the same. For example, two samples whose melting points differed by five degrees when determined at the same time could not be called identical. However, a five-degree discrepancy between an experimental melting point and a literature value would not usually *in itself* cause you to conclude that the samples were different. Factors such as different apparatus, different thermometer, different rates of heating, and different experimenters could easily account for the disagreement; the samples could be identical.

It should now be apparent that the question "Is this sample pure?" can never be answered with an unqualified Yes. The strongest statement that can be made regarding purity is "The sample was shown to be homogeneous by separation procedures A, B, and C," or "The sample had the same properties as a sample prepared by method X and purified by procedures Y and Z."

7 Filtration

Filtration involves the separation of insoluble solid materials from a liquid. In this operation, the liquid passes through a porous barrier (sintered glass or filter paper) and the solid is retained by the barrier. The liquid can be made to pass through the barrier by gravity alone, in which case the procedure is called a *gravity filtration.* Alternatively, the liquid can be caused to pass through by a combination of gravity and air pressure. Such an operation is called a *vacuum* or *suction filtration.*

Figure 7.1. Folding of filter paper for gravity filtration: (a) Fold the filter paper circle (11 cm diameter) in half. (b) Crease the half to divide it into eight equal pie-shaped sections; it is easiest to make the creases in the numerical order shown. (c) Turn the piece over and pleat it into a fan by folding each pie-shaped section in half in the direction opposite to the previous creases. (d) Pull the two sides apart.

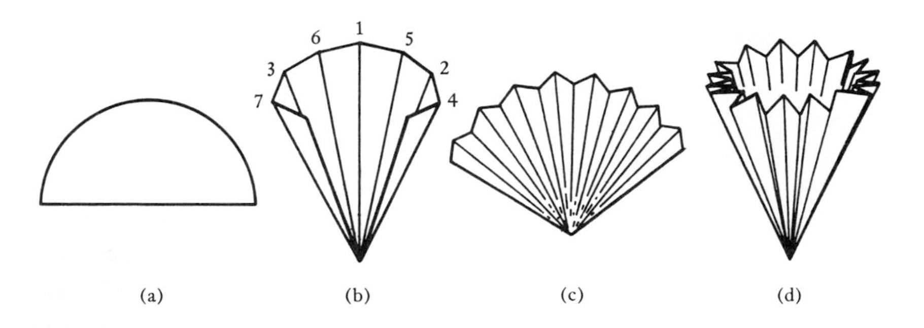

7.1 GRAVITY FILTRATION

A piece of filter paper and a conical glass funnel to support it are all that are required for gravity filtration. In order to maximize the rate at which the liquid flows through the filter paper, the paper should be folded as indicated in Figure 7.1. The folded paper is then dropped into the funnel (see Figure 7.2). The funnel is best supported in an iron ring, as shown in the figure. The material to be filtered is poured into the filter paper cone, in portions if necessary. This operation is used with a stemless funnel in hot filtration during recrystallization (Section 8.4), or in removal of a drying agent from a solution (Section 15.2).

7.2 VACUUM OR SUCTION FILTRATION

In vacuum or suction filtration, a partial vacuum is created below the filter, causing the air pressure on the surface of the liquid to increase the rate of flow through the filter. A typical apparatus is illustrated in Figure 7.3.

Don't make this mistake

A circle of filter paper just large enough to cover the holes in the bottom of the Hirsch or Buchner funnel should be used. A common error is to try to use a piece of filter paper so large that it must be turned up at the edges. If this is done, it is almost impossible to create a vacuum in the suction flask. Not only will the filtration take much longer, but any material that flows over the edge of the filter paper will run down into the suction flask without being filtered.

Figure 7.2. Arrangement of filter paper, funnel, and flask for gravity filtration.

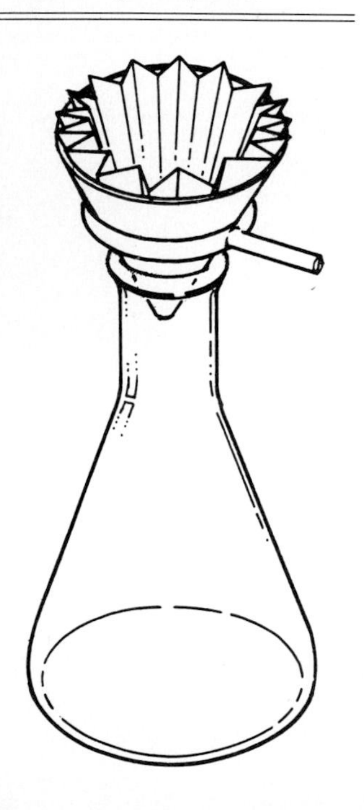

Filtration is carried out by connecting the side arm of the suction flask to the source of vacuum, which is almost always the water aspirator. When a water aspirator is used, the flask should be connected to the aspirator through a trap, as shown in Figure 7.3. The trap prevents water from the aspirator from being sucked back into the filter flask. Turn the aspirator on just a little at first so as to create a gentle vacuum, wet the filter paper with a small portion of the same solvent used in the solution being filtered while making sure that the paper is being pushed down over the holes, and pour the mixture to be filtered onto the center of the paper. Once the mixture has been added, the vacuum may be increased. When using the water aspirator, be sure to break the vacuum by disconnecting the tubing attached to the side arm of the filter flask before turning off the water. Suction filtration is used to collect a solid after recrystallization.

8 Recrystallization

Purification of a solid by recrystallization from a solvent depends upon the fact that different substances are soluble to differing extents in various solvents. In the simplest case, all the unwanted materials are

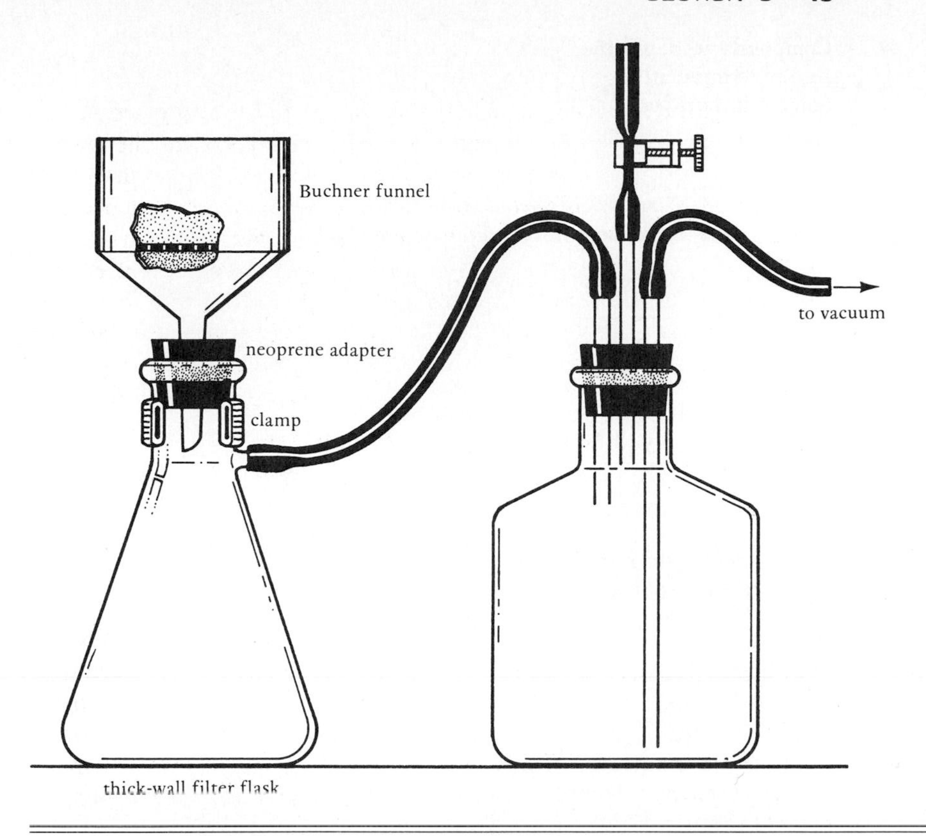

Figure 7.3. Apparatus for vacuum filtration.

much more soluble than the desired compound. In this case, the sample is dissolved in just enough of the hot solvent to form a saturated solution, the solution is cooled, and the crystals, which will have separated upon cooling, are collected by suction filtration (Section 7.2). The soluble impurities remain in solution after cooling and pass through the filter paper with the solvent upon suction filtration.

If insoluble impurities are present in the sample, they are removed by filtering the hot solution by gravity (Section 7.1) before it is allowed to cool.

These procedures are summarized in the flow chart in Figure 8.1.

8.1 CHOICE OF SOLVENT

The choice of solvent is crucial in purification by recrystallization, but there is no easy way to know which solvent will work best. If you wish to recrystallize a known compound, the chemical literature (perhaps a

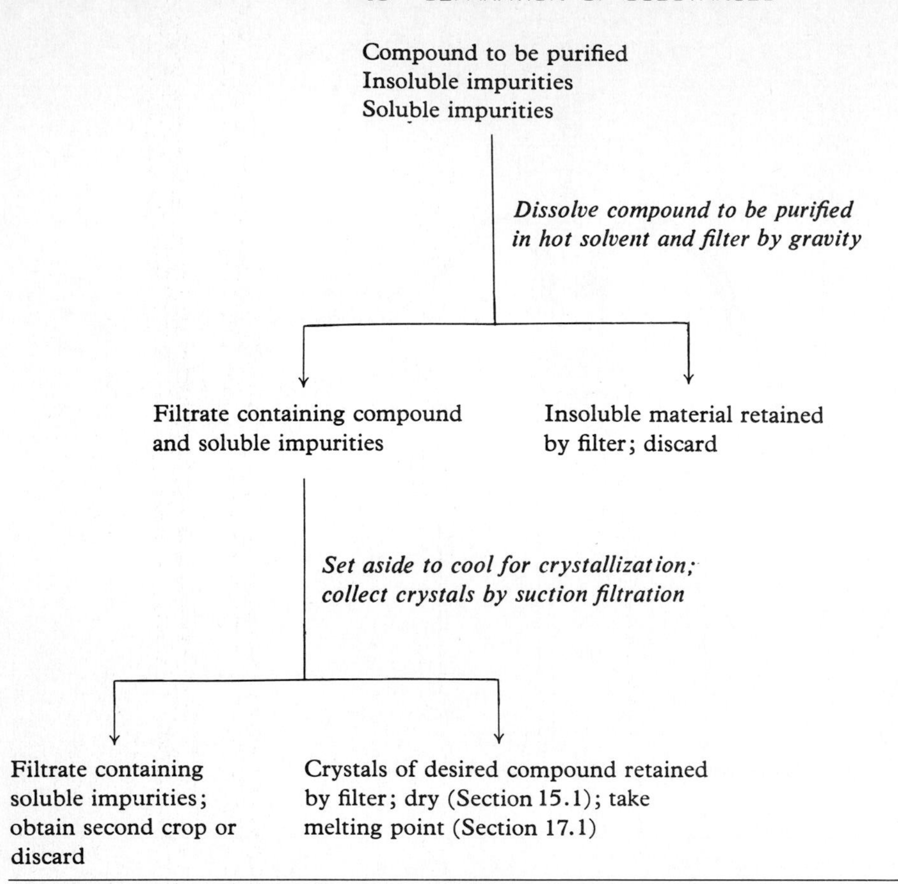
Compound to be purified
Insoluble impurities
Soluble impurities
Dissolve compound to be purified in hot solvent and filter by gravity
Filtrate containing compound and soluble impurities
Insoluble material retained by filter; discard
Set aside to cool for crystallization; collect crystals by suction filtration
Filtrate containing soluble impurities; obtain second crop or discard
Crystals of desired compound retained by filter; dry (Section 15.1); take melting point (Section 17.1)

Figure 8.1. Flow chart for the recrystallization of a solid.

lab manual) may report or recommend the use of certain solvents. However, these solvents may not be useful if different impurities are involved. Quite often, several solvents must be tried.

Test the solvent

One essential characteristic of a useful solvent is that the desired compound must be considerably more soluble in the solvent when it is hot than when it is cold. A way of testing whether or not this requirement is met is to add as much of the compound as will cover the tip of a spatula (100 mg) to a small test tube, add a few drops of solvent, and (if the substance does not dissolve in the cold solvent) heat the mixture to boiling on the steam bath or over a low burner flame. If the material does not go into solution at this point, more solvent may be added, a little at a time, with continued heating, until it does. Then the solution is cooled by placing the test tube in a beaker of cold water to see whether or not the compound will crystallize from solution.

There are several points at which you may decide that a particular solvent is not suitable for use in recrystallizing a given compound. If a very small amount of solvent (say 1 mL per gram) serves to dissolve the compound when it is cold, the amount of solvent that can be used will be so small that at the point of suction filtration you will be working with a damp mush rather than a suspension of crystals. When the compound dissolves this readily in the cold solvent, it will not be possible to use fresh cold solvent either to help in transferring the mush to the funnel for suction filtration or to wash the crystals free of the remaining solution containing the impurities, without involving unacceptable losses.

If a very large amount of solvent is required to dissolve the compound (say 100 mL per gram), you may not have large enough flasks and may not be able to afford the amount of solvent required. If only 1 gram needs to be recrystallized, there is no problem, but 100 grams would require 10 liters of solvent!

Finally, of course, the solvent might have to be rejected because the compound is not much more soluble in the hot solvent than in the cold solvent.

The relatively fast, qualitative test just described will be satisfactory in many cases. Sometimes you will want to be more quantitative and to weigh the amounts of compound used and recovered, as well as to measure the amount of solvent used. In order to obtain a valid measurement of the amount of solvent used, you will have to fit the flask or test tube used for the trial recrystallization with a reflux condenser that will condense the vapors of the boiling solvent and return them to the flask.

With experience, you will find that this quick test of the suitability of a solvent for recrystallization often works satisfactorily. But at first, it may give you misleading indications. There are several mistakes that you should avoid. If the compound dissolves only slowly in the boiling solvent, you may prematurely reject the solvent as incapable of dissolving the compound. Or you may decide that more solvent will be required than is actually necessary. As a result, you will add too much solvent; your estimate of the amount of solvent needed will be too big, and your estimate of the percent of compound that can be recovered on cooling will be too small, since the additional solvent also will dissolve more compound when it is cold. As an extreme case, you may use so much extra solvent that the solution does not become saturated on cooling and nothing is recovered.

Don't be fooled

Sometimes a compound fails to separate immediately from a cooled solution and does so only after a considerable length of time. When this occurs, the solution is said to be *supersaturated,* and it is possible to mistakenly decide that the substance is as soluble in the cold solvent as in the hot solvent, or that too much solvent was used to dissolve the sample. Since supersaturation occurs when the process of forming new crystal nuclei is unusually slow, supersaturation can be relieved by

Supersaturation

Seeding; scratching

adding a powdered crystal of the substance (seed crystals) or by scratching the walls of the flask below the surface of the solution with a stirring rod. Scratching the glass relieves supersaturation by producing burrs and bits of glass that, by chance, act as points of crystal growth.

Although crystallization is sometimes essentially complete in a few minutes and often within a half hour, it *can* take several hours or even several days. When this happens, it is easy to mistakenly decide that only a small percent of the sample will ever separate from solution, simply because you did not wait long enough for crystallization to become complete.

Allow equilibrium to be established

The mistakes that can be made in estimating the suitability of a solvent can be summarized by saying that you mistakenly assume the equilibrium has been established when actually it has not. Either not all that will dissolve has yet dissolved, or not all that will crystallize has yet crystallized.

Problem of insoluble impurities

So far, we have assumed that the sample of the compound being used to determine the suitability of a solvent for recrystallization is relatively pure. If the sample contains appreciable amounts of "insoluble" impurities, another problem arises. At some point in your attempt to dissolve the sample in the hot solvent, the residue will be the insoluble impurity, not the desired compound. What you want to do is stop at this point, separate the solution from the residue, and allow the solution to cool. The problem is to distinguish this situation from the one in which the residue is composed of both the desired compound and the impurity, or is entirely the desired compound. A careful observer may be able to detect a different appearance of the residue, or a slowing in the process of dissolution since the more soluble material is already in solution and since the addition of more solvent does not dissolve as much material now as a similar amount did earlier. If this situation is not recognized, you may succeed only in recrystallizing the relatively insoluble impurity.

Probably the best approach in such a situation is to use sufficient solvent to dissolve the desired compound and to leave as residue what appears to be the insoluble impurity, then to separate the solution from the residue (1) by allowing the latter to settle and decanting the supernatant solution or (2) by removing the solution with a medicine dropper or (3) by carrying out a gravity filtration on the hot mixture. The properties of the residue (melting point, infrared spectrum) can be compared with those of the original sample.

Even if a solvent is found, which in convenient amounts (say 5–10 mL per gram) quickly dissolves the compound when hot, and from which the compound separates quickly on cooling as nice crystals with 80–90% recovery, the solvent still may not be suitable for the purification of a particular sample. It may be that certain impurities are neither much more soluble nor much less soluble than the desired substance; they neither remain in solution upon cooling nor remain "entirely undissolved" in the hot solvent so that they can be removed by gravity filtration of the hot solution. Instead, there can be intermediate situations in which the relative solubilities of the desired compound and an impurity are such that it is only the impurity that can be purified by recrystallization. When this occurs, another solvent must be used or this particular impurity must be removed in another way.

If you are working with a new compound or if you can find no recommendation for the use of a particular solvent in the recrystallization of a compound, you must simply test various solvents. A guiding principle is that solvents of molecular structure similar to that of the compound may be good solvents for that compound.

In some cases, mixtures of solvents may be necessary or useful (Section 8.9). Table 8.1 lists certain properties of solvents commonly used for crystallization.

Table 8.1. Properties of common solvents

Solvent	*Boiling Point (°C)*	*Density (g/mL)*	*Solubility in Water*	*Flammability*
Acetic acid	118[a]	1.05	∞[e]	+
Acetone	57	0.79	∞	+++
Acetonitrile	80	0.79	∞	++
Benzene	80[b]	0.88	i[f]	+++
Carbon disulfide	45	1.26	i	++++
Carbon tetrachloride	77	1.59	i	−
Chloroform	61	1.49	i	−
Cyclohexane	81	0.78	i	+++
Dichloromethane	41	1.34	i	−
Dimethylformamide	153	0.94	∞	+
Dimethylsulfoxide	189[c]	1.10	∞	+
Dioxane	101[d]	1.03	∞	+++
Ethanol, 95%	78	0.81	∞	++
Ether, diethyl	35	0.71	7[g]	++++
Ethyl acetate	77	0.90	8[g]	+++
Hexane	68	0.66	i	+++
Ligroin	60–90	0.67	i	+++
Methanol	65	0.79	∞	+
Pentane	36	0.63	i	++++
Petroleum ether	30–60	0.64	i	++++
Tetrachloroethane	146	1.60	i	−
Tetrahydrofuran	65	0.89	∞	+++
Toluene	111	0.87	i	++
Water	100	1.00	∞	−

[a] Freezes at 17°C
[b] Freezes at 5°C
[c] Freezes at 18°C
[d] Freezes at 12°C
[e] Miscible in all proportions
[f] "Insoluble"
[g] Grams per 100 mL

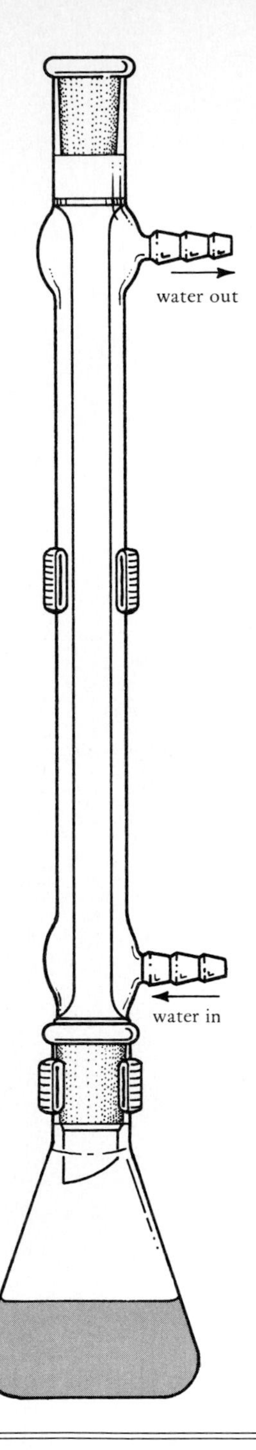

Figure 8.2. Apparatus for dissolving a sample for recrystallization.

8.2 DISSOLVING THE SAMPLE

After a solvent has been chosen, either through a recommendation in the literature or by the test procedure just described, a solution of the sample in the hot solvent must be prepared. Figure 8.2 shows a typical apparatus.

Choose an Erlenmeyer (conical) flask of such a size that it will be less than half filled with solution. Place the solid in the flask, add about 75% of the amount of solvent thought to be required, fit the flask with a reflux condenser, and bring the mixture to a boil. For solvents boiling below 95°C, use the steam bath; for higher-boiling solvents, a burner or an oil bath is appropriate (Section 32.1). If a little more solvent is needed, add it down the condenser (if a flammable solvent is being heated with a burner, extinguish the burner before adding the solvent). Use of a boiling stone is recommended (Section 9.6).

If the amount of solvent needed is not specified or has been determined only roughly, place only a portion of the sample in the flask (for example, 1.0 gram in a 125-mL Erlenmeyer flask) and add measured amounts of solvent down the condenser. From the amount of solvent required to dissolve this first portion, the amount needed for the whole sample can be calculated. If the total will fill the flask much more than half full, transfer the first solution to a larger flask of appropriate size and prepare the rest of the solution in the larger flask. If no transfer is necessary, add the remainder of the sample (remove the flask from the heat and remove the condenser from the flask) followed by the rest of the solvent, using the solvent to wash the solid down from the neck of the flask as necessary.

If all the sample is added to the flask at once and then it turns out that more solvent is required to dissolve the sample than will fit in the flask, a very messy transfer will be necessary in which much material can be lost.

You should keep in mind the possibilities that the compound may dissolve only slowly in the boiling solvent and that "insoluble" impurities may be present in your sample.

8.3 DECOLORIZING THE SOLUTION

The presence of colored impurities in a sample or in a solution of a colorless compound is obvious. Sometimes during recrystallization, the colored substances remain in solution upon cooling and are removed with the solvent upon suction filtration. Often the colored substances are adsorbed by the crystals as they are formed, giving an obviously impure product. Since the types of molecules that impart the color to the solution are often the types that are preferentially adsorbed by activated charcoal (decolorizing carbon),* the addition of a small amount of activated charcoal (1–2% by weight of the sample; or 1 mg per milliliter of solution; or enough to cover the tip of a spatula) in 50 mL of solution, followed by gravity filtration of the hot solution, can serve to remove some if not most of the color. The molecules responsible for the color will be adsorbed by the carbon and will be separated along with it during the filtration.

The carbon should be added only after the solution has cooled a bit below the boiling point; otherwise, the solution will boil over when the addition is made. After the addition, the solution may be reheated to boiling. In general, it is quite all right to use decolorizing carbon in any recrystallization unless it is obvious from the lack of color of the solution that this step is not necessary.

Warning: cool below b.p. before adding carbon

Activated charcoal can also serve to adsorb small amounts of resinous material or very finely divided solid impurities that might not otherwise be removed by filtration.

Adsorption is not as efficient at higher temperatures, and sometimes it may be preferable to prepare an ethereal solution of the substance at room temperature, treat this solution with activated charcoal, filter it by gravity, and remove the ether by evaporation. This process is very similar to that of chromatography (Section 14), and the principle is just the same.

8.4 HOT FILTRATION

When the desired substance is in solution in the hot solvent, insoluble impurities (including dust, pieces of filter paper, glass, or cork) and decolorizing carbon, if used, can be separated by filtering the hot solution by gravity. Vacuum filtration cannot be used, because the reduced pressure in the suction flask will cause the filtrate to boil and material in solution will be deposited over the walls of the flask.

The main problem in hot filtration is that the hot solution cools a little before it runs through the filter. This means that some crystallization can take place in the filter. You can try to avoid this undesired crystallization by warming the funnel (with steam, quickly wiping it dry with a towel; or with a flame) and pouring only a little of the solution into the filter at a time, keeping the remainder at the boiling point. If a large volume of solution must be filtered, or if the solubility of the substance decreases greatly in the range just below the boiling point, a heated funnel is a necessity. Often it is a good idea to use a small excess (10–25%) of solvent so that the solution will not become

* Activated carbon prepared from wood is desirable, having a surface area of hundreds of square meters per gram. Animal charcoal (bone black) contains a large proportion of inorganic salts and has less adsorptive power. Excellent decolorizing carbons are sold under the trade names of Norit and Nuchar.

saturated until the temperature has fallen somewhat below the boiling point. A stemless funnel is always recommended: with a stemless funnel, you can avoid the problem of crystallization and subsequent clogging in the stem.

Use a stemless funnel

Although usually the filtrate should be collected directly in the Erlenmeyer flask in which the cooling for crystallization is to take place, an arrangement such as that illustrated in Figure 8.3 can be very useful in cases where there is trouble with crystallization of the compound in the filter. The filtrate in the beaker can be boiled up around the funnel, keeping it hot. Some care must be used with a flammable solvent if the heating must be done with a flame. After filtration is complete, the filtrate can be transferred to an Erlenmeyer flask for crystallization.

If an excess of solvent has been used to minimize the problem of crystallization in the filter paper, it may be removed at this point by distillation. This will also serve to bring back into solution any crystals that may have separated during filtration.

Reheat filtrate

In all cases, the filtrate should be reheated, if necessary, to dissolve any crystals that may have formed and to give a clear solution. Crystals that form when the hot filtrate hits the cold flask are likely to be less pure than those that separate more slowly from solution.

8.5 COOLING

After the filtrate from the hot filtration has been adjusted to the desired volume and all solid has been brought back into solution, the filtrate is allowed to cool.

Don't cool too fast!

The rate of cooling determines the size of the crystals. Slow cooling tends to favor fewer and larger crystals; fast cooling tends to favor more and smaller crystals. Very large crystals are to be avoided since they often occlude the solvent and its dissolved impurities. Very small crystals are undesirable because it is difficult to wash them free of the solvent and the soluble impurities, and it takes longer to dry them. Needles between 2 and 10 mm in length are fine, as are prisms 1 to 3 mm in each dimension.

Usually the best compromise of speed, convenience, and quality of crystals is reached by allowing the solution to cool to room temperature on a non-heat-conducting surface such as a cork ring. Sometimes it will be possible to hurry the cooling without getting overly small crystals by swirling the flask in a beaker of water at room temperature, or even by placing the flask directly in an ice bath. The rate of cooling can be slowed greatly by supporting the flask in a beaker of water at the temperature of the solution and allowing both to cool spontaneously to room temperature. The fastest and most satisfactory procedure can only be determined by experimentation.

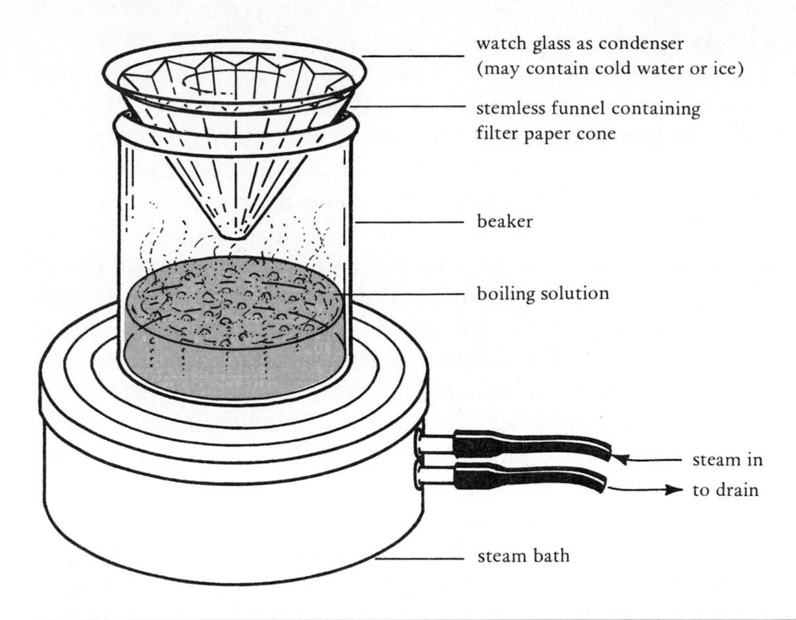

Figure 8.3. A way to prevent crystallization in the funnel during hot filtration.

Since solubility decreases with decreasing temperature, it is often a good idea to finally cool the mixture from room temperature to 0°C (or to the freezing point of the solvent if it is above 0°C) in a mixture of ice and water. Cooling below 0°C is not often done, since colder baths are not so easily prepared and other problems related to the condensation of water vapor from the air in the form of water or frost require special techniques and apparatus.

If the crystals are collected too soon, some material will be lost, which would have separated from solution on further standing. The minimum acceptable time for crystallization—varying from a few minutes for some substances to days for others, as mentioned in Section 8.1—can be determined only by experiment. In order to tell whether one-half hour is enough, two identical samples must be prepared. One is filtered after standing for half an hour at room temperature, and the other perhaps the next day. A comparison of the amount recovered in the two cases will tell whether anything is to be gained by letting the sample stand for more than half an hour.

When is crystallization complete?

Similar experiments can be performed to answer the question of whether it is worth cooling the mixture in an ice bath before the filtering operation.

8.6 COLD FILTRATION

When crystallization is complete, the product is collected by suction filtration. The size of the funnel used should be such that it will not be more than half filled with crystals. The suction flask should be large enough so that the solution will not fill it above the tip of the funnel or the side arm.

Sometimes it is possible to decant the solvent through the funnel until the mass of crystals remaining in the flask is just covered with solvent, and then to pour the entire mass of crystals into the funnel in one smooth operation. More often, it is necessary to suspend the crystals in the solvent by swirling, and then to quickly pour part of the suspension into the funnel, wait until the liquid level has fallen almost to the level of the crystals in the funnel, and then repeat the swirling and pouring operation.

Crystals will often remain in the flask after all the solvent has been poured out. These must be scraped into the funnel, rinsed into the funnel with fresh cold solvent (if the compound is relatively insoluble in the cold solvent), or rinsed into the funnel with portions of the filtrate. The last process can easily lead to a mess. It is far better to work hard to pour the crystals out with the liquid in the first place!

During the entire operation of getting the crystals into the funnel, it is best if the level of the liquid in the funnel does not fall below that of the crystals. Air should never be drawn through the crystals until they have been rinsed with fresh solvent (see next section). Sometimes it is necessary to break the vacuum in the flask by disconnecting the hose at the side-arm in order to slow the rate of filtration sufficiently.

8.7 WASHING THE CRYSTALS

After the crystals have been transferred to the funnel for suction filtration and the liquid has been drawn off, some fresh solvent should be poured over the crystals in order to wash off the liquid that contains the soluble impurities. If this is not done, the soluble impurities will be deposited on the crystals when the solvent evaporates. If the product is relatively soluble in the cold solvent, one washing will usually have to suffice. Two washes are generally appropriate.

If the crystals are relatively soluble, a minimum amount of solvent must be used, and it should be cooled thoroughly in an ice bath. If the crystals are not very soluble, larger amounts of solvent may be used and it need not be chilled. In either case, it is possible to use some of the wash liquid to rinse any remaining crystals out of the flask in which crystallization took place.

Don't use too much solvent

If the crystals are not matted down tight into a solid cake, the vacuum can be released and the wash liquid poured evenly over the crystals and then drawn off by reestablishing the vacuum. If the crystals do form a solid cake, the wash liquid will have to be added to the crystallized matter in the funnel and the product mass carefully pulled apart and suspended evenly by means of a small spatula. Some care is required to suspend all of the product and not tear or dislodge the filter paper. It is best if a very slight vacuum can be maintained during this operation without drawing the wash liquid through too fast. When a procedure says "wash thoroughly," it is probably calling for this rather tedious and delicate operation of carefully breaking up and suspending the filter cake in the wash liquid. After this has been done, the wash liquid is drawn off by suction. An alternative procedure that is sometimes appropriate is to transfer the filter cake to a beaker, add solvent, and break up and suspend the material in the beaker. Following this, the product is again collected by suction filtration.

Sometimes, if a relatively nonvolatile solvent such as acetic acid or nitrobenzene is used for recrystallization, it may be possible to wash this solvent off with a more volatile solvent so as to speed the drying of the crystals. It should hardly be necessary to mention that the crystals should not be soluble in the more volatile solvent.

8.8 DRYING THE CRYSTALS

After you have collected the crystals by suction filtration, remove as much solvent from the product as possible by continuing to draw air through the crystals while they remain on the filter paper in the funnel. The last traces of solvent will evaporate when the crystals are removed from the funnel and spread out to dry.

If the crystals collect as a solid cake, air should be drawn over them in the funnel until the solvent has almost stopped dripping. Failure to suck the filter cake as dry as possible before breaking it up and spreading it out to dry is a very common mistake. The filter cake should be a damp, friable solid, not a paste or mush, when spread out to dry.

Suck crystals as dry as possible

More ways of drying solids are described in Section 15.1.

8.9 MORE TECHNIQUES OF CRYSTALLIZATION

The preceding sections describe the normal techniques for recrystallization. There are, however, various problems that can be encountered during a recrystallization. This section describes some of these difficulties and suggests some possible solutions.

Solvent Pairs

Occasionally you will wish to recrystallize a compound that is too readily soluble in some of the available solvents and not soluble enough in others. When this is necessary, a mixture of solvents can be useful. To test the suitability of a pair of solvents, prepare a hot solution of a sample of the compound in a small amount of the better solvent, and add slowly, while keeping the mixture hot, some of the poorer solvent. When a cloudiness is produced by slight crystallization, add a little of the better solvent, still keeping the mixture hot, until the cloudiness has been dispelled, and then allow the solution to cool. Although the solubility of a substance in a mixture of solvents usually changes gradually with the proportion of the solvents, the solubility of a substance is sometimes greatly increased or decreased by the addition of only a small amount of a better or a poorer solvent.

Solvent pairs that are most often used include toluene/hexane, acetic acid/water, and alcohol/water. Any pair of miscible liquids can be used. Although alcohol/water mixtures are often used, they seem to promote separation of the product as a liquid.

When a recrystallization is carried out using a pair of solvents, it is often a good idea to do the hot filtration before adding the poorer solvent. The addition of the poorer solvent should then be carried out as just described. A very common error is to add more of the poorer solvent than is needed to just achieve saturation of the hot solution. In extreme cases, this results in the precipitation of everything in solution, impurities as well as the desired compound.

Oiling Out

Sometimes, during cooling for crystallization, the product separates not as crystals but as a liquid (an oil). This may be indicated first by the formation of a cloudiness or opalescence, and then by the formation of visible droplets. It is undesirable to allow the product to separate as an oil because often the oil is an excellent solvent for impurities. When (or if) the oil finally freezes, the impurities that have dissolved in the oil will be in the crystals.

Separation of the product as an oil occurs most often in the recrystallization of low-melting substances, or when mixtures of alcohol and water are used as solvent.

Sometimes oiling out can be prevented by using a little more solvent (or more of the better solvent of a pair) so that the solution will become saturated at a lower temperature. The lower the temperature at which the product separates, the more likely it is to separate as a solid rather than as a liquid.

If the first traces of oil can be caused to solidify by the addition of seed crystals, by vigorous stirring or swirling of the mixture, or by scratching the walls of the flask with a stirring rod, the remainder of the product will usually separate as crystals if the rate of cooling is not too great. If oiling out cannot be prevented—that is, if most of the product separates as an oil before it can be caused to solidify—you can hope that recrystallization of the solidified oil will give a better result, you can try a different solvent, or, probably best, you can purify the product by another method before attempting to recrystallize it.

We have assumed so far that, at equilibrium, crystals are present rather than oil. In some cases, oil formation cannot be prevented, since oil is present in the equilibrium state. This happens, for example, in the recrystallization of *pure* acetanilide from water at acetanilide concentrations of greater than 5.2% by weight. The only possible remedy in these cases is to change the composition of the system by adding more water or by changing the solvent.

Failure to Crystallize

Occasionally, crystallization will not occur when a solution is cooled, even though it is supersaturated. The most stubborn cases are the result of impurities, or "tar," acting as a protective colloid. If the normal expedients of adding a seed crystal or scratching the flask with a stirring rod fail, crystallization can sometimes be initiated by cooling the mixture in a salt/ice bath (about −10°C) or a Dry Ice/acetone bath (about −70°C), depending upon the freezing point of the solvent.

Since the rate of crystal growth is lower at low temperatures and in the more viscous solutions that are obtained at low temperatures, a higher temperature is needed for a good rate of crystal growth than for crystal initiation. For this reason, it sometimes works to cool the mixture for a while to −70°C to initiate crystal formation, and then to allow it to warm slowly to room temperature. If crystals have been initiated at the low temperature, they may have an opportunity to grow at an optimum rate in an intermediate temperature range attained during the warming process. Placing the flask on a piece of Dry Ice for a few minutes is sometimes helpful. Alternatively, bits of Dry Ice may be dropped into the solution.

For crystallization to occur in some cases, the solution must be stored in a refrigerator or freezer for long periods of time, even years.

Solubility Differences Caused by Impurities

As pointed out, impurities frequently cause oiling out and can inhibit crystallization. Often, when the level of impurity has been reduced by one recrystallization, the sample will behave appropriately in succeeding crystallizations. Occasionally an impurity is present that greatly increases the solubility of the substance. As this material is removed by recrystallization, the solubility of the substance may decrease dramatically, and in some cases the compound will become "insoluble" in the solvent previously used for recrystallization.

Wet Samples

A solid will often be isolated by pouring a reaction mixture into water and collecting the resulting precipitate by suction filtration. When, in this procedure, the product separates as a fine powder, as it frequently does, it is difficult to suck it dry on vacuum filtration. The crude, damp material can easily be more than half water. Recrystallization of such damp material from a water-miscible solvent will often require the use of more solvent than expected, since the water in the crude product will reduce its solubility. If the damp material is recrystallized from a water-immiscible solvent, an extra water phase will be present along with the hot solution. The water should be removed with a medicine dropper or pipet before hot filtration. The solution will also be saturated with water at this point.

Low-Melting Compounds

Recrystallization of low-melting compounds is not easy. Low melting point and high solubility in nonpolar solvents usually go together, as explained in Section 21.2. This gives you the choice of using very small volumes of nonpolar solvents or using solvent pairs that include water. If water is used, the product will often separate as a liquid upon cooling. Neither alternative is very attractive.

Furthermore, since crystal formation is impossible above the melting point of the compound, all useful cooling of the hot solution must take place from a maximum temperature somewhat below the melting point of the substance. The lower the melting point, the smaller the range of cooling and the smaller the difference in solubility at the higher and lower temperatures. This leads to relatively large losses on recrystallization unless the mixture can be cooled below 0°C.

Small Samples

When the amount of material is less than about 50 mg or the volume of the hot solution is less than about 5 mL, the usual techniques of recrystallization give unacceptably large losses in the two filtration steps.

With small amounts, the hot solution should be prepared in a small flask or a short test tube so that it can be withdrawn by a medicine dropper for the hot filtration. With more than a couple of milliliters, you can use a very small funnel and, instead of using filter paper, put a small plug of cotton or glass wool in the stem of the funnel. It is easy to use so much plugging material that the liquid will not run through, and it is very wise to test the funnel and the plug with a sample of the pure hot solvent. With less than 1 or 2 mL, you can put a small plug of cotton in the tip of a medicine dropper, draw the solution into the dropper through the cotton, remove the cotton with tweezers while holding the tip of the dropper over the test tube to which the filtrate is to be added, and then release the hot solution into the test tube.

An advantage in cooling the solution for crystallization in a test tube is that the crystals can be collected by centrifugation followed by decanting of the solvent. Washing can be done by suspending the crystals in a little cold solvent, centrifuging, and decanting. Alternatively, the solvent can be removed with a dropper that has had its tip drawn down to a capillary of about 1 mm. The wash liquid can be removed in the same way.

The crystals can be dried in the test tube by laying the tube on its side with the bottom slightly raised or by connecting it to the vacuum as shown in Figure 8.4.

When the crystals are dry, they can be removed from the test tube by inverting it over a piece of filter paper and tapping it with a stirring rod.

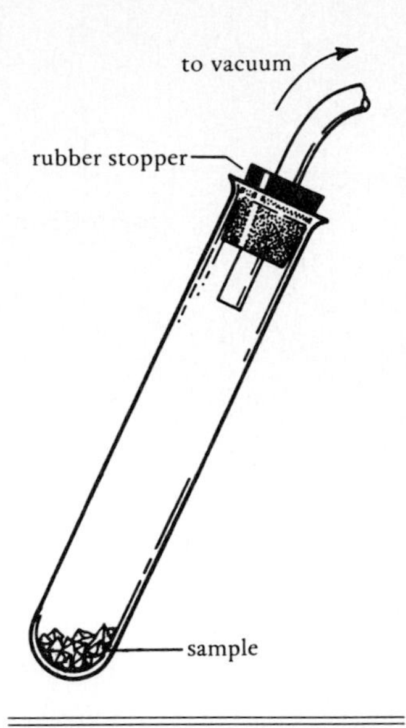

Figure 8.4. A way to dry crystals in a test tube.

Second Crops

The filtrate that is removed when the crystals are collected by suction filtration is saturated with respect to the compound. It is often possible to concentrate the filtrate by distilling off some of the solvent, and then to obtain a further crop of crystals by allowing the concentrated filtrate to cool. The second crops are usually not as pure as the first, and this technique is most useful when there is a relatively small difference in solubility of the compound in the hot and cold solvent.

Questions

1. Suppose you are recrystallizing a sample of benzoic acid from water. The original sample is contaminated with sand and salt.
   a. Briefly describe the procedure that you would use.
   b. Explain how the two impurities would be separated from the benzoic acid by your procedure.
2. Explain what effect each of the following mistakes would have on the success of a recrystallization:
   a. Too much solvent was used.
   b. Too little solvent was used.
   c. The hot solution was filtered by suction.
   d. The decolorizing carbon was not completely removed by the hot filtration.
   e. The hot solution was immediately placed in an ice bath.
   f. The crystals were washed with warm solvent.
   g. The crystals were not washed at all.
   h. There was inadequate suction during the suction filtration.

Problems

1. What is the expected percent recovery upon recrystallization of 2.00 grams of benzoic acid from 100 mL of water, assuming the solution cools to 18°C for crystallization?

    Solubility of benzoic acid in water:
    2.2 g/100 mL water at 75°C
    0.27 g/100 mL water at 18°C

2. What is the expected percent recovery upon recrystallization of 30 grams of benzoic acid from 100 mL of carbon tetrachloride, assuming that the solution will be cooled to 20°C for crystallization?

    Solubility of benzoic acid in carbon tetrachloride:
    32 g/100 mL CCl_4 at 60°C
    5.6 g/100 mL CCl_4 at 20°C

3. Compare the percent recovery to be expected upon recrystallization of 2.00 grams of benzoic acid from 200 mL of water with that to be expected if only 100 mL of water is used, cooling to 18°C in either case.
4. If the percent recovery to be expected increases when less solvent is used, other things being equal, would it be a good idea to try to recrystallize 2.00 grams of benzoic acid from less than 100 mL of water? Explain.
5. **a.** Calculate the percent recovery to be expected upon recrystallization of 2.00 grams of benzoic acid from 100 mL of water if the solution is allowed to cool to 4°C before suction filtration.

    Solubility of benzoic acid in water at 4°C:
    0.18 g/100 mL

    b. Do the same for the recrystallization of 30 grams of benzoic acid from carbon tetrachloride, assuming that the solution can be cooled to 0°C before suction filtration.

    Solubility of benzoic acid in carbon tetrachloride at 0°C:
    1.5 g/100 mL

6. **a.** Would you choose to recrystallize a 100-milligram sample of benzoic acid from water or from carbon tetrachloride? Present the reasons for your choice.
    b. Would you choose to recrystallize a 100-gram sample of benzoic acid from water or from carbon tetrachloride? Present the reasons for your choice.
7. A mixture contains 95% by weight *A* and 5% by weight *B*. Assume that you must obtain pure *A* by recrystallization of a 100-gram sample of the mixture.
    a. What is the minimum amount of solvent necessary for the recrystallization? What percent of *A* in the sample should crystallize out upon cooling of the hot solution? Assume that the solubilities are as follows.

	Hot	*Cold*
A	10 g/100 mL	2 g/100 mL
B	10 g/100 mL	2 g/100 mL

b. Answer the same two questions as in **a**, but assume the solubilities of B are

5 g/100 mL (hot) and 1 g/100 mL (cold)

c. Answer the same two questions as in **a**, but assume the solubilities of B are

2 g/100 mL (hot) and 0.4 g/100 mL (cold)

d. Answer the same two questions as in **a**, but assume the solubilities of B are

1 g/100 mL (hot) and 0.2 g/100 mL (cold)

e. Answer the same two questions as in **a**, but assume the solubilities of B are

0.5 g/100 mL (hot) and 0.1 g/100 mL (cold)

Exercises

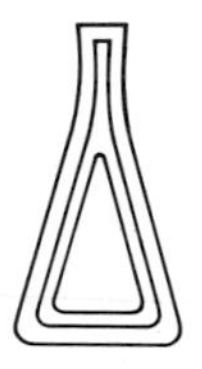

Any skill improves with practice. The purpose of an exercise is to provide practice. For this reason, I have included a few suggestions for practice in recrystallization.

1. Recrystallization of *endo*-5-norbornene-2,3-dicarboxylic acid.

   This substance can be recrystallized very nicely from water. Its tendency to remain supersaturated for a little while makes it easier to complete the hot filtration before crystallization begins. If the solution is allowed to cool slowly, the crystals separate in beautiful long spars. Since the corresponding anhydride is much less expensive and is converted to the diacid on boiling with water, the instructions specify that you start with it rather than the acid. The anhydride can be prepared according to the procedure of Section 69.2.

   Procedure. Place 4.0 grams of the anhydride in a 125-mL Erlenmeyer flask and add 50 mL of water. Heat the mixture to boiling and continue to heat until the oily liquid goes into solution. Filter the hot solution by gravity and allow it to cool. Collect the resulting crystals by suction filtration and wash them with a little water.

H H H H H COOH H H H COOH

endo-5-norbornene-2,3-dicarboxylic acid

2. Recrystallization of acetanilide. Water is often recommended for the recrystallization of acetanilide. Solubility information for acetanilide can be found in the *Handbook of Chemistry and Physics* (older editions).
3. Recrystallization of *m*-nitroaniline. Both water and 75% aqueous ethanol have been recommended for the recrystallization of this compound. It usually forms lovely yellow needles.
4. Recrystallization of *p*-nitroaniline. Water at 100 mL per gram has been recommended for the recrystallization of this substance.
5. Recrystallization of benzil. If benzil is dissolved in 95% ethyl alcohol at the rate of 6.5 mL per gram in a relatively large flask, and if care is taken to leave no traces of solid in any part of the flask, the solution can be cooled to room temperature without crystallization. When a minute seed crystal is added, a very beautiful phenomenon of crystal growth may be observed. If no seed crystal is added, crystallization may take a long time to occur, but sometimes the entire sample of benzil will separate as a single crystal.

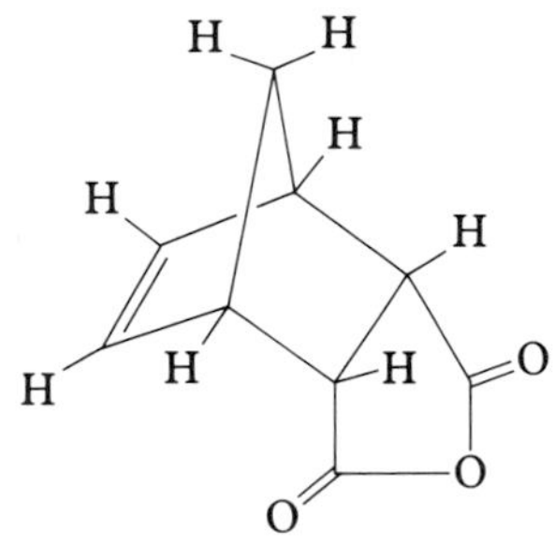

endo-5-norbornene-2,3-dicarboxylic anhydride

References

Sources from which more information about recrystallization processes can be obtained include:

1. K. B. Wiberg, *Laboratory Technique in Organic Chemistry,* McGraw-Hill, New York, 1960, p. 98.
2. A. I. Vogel, *Practical Organic Chemistry,* 3rd edition, Wiley, New York, 1956, p. 122.

9 Distillation

In a distillation, a liquid is heated to the temperature at which it changes to a vapor. The vapor is then cooled, and thus liquefied, in another part of the apparatus. Separation of the components of a mixture by distillation takes advantage of the fact that different substances can differ in the degree to which they can be vaporized under the conditions of the experiment.

Figure 9.1. Graph of the equilibrium vapor pressure versus temperature for several liquids: (1) diethyl ether, (2) acetone, (3) ethyl alcohol, (4) carbon tetrachloride, (5) water, (6) bromobenzene.

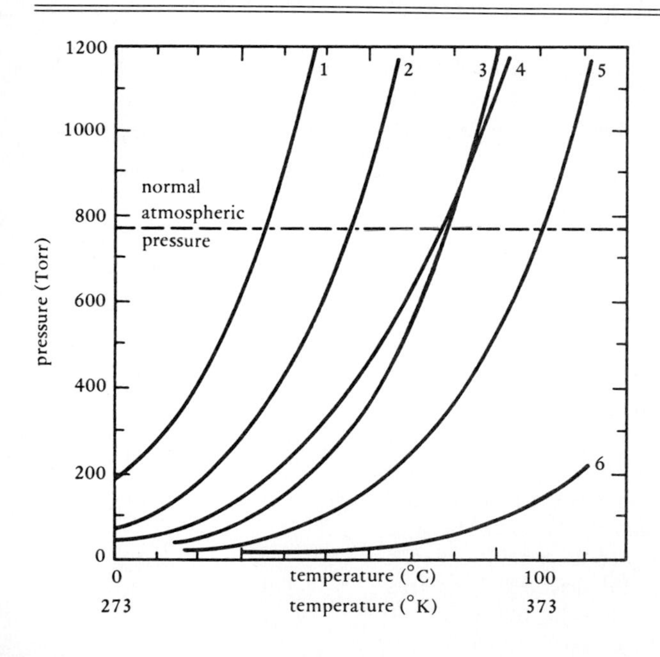

9.1 VAPOR PRESSURE

If a sample of a liquid is placed in an otherwise completely empty space, some of the liquid will vaporize. As this happens, the pressure in the space above the liquid will rise and will finally reach some constant value. The pressure under these conditions is due entirely to the vapor of the liquid and is called the *equilibrium vapor pressure*.

Equilibrium vapor pressure

The equilibrium vapor pressure increases with temperature according to Equation 9.1-1 (see Figure 9.1), where C is a constant and T is the absolute temperature:

$$P \propto e^{-C/T} \quad \text{or} \quad P \propto \frac{1}{e^{C/T}} \tag{9.1-1}$$

This relationship between vapor pressure and temperature can be rewritten in a logarithmic form:

$$\ln P = \frac{-C}{T} + \text{constant} \tag{9.1-2}$$

or, using common logarithms:

$$2.3 \log P = -\frac{C}{T} + \text{constant} \tag{9.1-3}$$

$$\log P = -\frac{C}{2.3T} + \frac{\text{constant}}{2.3}$$

In these last equations, "constant" is the natural log of the proportionality constant implied in Equation 9.1-1. The logarithmic form of the equation makes it apparent that a graph of log P versus $1/T$ should give a straight line of slope $-C/2.3$ (see Figure 9.2).

The existence of vapor pressure is explained by molecules of liquid escaping into the empty space above the liquid. As the number of molecules in the vapor space above the liquid becomes larger, the rate of return of molecules from the vapor space to the liquid increases until the rate of return has risen to equal the constant rate of escape. This is the equilibrium condition, and the corresponding concentration of molecules in the vapor space gives rise to the equilibrium vapor pressure. At higher temperatures, the greater kinetic energy of the molecules in the liquid results in a greater constant rate of escape. Equilibrium is established at higher temperatures, then, with larger numbers of molecules in the vapor phase, and at correspondingly higher pressures.

Molecular interpretation

The equilibrium vapor pressure will be exerted in a closed container whether or not there are other molecules present in the gas phase

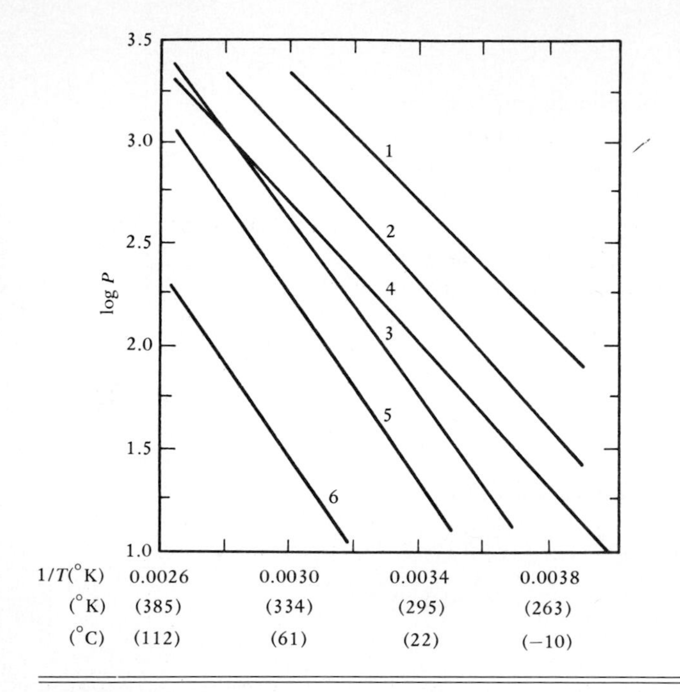

Figure 9.2. Graph of log (equilibrium vapor pressure) versus $1/T$, in degrees K, for several liquids. The liquids are the same as those of Figure 9.1: (1) diethyl ether, (2) acetone, (3) ethyl alcohol, (4) carbon tetrachloride, (5) water, (6) bromobenzene.

(air, for example). If there are other molecules present, the equilibrium vapor pressure will not be equal to the total pressure as in the previous example. The total pressure will be the sum of the equilibrium vapor pressure of the liquid (its *partial pressure*) plus the pressure due to the other molecules (the sum of their various partial pressures).

Partial pressure

Different substances have different vapor pressures at any given temperature, as illustrated by the examples in Figures 9.1 and 9.2. This is equivalent to saying that the values for the constants C and "constant" in Equations 9.1-2 and 9.1-3 are different for different molecules. In Section 16.3, boiling points will be interpreted in terms of the heat of vaporization (ΔH_{vap}; the amount of heat required to convert a mole of the liquid to a vapor at the normal boiling point; always positive) and the entropy of vaporization (ΔS_{vap}; the entropy increase that accompanies the conversion of a mole of liquid to a vapor at the normal boiling point; always positive). Since it can be shown that in Equation 9.1-3

$$C = \frac{\Delta H_{vap}}{R} \quad \text{and} \quad \text{Constant} = \frac{\Delta S_{vap}}{R}$$

where R is the ideal gas constant, then

$$\log P = -\frac{\Delta H_{vap}}{2.3RT} + \frac{\Delta S_{vap}}{2.3R}$$

A high vapor pressure at any given temperature is thus the result of either a small heat of vaporization or a large entropy of vaporization, or both, and to say that a compound generally has a high vapor pressure is equivalent to saying that it has a low boiling point.

Figure 9.3. A simple apparatus for distillation at atmospheric pressure.

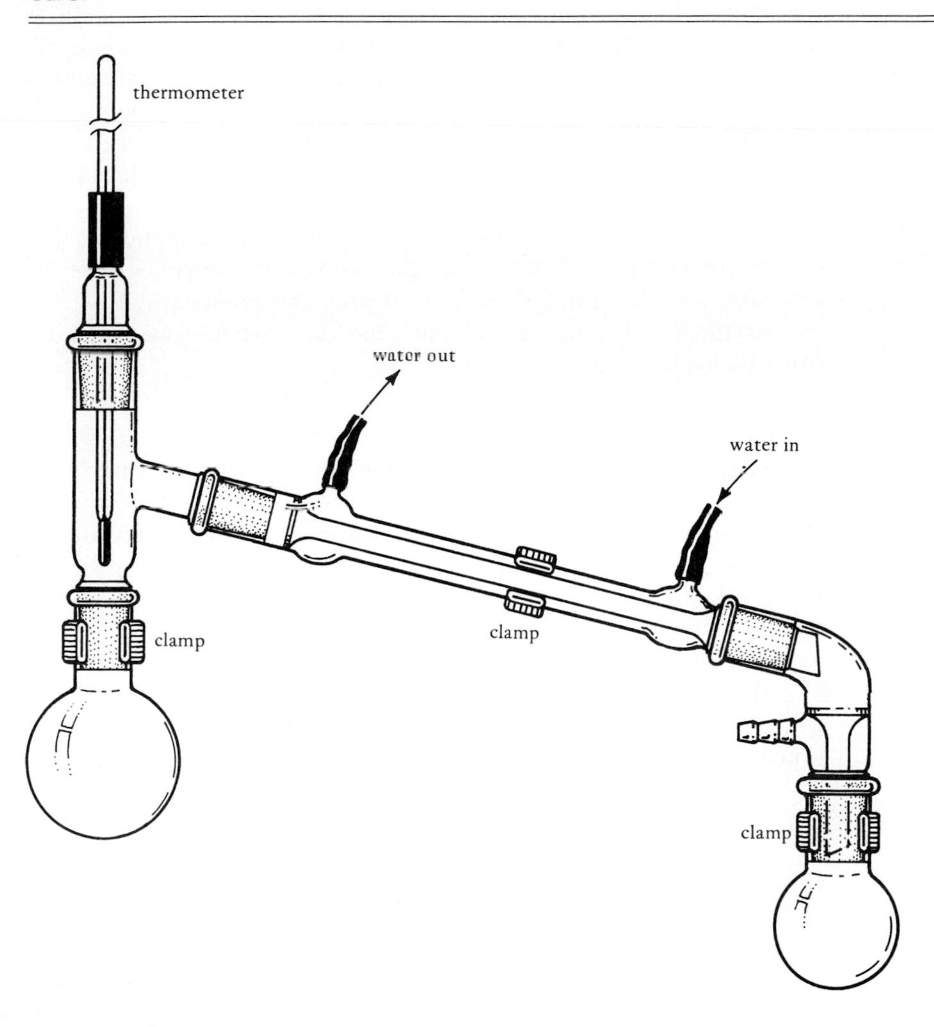

9.2 DISTILLATION OF A PURE LIQUID

If a pure liquid is heated in a flask connected to a condenser that is open to the atmosphere at the other end (see Figure 9.3), its vapor pressure will rise, as explained in the preceding section. When the temperature of the liquid becomes sufficiently high, the vapor pressure of the liquid will slightly exceed that of the atmosphere, and the vapor will start to expand out of the flask and into the condenser. It is the function of the condenser to cool the vapors and reconvert them to liquid. In a distillation, the condenser is arranged so that the condensate does not return to the flask, in contrast to its use as a reflux condenser. As long as liquid remains in the flask, the temperature of the distilling vapor will not rise. The continuing input of heat serves only to supply the required heat of vaporization, and thus to convert more liquid to vapor.

Boiling

The temperature at which distillation takes place at a total pressure of 1 atmosphere is called the *normal boiling point* of the liquid. If the pressure in the apparatus, which is usually open to the atmosphere, is not equal to exactly 1 atmosphere, the temperature at which boiling will begin (the temperature at which the equilibrium vapor pressure equals external pressure) will be different from the normal boiling point. It is sometimes desirable to distill a substance at as low a temperature as possible, using an apparatus in which the pressure can be reduced (see Section 10).

Normal b.p.

The significant features in the distillation of a pure liquid are that (1) the compositions of the liquid, the vapor, and the condensate (or distillate) are identical and constant during the process, and (2) the temperatures of the liquid and the vapor are constant and, ideally, equal throughout the distillation.

Distillation of a pure liquid

9.3 MISCIBLE PAIRS OF LIQUIDS

When two liquids that are completely soluble in one another are mixed, the vapor pressure of each liquid at a particular temperature is diminished by the presence of the other liquid. Such mixtures can be characterized according to the contribution of each component to the total vapor pressure as a function of the composition of the mixture.

Miscible pairs of liquids are said to behave ideally if the contribution of each component to the total vapor pressure is directly proportional to its mole fraction. That is,

Ideal behavior

$$P_A = x_A P_A^0 \qquad P_B = x_B P_B^0 \tag{9.3-1}$$

$$P_{\text{total}} = P_A + P_B = x_A P_A^0 + x_B P_B^0 \tag{9.3-2}$$

where P_A and P_B are the vapor pressures of A and B above a solution of mole fraction x_A and x_B and where P_A^0 and P_B^0 are the vapor pres-

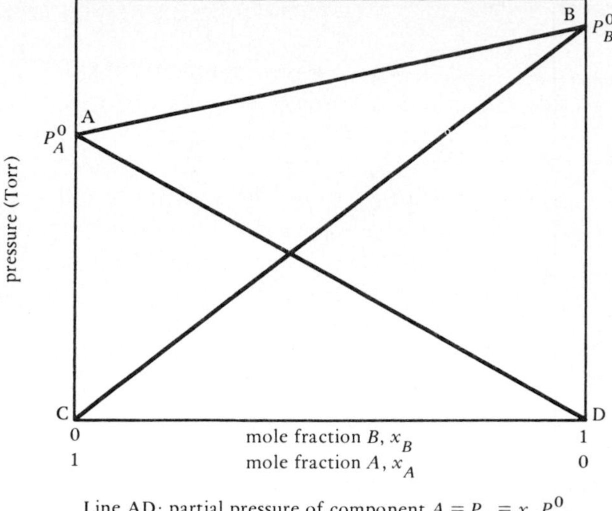

Line AD: partial pressure of component $A = P_A = x_A P_A^0$.

Line BC: partial pressure of component $B = P_B = x_B P_B^0$.

Line AB: total vapor pressure P_{total} = line AD + line BC.

Figure 9.4. Vapor pressure–composition diagram for the ideal system A/B at a particular temperature.

sures of pure A and pure B at that particular temperature. This type of behavior, often referred to as behavior according to *Raoult's law*, can also be represented graphically as in Figure 9.4.

Raoult's law

Ideal behavior is approximated by mixtures such as benzene/toluene, n-hexane/n-heptane, carbon tetrachloride/silicon tetrachloride, and n-butyl bromide/n-butyl chloride, in which the mixture is composed of molecules of similar size and type of intermolecular interaction. A vapor pressure–composition diagram is given for the system benzene/toluene in Figure 9.5.

Boiling point diagrams

For discussing the separation of a pair of miscible liquids by distillation, a boiling point diagram is very helpful. This is a diagram that shows the temperature at which mixtures of various composition boil (at a given total external pressure, usually 1 atmosphere) and the compositions of the liquid and vapor that are in equilibrium at this temperature. The way this information is stored in a boiling point diagram is best shown by considering how one is constructed. Suppose, for example, that you prepare mixtures of benzene and toluene of 0.1, 0.3, 0.5, 0.7, and 0.9 mole fraction benzene and heat each mixture to boiling in an apparatus open to 1 atmosphere of pressure and in which the condensed vapor is returned to the boiler. A thermometer can be used to determine the temperature at which the mixture boils, and a sample of the condensate can be removed to determine the composition of the

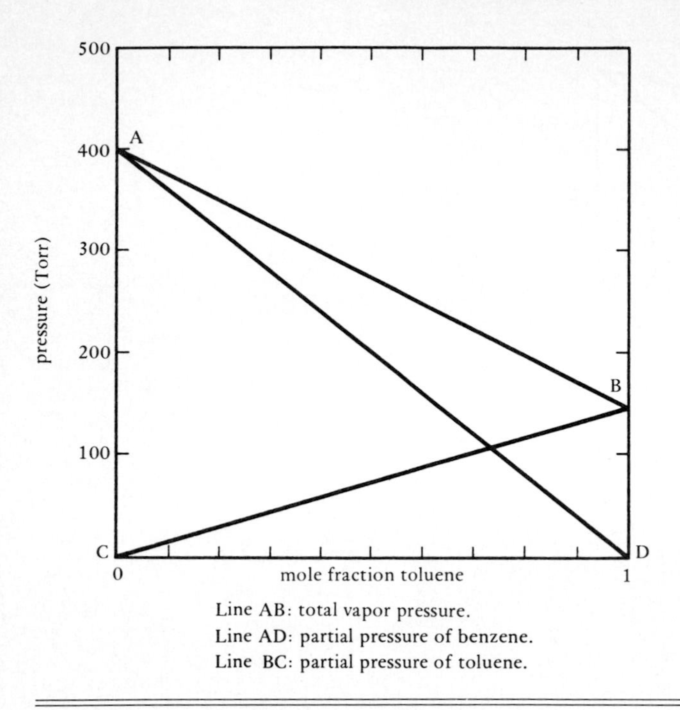

Figure 9.5. Vapor pressure–composition diagram for the system benzene/toluene at 60°C.

vapor. (The composition of the liquid is known since the mixture was made up of known amounts of the components; it could also be determined experimentally.) Suppose the results are as shown here:

b.p. (°C)	*Liquid Composition (mole fraction benzene)*	*Vapor Composition (mole fraction benzene)*
110.6	0.00	0
105.7	0.10	0.21
98.3	0.30	0.51
92.4	0.50	0.71
87.3	0.70	0.86
82.6	0.90	0.96
80.0	1.00	1.00

These data can be plotted as shown in Figure 9.6. A smooth line can then be drawn to interpolate all the other possible experimental points, and the result is shown in Figure 9.7. Now, from Figure 9.7,

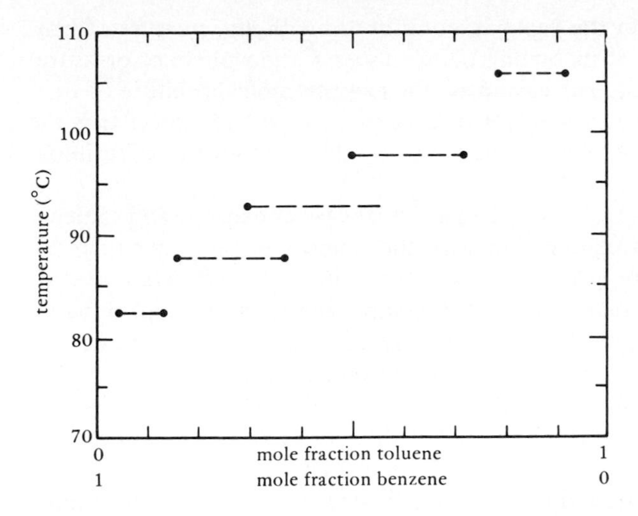

Figure 9.6. Construction of the boiling point diagram for the system benzene/toluene at 1 atmosphere. Each pair of points indicates the boiling point of a mixture of benzene and toluene and the compositions of the liquid and the vapor that are in equilibrium.

Figure 9.7. Boiling point diagram for the system benzene/toluene at 1 atmosphere.

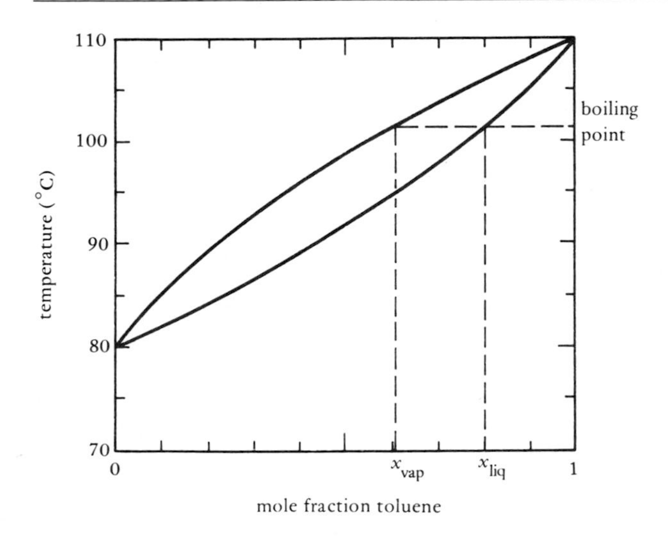

the composition of the vapor in equilibrium with any mixture of benzene and toluene at its boiling point (under 1 atmosphere of pressure) can be determined. For example, if a 1 : 4 (by moles) mixture of benzene and toluene is heated for distillation, you would expect that the mole fraction of benzene in the vapor would be about 0.4 at an initial boiling point of about 102°C.

It is generally true, as in the particular case of benzene and toluene, that the vapor in equilibrium with the liquid will be richer than the liquid in the more volatile component. This seems intuitively reasonable in that the molecules of the component with the higher vapor pressure at any given temperature should tend to escape more frequently and thus be overrepresented in the vapor phase.

As the distillation of a mixture of miscible liquids progresses, the mixture will gradually be depleted of the more volatile component. As this happens, according to the distillation diagram, *the boiling point will gradually rise,* and the distillate, though always richer than the residue in the more volatile component, will contain a continually decreasing proportion of the more volatile component. You can think of this process as being represented by the gradual movement of line AB in Figure 9.8 upward and to the right to, say, A′B′. Thus, the significant features in the distillation of a miscible pair of liquids that serve to distinguish it from the distillation of a pure liquid are that (1) the compositions of the liquid and vapor (or distillate) are not the same, and (2) the boiling

Distillation of a miscible pair of liquids

Figure 9.8. Boiling point diagram for the system benzene/toluene at 1 atmosphere. The horizontal lines AB and A′B′ connect points on the two curves that represent the compositions of liquid and vapor that are in equilibrium at two different temperatures.

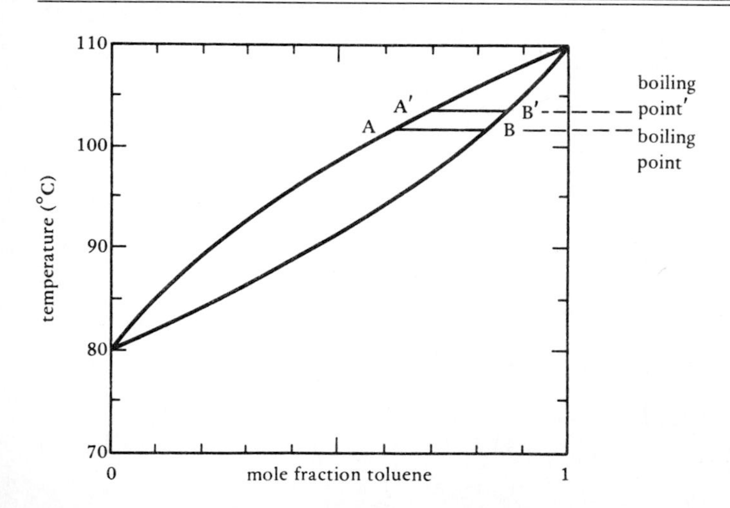

point of the liquid will gradually rise during the distillation. Exceptions to this generalization are discussed in the section on azeotropic mixtures (Section 9.5).

9.4 FRACTIONAL DISTILLATION

From Figure 9.7 and the example cited in the previous section, you can see that if you started to distill a mixture of benzene and toluene that was 0.2 mole fraction in benzene, the vapor initially in equilibrium with the mixture would be about 0.4 mole fraction in benzene. If you started with a large sample, this would be the composition of the first part of the distillate. If the first part of the distillate were then redistilled, the vapor in equilibrium with it would be approximately 0.6 mole fraction in benzene and the first few drops of the distillate would have this composition. You can see that if you started with a large enough sample and repeated this process several times, collecting only the very first part of the distillate each time, you could obtain a very small sample of fairly well-purified benzene. A four-step process of this type could be represented by the movement of a point from A to B . . . to I in Figure 9.9.

Figure 9.9. Boiling point diagram for the system benzene/toluene at 1 atmosphere. The series of four horizontal and four vertical lines connecting the points A and I indicate that after four successive steps of vaporization followed by condensation, a large sample of liquid whose boiling point and composition correspond to point A would yield a small sample of a liquid whose boiling point and composition would correspond to point I.

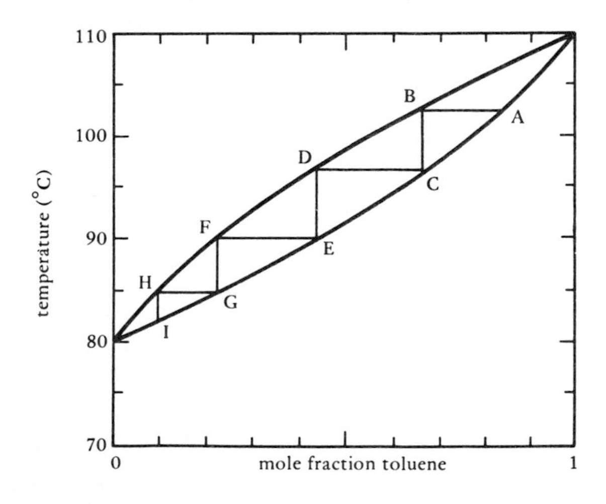

The process just described would be too inefficient to be practical. If 10% of the material were distilled each time, then, starting with 1 liter, only 0.1 milliliter would be obtained at the end of the fourth stage. Also, if a finite amount were collected in each step, the final purity would be less than that indicated in Figure 9.9.

A more practical procedure, which was used in the past, is to distill the entire sample, collecting the distillate in several portions, or *fractions*. The boiling ranges of these fractions might be, for example, 80–85°, 85–90°, 90–95°, etc. The first fraction, the most volatile fraction, is then redistilled in a similar manner until the temperature of the distillate rises to the lower temperature of the boiling range of the second fraction, or, in this example, 85°. The second fraction is then added to the boiler and distilled in a similar manner until the temperature of the distillate rises to the lower temperature of the third fraction, and so on. The entire process may be repeated four or five times. In each series of redistillations, the distillate is collected in fractions, and in the later series, the temperature ranges for the initial and final fractions are made smaller while those of the intermediate fractions are made larger. In a successful fractional redistillation process such as this, most of the distillate in the last stage will be in the first and last fractions.

Fortunately, an apparatus called a *fractionating column* has been developed that can effect a high degree of purification much more quickly and easily, and with little loss or rejection of material. Some different types of fractionating columns are illustrated in Figure 9.10.

Fractionating columns

The significant feature of a fractionating column is that it provides for efficient exchange of heat and material between the condensate flowing down the column and the vapors flowing up. The material finally coming out of the top of the column as a vapor has been subjected to multiple condensations and evaporations on the way up, each of which has served to enrich the vapor in the more volatile component. A good column can produce a distillate in which the enrichment corresponds to between 25 and 100 steps like the four in Figure 9.9. Since some industrial fractionating columns are built up of units called plates, each of which theoretically provides enrichment corresponding to 1 step, the efficiency of enrichment of a column is expressed as *theoretical plates*, rather than as steps. Mixtures for which the boiling point diagram is known are used to determine the efficiency of a column.

Theoretical plates

While it might seem that the more theoretical plates a fractionating column has, the better, there are other factors that need to be considered in choosing a column for a particular distillation. *Column holdup* is the volume of material that would not flow out the bottom of the column if poured into the top—the volume of liquid required to wet the column. It would not be possible to distill a sample whose volume is less than the volume of the holdup; the loss on distillation will always be at least equal to the holdup. *HETP* is the height (length of column) equivalent to one theoretical plate. A low HETP is desirable both to

Holdup

HETP

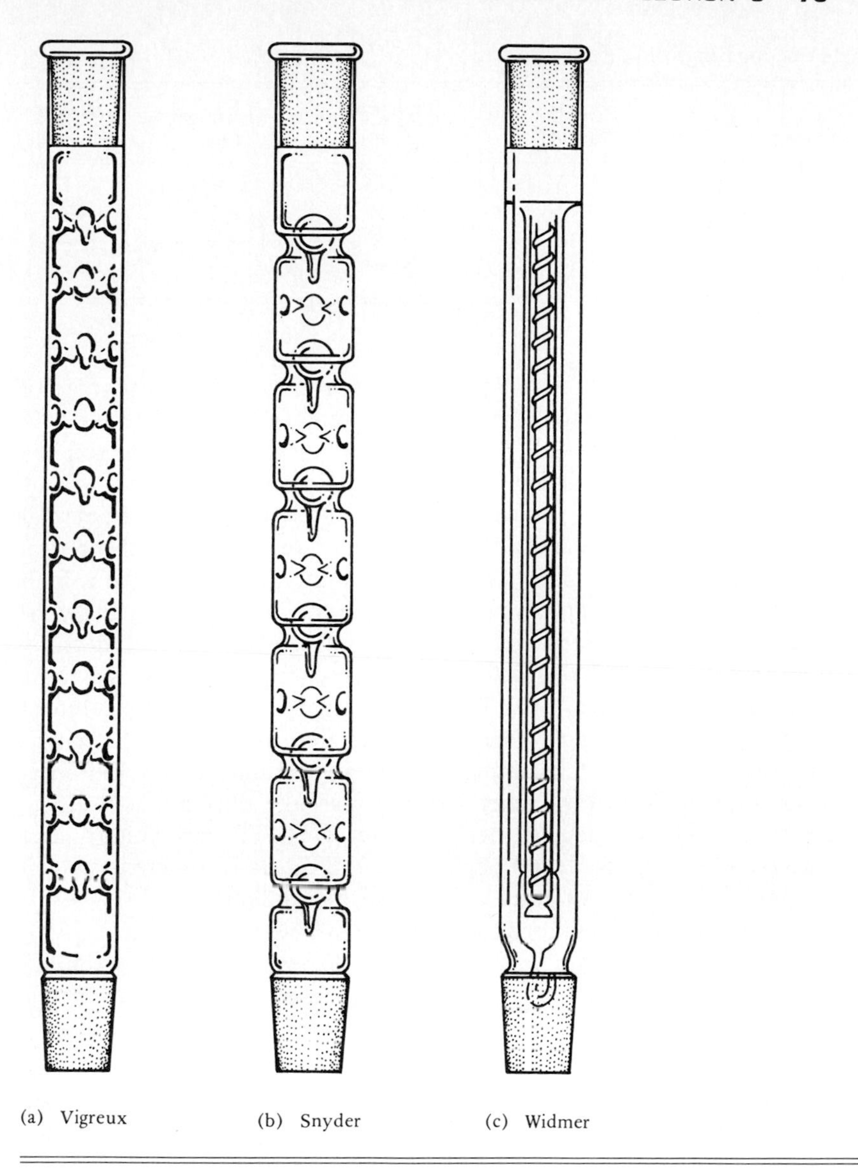

Figure 9.10. Some different types of fractionating columns.

keep columns that are highly efficient from being also inconveniently tall and to minimize holdup. *Throughput* is the maximum volume of liquid that can be boiled up through the column per unit of time while still maintaining equilibrium throughout the column. A high throughput is desirable so that a separation can be done quickly. The *reflux ratio, R,* is the ratio of the ratio of the amount of condensate formed at the top of the column and returned to the column to the amount

Throughput

Reflux ratio

Table 9.1. Comparison of simple fractionating columns

Type	*Diameter (mm)*	*Throughput (mL/min)*	*HETP (cm)*	*Holdup (mL/plate)*
Glass tube	6	2	~40	~1.3
Vigreux	12	2–5	~7	~2/3
Glass tube with $\frac{1}{8}$-in. helices	10	3–8	~4	~2
Glass tube with stainless steel sponge	12	2–5	~4	~1.5

removed as distillate. A reflux ratio of 19 to 1 would mean that 5% of the condensate at the top of the column is removed as distillate and 95% is returned to the column. The higher the reflux ratio, the greater the operating efficiency of any given column, since the column will be operating more nearly at the equilibrium conditions, which exist at total reflux (infinite R)—the conditions used in determining the theoretical plate rating of a column. The ideal column, then, will have a high theoretical plate rating, a low HETP, low holdup, and high throughput, and will not suffer a great loss in operating efficiency with decreasing reflux ratio. All real columns involve compromises among these factors and, of course, cost. Table 9.1 compares some simple columns in terms of these criteria. These numbers are estimates of what might be expected from certain simple fractionating columns with a minimum of insulation. A vacuum jacket would increase the efficiency about 25%. A larger column diameter would increase HETP and holdup, but would allow a greater throughput. The HETP was estimated for total reflux and will increase with a finite rate of distillation.

Since column height and holdup increase with increasing theoretical plates, it is undesirable to use a more efficient fractionating column than is necessary to effect the desired separation.

9.5 AZEOTROPIC MIXTURES

Nonideal behavior

In many mixtures of pairs of miscible liquids, the vapor pressure of each component cannot be represented by Equation 9.3-1, and a plot of the total vapor pressure of the mixture versus composition at any given temperature will not give a straight line such as the line AB in Figures 9.4 and 9.5. As long as the total vapor pressure of any mixture at a certain temperature is neither greater than nor less than the vapor pressure of either pure component at that temperature, mixtures of these substances will approximate the expected behavior of ideal mixtures upon distillation. The system water/methanol is such a system; its vapor pressure–composition diagram is given in Figure 9.11, and its distillation diagram in Figure 9.12.

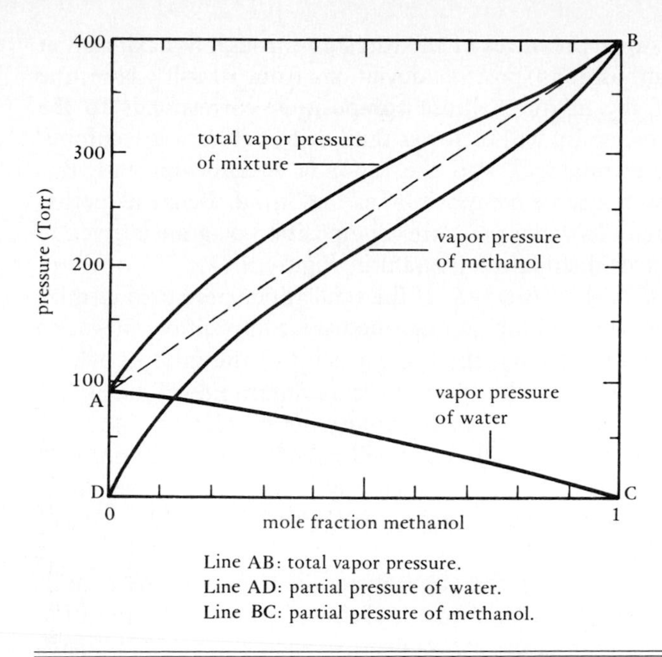

Line AB: total vapor pressure.
Line AD: partial pressure of water.
Line BC: partial pressure of methanol.

Figure 9.11. Vapor pressure–composition diagram for the system water/methanol at 49.76°C.

Figure 9.12. Boiling point diagram for the system water/methanol at 1 atmosphere.

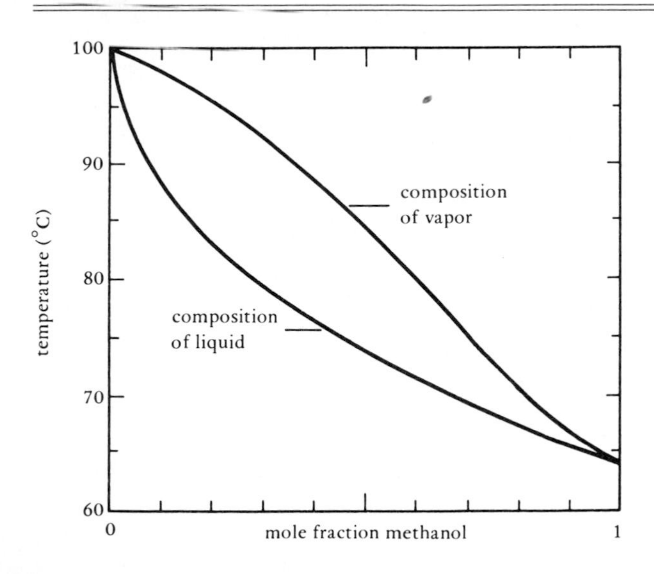

If the total vapor pressures of mixtures go through a maximum at intermediate compositions (positive deviations from Raoult's law), the boiling point of the mixture whose composition corresponds to the vapor pressure maximum will be lower than that of either pure component (or any other mixture), and the vapor in equilibrium with this mixture will have the same composition as the liquid. Benzene/methanol is such a system; its vapor pressure–composition diagram is given in Figure 9.13, and its distillation diagram in Figure 9.14.

The converse is also observed. If the total vapor pressures of mixtures go through a minimum at intermediate composition (negative deviations from Raoult's law), the boiling point of the mixture whose composition corresponds to the vapor pressure minimum will be higher than that of either pure component (or any other mixture), and the vapor in equilibrium with this mixture will have the same composition as the liquid. Chloroform/acetone is such a system; its vapor pressure–composition diagram is given in Figure 9.15, and its distillation diagram in Figure 9.16.

Since the composition of the vapor in equilibrium with these minimum- or maximum-boiling mixtures is identical with the composition of the liquid, separation of the components of such a mixture by distillation is impossible. Such mixtures are called *azeotropes*.

Azeotrope

It is invariably true that if the vapor pressure versus composition diagram shows a maximum at an intermediate composition, a minimum-boiling azeotrope will exist at that composition, at the same temperature and pressure. That the mixture is minimum boiling is easily seen, since high vapor pressure corresponds to low boiling point. That

Figure 9.13. Vapor pressure–composition diagram for the system benzene/methanol at 35°C. The vapor pressure maximum occurs at 0.567 mole fraction methanol.

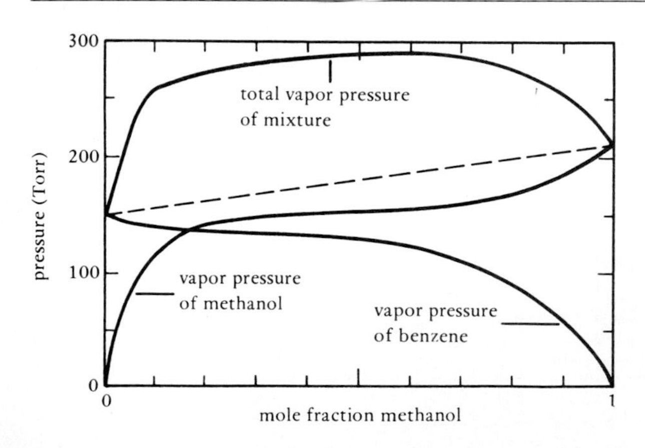

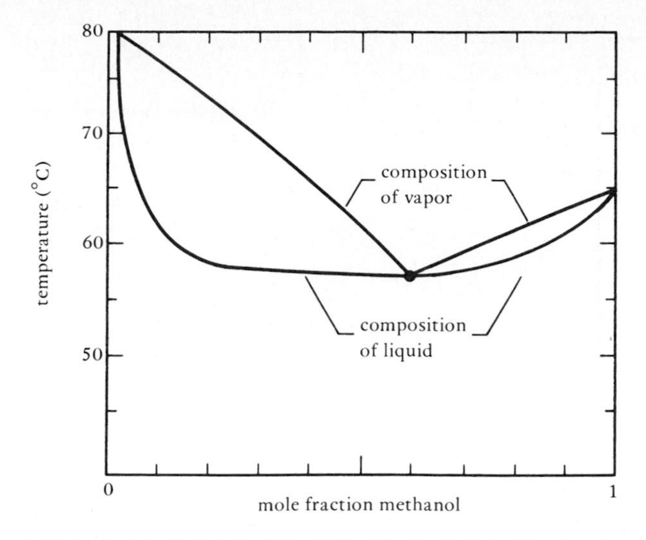

Figure 9.14. Boiling point diagram for the system benzene/methanol at 1 atmosphere. The composition of the azeotropic mixture is 0.609 mole fraction methanol, and it boils at 57.6°C. The point • indicates boiling point and composition of minimum-boiling azeotrope.

Figure 9.15. Vapor pressure–composition diagram for the system chloroform/acetone at 35.17°C. The vapor pressure minimum occurs at 0.383 mole fraction acetone.

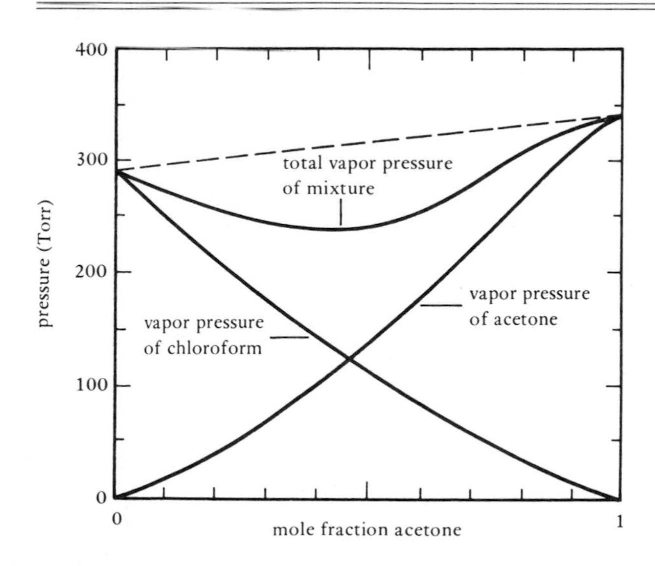

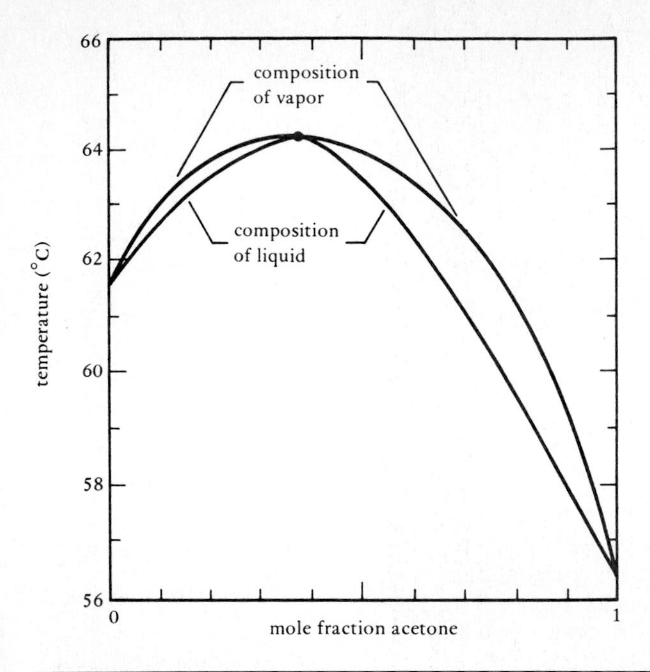

Figure 9.16. Boiling point diagram for the system chloroform/acetone at 1 atmosphere. The composition of the azeotropic mixture is 0.360 mole fraction acetone, and it boils at 64.4°C.

the mixture must be azeotropic can best be deduced by seeing from the boiling point diagram (Figure 9.14) that the situation cannot be otherwise. If the upper curve (representing vapor composition) does not touch the lower curve (representing liquid composition) at the minimum point, the implication would be that there is no vapor in equilibrium with this liquid. Therefore, the curves must touch at the minimum point, which is to say that the same point represents the composition of both liquid and vapor. A similar argument can be applied to the case of a maximum-boiling azeotrope as well.

The practical consequences of distillation of mixtures that form either minimum- or maximum-boiling azeotropes are best deduced from the boiling point diagrams. If a pair of liquids can form a minimum-boiling azeotrope, a perfectly efficient fractionating column will produce a distillate of the composition of the azeotrope, no matter what the initial composition of the mixture.

The residue in the boiler after complete removal of the other components as the azeotropic mixture will be pure A or pure B, depending upon whether the mixture initially contained more A than the azeotropic composition, or more B than the azeotropic composition. If, on the other hand, a pair of liquids can form a maximum-boiling azeo-

Table 9.2. Azeotropic mixtures of maximum boiling point

Component A	b.p. (°C)	Component B	b.p. (°C)	Azeotropic Mixture b.p. (°C)	Azeotropic Mixture Weight % B
Water	100	Hydrofluoric acid	19.5	111.4	35.6
Water	100	Hydrochloric acid	−80	108.6	20.22
Water	100	Hydrobromic acid	−73	126	47.5
Water	100	Hydriodic acid	−34	127	57
Water	100	Nitric acid	86	120.5	68
Water	100	Sulfuric acid	dec.	338	98.3
Water	100	Perchloric acid	110	203	71.6
Water	100	Formic acid	100.8	107.2	22.6
Water	100	Ethylenediamine	116	118	20–25
Acetic acid	118.2	Pyridine	115.5	140	53
Acetone	56.1	Chloroform	61.2	64.4	78.5

trope, a perfectly efficient fractionating column can produce a distillate of either pure A or pure B, depending upon whether the mixture initially contained more A or more B than the azeotropic composition. The residue, after complete removal of the amount of either A or B that was initially present in excess over the material of azeotropic composition, will be the pure azeotrope.

Possibly the easiest way to interpret the boiling point diagrams of mixtures involving azeotropes is to consider them to be made up of two simple distillation diagrams side by side, one of which involves pure A and the azeotrope of A and B, and the other the azeotrope of A and B, and pure B. Then you can apply the same stepping process to either side of the diagram as was used in the case of the ideal mixture.

The existence of maximum-boiling azeotropes (negative deviations from Raoult's law) can be seen through the existence of energetically more favorable interactions (lower potential energy) between the unlike molecules than between the like molecules. It is tempting to extend this idea to that of the formation of a "complex" or weakly bonded compound. However, since the compositions of most azeotropes do not correspond to nice molecular ratios and, even worse, show slight variations with temperature and pressure, this interpretation must be ruled out. Table 9.2 lists some maximum-boiling azeotropes and some of their properties.

The existence of minimum-boiling azeotropes (mixtures of maximum vapor pressure) can be seen through the presence of energetically less favorable interactions (higher potential energy) between the unlike molecules than between the like molecules. This has been described as "incipient insolubility" in which the like molecules tend to squeeze out

Table 9.3. Azeotropic mixtures of minimum boiling point

				Azeotropic Mixture	
Component A	*b.p.* (°C)	*Component B*	*b.p.* (°C)	*b.p.* (°C)	*Weight % B*
Water	100	Acetonitrile	81.5	76.0	85.8
Water	100	Ethanol	78.3	78.1	96
Water	100	*n*-Propanol	97.3	87	71.7
Water	100	*i*-Propanol	82.3	80.3	87.4
Water	100	Propionic acid	141.4	99.1	17.8
Water	100	*t*-Butanol	82.5	79.9	88.2
Water	100	Pyridine	115	94	43
Water	100	Dioxane	101.3	87.8	82
Carbon tetrachloride	76.8	Methanol	64.7	55.7	20.6
Carbon tetrachloride	76.8	Ethanol	78.3	65.1	15.8
Carbon tetrachloride	76.8	Acetone	56.2	56.1	88.5
Chloroform	61.2	Methanol	64.7	53.4	12.6
Chloroform	61.2	Ethanol	78.3	59.4	7
Chloroform	61.2	Hexane	69.0	60.0	28
Methanol	64.7	Acetone	56.2	55.5	88
Methanol	64.7	Benzene	80.1	57.5	60.9
Methanol	64.7	Cyclohexane	80	54	72
Ethanol	78.3	Ethyl acetate	77.1	71.8	69
Ethanol	78.3	Benzene	80.1	68.2	67.4
Ethanol	78.3	Cyclohexane	80	64.9	69.5
Acetone	56.2	Hexane	69	49.7	46.5
Benzene	80.1	Cyclohexane	80.6	77.7	48.2

the unlike. The extreme case is found when the two liquids are immiscible (see Steam Distillation, Section 11) and each exerts its own vapor pressure independently. Table 9.3 lists some minimum-boiling azeotropes and some of their properties.

9.6 TECHNIQUE OF DISTILLATION

Heating

A constant rate of heating is essential for optimum performance of any distillation apparatus. Of the methods of heating described in Section 32.1, an electrically heated oil bath or an electric heating mantle gives the most constant and most easily regulated rate of heat input. If the separation is very easy and high efficiency is not needed, a steam bath can be used at lower temperatures with flammable materials (for example, to remove the volatile solvent from a high-boiling product), or a flame can be used with higher-boiling substances (for example, to complete the distillation after the solvent has been removed by heating with the steam bath).

The rate of distillation is controlled by the rate of heating. The higher the temperature of the bath above the boiling point of the mixture, the greater the rate of distillation. The lower the rate of distillation, the greater the efficiency of any distillation apparatus. With an easy separation, a collection rate for the distillate of one to three drops per second (three to ten mL per minute) will often be a reasonable compromise between speed and efficiency. In careful fractionations, the reflux ratio may be quite high, and the collection rate much lower than this. Since the required rate of heating for a given rate of distillation depends upon how well the column is insulated against heat loss, the rate must be determined by experimentation.

Rate of distillation

Even with a constant source of heat, the rate of boiling will not always be constant. Sometimes alternate periods of superheating followed by vigorous boiling will occur. More than being just an annoyance, this phenomenon, called *bumping,* will prevent equilibrium from being established within the fractionating column. Smooth boiling can usually be promoted by adding several *boiling stones*. These may be bits of unglazed porous clay plate, pieces of carborundum, or bits of a specially prepared anthracite coal. The tiny air bubbles trapped in the pores of such materials prevent superheating by providing the nuclei for bubble formation. The use of several boiling stones is recommended whenever a liquid is to be boiled, in recrystallizations as well as in distillations. Should you forget to add the boiling stones, let the mixture cool a bit before you drop them in; if you do not allow a brief cooling time, you run the risk of having the mixture boil over.

Bumping

Boiling stones

In order for a distillation column to operate at maximum efficiency, *it must be insulated against heat loss*. The most usual form of insulation is a high-vacuum jacket (like a Thermos bottle) that is an integral part of the column. Sometimes provision is made for electrical heating, with or without the vacuum jacket. No insulation at all is provided with the simple columns illustrated in Figure 9.10. The simplest way of providing some insulation for these columns is to wrap them, not too smoothly, with a couple of layers of aluminum foil. Asbestos tape, glass wool, and many other materials have been used, either singly or in combination. The higher boiling a substance is, the more easily it will be condensed by the column. If the column is tall and poorly insulated, it may be impossible to put as much heat in at the boiler as can be lost in the column. In this case, it will not be possible to drive the material over.

Insulation

When you are heating with a flame, it is easy to superheat the vapor, upset the equilibria within the column, and lower column efficiency. Superheating can be minimized by heating through an abestos board, and by wrapping the upper part of the boiling flask and the column in a layer of aluminum foil.

Avoiding superheating

The flask used as the boiler in a distillation should not be more than half full at the start. If a large amount of solvent must be removed from a small amount of a high-boiling compound, it is best to remove most of

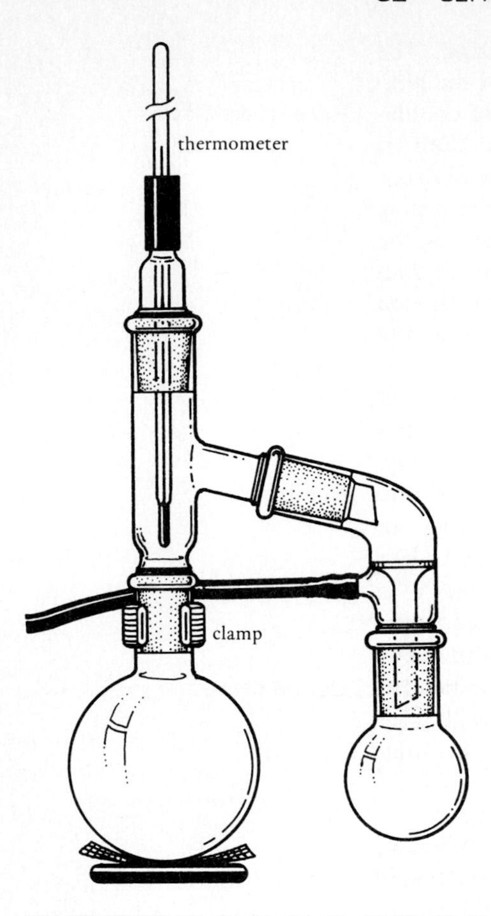

Figure 9.17. An apparatus for distillation. The receiver and vacuum adapter should each be held on with a spring clamp or a rubber band. The hose should lead over the edge of the desk in order to keep flammable fumes from the burner, or to the vacuum pump in the case of a vacuum distillation. A small piece of stainless steel sponge may be placed just below the thermometer as a packing for better fractionation. This set-up, which does not make use of a condenser, should not be used with low-boiling compounds.

Size of boiling flask

the solvent in a first distillation and then to transfer the residue to a smaller flask to complete the distillation. Losses in transfer can be prevented by rinsing with a small amount of the solvent that was just removed. If you attempt to complete the distillation from a large flask, not only will the losses be greater because of the larger vapor volume and surface area, but the walls of the flask will act as a condenser and may make it impossible to drive the higher-boiling material over.

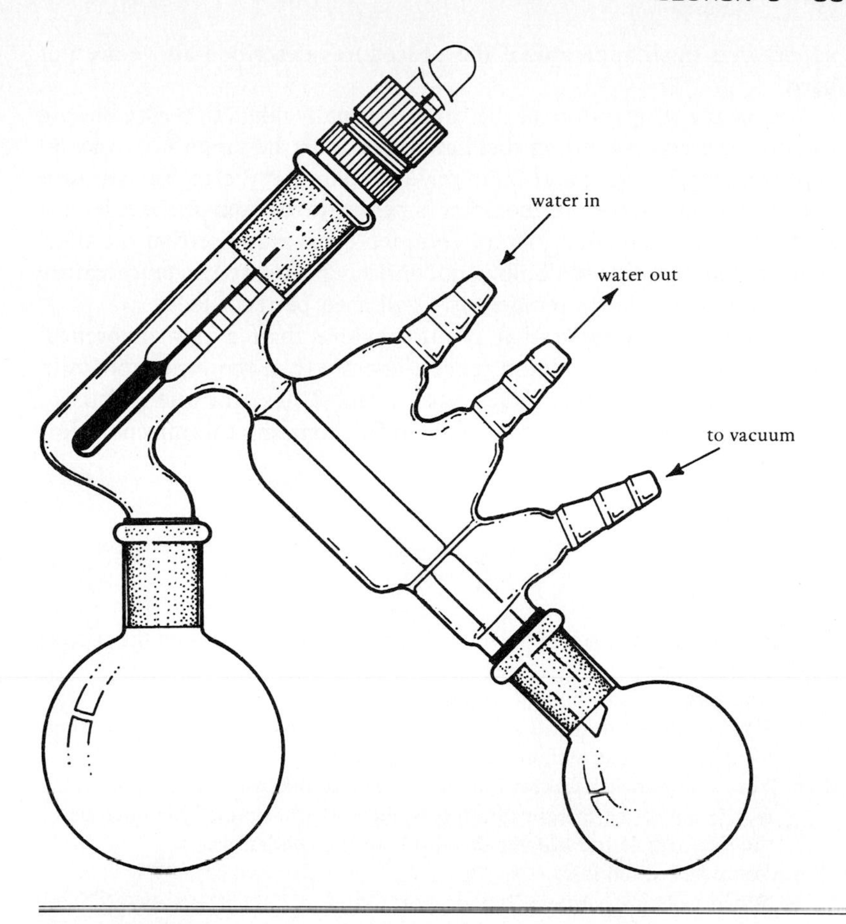

Figure 9.18. A small-scale apparatus for distillation.

Avoiding loss of material

There are several reasons for loss of material upon distillation. The vapor volume of the boiling flask will result in a loss of 0.3 to 0.6 g per 100 mL; the walls of the boiling flask, adapters, and condenser will be wet with liquid at the end of the distillation; and the column will hold up a certain amount. The distillation apparatus illustrated in Figure 9.3 will result in losses of approximately 1 to 3 mL, depending on the size of the flask, and if a fractionating column is used, the total may be between 3 and 5 mL. It should be obvious now that if the volume of material to be distilled is less than 15 or 20 mL, a smaller apparatus should be used. If the separation problem is very easy, the apparatus of Figure 9.17 can be used. If a substance is to be distilled without using a condenser, it must boil at a relatively high temperature, or there must be very little of it, or the distillation must go slowly; the receiver may be cooled occasionally in a cold water bath. A better but more expensive apparatus is shown in Figure 9.18. It is very easy to superheat the vapors in a small apparatus if the procedures described above are not used.

Position of thermometer

Since the temperature of the vapor is usually taken to be the boiling point of the corresponding distillate, care must be taken (1) to avoid superheating the vapor and (2) to make certain that the entire mercury-containing bulb of the thermometer is below the bottom of the side-arm leading to the condenser. A very common error is to position the thermometer so high that the bulb is not entirely heated to the temperature of the vapor; too low a temperature will then be recorded.

Don't make this mistake

If the desired separation requires more than a few theoretical plates, a fractionating column more elaborate than the types shown in Figure 9.10 is desirable. References 1 and 2 describe the details of operation of a number of highly efficient fractionating columns of different designs.

Questions

1. Explain what effect each of the following mistakes would have on the success of a distillation:
   - **a.** You forgot to add boiling stones.
   - **b.** You attempted to distill a low-boiling, flammable liquid (such as ether) using a flame but without using a condenser.
   - **c.** Your thermometer was positioned too high in the distilling adapter. That is, the top of the mercury bulb was above the bottom of the opening of the side-arm of the adapter that leads to the condenser.
   - **d.** You distilled too fast.
   - **e.** When isolating carvone (boiling point around 230°C; Section 47) from either oil of spearmint or oil of caraway, you collected as product the material boiling from 110°C to 231°C.

Problems

1. **a.** Estimate from Figure 9.12 the boiling point of a water/methanol mixture 0.6 mole fraction in water.
   b. What would be the composition of the vapor in equilibrium with this mixture at the boiling point?
2. **a.** Estimate from Figure 9.14 the boiling point of a mixture of benzene and methanol that is 0.1 mole fraction in methanol. What would be the composition of the vapor in equilibrium with this mixture at the boiling point?
   b. Answer the same questions as in **a**, for the case of a mixture that is 0.1 mole fraction in benzene.
3. Referring to Figure 9.12, and assuming the use of a magic fractionating column with an infinite number of theoretical plates and zero holdup,
   a. What would be the boiling point and the composition of the initial distillate, starting with an equimolar mixture of methanol and water?

   b. What mole percent of the total mixture would distill with this composition?
   c. What would be the boiling point and composition of the remainder?
   d. Present the results of your calculations for **b** and **c** as a plot of boiling point of the distillate (vertical axis) versus mole percent distilled (horizontal axis).
4. Do the same as in Problem 3, but substitute benzene and methanol (Figure 9.14).
5. Do the same as in Problem 3, but substitute acetone and chloroform (Figure 9.16).
6. Interpret the existence of a minimum-boiling azeotrope in the system acetone/chloroform in terms of the attractive forces between the molecules. Does your hypothesis explain the composition of the azeotrope?

Exercises

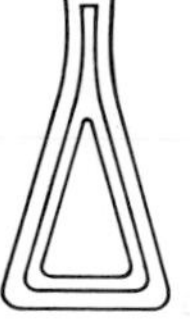

1. Distillation of water. The distillation of water provides an example of the distillation of a pure liquid.

   Procedure. Using a clamp and a ring stand, support a 100-mL round-bottom boiling flask about 4 cm above the top of your Bunsen burner. Place a wire gauze on an iron ring, and clamp the ring to the ring stand so that the wire gauze presses up against the bottom of the flask. Add 50 mL of water and 1 boiling stone to the flask. Fit the flask with a distilling adapter, as illustrated in Figure 9.3. Support a thermometer in the top joint of the adapter, and connect a condenser, with distillation adapter and receiver, to the side-arm of the distillation adapter. Using 2 lengths of rubber tubing, provide for a modest flow of water from the cold water tap to the lower hose connection of the condenser, and from the upper hose connection of the condenser to the drain. Adjust the height of the thermometer so that the top of the bulb of mercury is just below the bottom of the opening of the side-arm.

   Heat the contents of the flask with the Bunsen burner, vigorously at first and then more gently as the contents of the flask appears to approach the boiling point. Slowly distill the contents, keeping a record of the temperature of the vapor, as indicated by the thermometer, versus the volume of the distillate. Remove the burner and extinguish the flame when the boiling flask approaches dryness.

   After the distillation is complete, make a plot, in your lab notebook, of the temperature (vertical) versus the volume of distillate (horizontal).
2. Distillation of methanol. The distillation of methanol provides a second example of the distillation of a pure liquid. The procedure of Exercise 1 can be used, or a steam bath (Section 32) can be used for heating instead of the burner. The steam bath can be supported on the iron ring, or the flask can be clamped in position over the steam bath.
3. Distillation of an azeotropic mixture. Distill (as described in Exercise 1) 50 mL of a mixture of *n*-propyl alcohol and water that is 0.57 mole fraction water. How can you tell whether you are distilling a pure substance or an azeotropic mixture?

4. Distillation of a mixture of methanol and water. Distill 100 mL of an equimolar mixture of methanol and water, using a simple fractionating column as illustrated in Figure 9.10. Keep a record of the temperature of the vapor at the top of the column versus the volume of the distillate collected. When the distillation is complete, plot your data as in Problem 3 and compare the actual result with the ideal.
5. Isolation of (R)-(−)- or (S)-(+)-carvone from oil of spearmint or oil of caraway by distillation (Section 47).

References

More information about simple distillation and fractional distillation can be obtained from References 1 and 2.

1. K. B. Wiberg, *Laboratory Technique in Organic Chemistry*, McGraw-Hill, New York, 1960, p. 24.
2. *Technique of Organic Chemistry*, Vol. IV, A. Weissberger, editor, Interscience, New York, 1951.

10 Reduced-Pressure Distillation

A liquid will boil when its vapor pressure equals the pressure on its surface. Thus, you can make a substance boil at a temperature lower than its normal boiling point by distilling it in an apparatus in which the pressure can be reduced. Because many substances undergo noticeable decomposition at elevated temperatures, distillation under reduced pressure is desirable for these substances in order to minimize decomposition. With substances whose thermal stability is not known, distillation is usually carried out under reduced pressure if it appears that the normal boiling point will be greater than 125–175°C.

10.1 ESTIMATION OF THE BOILING POINT AT REDUCED PRESSURE

A useful generalization is that the boiling point of a compound will decrease by about 20–30°C each time the external pressure is reduced by a factor of two. Thus, the boiling point of a compound in an apparatus in which the pressure is maintained at 45–50 Torr (1/16 atmosphere) would be expected to be 80–100°C lower than its normal (atmospheric pressure) boiling point.

The *nomograph* in Figure 10.1 is useful for estimating both expected reduced-pressure boiling points from normal boiling points and normal boiling points from observed reduced-pressure boiling points. The nomograph applies to liquids that are not associated in the liquid

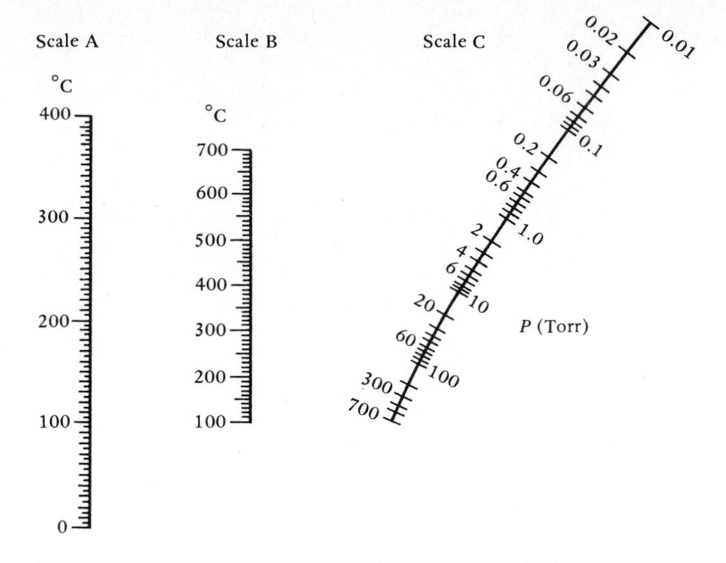

Figure 10.1. A nomograph for estimating the boiling point as a function of pressure. The nomograph relates the normal boiling point of a substance (scale B) to boiling points at reduced pressures (scales A and C). A line connecting points on two scales will intersect the third scale at some point. Thus, from values for A and C, you can estimate B. Knowing B (or having estimated B), you can estimate the boiling point A′ at a reduced pressure C′. A transparent plastic ruler works as well as anything for connecting points and noting intersections.

phase. The variation of boiling point with pressure for associated liquids, such as alcohols, is 10–20% less than for nonassociated liquids.

Reference 1 describes a more accurate method of calculating changes in boiling point with variations in pressure.

10.2 APPARATUS

It is possible to carry out distillations at reduced pressure in a regular distillation apparatus such as those illustrated in Figures 9.3 and 9.17 by connecting a source of vacuum to the side-arm of the distillation adapter. Other apparatus that are suitable for vacuum distillation, if the separation requirements are modest, are illustrated in Figure 10.2. For smaller-scale work, an arrangement such as that shown in Figure 9.18 could be used.

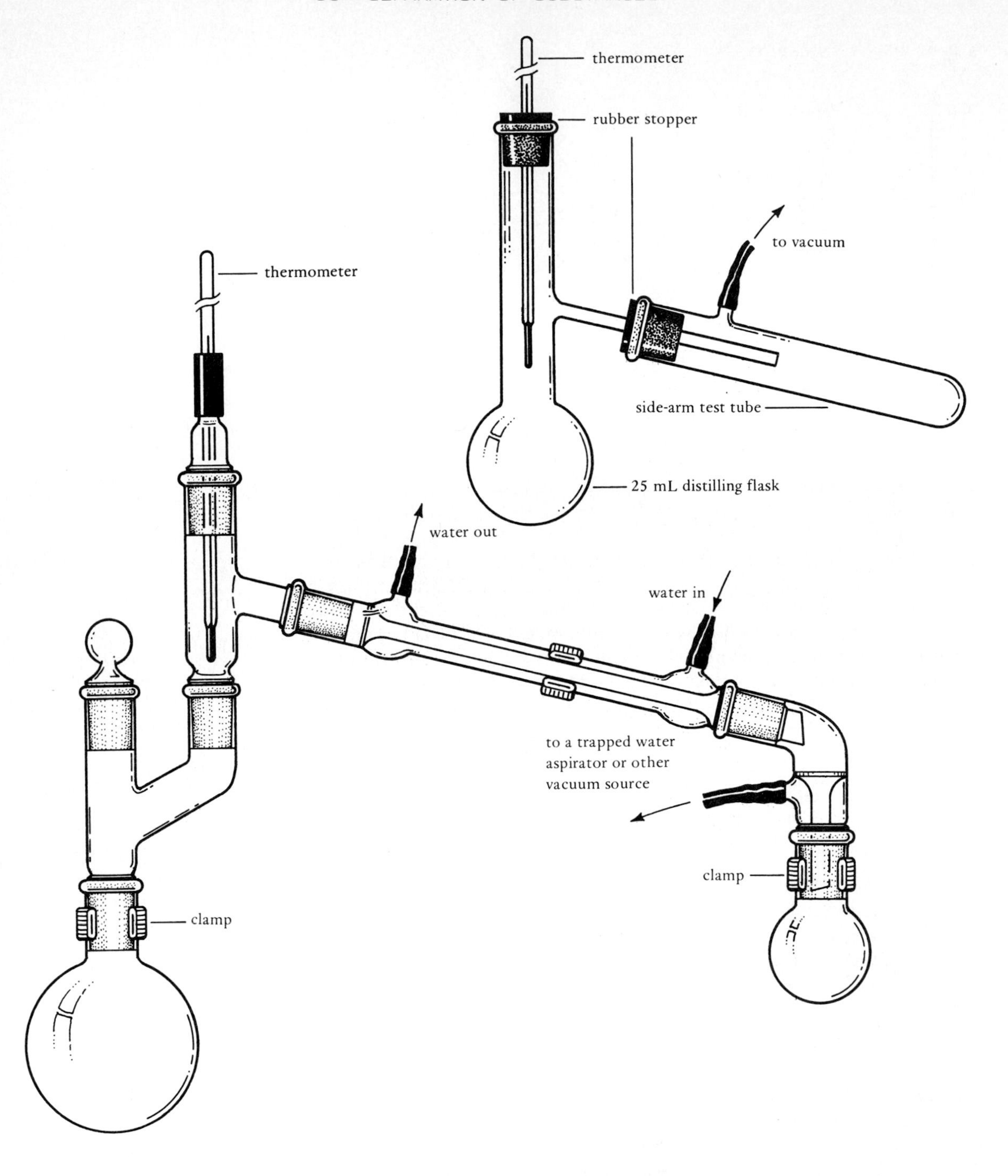

Figure 10.2. Two examples of apparatus for vacuum distillation.

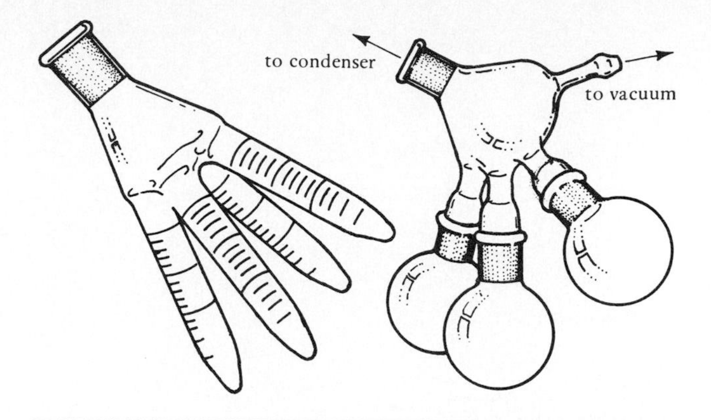

Figure 10.3. Two simple fraction collectors ("cows").

For separations that require more than simply removing a volatile solvent and then vacuum distilling the product, or vacuum distilling the product from a nonvolatile residue, a more specialized apparatus must be used. This apparatus must allow the collection of fractions of different boiling ranges without variation of the internal pressure or interruption of the distillation process. The simplest type of so-called *fraction collector* uses a distillation adapter that can be rotated in order to collect the distillate in different receivers. Some examples are illustrated in Figure 10.3. The apparatus illustrated in Figure 9.18 can also be used

Figure 10.5. Fraction cutter for vacuum distillation. Rotation of the stopcock to different positions allows changing of receivers without breaking the vacuum to the system.

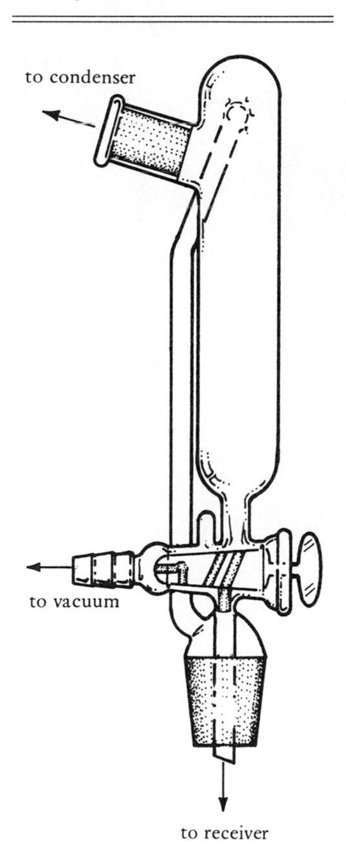

Figure 10.4. Small-scale vacuum distillation apparatus with rotating fraction collector.

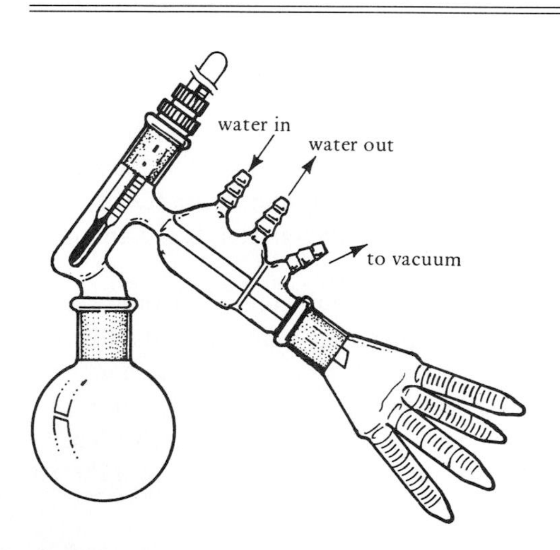

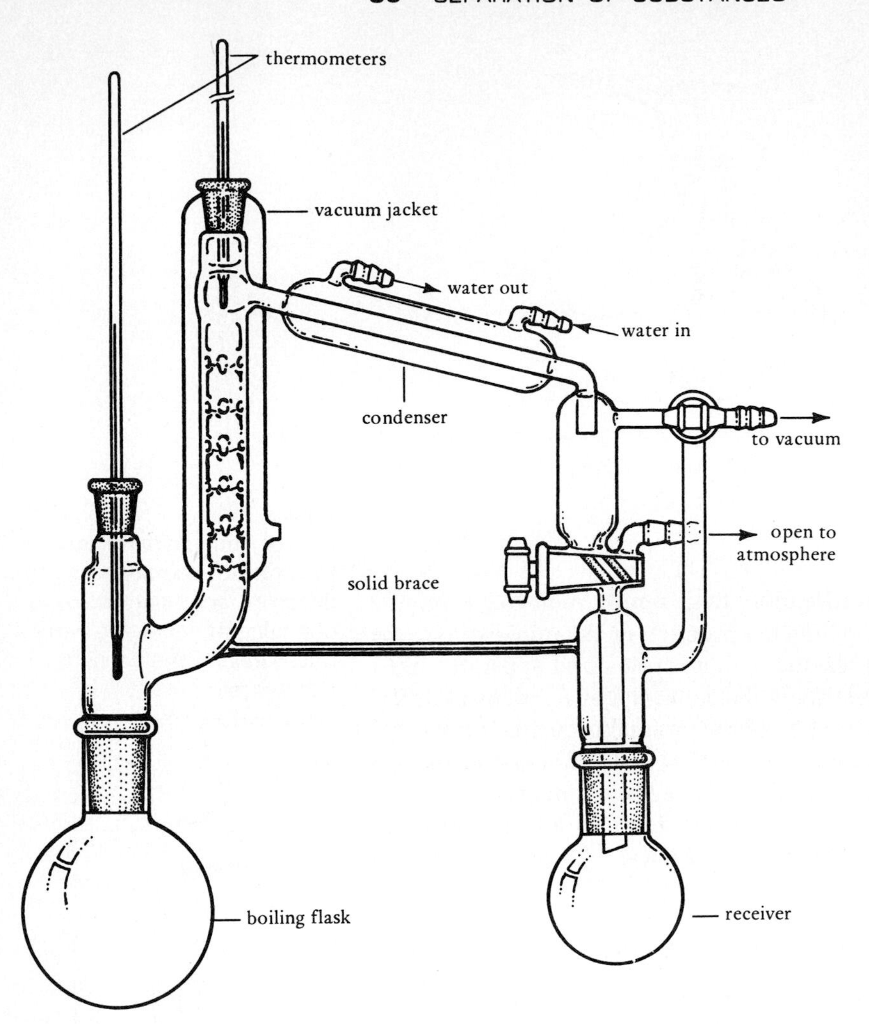

Figure 10.6. Apparatus for fractional distillation under vacuum.

for vacuum distillation of small samples when a rotating fraction collector of this type is attached to it, as shown in Figure 10.4. The disadvantages of this type of fraction collector are the limited number of fractions allowed and the fact that the more volatile fractions can distill into the less volatile ones during the distillation.

Fraction cutter

Another approach to the problem of how to collect fractions without disturbing the equilibria in the apparatus is to use a distillation adapter (*fraction cutter*), which will allow removal and replacement of a receiver without breaking the vacuum in the distillation apparatus. An example of an adapter of this type is shown in Figure 10.5. Distillation apparatus have been designed that include a short vacuum-jacketed

Vigreux column, condenser, and fraction cutter in one compact piece. An example is illustrated in Figure 10.6.

10.3 SOURCE OF VACUUM

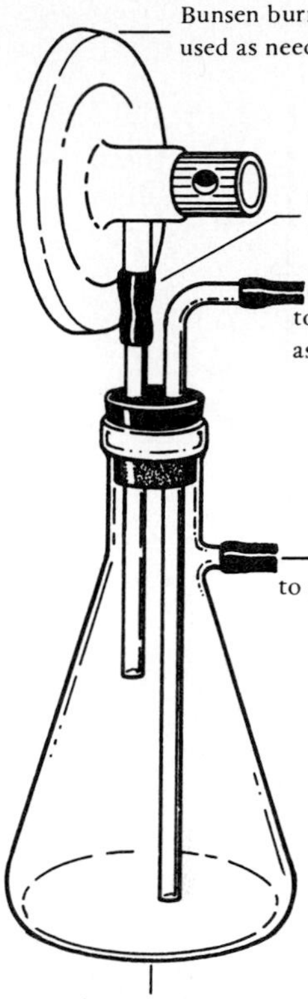

The most common and convenient source of vacuum is the water aspirator. The ultimate pressure attainable with a water aspirator equals the vapor pressure of water at temperature of flow. Between 5 and 30°C, the vapor pressure of water, in Torr, is numerically equal to the temperature, in degrees C, within 2.5 Torr. Thus, in the winter, you might hope to attain a pressure below 15 Torr, while in the summer, 25 Torr might be the ultimate pressure. An arrangement such as that illustrated in Figure 10.7 is suitable for reduced-pressure distillation when the desired pressure is equal to or above the pressure attainable with the water aspirator. The suction flask not only serves as a trap to prevent water from being sucked back into the apparatus in the event that the water pressure decreases momentarily, but also buffers the system against changes in pressure that result from variations in the flow rate of the water or from changing distillation receivers. The Bunsen burner valve can serve to provide a controlled rate of leakage of air into the system in case you wish to operate at a pressure above the ultimate pressure that you can attain with the aspirator.

Figure 10.7. Pressure regulation using the water aspirator.

For pressures below 10–25 Torr, a mechanical vacuum pump is generally used. A pump in average condition can give an ultimate pressure of 3–5 Torr, and 1 Torr if in good condition. A mechanical vacuum pump should always be connected to the system through a *cold trap*, as shown in Figure 10.8, in order to keep volatile (or corrosive) materials from reaching the pump, since the ultimate pressure attainable by the pump will be no lower than the vapor pressure of the liquid (oil) in the pump. Often, the performance of a mechanical pump can be restored merely by changing the oil. If it is necessary to carry out a reduced-pressure distillation with a mechanical vacuum pump at a pressure greater than the ultimate pressure attainable by the pump, a *manostat*, such as that described in *Laboratory Technique in Organic Chemistry* (2), should be used, and *not* an air leak. Drawing air through the hot oil of the pump would lead to degradation of the oil and an increase in its vapor pressure.

10.4 PRESSURE MEASUREMENT

Manometer; McLeod gauge

For measuring pressures in excess of 5–10 Torr, you will find a simple *manometer* such as that illustrated in Figure 10.9 to be quite adequate. For determination of pressures down to 1 Torr, a tilting *McLeod gauge* is useful (2).

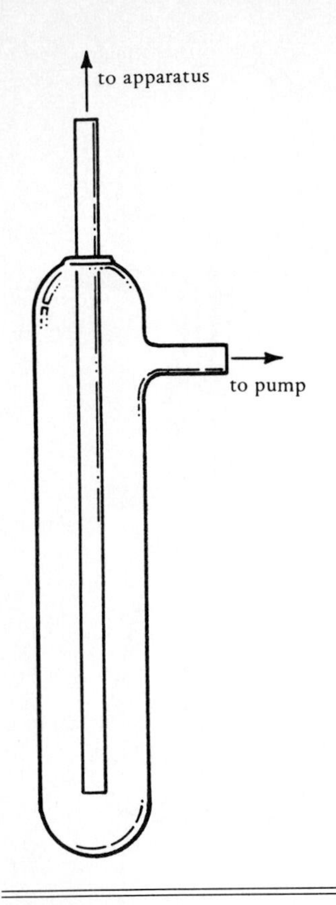

Figure 10.8. A cold trap. The lower part of the trap must be immersed in a Dry Ice/acetone bath or liquid nitrogen bath in order to condense vapors to keep them from getting to the pump.

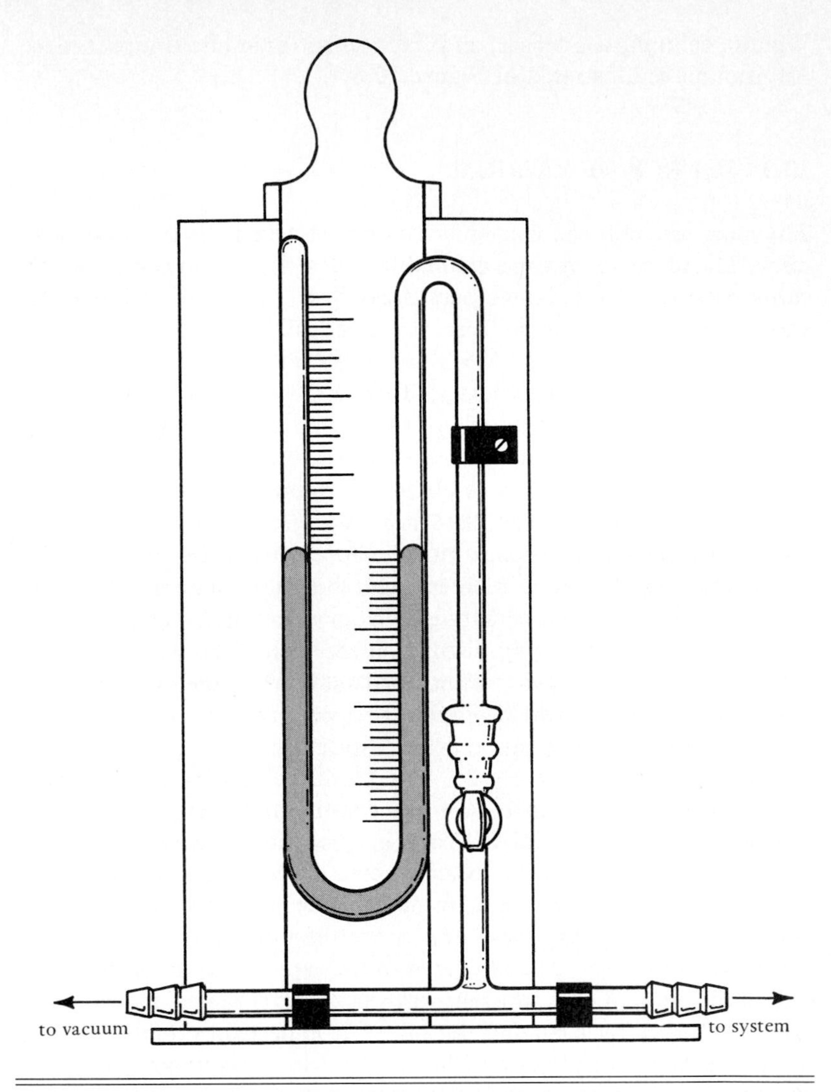

Figure 10.9. A simple manometer. The pressure equals the difference in height of the two columns of mercury. This height difference may be read by means of the movable scale.

10.5 TECHNIQUE OF DISTILLATION UNDER REDUCED PRESSURE

The volume of the vapor that results from the vaporization of a given amount of substance is many times larger at a reduced pressure than at 1 atmosphere (100 times larger at 7.6 Torr; 760 times larger at 1 Torr). Thus, maintenance of smooth boiling and prevention of bumping are even more important in vacuum distillations than in distillations at atmospheric pressure: a burst of vapor produced in a bump during vacuum distillation can easily splash material over into the receiver and can upset the vapor–liquid equilibria as well.

Boiling stones are often used to prevent bumping in a brief distillation, but they usually lose their effectiveness fairly quickly. If, for some reason, the pressure in the apparatus rises momentarily, the boiling stones often cease to function afterwards. Microporous carbon boiling chips (available from Fisher Scientific Company) are often recommended as being most suitable.

One of the most convenient methods of heating the boiling flask during distillation is to use a magnetically stirred and electrically heated oil bath (see Figure 33.1). If this method is used, bumping can usually be eliminated by placing a small magnetic stirring bar in the boiling flask and adjusting the level of the heating bath until it is somewhat higher than the level of the liquid in the flask. If the temperature of the bath is not too much above the boiling point of the liquid, and the liquid and bath are well stirred (the stirring bar in the bath turns the smaller one in the flask), boiling will usually be very smooth.

Experimental measurement of the boiling point of a substance in a reduced-pressure distillation is much less certain than in distillations at atmospheric pressure, and extrapolations of the reduced-pressure boiling points to atmospheric-pressure boiling points will contain these uncertainties. Since, for a given heating rate, the vapor velocity in a reduced-pressure distillation may be several hundred times greater (because of the greater volume) than in an atmospheric-pressure distillation, it is much easier for superheated vapors to reach the thermometer. In addition, the rapid flow of large amounts of vapor through the apparatus may result in a large pressure gradient. That is, the pressure may be considerably higher in the boiler than at the manometer. Both of these effects can be minimized by using a heating bath, and using it at a temperature that provides a minimum rate of distillation. An additional uncertainty is introduced when an air bleed is used to promote even boiling, because the air will contribute an unknown amount to the total pressure of the system.

Other techniques of distillation, which also apply to distillation under reduced pressure, are described in Section 9.6.

Problems

1. **a.** Benzyl alcohol is reported to boil at 93°C at 10 Torr. At what temperature would it be expected to boil at atmospheric pressure?
   b. 1,2-Dibutoxybenzene is reported to boil at 135–138°C at 12 Torr. What would you expect its normal boiling point to be?
2. **a.** α-Phenylethylamine boils at 187°C at atmospheric pressure. At what temperature would you expect it to distill under the vacuum that can be obtained with a water aspirator (assume 20 Torr)?

b. The normal boiling point of nitrobenzene is 211°C. At what temperature would it distill under a pressure of 15 Torr?

3. a. Diethyl ether has a normal boiling point of 34.6°C. Calculate the vapor pressure of ether at 0°C.

b. Diethyl phthalate is reported to have a normal boiling point of 296°C. How low would the pressure have to be in order to distill this substance at a temperature below 100°C?

Exercises

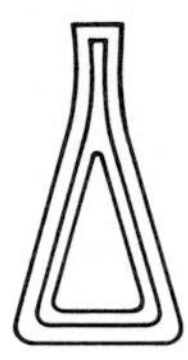

1. In the isolation of eugenol from oil of cloves (Section 45), the procedure can be extended to include vacuum distillation of the eugenol.

2. (R)-(−)- or (S)-(+)-carvone can be isolated from oil of spearmint or oil of caraway by distillation under reduced pressure (Section 47).

3. In the preparation of dibenzylketone (Section 80.3), the final distillation of the product can be carried out much more easily under reduced pressure.

References

1. H. B. Hass, *J. Chem. Educ.* **13**, 490 (1936).
2. K. B. Wiberg, *Laboratory Technique in Organic Chemistry*, McGraw-Hill, New York, 1960.

11 Distillation of Mixtures of Two Immiscible Liquids; Steam Distillation

In a mixture of two completely immiscible liquids, each exerts its own vapor pressure independently of the other. As the temperature of such a mixture in an apparatus open to the atmosphere is raised, the vapor pressure of each substance increases until the total vapor pressure equals the pressure of the atmosphere. Since the total vapor pressure is the sum of the individual vapor pressures, the total vapor pressure must become equal to atmospheric pressure at a temperature below the boiling point of either pure substance. The mixture thus distills at a temperature below the boiling point of either pure component.

Since organic compounds are generally miscible with one another, this phenomenon is usually observed only when one of the liquids is water; in these cases, the distillation process is called *steam distillation*. Steam distillation is a useful technique for effecting certain separations and is a useful method for distilling certain high-boiling compounds at a temperature no greater than 100°C.

11.1 THEORY OF STEAM DISTILLATION

In Section 9.5, the occurrence of minimum-boiling azeotropes was explained by incipient insolubility. If insolubility persists to high enough temperatures, two insoluble liquids will be present at the boiling point of the mixture. The gradual change from the one situation to the other can be imagined to occur by the process outlined in Figures 11.1 and 11.2.

Composition of the Distillate

The composition of the vapor at the boiling point, and therefore of the distillate, can be calculated by realizing that the ratio of the number of

Figure 11.1. A hypothetical boiling point diagram for a system involving a minimum-boiling azeotrope. If A and B are not completely miscible at all temperatures below the boiling point, certain mixtures of A and B will exist as two phases below certain temperatures. The area of the diagram under the broken curve XYZ represents the temperatures and compositions at which two phases exist in this hypothetical example. If you imagine that immiscibility could persist to temperatures above boiling point (XYZ moves up to X′Y′Z′), the boiling point diagram could be imagined to change to that shown in Figure 11.2.

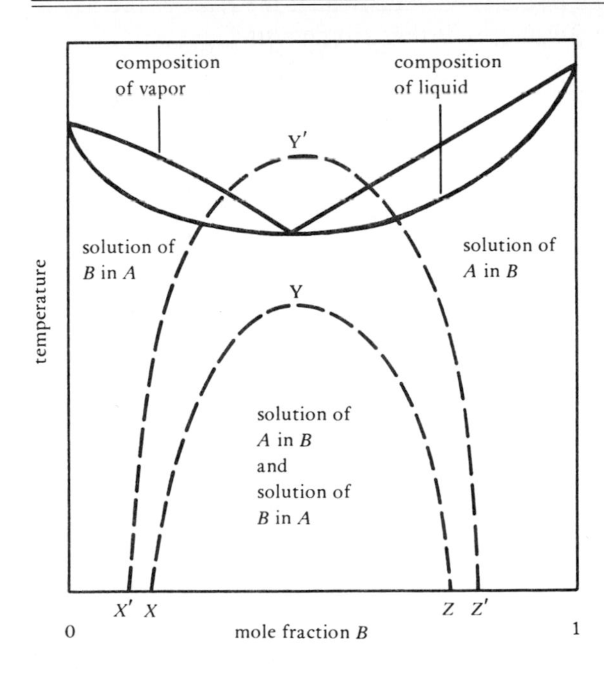

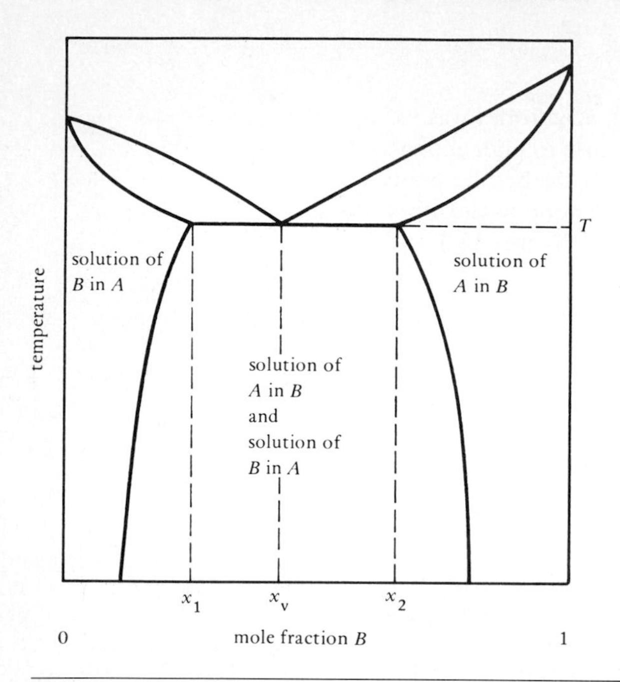

Figure 11.2. A hypothetical boiling point diagram of a system involving two "immiscible" liquids. At all compositions between x_1 and x_2, two phases will exist at the boiling point. The boiling point of all such mixtures will be $T°$, and the composition of the vapor will be x_v.

molecules of each type in the vapor must equal the ratio of the vapor pressures of the pure substances at that temperature:

$$\frac{P_B}{P_A} = \frac{\text{molecules of } B}{\text{molecules of } A} = \frac{\text{moles of } B}{\text{moles of } A} = \frac{n_B}{n_A} \quad (11.1\text{-}1)$$

For example, a mixture of the immiscible liquids bromobenzene and water steam distills at 95.3°C at 760 Torr. The vapor pressure of water at this temperature is 641 Torr, and therefore the vapor pressure of bromobenzene must be 760 − 641 = 119 Torr. The vapor pressure ratio of 119/641 must equal the mole ratio of the vapor and thus of the distillate, as in Equation 11.1-2:

$$\frac{119}{641} = \frac{n_{\text{brom.}}}{n_{\text{water}}} = 0.186 \text{ mole bromobenzene/mole water} \quad (11.1\text{-}2)$$

Since water is usually one of the substances involved in the distillation of mixtures of two immiscible liquids, we can substitute water for one of the substances in Equation 11.1-1. Substituting water for substance A gives us

$$\frac{P_B}{P_{H_2O}} = \frac{n_B}{n_{H_2O}} \quad (11.1\text{-}3)$$

Remembering that weight divided by molecular weight equals moles, we can write

$$\frac{P_B}{P_{H_2O}} = \frac{wgt_B/M_B}{wgt_{H_2O}/M_{H_2O}} = \frac{wgt_B}{wgt_{H_2O}} \cdot \frac{18}{M_B} \qquad (11.1\text{-}4)$$

Solving for wgt_B/wgt_{H_2O}, we get

$$\frac{wgt_B}{wgt_{H_2O}} = \frac{M_B}{18} \cdot \frac{P_B}{P_{H_2O}} \qquad (11.1\text{-}5)$$

where M_B is the molecular weight of compound B, which is insoluble in water, and where P_B/P_{H_2O} is the ratio of the vapor pressure of B to the vapor pressure of water at the temperature at which the sum of the two vapor pressures equals atmospheric pressure. Using again the example given above,

$$\frac{wgt_{brom.}}{wgt_{water}} = \frac{157}{18} \cdot \frac{119}{641} = 1.62 \text{ g bromobenzene/g water}$$

The higher the ratio of organic compound to water the better, since less distillate will have to be collected (and hence it will take less time) in order to collect a given amount of the organic compound. Equation 11.1-5 indicates that for a compound of a given molecular weight, the higher the vapor pressure, the more efficient the steam distillation. Solids with sufficiently high vapor pressures can also be steam distilled.

Because all real liquids are at least very slightly soluble in one another, the analysis given above can be only an approximation to any real situation. In the example involving bromobenzene and water, the two "immiscible" liquid layers would be a solution of a very small amount of water in bromobenzene and a solution of a very small amount of bromobenzene in water. However, according to Raoult's law (Section 9.3), which accurately accounts for the vapor pressure of the *solvent* in dilute solutions, the vapor pressure of bromobenzene in the bromobenzene-rich solution would be approximately equal to that of pure bromobenzene at that temperature, and the vapor pressure of water in the water-rich solution would be approximately equal to the vapor pressure of pure water at the same temperature.* Therefore, the analysis given for "completely immiscible" liquids will apply quite well if the two liquids are not very soluble in one another.

* It is interesting to realize that since both solutions are in equilibrium with vapor of the same composition, and thus must be in equilibrium with each other, the vapor pressure of water in the bromobenzene-rich solution (less than 1% water) must be equal to the vapor pressure of water in the water-rich solution (more than 99% water). A similar statement can be made about the vapor pressure of bromobenzene in the two solutions.

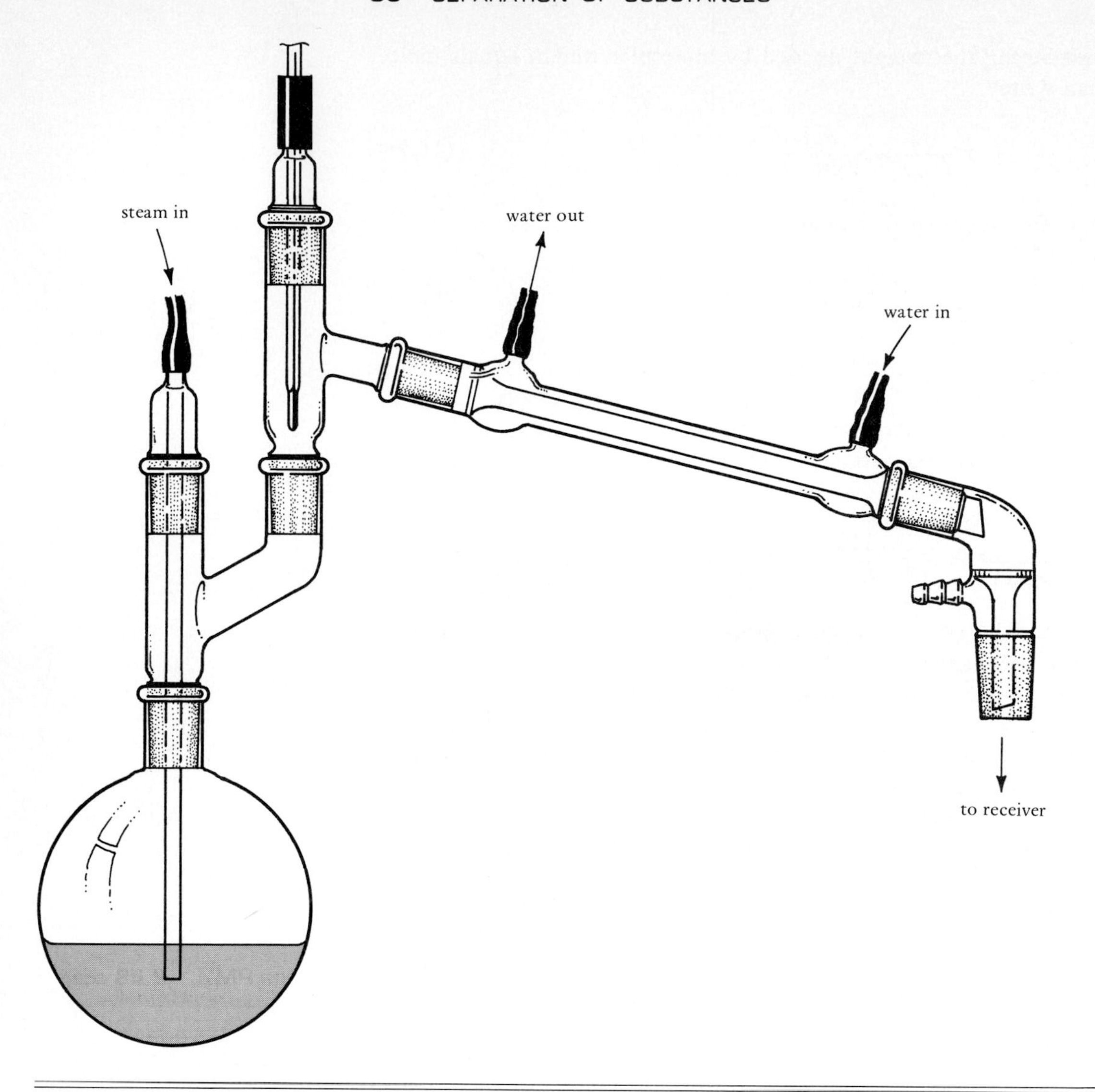

Figure 11.3. An apparatus for steam distillation. Replace the adapter and steam inlet tube with a separatory funnel if steam is generated internally.

11.2 TECHNIQUE OF STEAM DISTILLATION

Internal generation of steam

A substance can be distilled with steam by adding water to a mixture in which the substance (for example, a crude product from a reaction) is present, and boiling the resulting suspension. If the desired compound has a significantly higher vapor pressure at about 100°C than any of the other components of the mixture, it can be selectively removed by this

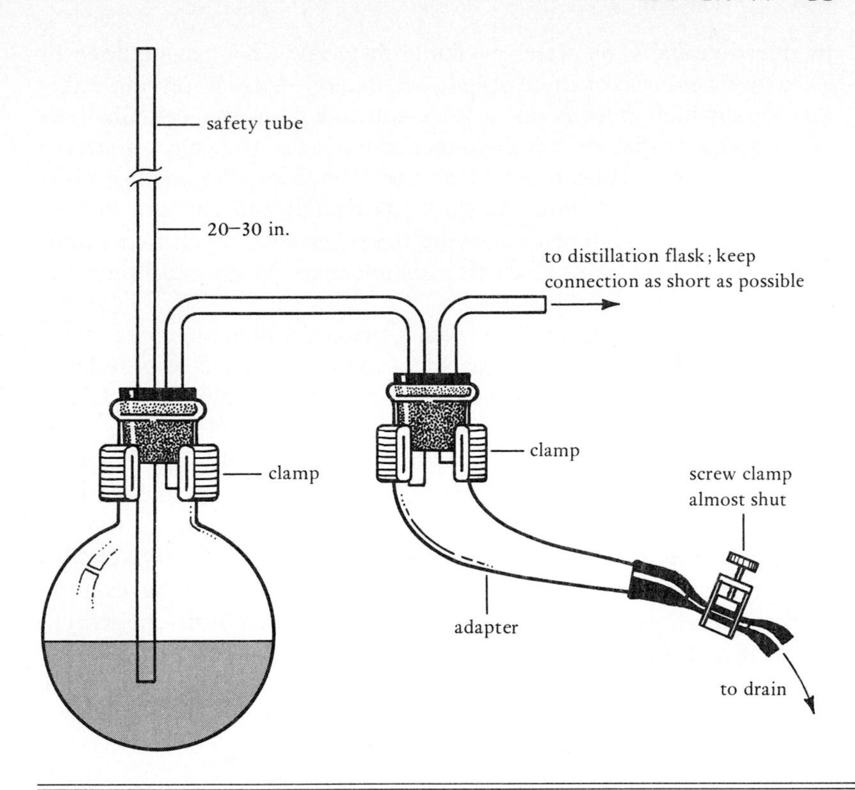

Figure 11.4. An apparatus for generating steam.

process. An apparatus suitable for this kind of operation is illustrated in Figure 11.3.

Sometimes the presence of a solid in a mixture will cause unavoidable superheating and violent bumping if the flask is heated directly. In these cases, the steam must be passed in from an external source. When steam is supplied externally, a satisfactory rate of steam distillation can be attained without any danger that the contents of the flask will be overheated by the vigorous heating necessary to supply the large molar heat of vaporization of water. If a large amount of water will be required to distill the substance, this can be an important consideration, since a maximum rate of distillation will probably be desired. An arrangement for the production and use of external steam is shown in Figure 11.4. When an external source of steam is used, water can condense in the distillation flask. This can be prevented by gently heating the flask with a Bunsen burner, an oil bath, or a heating mantle.

External generation of steam

Test for Completion of Distillation

There are two useful ways to determine when a steam distillation is complete. The first is to determine that the distillate currently being produced contains no water-insoluble material. This can be done by collecting a sample of the distillate separately, or by observing either that the distillate flowing down the condenser no longer contains little oily droplets or that solid is no longer collecting in the cold condenser. (If solid tends to collect and block the condenser, the cooling water should be turned off until the solid has melted and run into the receiver.) No harm is done in allowing the steam distillation to continue somewhat past the point at which insoluble material can be observed in the newly formed distillate.

Look for oily droplets

Another way of estimating when a steam distillation can be terminated is by calculating how much water should be collected in order to distill a given amount of substance. The amount of water collected can be estimated by collecting the distillate in a graduated cylinder or an approximately calibrated Erlenmeyer flask. According to Equation 11.1-5, if you know the molecular weight of the substance and the ratio of the vapor pressure of the substance to the vapor pressure of water at the temperature of distillation, the weight of water required to distill a given amount of the substance can be calculated. In the example worked out above, it was calculated that 1.62 grams of bromobenzene could be distilled by 1 gram of water. Since most steam distillations occur at approximately 100°C, it is satisfactory to use the ratio of vapor pressures at 100°C, rather than at the (usually) unknown temperature of distillation:

Calculate the volume of water required

$$\frac{\text{wgt}_B}{\text{wgt}_{H_2O}} = \frac{M_B}{18} \cdot \frac{P_B^{100°}}{760} \qquad (11.2\text{-}1)$$

The vapor pressure of B at 100°C may be estimated according to the nomograph of Figure 10.1 as long as the boiling point at atmospheric pressure is known. Employing the case of bromobenzene again as an example, and using 140 Torr as the value of the vapor pressure that can be estimated for bromobenzene at 100°C, we get

$$\frac{\text{wgt}_{\text{brom.}}}{\text{wgt}_{\text{water}}} = \frac{175}{18} \cdot \frac{140}{760}$$

$$= 1.6 \text{ g bromobenzene/g water}$$

This is the same value as that calculated using the ratio of vapor pressures at the actual temperature of distillation. The less the mutual solubilities of the substances and the closer the temperature of distillation to 100°C, the better this approximation.

Problems

1. In Section 44, oil of cloves is isolated from cloves by steam distillation. Assuming oil of cloves to be composed mostly of eugenol, which has a normal boiling point of 252°C,

CH_2 CH CH_2

HO

OCH_3

eugenol

a. Estimate the vapor pressure of eugenol at 100°C.

b. Calculate how many grams of water would be required to steam distill a gram of eugenol (oil of cloves).

Exercises

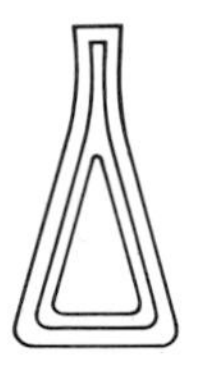

1. Isolation of oil of cloves from whole cloves (Section 44).
2. Other experiments that involve a purification by steam distillation include the oxidation of cyclohexanol to cyclohexanone (Section 54), the preparation of *n*-butyl bromide (Section 56.1), the preparation of aniline (Section 62), the preparation of chlorobenzene and the chlorotoluenes (Section 67), and the synthesis of 1-bromo-3-chloro-5-iodobenzene (Section 83.4).

12 Sublimation

Small samples of volatile solids can be purified by *sublimation*, or vaporization followed by condensation.

12.1 THEORY OF SUBLIMATION

The processes of sublimation and distillation are closely related. They can be compared with reference to the pressure-temperature diagram in Figure 12.1. In the process of *distillation of a liquid*, the liquid is heated until it reaches the temperature at which its vapor pressure equals the total pressure of the system; the state of the system can be represented during this process by a point moving from B to C in Figure 12.1.

The vapors that expand to a cooler part of the apparatus are then recondensed, the point now moving from C back to B. The distillation of a solid involves first melting and then boiling, condensation, and resolidification (a point moving from A to C and back again); it is just like distillation of a liquid except for melting and solidification. However, if a solid is heated under a total pressure less than that corresponding to the triple point (T in Figure 12.1), it will reach a temperature at which its vapor pressure equals the total pressure and will "distill" without melting, a point moving from D to E and back. Such behavior is called *sublimation*. The loss of water from food that has been stored in

Distillation versus sublimation

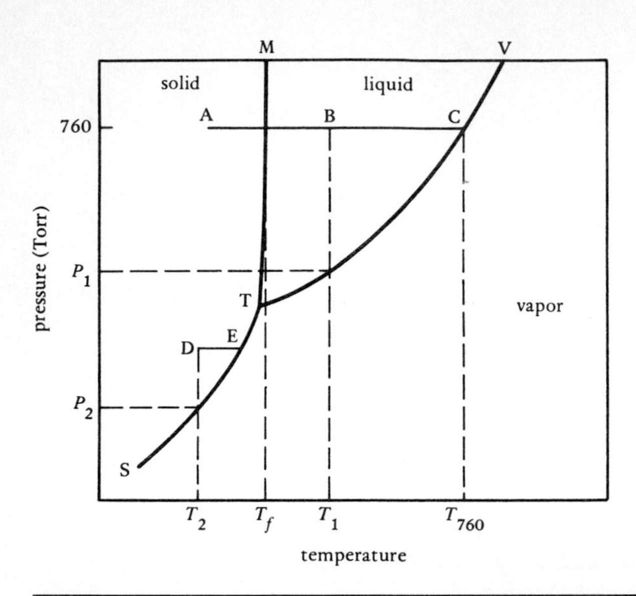

Figure 12.1. A general pressure-temperature diagram showing equilibria between solid, liquid, and vapor phases.

Table 12.1. Some substances that are easily sublimed under laboratory conditions

Compound	*Melting Point* (T_f, °C)	*Vapor Pressure at* T_f *and at* T[a]
Hexachloroethane	185	780 Torr
Perfluorocyclohexane	59	950 Torr
Camphor	179	370 Torr
Anthracene	218	41 Torr
Naphthalene	80	7 Torr
Benzene	5	36 Torr
p-Dibromobenzene	87	9 Torr
p-Dichlorobenzene	53	8.5 Torr
Phthalic anhydride	131	9 Torr
Benzoic acid	122	6 Torr
Carbon dioxide	−57	5.2 atm
Iodine	114	90 Torr
Many aromatic amines and phenols are easily sublimed.		

[a] Since the volume change upon melting is very small, the melting point of a substance is practically independent of pressure; therefore, the temperature at the triple point will be practically the same as at the normal melting point. Thus, since the temperature of the solid at the triple point and at the melting point is practically the same, the vapor pressure of the substance at the melting point should be the same as at the triple point.

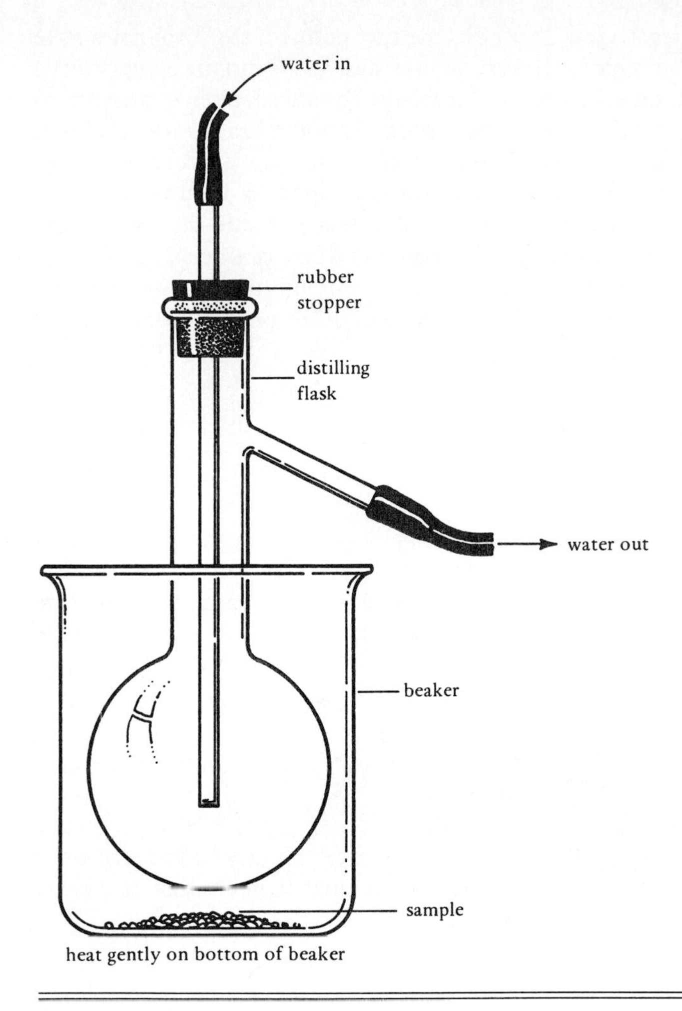

Figure 12.2. An apparatus for sublimation at atmospheric pressure.

a freezer for a long time takes place by sublimation ("freezer burn"). Similarly, wet clothes hung out to dry on a winter day first freeze and then dry by sublimation.

It is very rare for a substance to have a vapor pressure of 760 Torr or greater at the temperature of the triple point, which, as Figure 12.1 indicates, is approximately the same temperature as the normal melting point T_f, but many substances have a large enough vapor pressure near their melting point to be sublimed at the reduced pressures easily obtained in the laboratory. Table 12.1 lists a number of substances and their vapor pressures at the melting point.

Sublimation is a useful procedure for purification if the impurities are essentially nonvolatile and if the desired substance has a vapor

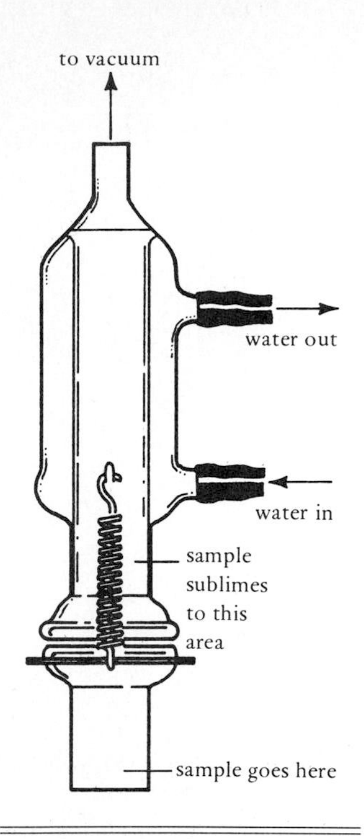

Figure 12.3. An arrangement for sublimation under reduced pressure.

pressure of at least a few Torr at its melting point. If the impurities have about the same vapor pressure as the desired compound, very little separation will be achieved, and it would be much better to attempt to purify the substance by recrystallization. Sublimation is most useful in the case of small samples, as mechanical losses can be kept very low. Several apparatus for sublimation are illustrated in Figures 12.2 and 12.3. Substances like camphor whose vapor pressure at the melting point is substantial, although less than 760 Torr, can be sublimed successfully at atmospheric pressure. The rate of sublimation will be relatively low and will depend upon the rate of diffusion of the vapor to the cold condensing surface.

12.2 TECHNIQUE OF SUBLIMATION

In a sublimation at atmospheric pressure, spread out the material to be sublimed in the container to be heated and arrange for the condensation of the vapors (Figure 12.2). Heat the container cautiously without melting the solid until crystals can be seen to appear on the cold surface of the condenser. At this point, only a small amount of heat is needed to keep the sublimation going. If you are heating with a burner, the flame will probably have to be applied intermittently. The rate of sublimation is much easier to control if an oil bath is used for heating (Section 32.1). When only a nonvolatile residue remains, carefully remove the condenser and scrape off the product.

In sublimation at reduced pressure, the procedure is the same except that a closed system in which the pressure can be reduced must be used (Figure 12.3). The source of vacuum can be either the water aspirator or the oil pump (Section 10.3).

Problems

1. Whether or not a substance can be sublimed depends upon its vapor pressure at the triple point. To be easily and conveniently sublimed, a substance should have a vapor pressure of at least 5–25 Torr. In Table 12.1, it was pointed out that the vapor pressure at the melting point should be a good approximation of the vapor pressure at the triple point. Since at the melting point the vapor pressure of the solid is equal to the vapor pressure of the liquid, and since it is possible to estimate the vapor pressure of a liquid at any temperature (even at its freezing point), when its vapor pressure at any other temperature is known (for example, by using the nomograph of Figure 10.1), knowledge of the boiling point of the substance at any pressure and the melting point of the substance should make it possible to estimate the vapor pressure at the triple point.
   a. Knowing that benzene boils at 80°C, calculate the vapor pressure to be expected for benzene at its freezing point of 5°C. Compare your result with the value in Table 12.1.

b. Do the same for benzoic acid, given that its normal boiling point is 249°C.

2. The vapor pressure, in Torr, of a substance at its melting point (P_f) can be related to its heat of vaporization ΔH_{vap}, its boiling point T_b, its melting point T_f, and atmospheric pressure (760 Torr) by the following equation:

$$\log P_f = \log 760 - \frac{\Delta H_{vap}}{2.3RT_b}\left(\frac{T_b - T_f}{T_f}\right)$$

You can see that compounds with relatively high melting points and short liquid ranges should be relatively easily sublimed.

a. Which of the isomeric octanes would be expected to sublime most easily? (See Table 17.2.)

b. In Table 12.1, it was stated that in general aromatic amines and phenols are relatively easily sublimed. Why does this make sense in terms of the structures of these molecules?

Exercises

1. Sublime a sample of hexachloroethane, using an arrangement such as that shown in Figure 12.2.
2. Sublime a sample of camphor. This can be done at atmospheric pressure or in an apparatus such as that illustrated in Figure 12.3.
3. Benzoic acid has a vapor pressure of 6 Torr at the melting point and can be purified by vacuum sublimation as well as by crystallization. A mechanical vacuum pump should be used so that the process will not take too long.
4. In the procedure of Section 42, caffeine is purified by recrystallization from ethanol. An alternative method should be purification by sublimation, since caffeine is reported to sublime slowly at atmospheric pressure at temperatures above 120°C, to sublime quickly at 1 Torr at 160–165°C, and to sublimc at 89°C and 15 Torr.

13 Extraction by Solvents

The fact that different substances may differ in solubility in various liquids can serve as a basis for separating them. Purification by recrystallization involves one application of this phenomenon, and solvent extractions and chromatographic methods (Section 14) are others.

13.1 THEORY OF EXTRACTION

If a mixture of substances, say A and B, is dissolved in an organic solvent, such as diethyl ether, and the solution is mixed thoroughly with the immiscible solvent water, an essentially complete separation of

A and *B* can be effected if one of the substances, say *A*, is much more soluble in water than in ether, and the other is much more soluble in ether than in water. If, at equilibrium, most of the *A* molecules are in the water phase and most of the *B* molecules are in the ether phase, a physical separation of the layers by means of a separatory funnel (Figure 13.1) will result in a separation of the two kinds of molecules. The material in each layer can be recovered, typically, by distillation or evaporation of the solvent.

Table 13.1 indicates in a general way the relative solubility of different types of compounds in water and organic solvents.

Table 13.1. Estimated relative solubility of different types of compounds in organic solvents and water

Solubility in all solvents decreases with increasing molecular weight.

Type of Compound	*Estimated Ratio of Solubility in Organic Solvent to Solubility in Water*
Covalent substances containing only carbon, hydrogen, and halogen	Very much greater than 1
Covalent substances containing oxygen and/or nitrogen in addition to carbon, hydrogen, and halogen	
a. 5 carbon atoms per functional group	10:1
b. 2 carbon atoms per functional group	1:1
c. 1 carbon atom per functional group	1:10
Salt of an organic acid	Very much less than 1
Salt of an organic base; amine salt	Very much less than 1
Inorganic salts	Very much less than 1

13.2 EXTRACTION OF ACIDS AND BASES

The extraction of an acidic or basic substance, either the product of the reaction or an undesired side product, from an organic solvent into water can be effected by mixing (shaking) the solution in the separatory funnel with an aqueous solution of base or acid, respectively.

Extraction of acids by bases

In the extraction of an acidic material with aqueous base, the acid in the organic solution dissolves in the water and is immediately converted to its salt. The salt is very much more soluble in water than in the organic solvent. Thus, the concentration of free acid in the aqueous phase will remain very low as long as the aqueous phase remains sufficiently basic to convert the acid to its salt. After equilibrium has been established, the *ratio* of free acid dissolved in the organic solvent to free acid dissolved in water may be much greater than 1 (because of greater solubility in the organic solvent than in water). Still, the fact that the vast majority of acid molecules are in the form of their corresponding salt dissolved in water means that only a very small *amount* of the acid originally present in the organic solvent is still there. From this analysis, you can see that the degree of completeness with which an acid can be extracted from an organic solvent depends upon the basicity of the extracting aqueous solution: the solution must be basic enough to convert the acid "completely" to its salt. A useful approximation is that the pH of the extracting solution should be at least 4 pH units on the basic side of the pK_a of the acid to be extracted. Thus, aqueous sodium carbonate solution (pH $\approx$ 11) should "completely" extract acids stronger than $pK_a \approx 7$ (see Table 13.2).

Table 13.2. pH required for extraction of different acids and bases

Approximate pH of Aqueous Solution Needed to Extract an Acid from an Organic Solvent		
Compound	pK_a	pH *as Basic or More Basic Than:*
Mineral acids	>1	~4
$Ar-C(=O)-OH$	~4	~8
$R-C(=O)-OH$	~5	~9
Phenols	~10	~14
Approximate pH *of Aqueous Solution Needed to Extract a Base from an Organic Solvent*		
Compound	pK_a	pH *as Acidic or More Acidic Than:*
Anilines	~5	~1
Pyridines	~6	~2
Aliphatic amines	~11	~7

Approximate pH *of Aqueous Solutions, 5–10% by Weight*	
Compound	*Approximate* pH
HCl; H_2SO_4	0
Acetic acid	3
$NaHCO_3$	8
Na_2CO_3; K_2CO_3	11
NaOH; KOH	14

Extraction of bases by acid

Similarly, you could expect to "completely" extract an amine from an organic solvent by means of an aqueous acid solution whose pH is at least 4 pH units on the acidic side of the pK_a of the conjugate acid of the amine (see the data of Table 13.2).

It should also be apparent that if a basic aqueous solution of an organic acid is acidified with a mineral acid ($pH \approx 1$; $[OH^-] \approx 10^{-13}$), the acid should be extractable from water into an organic solvent if the free acid is less soluble in water than in the organic solvent. Of course, if the acid is not very soluble in water, it will separate or precipitate from the acidified solution before the organic solvent is added.

In a similar way, it is possible to extract an organic base from an acidic aqueous solution by making the solution basic and then extracting the mixture with an organic solvent.

13.3 TECHNIQUE OF EXTRACTION

The objective of a simple extraction is to partition one or more substances between two immiscible solvents. This is usually accomplished with the use of a separatory funnel (Figure 13.1). If the separatory funnel has a glass stopcock, prepare the funnel for use by making sure that the stopcock is lightly greased and will turn without difficulty; Teflon stopcocks need not be greased. With the separatory funnel supported in a ring (Figure 13.1), check to make sure that the stopcock is closed and pour in the solution to be extracted. Then add the extracting solvent (the funnel should not be filled to more than about three-fourths of its height), replace the stopper after wetting it with water (to keep the organic solvent from creeping out around the stopper), and swirl or shake the contents to mix them. With vigorous shaking, a total mixing period of ten to thirty seconds is usually considered adequate to establish equilibrium. After allowing the mixture to stand in the separatory funnel until the two immiscible layers have separated cleanly, remove the stopper at the top and draw off part of the lower layer through the stopcock at the bottom. Wait a little while for the remainder of the lower layer to drain down (gentle swirling of the separatory funnel can speed this up), and draw this off also. Then pour the upper layer out the top.

Close the stopcock

Mix thoroughly

Let the air in

If a glass stopcock plug is used, it should be removed from the separatory funnel, cleaned, and stored out of the funnel. If this is not done, you may find it impossible to turn the stopcock next time you use the funnel. (Teflon stopcock plugs are much better suited for separatory funnels than are glass plugs since they will not freeze in place, do not need to be greased, and do not need to be removed for storage.)

Remove stopcock for storage

When a volatile solvent is involved in an extraction, the establishment of the equilibrium vapor pressure of the solvent will cause the

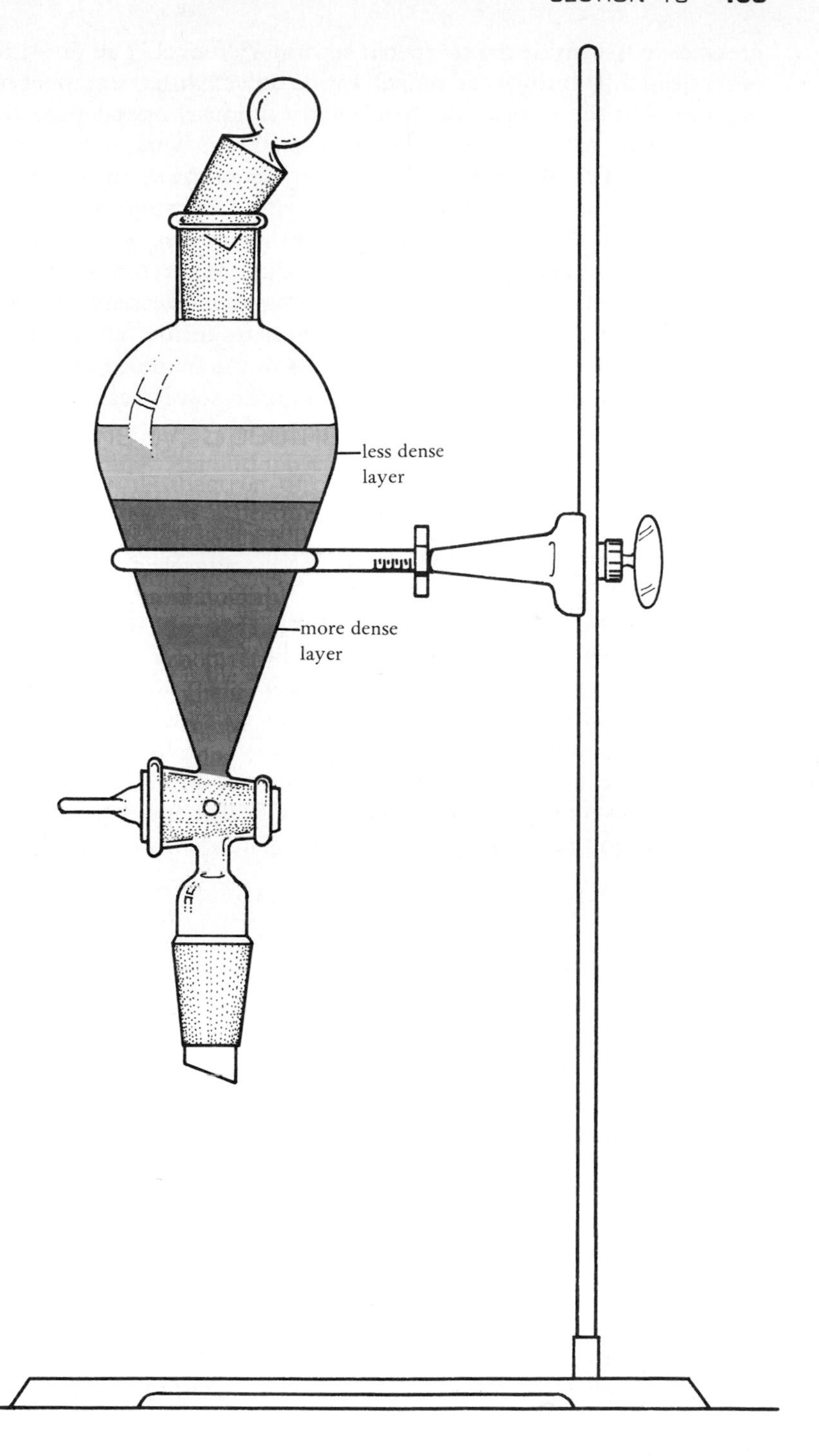

Figure 13.1. A separatory funnel.

pressure to rise inside the stoppered separatory funnel. The pressure is best released by turning the funnel upside down (with the stopper held in place with the palm of the hand) and cautiously opening the stopcock. When a very volatile solvent such as ether is being used, the first mixing should consist only of a slow inversion of the separatory funnel followed by release of the pressure. After alternate cautious sloshing of the contents of the separatory funnel and then release of pressure, the sound of the escaping vapors will indicate that the pressure is not being built up so fast and the periods of mixing can be longer and more vigorous. If you neglect to release the pressure inside the funnel, the stopper may be forced out. If the contents of the funnel are forced out as well, the escape of a volatile and flammable solvent can result in a dangerous fire.

Relief of pressure

Warning: beware of boiling solvent

It should be obvious that it is dangerous to attempt to extract a solution if its temperature is near or above the boiling point of the extracting solvent. This means that if an extraction is to be carried out with pentane, ether, or dichloromethane, it may be necessary to cool the solution to below room temperature. When these solvents are used, it is also a good idea to hold the separatory funnel by the ends so that the contents will not be warmed by your hands.

If a strong acid is to be extracted with carbonate or bicarbonate solution, the carbon dioxide produced can cause a large buildup in pressure unless mixing is done very cautiously with frequent release of pressure. In cases where much carbon dioxide production is anticipated, it is best to do the mixing in a flask or beaker and then to transfer the mixture to the separatory funnel for separation.

Warning: beware of CO_2 evolution

Choice of Solvent

If an organic product is to be purified by dissolution in an organic solvent followed by extraction of the solution with two or more portions of aqueous solution, the whole process will be much faster and easier, and will involve less loss, if the organic solution is less dense than water. In this case, the water layer can be drawn off through the stopcock, and the organic solution is retained in the funnel, ready for the next extraction. If the organic phase is heavier than water, it will have to be drawn off through the stopcock, the aqueous layer poured out, and the organic layer returned to the separatory funnel for the next extraction. Each such transfer will take time and may involve a loss of material. Conversely, if it is necessary to extract an aqueous solution with several portions of solvent in order to achieve the maximum recovery of a substance, it will be more convenient to extract with a solvent heavier than water so that the solvent can simply be drawn off each time through the stopcock without removal of the water layer first. The densities of a number of solvents are listed in Table 8.1.

Extraction in Batches

If it is necessary to extract a larger volume of material than will fit into the available separatory funnel, the extraction may, of course, be done in batches. Small amounts of water-insoluble material may be efficiently removed from large amounts of water by adding a little ether (or other solvent less dense than water) to the flask containing the water/product mixture, swirling the mixture well, and then adding it in portions to the separatory funnel, drawing off most of the water layer after each addition. The converse of this procedure can be used to wash a large amount of an organic solution with a little water, if the organic solution is heavier than water.

More Suggestions

Distinguishing the layers

Often it is possible to tell which layer in the separatory funnel is organic and which is aqueous from knowledge of the relative volumes used or relative densities of the two solvents. Sometimes, however, the transfer of material from one layer to the other or the presence of several substances of different densities can make the identification of the layers uncertain. It is not possible, of course, to tell by smell which layer is which, since the vapor pressure of each component in each phase will be the same. But sometimes you can determine which is which by adding a little water and seeing with which layer the added water combines. Or you can withdraw a little of the lower layer and see if it is miscible with water. If doubt still remains, each layer should be carried on in the procedure as if it were the desired one until it becomes obvious that one of the two cannot be the right one. It is very common to discard the product layer through error or ignorance; it is always advisable to *save everything* until the product has been safely isolated.

Smell won't tell

Don't throw away the wrong layer

Insoluble material

In most extractions, at least a trace of insoluble material collects at the interface between the two immiscible layers. It is often impossible to separate the layers without taking along some of this insoluble material. This is not a crucial matter, since whatever is picked up can always be removed by filtration at the end of the extraction or at some later stage in the purification. For example, in the very common case in which the organic product of a reaction is isolated by dissolving it in an organic solvent, extracting ("washing") the solution with one or more aqueous solutions in order to remove certain undesired materials, drying it (Section 15.2), and removing the drying agent by gravity filtration, any insoluble impurity that may have been carried along in the organic layer will be removed along with the drying agent in the gravity filtration.

If the distribution coefficient for a substance to be extracted from water is much less than 1—that is, if the ratio at equilibrium of the concentration in the organic solvent to the concentration in water is much less than 1—a simple extraction process will not give a satisfactory recovery. The distribution coefficient can sometimes be increased by adding sodium chloride or sodium sulfate to the aqueous solution, since the solubility of most organic compounds is less in salt solutions than in water. (The interpretation for this phenomenon, known as *salting out*, is given in Section 22.1.) Alternatively, the distribution coefficient can be increased by using an organic solvent that is a better solvent for the type of compound being extracted. For substances with oxygen-containing functional groups, ethyl acetate and *n*-butyl alcohol are probably better solvents than non-oxygen-containing solvents. Chloroform is an especially good solvent for amines. (The factors that determine solubility are also discussed in Section 22.1.)

Salting out

Sometimes the mixture in the separatory funnel does not separate into two phases, but forms a single, homogeneous solution. This can happen when large amounts of methanol, ethanol, or tetrahydrofuran are present, as these liquids are good solvents for *both* water *and* organic materials. Sometimes separation into two liquid phases can be brought about by the addition of more water and more of the organic solvent, or by the addition of saturated, aqueous, sodium chloride solution. It is best, however, to avoid this problem by removing most of the methanol, ethanol, or tetrahydrofuran, possibly by distillation, *before* doing the extraction.

O

tetrahydrofuran
THF

Occasionally, when the organic substance being purified is a solid, it will begin to crystallize during the extraction. This happens when the amount of organic solvent is reduced below that required to dissolve all of the solid, the loss of the organic solvent occurring because of its slight solubility in the aqueous liquids that are used to wash the organic solution. Or this can happen because another substance, perhaps an alcohol, that helped to dissolve the solid product in the organic layer has itself been removed by extraction. No matter what the cause, addition of more of the same solvent or of a better solvent should bring the solid material back into solution.

Emulsions

A common problem in extraction is failure of the immiscible solutions to separate completely and cleanly into two layers; a certain volume of the mixture at the interface sometimes consists of droplets of one solution suspended in the other (an *emulsion*). In some cases, the separation becomes clean and complete if the separatory funnel and its contents are allowed to stand undisturbed for a few minutes. At other times, the emulsion may persist for hours or days. If the volume of the emulsion is relatively small, it can sometimes be temporarily ignored and the extraction procedure continued in the hope that the emulsion will disappear in later extractions. But if most of the mixture is emulsified, it must either be allowed to stand (if the time is available) or be broken in some other way.

Breaking emulsions

If the emulsion is caused by too small a difference in density between the two layers, the addition of solvent to one or both layers may produce a larger density difference. Pentane will most efficiently decrease the density of an organic solution, while carbon tetrachloride will most efficiently increase the density. The addition of water may either increase or decrease the density of the aqueous phase, depending upon its composition. If the aqueous phase contains appreciable amounts of organic solvents (such as alcohols), it may have a density of less than one gram per milliliter; if, on the other hand, it is a solution of inorganic materials, its density will be greater than one gram per milliliter. The addition of salt or saturated sodium chloride solution can also increase the density of the aqueous phase.

Emulsions are most commonly encountered in extractions involving basic solutions. Presumably this is because traces of higher-molecular-weight organic acids are converted to their salts, and the resulting soap causes emulsification. The tendency of an extraction with a "neutral" aqueous solution to emulsify can sometimes be overcome by adding a few drops of acetic acid, thus suppressing soap formation.

Stubborn emulsions can sometimes be broken by centrifugation or suction filtration of the emulsified material. Suction filtration of an emulsion is a very messy operation and should be done only in desperation.

With emulsions, prevention is far better than cure. If the mixture is sloshed or shaken gently at first and then allowed to stand, the tendency toward emulsion formation can be estimated. When it appears that emulsion formation may be a problem, it is wise to mix the layers more gently for a longer time. In extreme cases, it may be desirable to carry out the mixing in a round-bottom flask by slow stirring with a magnetic stirrer. Since in this case the area of the interface will be much less than if the mixture were broken up into a suspension of tiny bubbles by vigorous shaking or stirring, it may take 30–60 minutes to approach equilibrium.

Problems

1. Estimate whether or not each of the following acids could be "completely extracted" with one portion of an excess of (a) aqueous sodium bicarbonate solution, (b) aqueous sodium carbonate solution, and (c) aqueous sodium hydroxide solution:

$Cl_3C{-}C({=}O){-}OH$
trichloroacetic acid
$K_a = 2 \times 10^{-1}$
$pK_a = 0.70$

$CH_3CH_2CH_2{-}C({=}O){-}OH$
butyric acid
$K_a = 1.5 \times 10^{-5}$
$pK_a = 4.52$

OH

phenol
$K_a = 1.3 \times 10^{-10}$
$pK_a = 9.89$

OH, NO_2, NO_2, NO_2

picric acid
$K_a = 1.6 \times 10^{-1}$
$pK_a = 0.80$

2. Using extraction procedures only, how would you separate a mixture of
   a. *p*-Dichlorobenzene, *p*-chlorobenzoic acid, and *p*-chloroaniline?
   b. *p*-Dichlorobenzene, *p*-chlorobenzoic acid, and *p*-chlorophenol?

Cl
Cl
p-dichlorobenzene

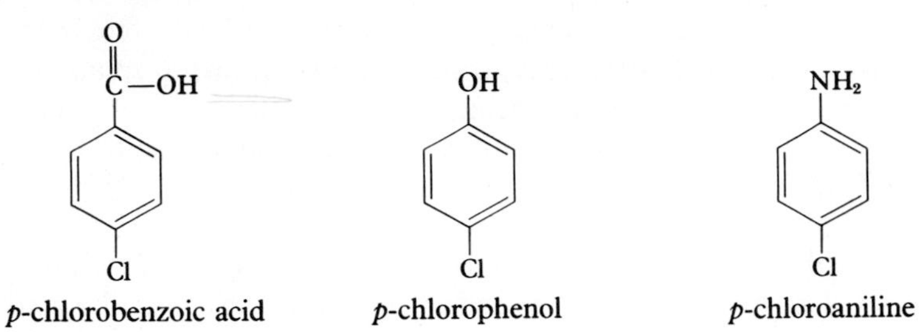

3. Why is carbon dioxide produced when a strong acid, such as sulfuric acid or hydrochloric acid, is neutralized with sodium bicarbonate or with sodium carbonate? Illustrate your answer with balanced equations.
4. The solubility of adipic acid in water is 1.5 g per 100 mL at 15°C, and 0.6 g per 100 mL in ether at the same temperature.
   a. What fraction of a sample of adipic acid could not be extracted from water into ether with one extraction with a volume of ether equal to that of the aqueous solution?
   b. What fraction of a sample of adipic acid could not be extracted from ether into water with one extraction with a volume of water equal to that of the ethereal solution?
   c. Repeat **a**, but use three portions of ether, each equal to one-third of the volume of the aqueous solution.
   d. Repeat **b**, but use three portions of water, each equal to one-third of the volume of the ethereal solution.
   e. Calculate the limiting fraction, which could not be extracted with an equal volume of ether.
   f. Calculate the limiting fraction, which could not be extracted with an equal volume of water.
   g. Repeat **a**, **c**, and **e**, but use a total volume of ether equal to five times the volume of the aqueous solution.

Exercises

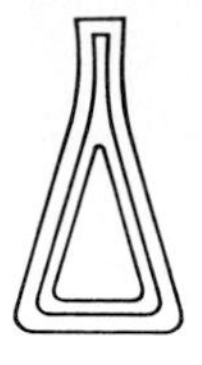

1. Extraction of caffeine from tea or NoDoz (Section 42).
2. Extraction of eugenol from oil of cloves (Section 45).
3. Separation of a mixture of an acid A—H, a base B:, and a neutral substance N.

Procedure. Dissolve about 5 grams of the mixture in 50 mL of diethyl ether and transfer the solution to a 125-mL separatory funnel. Add to the funnel 30 mL of 1 *M* HCl solution. Shake the mixture well in order to extract the basic substance as its hydrochloric acid salt into the water layer:

$$\mathrm{B{:}} + \mathrm{H_3\overset{+}{O}} + \mathrm{Cl^-} \longrightarrow \mathrm{\overset{+}{B}{-}H} + \mathrm{Cl^-} + \mathrm{H_2O}$$

Draw off the lower, aqueous, layer and save it.

Add to the ether solution in the separatory funnel about 30 mL of 1 *M* NaOH solution. Shake the mixture well in order to extract the acidic substance as its sodium salt into the water layer:

$$\mathrm{A{-}H} + \mathrm{HO^-} + \mathrm{Na^+} \longrightarrow \mathrm{A{:}^-} + \mathrm{Na^+} + \mathrm{H_2O}$$

Draw off the lower, aqueous, layer and save it. The ether layer should now contain only the neutral substance N.

Isolation of the neutral substance, N. Wash the ether layer by adding to the separatory funnel about 25 mL of water, shaking the mixture, allowing the layers to separate, and then drawing off and discarding the lower, aqueous, layer. Transfer the ethereal solution of the neutral compound to a small Erlenmeyer flask, dry the solution over 1–2 grams of anhydrous sodium sulfate for a few minutes (Section 15.2), remove the drying agent by gravity filtration (Section 7.1), and remove the ether by evaporation or distillation on the steam bath (Section 36). The infrared spectrum of the residue may be determined (Section 23); or, if the residue is a solid, it may be recrystallized (Section 8) and its melting point determined (Section 17).

Isolation of the basic substance, B:. Transfer the aqueous solution of the hydrochloric acid salt of the basic substance to a clean 125-mL separatory funnel. Make the solution strongly basic by adding about 2 mL of 50% aqueous sodium hydroxide. Make sure the solution is well mixed. The basic substance will now be present as the free base:

$$\mathrm{\overset{+}{B}{-}H} + \mathrm{Cl^-} + \mathrm{Na^+} + \mathrm{HO^-} \longrightarrow \mathrm{B{:}} + \mathrm{Na^+} + \mathrm{Cl^-} + \mathrm{H_2O}$$

Add 25 mL of ether to the separatory funnel. Shake the mixture well so as to extract the free base into the ether layer. Draw off the lower, aqueous, layer (which may now be discarded) and transfer the ethereal solution of the basic substance to a small Erlenmeyer flask. Dry the solution over 1–2 grams of anhydrous potassium carbonate for a few minutes, remove the drying agent by gravity filtration, and remove the ether by evaporation or distillation on the steam bath. The infrared spectrum of the residue may be determined; or, if the residue is a solid, it may be recrystallized and its melting point determined.

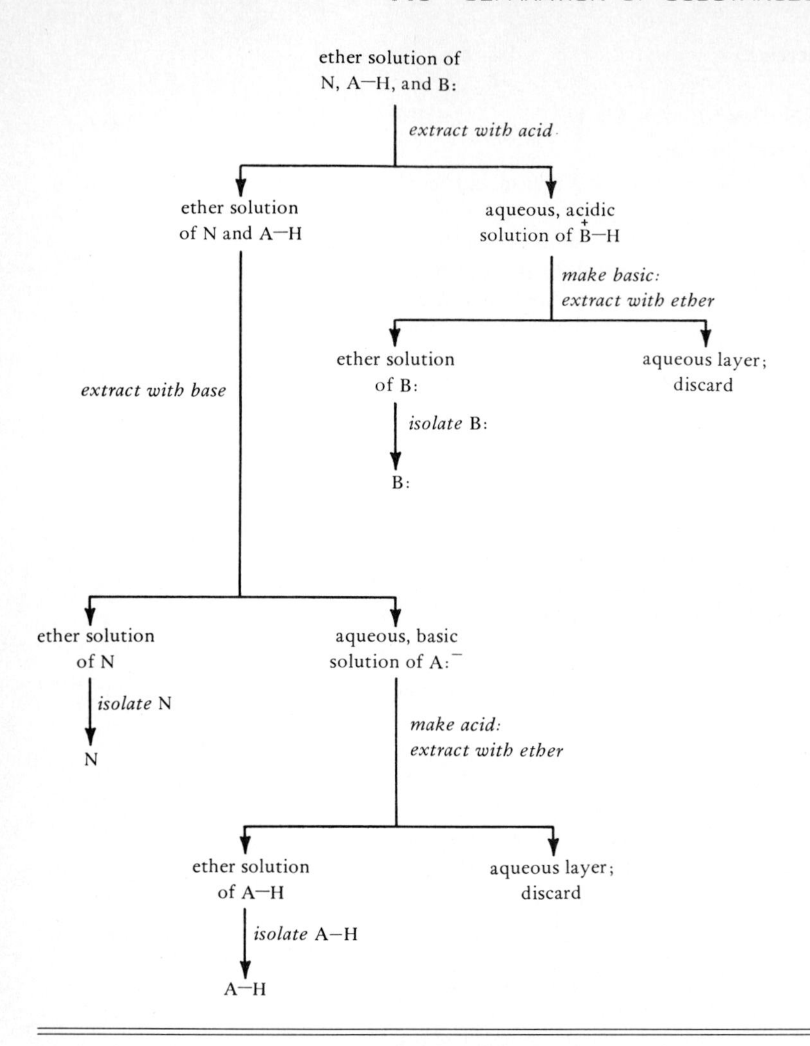

Figure 13.2. Flow chart for the separation of a mixture of an acid A—H, a base B:, and a neutral substance N.

Isolation of the acidic substance, A—H. Transfer the aqueous solution of the sodium salt of the acidic substance to a clean 125-mL separatory funnel. Make the solution strongly acidic by adding about 3 mL of conc. HCl. Make sure the solution is well mixed. The acidic substance will now be present as the free acid:

$$A{:}^- + Na^+ + H_3O^+ + Cl^- \longrightarrow A{-}H + Na^+ + Cl^- + H_2O$$

Add 25 mL of ether to the separatory funnel. Shake the mixture well so as to extract the free acid into the ether layer. Draw off the lower, aqueous, layer (which may now be discarded) and transfer the ethereal solution of the acidic substance to a small Erlenmeyer flask. Dry the solution over 1–2 grams of anhydrous sodium sulfate for a few minutes, remove the drying agent by gravity filtration, and remove the ether by evaporation or distillation on the steam bath. The infrared spectrum of the residue may be determined; or, if the residue is a solid, it may be recrystallized and its melting point determined.

The accompanying flow chart (Figure 13.2) summarizes these procedures.

4. Separation of the components of a commercial mixture of aspirin, phenacetin, and caffeine.

Some brands of headache or cold tablets, the so-called *APC tablets,* contain a mixture of acetylsalicylic acid (aspirin), phenacetin, and caffeine. It is possible to take advantage of the acid-base properties of these compounds so as to separate them by an extraction procedure. Caffeine, whose conjugate acid has a pK_a of -0.16, can just be extracted as the conjugate acid from chloroform into 4 *M* HCl. After the acidic extract has been neutralized, caffeine can be reextracted from the water with additional chloroform and isolated by evaporation of the chloroform. Aspirin, having a pK_a of 3.49, can be extracted from chloroform by 0.5 *M* aqueous sodium bicarbonate solution. After the aqueous extract is neutralized by addition of HCl, the precipitated aspirin can be isolated by reextraction with more chloroform and recovered by evaporation of the chloroform. Phenacetin, a substance that is neither acidic nor basic, remains in the original chloroform solution and is recovered by evaporation of the chloroform after the other two substances have been removed.

Procedure.

Isolation of caffeine. Crush 3 APC tablets (Note 1) and add them to a separatory funnel that contains 25 mL of chloroform. Then add to the funnel 20 mL of 4 *M* HCl (80 milliequivalents of acid). Shake the funnel in order to thoroughly mix the contents (Note 2). Allow the funnel to stand so that the layers will separate. Draw off and save the lower, chloroform, layer that contains the unextracted aspirin and phenacetin (Note 3).

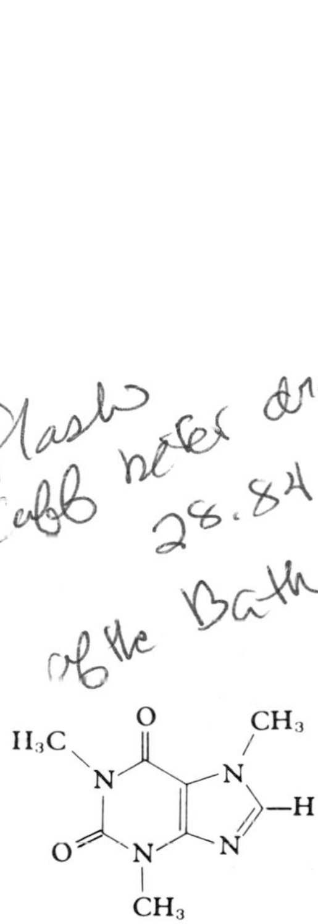

Recovery of caffeine. Neutralize the aqueous acidic extract of caffeine in the separatory funnel by adding 7.0 grams of solid sodium bicarbonate (83 meq. of base). Since a lot of carbon dioxide will be produced, you should add the sodium bicarbonate in portions. Swirl the contents of the funnel after each portion has appeared to react. After all the base has been added and the reaction appears to be complete, add 10 mL of chloroform, stopper the funnel, invert it, and release the pressure by opening the stopcock. Cautiously mix the contents of the funnel and frequently release the pressure by opening the stopcock of the inverted separatory funnel. Finally, when the pressure does not build up any more, shake the funnel to thoroughly mix the contents of the funnel (Note 2), allow the layers to separate (Note 4) and draw off the lower, chloroform, layer into a small Erlenmeyer flask labeled "caffeine." Reextract the neutralized acid solution with a second 5-mL portion of chloroform and add this further chloroform extract to the flask labeled "caffeine." Dry the combined extracts with a small portion of anhydrous magnesium sulfate (Section 15.2) and filter the mixture by gravity (Section 7.1) into a small, weighed Erlenmeyer flask containing a boiling stone (Note 5). Remove the chloroform by distillation on the steam bath (Note 6) and determine the weight of the residual solid by reweighing the flask. The recovered caffeine (Note 7) can be recrystallized from 10 mL of

caffeine

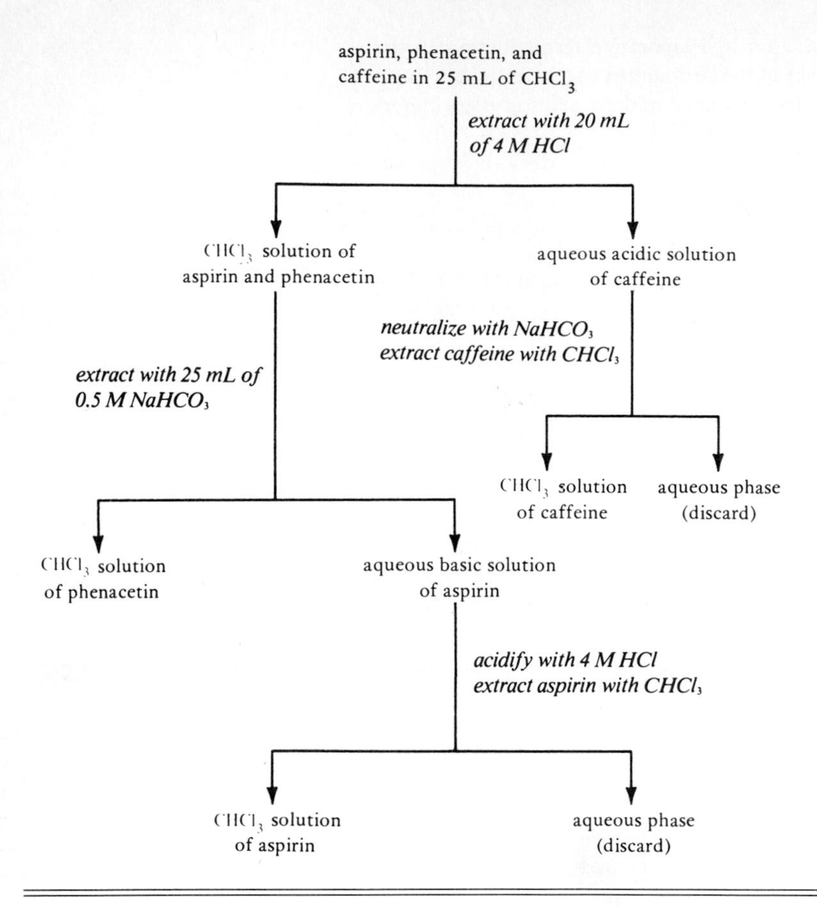

Figure 13.3. Flow chart for the separation of a mixture of aspirin, phenacetin, and caffeine.

acetylsalicylic acid

(aspirin)

carbon tetrachloride, if desired. The melting point of caffeine is reported to be 238°C, with sublimation starting at 170°C, and its infrared and NMR spectra are shown in Figures 42.1 and 42.2.

Isolation of aspirin. Place the original chloroform solution that was saved from the first extraction in a clean separatory funnel. Then add 25 mL of 0.5 *M* sodium bicarbonate solution (12.5 meq. of base) and thoroughly mix the contents of the funnel (Note 2) so as to extract the aspirin into the basic water layer. Since carbon dioxide will be formed, the funnel must be vented occasionally in order to prevent too great an increase in pressure. After mixing, allow the funnel to stand undisturbed so that the layers can separate. Draw off the lower, chloroform, layer into a small Erlenmeyer flask labeled "phenacetin" and add to this a small amount of anhydrous magnesium sulfate.

Recovery of aspirin. Add to the basic aqueous solution of aspirin in the separatory funnel 5 mL of 4 *M* HCl (20 meq. of acid). Mix the contents of the funnel by swirling it gently. Carbon dioxide will be evolved and the aspirin will separate as a solid. Recover the aspirin by extracting first with a

15-mL portion of chloroform and then with a 5-mL portion. Combine these two extracts in a small Erlenmeyer flask labeled "aspirin." Dry the aspirin extracts with a small portion of anhydrous magnesium sulfate and then filter the mixture by gravity into a small, weighed Erlenmeyer flask containing a boiling stone (Note 5). Remove the chloroform by distillation on the steam bath (Note 6) and determine the weight of the residue. The recovered aspirin (Note 8) can be recrystallized by adding 5 mL of water, heating the mixture on the steam bath, and adding slightly more than the minimum amount of 95% ethanol required to dissolve the solid; about 1.5 mL should be added. Aspirin is reported to melt at 135°C with rapid heating. Its IR and NMR spectra are shown in Figures 59.6 and 59.7.

Recovery of phenacetin. Remove the magnesium sulfate from the chloroform solution of phenacetin by gravity filtration, collecting the filtrate in a small, weighed Erlenmeyer flask containing a boiling stone (Note 5). Remove the chloroform by distillation on the steam bath (Note 6) and determine the weight of the residual solid. The recovered phenacetin (Note 9) can be recrystallized from a very small amount of 95% ethanol. The melting point of phenacetin is reported to be 134–135°C, and its IR and NMR spectra are shown in Figures 63.2 and 63.3.

The flow chart (Figure 13.3) summarizes these procedures.

O
H—N̈—C—CH_3
O—CH_2—CH_3

p-ethoxyacetanilide
(phenacetin)

Notes

1. Three APC tablets typically weigh 1.5 grams and contain 10.5 grains aspirin (680 mg), 7.5 grains phenacetin (486 mg), and 1.5 grains caffeine (97 mg).
2. One full minute of continuous shaking is sufficient.
3. If you have a second separatory funnel, this extract can be added directly to it and two parts of the experiment can be carried out simultaneously.
4. If an emulsion is formed, it can be broken by adding several drops of glacial acetic acid to the separatory funnel.
5. Weigh the flask with the boiling stone in it.
6. Put the recovered chloroform into the container reserved for it. The last of the chloroform may have to be removed by heating the flask without the adapters and condenser attached.
7. Only about half the caffeine is recovered, and NMR analysis shows that it is contaminated with phenacetin.
8. About three-fourths of the aspirin is recovered, and it appears to be pure by NMR analysis.
9. Almost all the phenacetin is recovered, but it contains a small amount of a less soluble material, perhaps part of the binder. Analysis by NMR shows it to contain between 5 and 10% caffeine, and something that contributes a single peak that overlaps the highest field component of the methyl triplet (2).

References

1. P. Haddad and M. Rasmussen, *J. Chem. Educ.* **53,** 731 (1976).
2. D. P. Hollis, *Analytical Chemistry* **35,** 1682 (1963).

14 Chromatography

Chromatographic separations, like extractions, take advantage of the fact that different substances are partitioned differently between two phases. In technique, however, the two methods differ greatly, and therefore chromatographic methods are discussed separately in this section.

14.1 THEORY OF COLUMN CHROMATOGRAPHY

In *column chromatography*, a solid phase (adsorbent) is held in a vertical tube, the mixture to be separated (A plus B) is placed on top of the column of adsorbent, and a solvent (eluant) is allowed to flow down through the column. At all times, a certain fraction of each component of the mixture will be adsorbed by the solid and the remainder will be in solution. Any one molecule will spend part of the time sitting still on the adsorbent with the remainder flowing down the column with the solvent. A substance that is strongly adsorbed (say, A) will have a greater fraction of its molecules adsorbed at any one time, and thus any one molecule of A will spend more time sitting still and less time moving. In contrast, a weakly adsorbed substance (B) will have a smaller fraction of its molecules adsorbed at any one time, and hence any one molecule of B will spend less time sitting and more time moving. Thus, the more weakly a substance is adsorbed, the faster it will get to the bottom of the column and flow out with the eluant. Since the eluant is collected in small portions (*fractions*), the early fractions will contain B and the later ones will contain A. The name *chromatography* was given to this process because the substances to which this method of separation was first applied were colored plant pigments.

Selectivity of adsorbent

Several factors determine the efficiency of a chromatographic separation. The adsorbent should show a maximum of selectivity toward the substances being separated so that the differences in rate of elution will be large. For the separation of any given mixture, some adsorbents may be too strongly adsorbing (holding all components near the top of the adsorbent) or too weakly adsorbing (allowing all components to move through the adsorbent almost as fast as the eluting solvent). Table 14.1 lists a number of adsorbents in order of adsorptive power.

Selectivity of eluting solvent

The eluting solvent should also show a maximum of selectivity in its ability to dissolve or desorb the substances being separated. The fact that one substance is relatively soluble in a solvent can result in its being eluted faster than another substance. However, a more important property of the solvent is its ability to be itself adsorbed on the adsorbent. If the solvent is more strongly adsorbed than the substances being separated, it can take their place on the adsorbent and all the substances will flow along rapidly together. If the solvent is less strongly adsorbed than

Table 14.1. Chromatographic adsorbents

	Most Strongly Adsorbent[a]
Alumina	Al_2O_3
Charcoal	C
Florisil	MgO/SiO_2 (anhydrous)
Silica gel	SiO_2
Lime	CaO
Magnesia	MgO
Magnesium carbonate	$MgCO_3$
Calcium phosphate	$Ca_3(PO_4)_2$
Calcium carbonate	$CaCO_3$
Potassium carbonate	K_2CO_3
Sodium carbonate	Na_2CO_3
Talc	MgO/SiO_2 (hydrous)
Sucrose	Carbohydrate (polyhydroxylic)
Starch	Carbohydrate (polyhydroxylic)
	Least Strongly Adsorbent

[a] The order in the table is approximate since it depends upon the substance being adsorbed and the solvent used for elution.

any of the components of the mixture, its contribution to different rates of elution will be only through its difference in solvent power toward them. If, however, it is more strongly adsorbed than some components of the mixture and less strongly than others, it will greatly speed the elution of those substances that it can replace on the column, without speeding the elution of the others.

Table 14.2 lists a number of common solvents in approximate order of increasing adsorbability, and hence in order of increasing eluting power. The order is only approximate since it is not independent of the nature of the adsorbent. Mixtures of solvents can be used, and, since increasing eluting power results mostly from preferential adsorbtion of the solvent, addition of only a little (0.5–2%, by volume) of a more strongly adsorbed solvent will result in a large increase in the eluting power. Because water is among the most strongly adsorbed solvents, the presence of a little water in a solvent can greatly increase its eluting power. For this reason, solvents to be used in column chromatography should be quite dry. (See Section 15.3 for methods of drying solvents.)

Solvents should be dry

The particular combination of adsorbent and eluting solvent that will result in the acceptable separation of a particular mixture can be determined only by trial. Alumina is the most commonly used adsorbent because it is readily available at relatively low cost and has a wide range of adsorptive power, depending upon how much water it has adsorbed. The adsorptive power of alumina can be decreased by adding up to 15% by weight of water. If alumina has become too heavily hydrated in storage, it can be "activated" by heating it at 200°C for

Table 14.2. Eluting solvents for chromatography

Least Eluting Power (alumina as adsorbent)[a]
Petroleum ether (hexane; pentane)
Cyclohexane
Carbon tetrachloride
Benzene
Dichloromethane
Chloroform
Ether (anhydrous)
Ethyl acetate (anhydrous)
Acetone (anhydrous)
Ethanol
Methanol
Water
Pyridine
Organic acids
Greatest Eluting Power (alumina as adsorbent)

[a] The order of eluting power of the solvents listed in the table will generally be observed with other highly polar adsorbents.

Table 14.3. Adsorbability of organic substances by functional group

Comparison of this table with Table 14.2 indicates that there is a relationship between the eluting power of a solvent and the tendency of the solvent to be adsorbed.

Least Strongly Adsorbed[a]
Saturated hydrocarbons; alkyl halides
Unsaturated hydrocarbons; alkenyl halides
Aromatic hydrocarbons; aryl halides
Polyhalogenated hydrocarbons
Ethers
Esters
Aldehydes and ketones
Alcohols
Acids and bases (amines)
Most Strongly Adsorbed

[a] The order depends upon the adsorbent.

about three hours. A trial to determine the conditions for a chromatographic separation might be made by preparing a small column using partially hydrated alumina, adding a sample of the mixture, and starting elution with a solvent of weak eluting power (hexane, for example). Solvents of successively greater eluting power (dichloromethane, dry ether, dry acetone, methanol; see Table 14.2) can then be tried until one is found that will move the material down the column. Further, more sensitive, trials can be made using different solvents, or mixtures of solvents, of similar eluting power. If the trial adsorbent is too strongly adsorbing, a more completely hydrated alumina can be tested or a weaker adsorbent can be tried; if it is too weakly adsorbing, a less hydrated grade of alumina can be tested.

If the substances in the mixture differ greatly in adsorbability, it will be much easier to separate them. Often, when this is so, a succession of solvents of increasing eluting power is used. One substance may be eluted easily while the other stays at the top of the column, and then the other can be eluted with a solvent of greater eluting power. Table 14.3 indicates an approximate order of adsorbability by functional group.

14.2 TECHNIQUE OF COLUMN CHROMATOGRAPHY

There are several types of tubes that can be used to support the adsorbent. One of the most satisfactory and least expensive is illustrated in

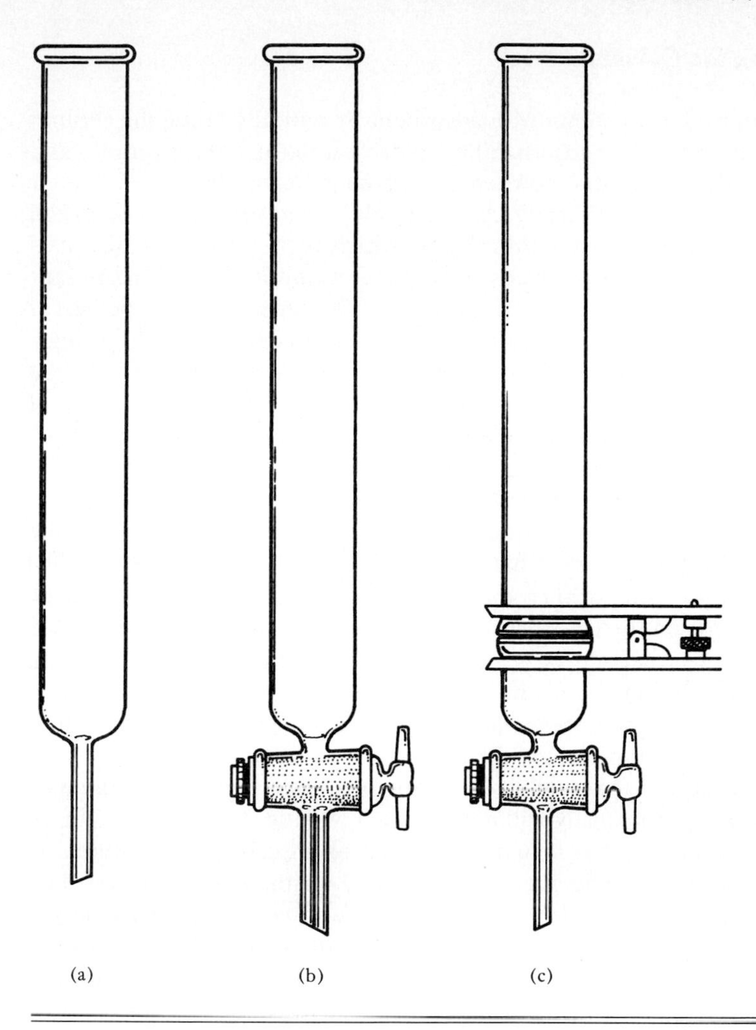

Figure 14.1. Chromatographic columns.

The column

Figure 14.1a. A chromatographic column like this, essentially a giant medicine dropper, can be made by heating, drawing down, cutting off, and fire-polishing a long piece of glass tubing 10–25 mm in diameter. The column should be at least 50 cm long so that the column of eluting solvent above the adsorbent can be high enough to contribute a sufficient hydrostatic head to achieve an acceptably large flow rate. If the tip is not too narrow, the flow of solvent through the column can be stopped by slipping a rubber policeman over the tip. Otherwise, a piece of rubber tubing and a clamp can be used. Burets and other columns with stopcocks (Figure 14.1b and c) are sometimes used, but they have the disadvantage that the stopcock grease will be eluted along with the other substances in the system, and the grease will contaminate the material being purified. Small trial chromatograms can be run in medicine droppers.

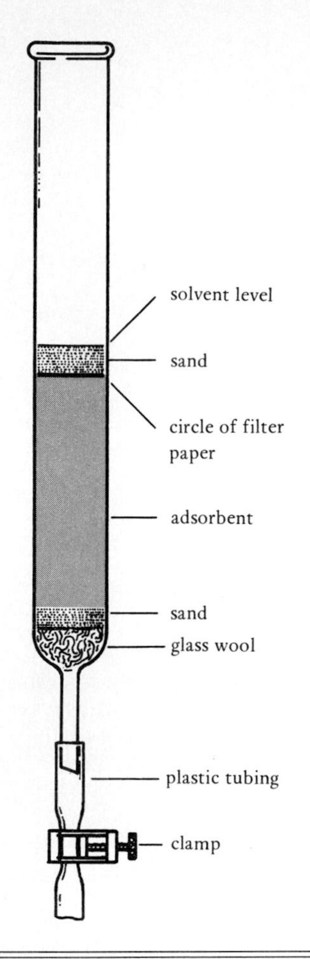

Figure 14.2. A simple chromatographic column ready for use.

Packing the Column

You can pack the column with adsorbent by partially filling the column with a liquid of low adsorbability (petroleum ether, for example) and slowly adding the dry powdered adsorbent down the top so that it settles evenly and uniformly. Alternatively, the adsorbent can be mixed with the liquid to form a thin slurry, which is then poured and rinsed into the column. The objective is to form a uniform bed of adsorbent without holes, channels, or air bubbles. The separation will be best if the bottom of the bed of adsorbent is flat. Therefore, if you are using a column that is tapered or has some other odd shape at the bottom, you should first partially fill the column with solvent, then push a plug of glass wool down to fill up the odd volume, and finally add a little sand, rinsing it down to form a flat base on which the adsorbent can come to rest. After all traces of adsorbent have been rinsed down (solvent will usually have to be drained off during this process) and the top of the bed of adsorbent has been flattened by jiggling the tube, a disk of filter paper just a little smaller than the diameter of the tube should be dropped flat on the adsorbent, and then enough sand to form a layer 2 or 3 mm deep should be sprinkled on the paper. The purpose is to protect the top of the column of adsorbent so that it will not be disturbed when the sample or the eluting solvent is poured onto it. Figure 14.2 shows a column ready for use.

The ratio of the weight of adsorbent to the weight of sample may vary greatly, but usually will not be much outside the range of 25 or 50 to 1. The ratio of the height of the bed of adsorbent to its diameter should normally be between 3 : 1 and 10 : 1. If the ratio is greater than this, the rate of flow of eluant may be too low. Thus, when larger samples are to be chromatographed, large-diameter columns must be used and not just deeper beds of adsorbent. The finer the adsorbent, the greater its surface area and hence its adsorptive capacity, but the lower the flow rate. Sometimes it is necessary to use a mixture of adsorbent plus diatomaceous earth (Celite) as a nonadsorbing diluent in order to attain a minimum acceptable flow rate.

Adding the Sample

When the column has been prepared, the solvent should be allowed to drain until it is just level with the top of the sand that covers the adsorbent. If the column is allowed to drain dry, the bed of adsorbent usually will crack, and its ability to separate will be greatly diminished because the solution will be able to flow through the cracks. It is a waste of time to try to use a column with cracked or channeled adsorbent.

The sample is then added. A liquid can be added directly; a solid should be dissolved in as little as possible of a solvent of low eluting power. After the addition, the column is allowed to drain again until the level of liquid has fallen just to the top of the adsorbent. If necessary, rinse the sample down from the walls of the column with additional *small* portions of solvent, draining the column each time. You will find that a medicine dropper is useful for these additions. The goal is to get the sample adsorbed in a minimum layer of adsorbent. The narrower the sample band, the better the separation, since narrow bands will overlap less.

Elution

After the sample has been added, the column is eluted using a solvent or series of solvents as recommended or as determined by trial. In a new situation, you should start with a solvent of low eluting power (petroleum ether, hexane) and work up through solvents of increasing eluting power (dichloromethane, ether, etc.) either as mixtures or as pure solvents. When the level of liquid in the column is low, the eluting solvent should be added very cautiously to avoid disturbing the top of the bed of adsorbent. During elution, the solvent should not be allowed to flow through any sort of rubber or plastic tubing, since material can be eluted from the tubing as well.

Collecting the Fractions

If the substances to be separated are colored, their progress down the column can be followed visually and the eluant that contains each component can be collected separately. If the substances are colorless, the eluant must be collected in successive fractions and the presence of components of the original mixture must be determined by some analytical procedure. The most common method is to collect fractions of approximately equal volume in tared (previously weighed) flasks, evaporate or distill the solvent (Section 36), and reweigh to determine the weight of the residue. If the residue is a solid, it will usually crystallize and can then be easily seen. The residue can be identified by determining one or more physical properties such as melting point or the infrared absorption spectrum. With substances that fluoresce (many aromatic compounds do), their progress down the column can be followed by illuminating the column with ultraviolet light ("black light"). Often, the presence of colorless substances in the eluant can be established by determining its ultraviolet absorption; negative fractions need not be evaporated.

Sometimes elution is carried out only until the components of the sample appear to be separated on the column. The adsorbent is then

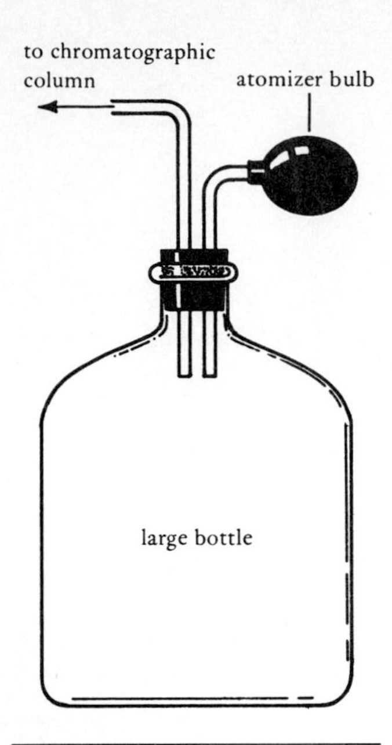

Figure 14.3. A way to apply pressure to the top of a chromatographic column.

pushed out of the tube (you must use a column such as (c) in Figure 14.1), and the parts of the adsorbent that contain the various components are separated and stirred with a solvent of high eluting power. Removal of the adsorbent by filtration and evaporation of the solvent yields the purified sample of the material.

The rate of flow through the column can be increased by filling the column with eluting solvent to a higher level. If the flow is still too slow, it can be hastened by applying a little air pressure to the top of the column with an atomizer bulb connected through a large ballast volume (Figure 14.3). Connecting the column to a water aspirator by means of a suction flask is not recommended because it is very easy to evaporate the solvent from the bottom of the column and thus form cracks and spoil the column.

Advantages and Disadvantages of Column Chromatography

Column chromatography is especially suited for purifying small amounts of material. Normally, it effects separations far more completely than distillation or recrystallization can. Its disadvantages are that it is a highly empirical method, often requiring several trials to establish acceptable conditions, and that it is not suited to the purification of large quantities since most columns of reasonable size have only a small capacity.

Exercises

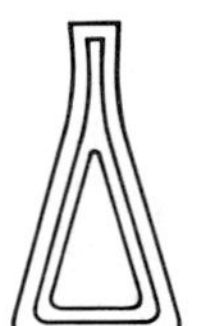

1. Purification of technical-grade anthracene.

   Procedure.

   Column. 1.5 to 2 cm diameter.

   Adsorbent. Alumina; use enough to make a bed of adsorbent 8–10 cm deep. Pack the column using hexane or petroleum ether.

   Sample solution. 0.3 g of technical anthracene in 150–200 mL of hexane.

   Eluting solvent. Hexane. Examination of the column in ultraviolet light will show a narrow, deep-blue fluorescent zone near the top, due to carbazole. Immediately below this should appear a nonfluorescent yellow band due to naphthacene. Anthracene forms a broad blue-violet fluorescent zone in the lower part of the column. Elution with hexane should be continued until anthracene begins to come off the column; the eluant collected before the anthracene begins to come off should be discarded. At this point, the elution should be continued with hexane: dichloromethane 1 : 1 (by volume), until the yellow band reaches the bottom of the column (1).

2. Isolation of (R)-(−)- or (S)-(+)-carvone from oil of spearmint or oil of caraway (Section 47).

14.3 THEORY OF THIN-LAYER CHROMATOGRAPHY

The theoretical basis for the separation of substances by *thin-layer chromatography* is exactly the same as that for column chromatography. The difference between the methods is that in thin-layer chromatography the adsorbent is used in the form of a thin layer (about 0.25 mm thick) on a supporting material, usually a sheet of glass or plastic. The sample is applied to the layer of adsorbent, near one edge, as a small spot of a solution. After the solvent has evaporated, the adsorbent-coated sheet is propped more or less vertically in a closed container, with the edge to which the spot was applied down. The solvent, which is in the bottom of the container, creeps up the layer of adsorbent, passes over the spot, and, as it continues up, effects a separation of the materials in the spot ("develops" the chromatogram) in the same way as the eluting solvent does in column chromatography. When the solvent front has nearly reached the top of the adsorbent, the sheet is removed from the container.

Since the amount of adsorbent involved is relatively small, and the ratio of adsorbent to sample must be high, the amount of sample must be very small, usually much less than a milligram. For this reason, thin-layer chromatography (TLC) is usually used as an analytical technique rather than a preparative method, although with thicker layers (about 2 mm) and large plates with a number of spots or a stripe of sample, it can be used as a preparative method. The separated substances are recovered by scraping the adsorbent off the plate (or cutting out the spots if the supporting material can be cut) and extracting the substance from the adsorbent.

Because the distance traveled by a substance relative to the distance traveled by the solvent front depends upon the molecular structure of the substance, TLC can be used to identify substances as well as to separate them. The relationship between the distance traveled by the solvent front and the substance is usually expressed by the so-called R_f value:

Identification through R_f value

$$R_f \text{ value} = \frac{\text{distance traveled by substance}}{\text{distance traveled by solvent front}}$$

The R_f values are strongly dependent upon the nature of the adsorbent and solvent. Therefore, experimental R_f values and literature values do not often agree very well. In order to determine whether an unknown substance is the same as a substance of known structure, it is necessary to run the two substances side by side in the same chromatogram, preferably at the same concentration.

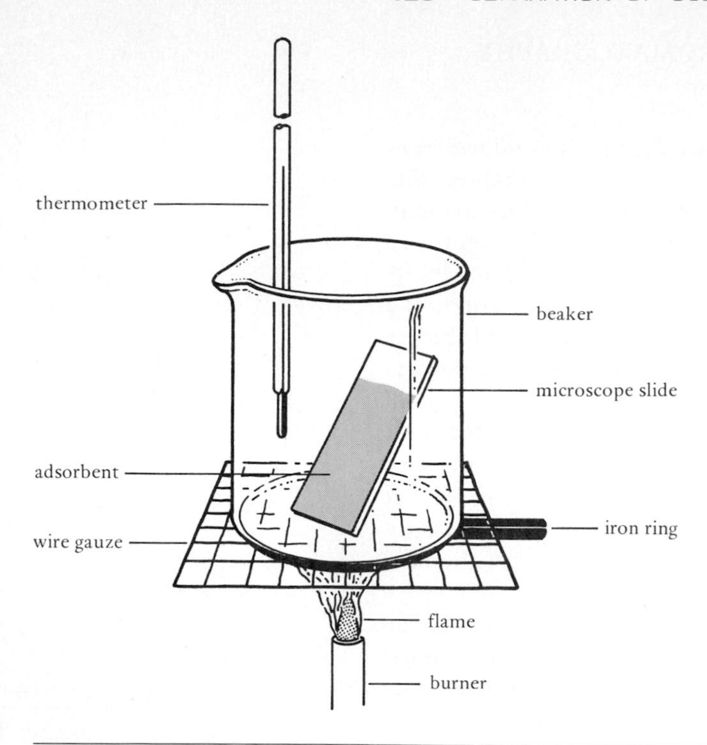

Figure 14.4. Activation of a plate for TLC.

14.4 TECHNIQUE OF THIN-LAYER CHROMATOGRAPHY

Preparation of Microscope Slide Plate

The two most widely used adsorbents in TLC are silica gel and alumina. Adsorbent-coated TLC plates may be either obtained commercially or prepared in the laboratory. Ready-made plates are more convenient but are relatively expensive. The preparation of large plates requires the use of fairly expensive apparatus, but small plates that use microscope slides as the support are easily prepared in the laboratory and are often used for small-scale work. The procedure consists of preparing a suspension of the adsorbent in a volatile solvent, dipping in a pair of clean microscope slides held tightly together until all but a handle of about 1 cm of their length is immersed in the slurry, and then drawing the slides out in one smooth, continuous motion. When the solvent has evaporated (in about a minute), the two slides may be separated. After the adsorbent has been activated by heating (Figure 14.4), the plates are ready for use.

Activation

Application of Sample

The sample to be separated is generally applied as a small spot (1 to 2 mm diameter) of solution about 1 cm from the end of the plate opposite the handle. The addition may be made with a microsyringe or with a micropipet prepared by heating and drawing out a melting point capillary. As small a sample as possible should be used, since this will minimize tailing and overlap of spots; the lower limit is the ability to visualize the spots in the developed chromatogram. If the sample solution is very dilute, make several small applications in the same place, allowing the solvent to evaporate between additions. Do not disturb the adsorbent when you make the spots, since this will result in an uneven flow of the solvent. The starting position can be indicated by making a small mark near the edge of the plate.

Development

The chamber used for development of the chromatogram can be as simple as a beaker covered with a watch glass, or a cork-stoppered bottle. The developing solvent (an acceptable solvent or mixture of solvents must be determined by trial) is poured into the container to a depth of a few millimeters. The spotted plate is then placed in the container, spotted end down; the solvent level must be below the spots (see Figure 14.5). The solvent will then slowly rise in the adsorbent by capillary action.

In order to get reproducible results, the atmosphere in the development chamber must be saturated with the solvent. This can be accomplished by sloshing the solvent around in the container before any plates have been added. The atmosphere in the chamber is then kept saturated by keeping the container closed all the time except for the brief moment during which a plate is added or removed.

Visualization

When the solvent front has moved to within about 1 cm of the top end of the adsorbent (after 15 to 45 minutes), the plate should be removed from the developing chamber, the position of the solvent front marked, and the solvent allowed to evaporate. If the components of the sample are colored, they can be observed directly. If not, they can sometimes be visualized by shining ultraviolet light on the plate or by allowing the plate to stand for a few minutes in a closed container in which the atmosphere is saturated with iodine vapor. Sometimes the spots can be visualized by spraying the plate with a reagent that will react with one or more of the components of the sample.

Figure 14.5. Development of a plate in TLC.

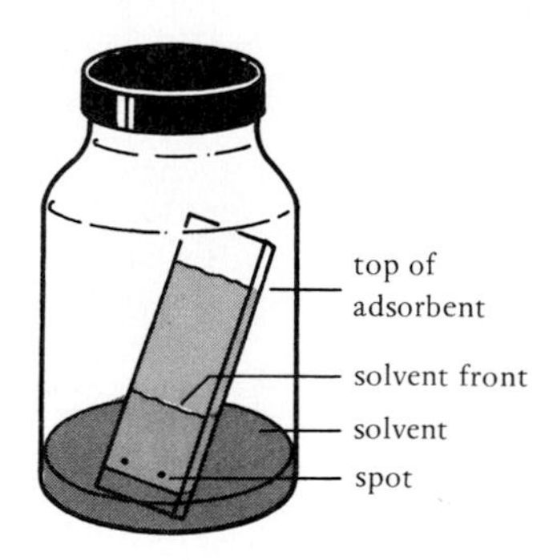

Exercises

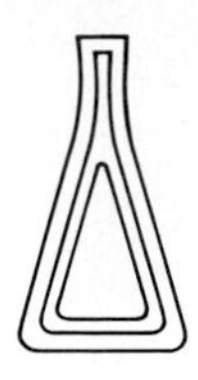

Silica gel–coated TLC plates can be obtained commercially from a number of sources, or can be prepared in the laboratory. Two procedures for coating microscope slides are described here.

a. Prepare a suspension of 35 g of silica gel G (Merck) in 100 mL of a 2 : 1 (by volume) mixture of chloroform : methanol. Coat the plates as described in Section 14.4.

b. Prepare a suspension of 5 g of silica gel (Biosil A-30B) in 12 mL of water. Pour 1 mL of the slurry on a clean microscope slide and spread it out with a stirring rod. Tap the slide gently to settle the slurry into an even film. After allowing it to set for a couple of minutes, dry it in an oven at 110°C for 15 minutes or longer.

The following separations have been described using silica gel TLC plates.

1. Separation of leaf pigments.

Procedure.

Sample solution. Crush green leaves in a mortar with a few mL of ethanol or acetone and twice this volume of petroleum ether or hexane. Filter the mixture, saving the filtrate. Extract the leaves again, filter, combine the two filtrates and transfer them to a separatory funnel. Wash the extract with three small portions of water and then dry it (Section 15.2) over anhydrous sodium sulfate.

Developing solvent. Hexane : acetone, 7 : 3 (by volume).

Visualization. The carotenes move most rapidly, then chlorophyll a, chlorophyll b, and the xanthophylls (2).

2. Analysis of mixtures of aspirin, phenacetin, and caffeine.

Procedure.

Sample solution. Use methanol; known solutions should be prepared to contain 5–10 mg per mL. Unknown mixtures can be prepared from materials available at drugstores.

Developing solvent. Ethyl acetate or acetone.

Visualization. Ultraviolet light or iodine vapors.

14.5 THEORY OF PAPER CHROMATOGRAPHY

The process of *paper chromatography* is very similar to that of thin-layer chromatography (TLC) except that a strip of paper replaces adsorbent and support. As with TLC, a very small amount of a dilute solution of the substance is applied as a spot near one end of a strip of filter paper, and the strip is supported in a closed container with the end containing the spot hanging down into a solvent. As the solvent rises up the paper by capillary action, it effects the separation ("develops" the chromatogram) as does the eluant in column chromatography.

Under ordinary conditions, about 18% by weight of filter paper consists of adsorbed water. This means that when the molecules of a substance are sitting still on the paper, it is possible that they should be considered to be dissolved in the adsorbed water rather than adsorbed by the paper.

As in the case of TLC, the amounts of sample must be so small that paper chromatography is used almost entirely as an analytical method by which you can demonstrate the homogeneity of a sample or qualitatively determine the composition of a mixture through determination of R_f values.

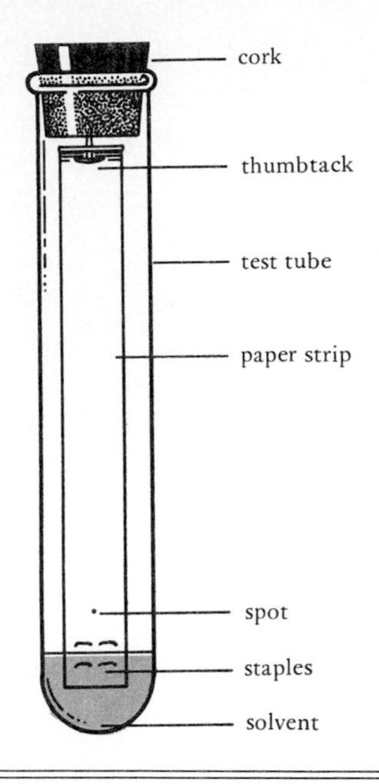

Figure 14.6. An arrangement for paper chromatography.

14.6 TECHNIQUE OF PAPER CHROMATOGRAPHY

Arrangement of Paper Strip

The following simple procedure can be used for the paper chromatographic analysis of a single sample spot. As a container for the development of the paper chromatogram, use a 25 mm × 125 mm or a 25 mm × 150 mm test tube. Cut a half-inch-wide strip of filter paper, and thumbtack one end of it to the bottom of the cork used to stopper the test tube. The strip should be just long enough that the lower end will dip into the half-inch layer of solvent in the bottom of the test tube. During the development of the chromatogram, the paper must not touch the walls of the test tube. A couple of staples fastened into the lower end of the strip will help it to hang straight. Figure 14.6 illustrates the suggested arrangement.

Several samples can be run simultaneously by using a sheet of filter paper bent around to form a short tube and held that way (without the edges touching) by two staples. Development is carried out with the tube standing on end in a beaker covered with a watch glass. The grain of the paper—that is, the direction of the longer axis of the ellipse formed when a drop of water is allowed to spread on the paper—must be vertical. This arrangement is illustrated in Figure 14.7.

Application of Sample

Figure 14.7. An arrangement for paper chromatography allowing several samples to be run at once.

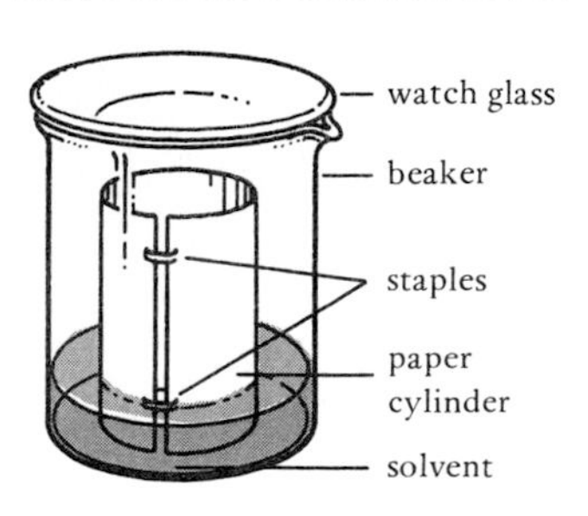

The solution of the sample of the material to be separated should be applied in as small a spot as possible. Use a micropipet, which can be made by heating and drawing out a melting point capillary or can be obtained commercially. If the solution is so dilute that a relatively large amount must be used, several small applications should be made, allowing time for the previous one to dry before the next one is added. As with TLC, the less sample used, the smaller the spots and the better the resolution. The only limitation is the ability to see the spots of separated material. The sample spot should be made far enough from the end of the paper that it will not dip into the solvent. Mark the starting line lightly with a pencil so that later on you will be able to more easily determine the R_f values.

Development

The chromatogram should be developed until the solvent front has risen about 10 cm. The paper should then be removed from the developing chamber, the line of the solvent front marked with a pencil, and the solvent rinsed off or allowed to evaporate.

Visualization

If the substances are colored, the spots can be observed directly. Ultraviolet light can sometimes be used to visualize the spots, or the paper can be sprayed with a reagent that will react with some or all of the substances of the sample in such a way as to transform them to a colored derivative.

R_f values

As in TLC, R_f values are fairly sensitive to the exact experimental conditions. Comparison of experimental R_f values of unknown substances with those of substances whose structures are known are best made by running a sample of the known substance alongside the unknown sample on the same strip of paper.

Exercises

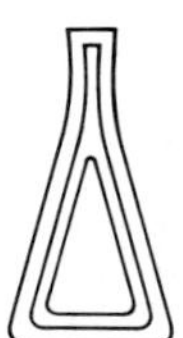

1. Analysis of spinach leaves.
   Procedure.
   Sample solution. Grind a few grams of fresh or frozen spinach leaves with about three volumes of acetone. Filter the mixture with suction and discard the filtrate. Grind the residue with a minimum volume of acetone and use the resulting solution.
   Developing solvent. Petroleum ether.
   Visualization. Carotenes move most rapidly, followed by xanthophylls and chlorophylls; a gray area due to decomposed chlorophylls may appear after the carotenes (3).
2. Analysis of carrots. Analysis of carrots as in Exercise 1 shows carotene to be predominant (3).
3. Analysis of food coloring.
   Procedure.
   Sample solution. Dilute the food coloring by a factor of five with isopropyl alcohol : water, 1 : 2 (by volume).
   Developing solvent. Isopropyl alcohol : water, 1 : 2 (by volume) (4).
4. Separation of ink pigments.
   Procedure.
   Sample solution. Dilute 1 mL of Script "washable" writing fluid in 5 mL of 95% ethanol. Use 0.005 mL of solution.
   Developing solvent. 95% ethanol (5).

14.7 THEORY OF VAPOR-PHASE CHROMATOGRAPHY

In *vapor-phase chromatography* (VPC), the stationary phase is a high-boiling liquid that is present as a coating upon an inert granulated support, and the mobile phase is a gas, usually helium. As in the other chromatographic methods, separations are possible when there are differences in the way in which different substances are partitioned between the stationary and mobile phases.

If the liquid phase does not preferentially dissolve molecules with certain functional groups, the order of elution is most volatile to least volatile (order of increasing boiling point). One would expect this order always to be observed with molecules of a homologous series and with structural isomers with the same functional group. Some liquids do appear to preferentially dissolve molecules with certain functional groups. They therefore display a relatively high selectivity toward these molecules and are especially well suited for the analysis of mixtures containing them. Table 14.4 lists some liquids that are commonly used as the stationary phase in VPC.

Table 14.4. Stationary phases most commonly used in vapor-phase chromatography

	Stationary Phase	*Maximum Useful Operating Temperature*	*Selectivity; Suitability*
Silicone oils	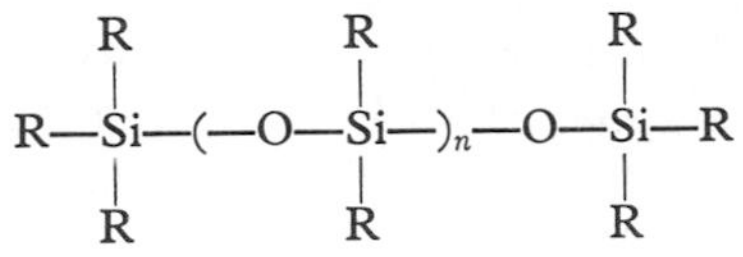 $R-Si(R)_2-(-O-Si(R)_2-)_n-O-Si(R)_2-R$	250°C	According to volatility; Generally useful
Apiezon grease	$CH_3-(-CH_2-)_n-CH_3$	250°C	According to volatility; Not suitable for hydroxylic compounds (tailing)
Polyethylene glycol	$HO-(-CH_2-CH_2-O-)_n-CH_2-CH_2-OH$	150°C	Relatively selective toward polar compounds (alcohols, amines, aldehydes, ketones); Best suited for polar compounds
Diisodecyl phthalate	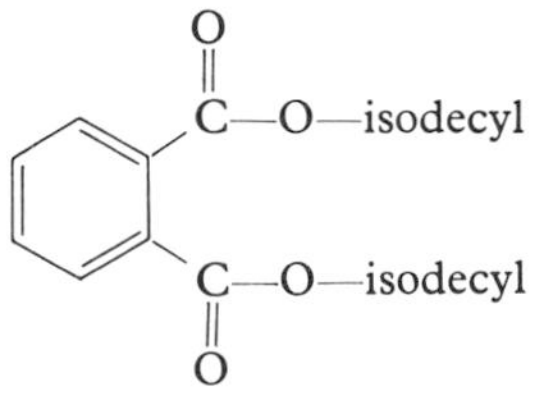 $C_6H_4(C(=O)-O-isodecyl)_2$ (benzene ring with two ortho C(=O)—O—isodecyl groups)	175°C	According to volatility; Generally useful

The apparatus required for VPC is considerably more complex and expensive than that needed for the other chromatographic methods. The essential parts, shown schematically in Figure 14.8, consist of a source of the carrier gas under pressure, a port through which the sample may be injected, the column (usually a 5- to 10-foot length of $\frac{1}{4}$-inch-diameter metal tubing packed with the liquid-coated support), the detector, and its recorder. The detector senses the presence of material in the carrier gas, and the recorder in effect records the output of the detector as a function of time.

Hot wire detector

A commonly used detector is an electrically heated wire (filament) positioned in the gas stream at the exit to the column. The wire is cooled to an equilibrium temperature by the flowing carrier gas, and at that temperature the wire has a certain resistance. When some substance other than the gas itself is present in the steam, the wire is not cooled as efficiently (helium has a very much larger thermal conductivity than any other gas but hydrogen); its temperature rises, and so does its resistance. The wire is part of a Wheatstone bridge circuit and, as the resistance of the wire varies, the resulting imbalance voltage developed in the bridge is compensated for by the recorder, which is an automatic recording null potentiometer. The position of the pen of the recorder corresponds to the position of the slide on the slide-wire of the potentiometer. The recorder output, which is a record of the pen position on the slide-wire as a function of time, is thus related to the variation of the thermal conductivity of the gas at the exit to the column. When nothing but helium is coming off the column, the pen of the recorder will draw a straight line (the base line). As a substance comes off the column, the pen will deflect and then return to the base line, thus drawing a "peak" (see Figure 14.9).

Flame ionization detector

A more sensitive detector is the flame ionization detector. In a system with a flame ionization detector, a portion of the eluant from the column is fed to a hydrogen/air flame. When a substance other than the

Figure 14.8. Schematic diagram of a vapor-phase chromatograph.

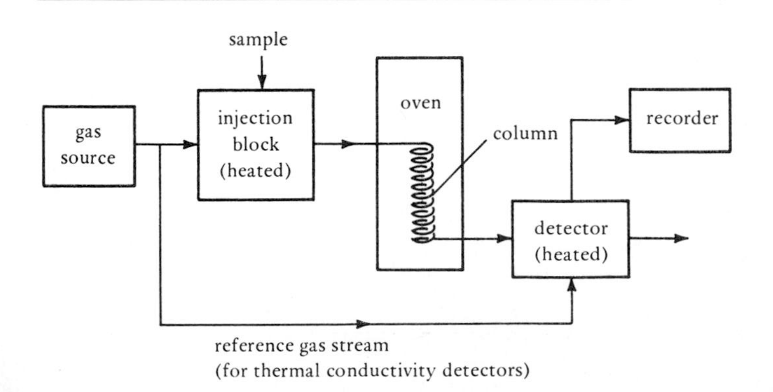

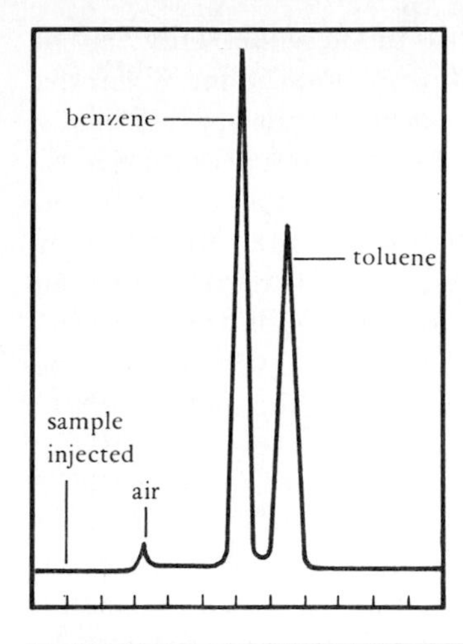

Figure 14.9. Vapor-phase chromatogram of a mixture of benzene and toluene.

carrier gas is present in the eluant, the electrical conductivity of the hot gas of the flame will increase because of the formation of ions as the substance in the carrier gas burns. This change in conductivity is sensed as a change in voltage between a wire in the flame and the burner, and this change in voltage is followed by the recorder operating as a potentiometer, as with the hot wire detector. Nitrogen gas can be used as the carrier gas when a flame ionization detector is used, rather than the much more expensive helium gas required when using a thermal conductivity detector.

Quantitative Analysis by VPC

It seems reasonable that the larger the amount of a given substance that comes off the column, the higher its concentration in the carrier gas. Thus, the area under the peak should be a measure of the amount of that substance present in the sample. If the sample is a mixture, the ratio of the areas of any two peaks should be a measure of the relative amounts of the two substances.

It is easy to determine which peak corresponds to which substance if authentic samples are available. A portion of the mixture being analyzed can be spiked with one of the substances known to be present in the mixture. Analysis of this sample will show that the area of one peak has increased relative to the others, thus indicating to which component of the mixture it corresponds. The more difficult problem is to determine exactly in what way the area ratio is a measure of the relative amounts of the two substances. That is, does the area ratio correspond to the weight ratio or to the mole ratio or to neither? The only way to know for sure is to analyze mixtures of known composition. When the area ratio has been determined, it generally corresponds approximately to the weight ratio, the agreement being better for similar compounds. If pure samples of the substances being analyzed are not available, you can only hope that the detector response is approximately the same for each. The more similar the substances are in molecular structure, the more likely it is that this will be true. If the compounds are structural isomers with the same functional groups, the area ratio will probably be very close to the weight ratio (which is the same as the mole ratio).

Qualitative Analysis by VPC

Vapor-phase chromatography is often used to roughly estimate the purity of a substance by comparing the area of the main peak, which corresponds to the compound of interest, to the area of the other peaks. VPC is often used to determine whether or not a certain substance is present in a mixture. This is done by running first a sample of the mixture and then a sample of the mixture plus a little of the compound of interest. If a peak present in the first chromatogram is larger in the second, the substance added probably has the same structure as a component of the original mixture. If the added substance results in a new peak in the second chromatogram, it must not be present in the original mixture. Similarly, as implied above, it is possible to identify the substance responsible for a given peak in the chromatogram of a mixture by spiking portions of the mixture with small amounts of various known substances and comparing the resulting chromatograms of the spiked mixtures with the chromatogram of the original mixture. Although this method of identification is fairly reliable, especially if the result is the same when several different columns are used, it requires that you be able to guess the identity of the substance responsible for the peak and that you have a pure sample of that material available. Sometimes it is necessary to collect the substance responsible for a certain peak as it leaves the gas chromatograph, determine some physical property such as its infrared or mass spectrum, and deduce from that the identity of the substance.

Preparative VPC

So far, the applications of VPC that have been described have been analytical. As in other chromatographic methods, it is possible to use VPC for preparative purposes by leading the carrier gas stream from the detector through a cooled tube in which the sample components of the gas stream can condense. If the tube is changed as each peak comes through, the components of the mixture can be collected separately. In order to purify larger amounts of material this way, it is not possible just to inject larger samples, since peak width increases with increasing sample size, and soon the point is reached where the impurity peaks overlap the peak of the desired substance. It is also not possible just to use a larger-diameter column, because the separating power of the column falls off very rapidly with increasing column diameter. It is therefore usually necessary to process successive samples of 0.01 to 0.25-mL volume, depending upon the difficulty of the separation, using quarter-inch or sometimes half-inch columns. Instruments are available commercially in which the repetitive processes of sample injection and fraction collection are carried out automatically.

Vapor-phase chromatography is one of the most widely used methods for separation or analysis of small amounts of reasonably volatile substances. While a very good fractionating column may have an efficiency corresponding to about 100 theoretical plates, an ordinary gas chromatographic column may have an efficiency corresponding to more than 1000 theoretical plates, and columns are available with efficiencies of greater than 10,000 theoretical plates. Holdup is essentially zero, but throughput is very small.

14.8 TECHNIQUE OF VAPOR-PHASE CHROMATOGRAPHY

Usually, the gas chromatograph is ready to use. The following things have been done for you:

1. The appropriate *column* has been installed, and the inlet port, column, and detector have been brought to their operating temperatures by their corresponding *heaters*.
2. The *carrier gas flow rate* has been set by adjustment of the *pressure regulator* and a *needle valve*.
3. The *filament current* has been set to the proper operating value.
4. The *sensitivity* has been set so that the pen of the recorder will not go off scale on the largest peak. If the pen does go off scale, the sensitivity or the sample size should be decreased; if the largest peak gives less than 50% of full-scale deflection, the sensitivity or the sample size should be increased.
5. The *balance* and *recorder zero* have been set to put the pen at the position desired for the base line.
6. The *recorder speed* has been set at the desired speed. One inch per minute is normal; if the sample comes through very quickly, a greater recorder speed is probably desirable.

Injection of Sample

What remains is for you to start the recorder and inject the sample. Liquid samples are injected as is; solids must be dissolved in a volatile solvent, such as ether. Normally, the special (expensive) microsyringe used for injection will be wet with the previous sample. If so, you can rinse it by drawing in some of your sample and then squirting it out into a waste solvent container. Two or three repetitions should be adequate. The next step is to draw your sample into the syringe without any air bubbles. Sometimes, the air can be removed by dipping the needle into the sample and pumping the plunger a few times. Sometimes, you may need to draw in some liquid, hold the syringe with the needle up, and tap the barrel with your finger until the bubbles rise to the needle end. They may then be forced out by pushing the plunger in, and then the rest of the syringe can be filled. The desired amount of sample—0.001 to 0.005 mL (1 to 5 microliters)—is taken by adjusting the end of the plunger so that the desired volume fills the barrel of the syringe. Now you are ready to inject the sample. Turn the recorder from Standby to On. Insert the needle of the syringe all the way into the rubber septum of the inlet port, slide in the plunger, and pull the needle straight back out.

Follow the progress of the chromatogram by watching the recorder. A chromatogram may take less than a minute or more than half an hour; a typical time is a few minutes. When the chromatogram is complete, turn the recorder from On to Standby. If the pen goes off scale, or if the peaks are too small, change either the sensitivity setting or the sample size and try it again.

It is not unusual to decide that a chromatogram is complete when there is still material in the column. If the gas chromatograph is not used again immediately, this residue may go undetected. If, however, it is used again soon, very shallow, broad peaks can sometimes be observed under the relatively sharp peaks of the new sample. The broad peaks can be attributed to the previous sample, since the longer a substance remains in the column, the more time it has to diffuse and the shorter and wider will be its peak, for a given area.

Determination of Peak Areas

If the recorder has either a mechanical or an electronic integrator built into it, it is possible to measure the area of a peak automatically. Otherwise, peak areas can be measured with a planimeter or by counting squares. If the peak is symmetrical, the product of the height of the peak times the width of the peak at half-height equals 93% of the area of the peak. If area ratios are to be used, as is usually the case, the product of peak height times width at half-height is just as useful as the area. Experimental error in measuring the width of the peak can be mini-

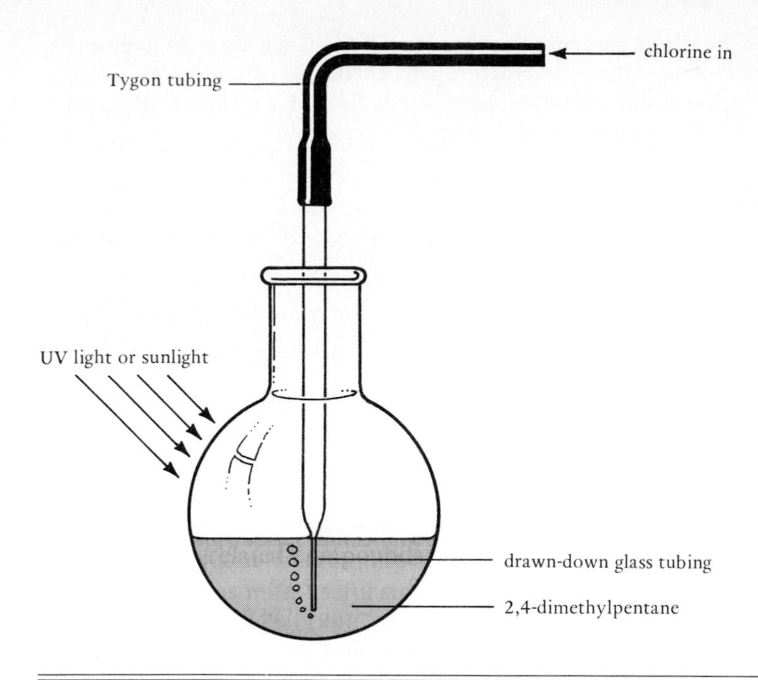

Figure 14.10. Arrangement for chlorination of 2,4-dimethylpentane.

mized by using a higher chart speed. If the peaks are not completely resolved but overlap somewhat, it is much more difficult to estimate their area. Resolution is increased by using a longer or narrower column, smaller sample size, and lower operating temperature. Two substances that cannot be resolved on one column may be resolved on another.

Exercises

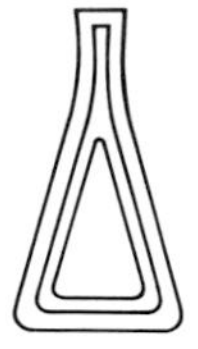

1. Chlorination of 2,4-dimethylpentane: determination of the relative reactivity of primary, secondary, and tertiary hydrogens in 2,4-dimethylpentane toward chlorine atoms. The experimental observations in the free-radical chlorination of alkanes are explained by the following mechanism:

1)	Cl—Cl	⟶	2 Cl·	initiation
2)	Cl· + H—R	⟶	Cl—H + R·	chain propagation
3)	R· + Cl—Cl	⟶	R—Cl + Cl·	
4)	2 R·	⟶	R—R	chain termination
5)	2 Cl·	⟶	Cl—Cl	
6)	R· + Cl·	⟶	R—Cl	

If the alkane has more than one kind of hydrogen atom, as in propane, products of more than one structure can be obtained. With propane, both 1-chloropropane and 2-chloropropane can be formed, depending upon which type of hydrogen is removed in Step 2:

2 (primary)

$$Cl\cdot + CH_3—CH_2—CH_3 \longrightarrow Cl—H + CH_3—CH_2—CH_2\cdot$$

3 (primary)

$$CH_3—CH_2—CH_2\cdot + Cl_2 \longrightarrow CH_3—CH_2—CH_2—Cl + Cl\cdot$$

1-chloropropane (45%)

2 (secondary)

$$Cl\cdot + CH_3—CH_2—CH_3 \longrightarrow Cl—H + CH_3—\dot{C}H—CH_3$$

3 (secondary)

$$CH_3—\dot{C}H—CH_3 + Cl_2 \longrightarrow CH_3—CH(Cl)—CH_3 + Cl\cdot$$

2-chloropropane (55%)

The relative amounts of the products equal the relative rates of formation of the radicals that are produced by removal of the different kinds of hydrogen atoms in Step 2. In the case of the light-catalyzed chlorination of propane at 25°C, the product is 45% 1-chloropropane and 55% 2-chloropropane. The ratio of the rate of Steps 2 (secondary) and 2 (primary) is thus inferred to be 55 : 45 or 1.22 : 1. Since there are three times as many primary hydrogen atoms as secondary hydrogen atoms, the ratio of the rate of Step 2 (secondary) to the rate of Step 2 (primary) *per hydrogen atom* is

$$\frac{1.22/1}{1/3} = 3.66:1$$

In the free-radical chlorination of 2,4-dimethylpentane, three products are possible, which correspond to the abstraction of a hydrogen atom from a primary, secondary, or tertiary position:

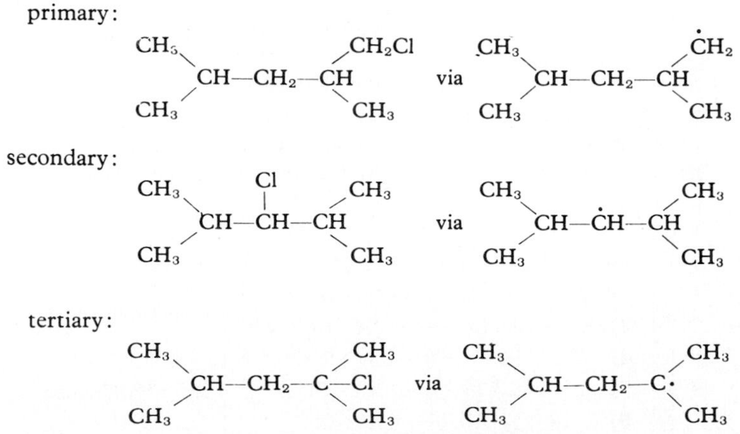

primary:

$$(CH_3)_2CH—CH_2—CH(CH_2Cl)(CH_3) \quad \text{via} \quad (CH_3)_2CH—CH_2—CH(\dot{C}H_2)(CH_3)$$

secondary:

$$(CH_3)_2CH—CH(Cl)—CH(CH_3)_2 \quad \text{via} \quad (CH_3)_2CH—\dot{C}H—CH(CH_3)_2$$

tertiary:

$$(CH_3)_2CH—CH_2—C(Cl)(CH_3)_2 \quad \text{via} \quad (CH_3)_2CH—CH_2—\dot{C}(CH_3)_2$$

Procedure. Pass a slow stream of chlorine gas through about 5 mL of 2,4-dimethylpentane in a 5-mL Erlenmeyer flask while irradiating the mixture with a weak ultraviolet light for about 5 minutes, as shown in Figure 14.10. The mixture can be analyzed by vapor-phase chromatography on a polyethyleneglycol column.

In addition to the large peak corresponding to unreacted 2,4-dimethylpentane, three peaks will be observed that correspond to the three isomeric monochlorination products. Assuming that the detector is equally sensitive to each isomer, the ratio of peak areas can be taken as the mole ratio of the isomers in the mixture. The order of appearance of the peaks is as follows: tertiary, secondary, primary (6).

2. The ratio of the products of the acid-catalyzed dehydration of 2-methylcyclohexanol (Section 50.2) can be determined by vapor-phase chromatography.

14.9 HIGH-PRESSURE LIQUID CHROMATOGRAPHY

High-pressure liquid chromatography, HPLC, also called high-performance liquid chromatography, is a widely used variation of column chromatography. In contrast to the particles in ordinary column chromatography, Section 14.1, the particles of the solid phase used for high-performance liquid chromatography are very finely divided, sometimes being as small as 5 microns in diameter. Furthermore, the solid phase often consists of a glass core coated with a thin layer of adsorbent. Therefore, since the depth of adsorbent is small, and the volume of liquid relative to the volume of adsorbent is also very small, equilibrium between the adsorbent and the liquid phase is established very rapidly, and columns about ½ meter long can contain the equivalent of several thousand theoretical plates. Thus, the resolving, or separating, power of HPLC, now available for samples of liquids and for solutions of solids, is similar to that provided by vapor-phase chromatography, Section 14.7, for samples of gases and vapors. The finely divided nature of the solid phase used in HPLC, however, provides great resistance to the flow of the eluant. High pressures, often several thousand pounds per square inch, are required to force the eluant through the 3- to 6-mm-diameter stainless steel columns, and the necessary pumps and valves are quite expensive.

14.10 BATCHWISE ADSORPTION; DECOLORIZATION

Sometimes it is possible to adsorb an undesired colored compound by treating a solution with an adsorbent such as activated charcoal or alumina and then filtering the mixture by gravity to remove the adsorbent and the adsorbed material (Section 8.3). The ideal conditions are that the impurity is much more strongly adsorbed than the desired substance, and that an amount of adsorbent just sufficient to adsorb the impurity is used.

Occasionally, it may be possible to decolorize a solution by filtering it through a short column of alumina, thus preferentially adsorbing the colored substance. A poorly adsorbed solvent (hydrocarbons, ether; not alcohols) must be used so that the solvent will not take the place of the colored material.

References

1. A. I. Vogel, *A Textbook of Practical Organic Chemistry,* 3rd edition, Wiley, New York, 1956.
2. C. Rollins, *J. Chem. Educ.* **40,** 32 (1963).
3. A. R. Patton, *J. Chem. Educ.* **27,** 574 (1950).
4. E. S. and D. Kritchevsky, *J. Chem. Educ.* **30,** 370 (1953).
5. L. F. Druding, *J. Chem. Educ.* **40,** 536 (1963).
6. A. Ault and W. Lambrecht, unpublished results; G. A. Russell and P. G. Haffley, *J. Org. Chem.* **31,** 1869 (1966).

Sources from which more information about chromatographic methods can be obtained include:

7. E. Lederer and M. Lederer, *Chromatography,* Elsevier, Amsterdam and New York, 1967.
8. E. Heftmann, *Chromatography,* 3rd edition, Reinhold, New York, 1975.
9. R. P. W. Scott, "Contemporary Liquid Chromatography," in *Techniques of Chemistry,* Vol. XI, Wiley, New York, 1976.

15 Removal of Water; Drying

One very common separation problem is that of removing water. Since the products of many reactions are separated by procedures that use water or aqueous solutions, water usually must be removed during the purification process. Occasionally, the presence of even small amounts of water in a reaction mixture is undesirable. If so, it is necessary to remove even the water that is present in the reagents and solvents as a result of exposure to water vapor in the air. The following sections describe some of the methods for removal of water from solids, liquids, and gases.

15.1 DRYING OF SOLIDS

A damp solid, such as that obtained when isolating a solid by suction filtration, can often be dried simply by spreading it out on a sheet of filter paper. If the solid is only slightly damp and the crystals are not so

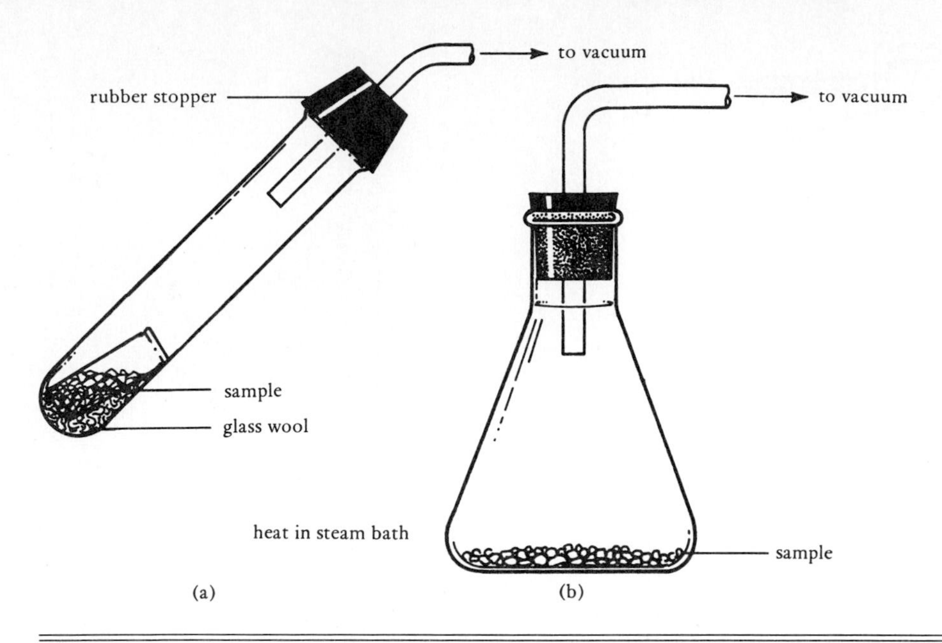

Figure 15.1. Arrangements for drying small samples.

small that the material is a damp powder or paste, it may dry within an hour to the point that the only water associated with the solid is that which must be adsorbed in order for it to be in equilibrium with the water vapor present in the air.

If the material is quite damp or pasty, the bulk of the water can often be removed by pressing the material between sheets of adsorbent paper (filter paper) or by spreading it out and pressing it down on a piece of unglazed porcelain plate or a block of plaster of Paris.

Increasing Rate of Drying

The overall *rate* of drying can be increased by increasing the rate of evaporation (by raising the temperature) and by minimizing the rate of recondensation (by decreasing the partial pressure of water in the atmosphere around the sample). To increase the rate of evaporation, the sample can be heated in an oven to a temperature somewhat below its melting point (but usually not above 110°C for organic compounds, in order to minimize the rate of reaction of the substance with oxygen in the air). Another very convenient method of heating is to shine an infrared heat lamp on the sample. In all cases, the thinner the layer in which the sample is spread, the faster it will dry. The rate of recondensation of water vapor on the sample can be decreased by providing for circulation of air over the sample, or by drying the sample in a container

Drying oven; heat lamp

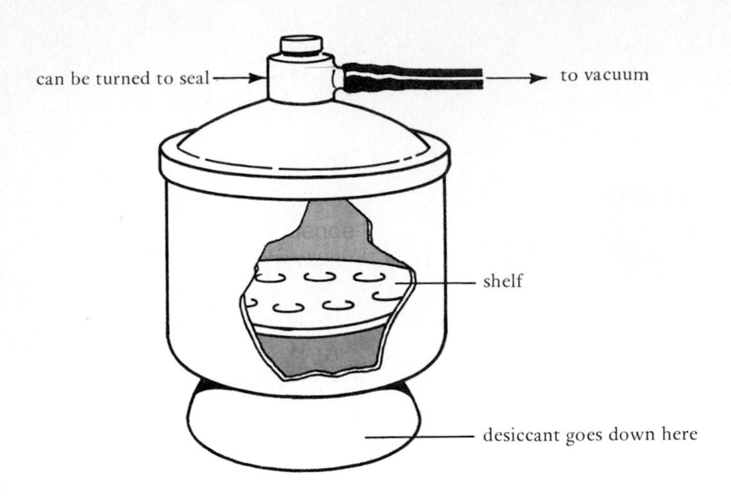

Figure 15.2. A vacuum desiccator.

Vacuum desiccator

in which the vapor pressure of water can be reduced. Figure 15.1 shows two simple arrangements for drying small samples. Figure 15.2 illustrates a vacuum desiccator. This device can be evacuated, and it also can contain a desiccant, a material that combines readily with water and has, in the hydrated form, a low equilibrium vapor pressure of water. The drying process inside an evacuated vacuum desiccator is evaporation of water from the sample, diffusion to the desiccant, and irreversible combination of the water with the desiccant. The absence of air in the evacuated space greatly increases the rate of diffusion of water, which is usually the slow step. Vacuum ovens are useful for drying larger quantities under vacuum at an elevated temperature.

Increasing Extent of Drying

The *extent* to which a substance can be dried depends upon the partial pressure of water in the atmosphere surrounding it. In the open air, it is determined by the partial pressure of water in the air. For example, at 25°C and 60% relative humidity, the partial pressure of water will equal 0.60 times the equilibrium vapor pressure of water at this temperature, or about 14 Torr. For many substances, the amount of adsorbed or combined water that will result in an equilibrium vapor pressure of this magnitude for the substance will be quite small, and drying in air will serve to remove practically all of the water. If a substance will adsorb or combine with an objectionable amount of water at the partial pressure of water in the atmosphere, it must be dried in a closed chamber that

Table 15.1. Desiccants commonly used in desiccators

Substance	*Hydrated Form*	*Equilibrium Vapor Pressure of Hydrated Form*
Aluminum oxide (alumina)	1% by weight of water (estimate)	0.001 Torr
Conc. sulfuric acid	95% sulfuric acid	0.001 Torr
	80% sulfuric acid	0.6 Torr
Potassium hydroxide	$KOH \cdot H_2O$	1.5 Torr
Sodium hydroxide	$NaOH \cdot H_2O$	0.7 Torr
Calcium chloride	$CaCl_2 \cdot H_2O$	0.04 Torr
Drierite	$CaSO_4 \cdot \frac{1}{2}H_2O$	0.004 Torr

Use of desiccant

contains a desiccant. The desiccant can be chosen to provide a sufficiently low equilibrium vapor pressure of water (see Table 15.1). Figure 15.3 illustrates an apparatus that is used to dry small samples quickly and thoroughly by providing for heating, evacuation, and the use of a desiccant.

A less generally useful method of drying a solid that has been collected by suction filtration is to wash it with a solvent in which water is at least somewhat soluble but the substance is not. Anhydrous methanol, ethanol, or ether can be used for this purpose, but the problem is that many organic substances are appreciably soluble in these solvents.

Figure 15.3. Apparatus for drying a small sample under vacuum in the presence of a desiccant: Abderhalden drying pistol. The temperature at which the sample is dried is determined by the boiling point of the solvent used.

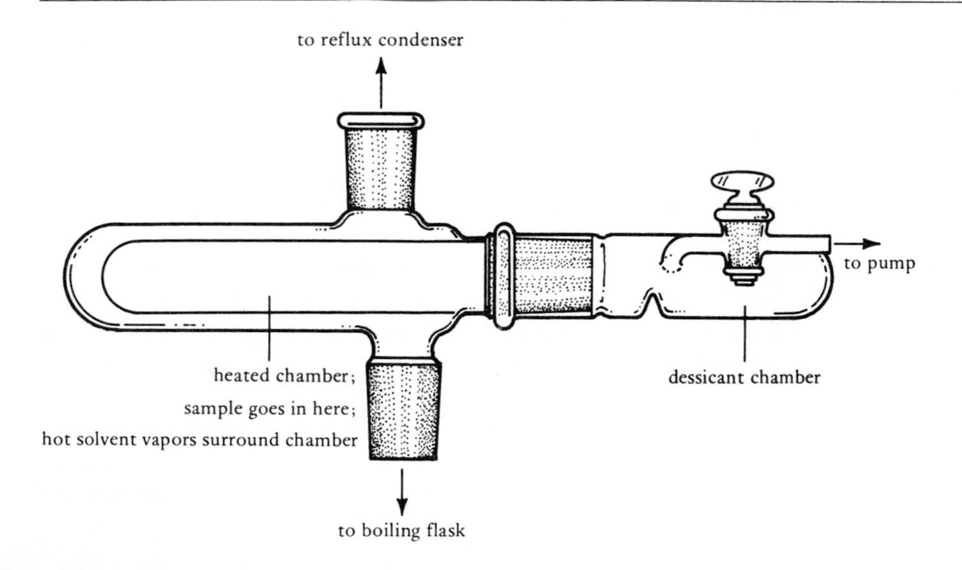

Removal of Organic Solvents

Although removal of water has been the objective of the methods previously described, they can also be used for the *removal of other solvents* from solids. Removal of other solvents by air drying is usually complete, since the partial pressure in air of solvents other than water should be zero. Of course, the drying process can always be speeded up by increasing the temperature or using a vacuum desiccator, or both. Occasionally, a substance will hold solvent quite tenaciously and will require long periods of drying at an elevated temperature. High-boiling solvents, acetic acid especially, are best removed directly after collection by suction filtration: the sample is washed on the funnel with small portions of a more volatile solvent that will dissolve the high-boiling solvent but not the compound.

15.2 DRYING OF SOLUTIONS

When an organic substance is isolated by extraction with an organic solvent, the aqueous phase from which the substance was extracted, or the aqueous solutions with which the extracts are washed, will leave the extract saturated with water. It is usually most convenient to remove the water by treating the extract with a solid that will combine with the water:

$$\underset{\text{anhydrous salt}}{\text{Solid} + nH_2O} \rightleftharpoons \underset{\text{hydrated salt}}{\text{Solid}\cdot nH_2O}$$

The hydrated salt is then removed by gravity filtration. The completeness of water removal by this method depends both upon how long the drying agent is left in contact with the extract and upon how low the equilibrium vapor pressure of water is for the particular hydrated salt.

The ideal material for this purpose is a solid that is insoluble in the organic solvent, is inert toward both the solvent and the substance in solution, can be used in small amounts, and will combine quickly and completely with the dissolved water to give a solid (the hydrate of the original material) that can be removed by gravity filtration. Table 15.2 lists the drying agents that are most commonly used in this way. The drying agents are characterized by their capacity (the amount of water that can be removed by a given weight of drying agent), speed (rate at which water combines to form the hydrate), intensity (degree of dryness ultimately achieved), suitability for use with different classes of compounds, and cost. Since the criteria are not independent of one another (for instance, the intensity depends upon the degree of hydration allowed, and thus upon the capacity), and since they also depend upon the solvent being dried (ether solutions dry faster than benzene or

Drying agents

Table 15.2. Drying agents commonly used for drying solutions in organic solvents

Substance	*Capacity*	*Speed*	*Intensity*	*Cost*	*Convenience*	*Suitability*
Calcium chloride	H	M	H	L	H	a
Calcium sulfate (Drierite)	L	H^+	H^+	M	H	b
Magnesium sulfate	H	H	M,H	L	M	c
Molecular sieves, 4Å	H	H	H		H	c
Potassium carbonate	M	M	M		M	d
Sodium sulfate	H^+	L	L	L	M	e

H: high M: medium L: low

a. Combines with alcohols, phenols, amines, amino acids, amides, ketones, and some aldehydes and esters. It should not be used to dry solutions containing compounds of these types unless it is desired to remove them also. Some calcium hydroxide may be present that will combine with acids. The hexahydrate is unstable above 30°C.
b. Generally useful. The hemihydrate is stable to at least 100°C.
c. Generally useful.
d. Combines with acids and phenols. It should not be used to dry solutions containing acids unless it is desired to remove them also.
e. Generally useful. The decahydrate is unstable above 32°C.

ethyl acetate solutions), the ratings can be only approximate and qualitative.

Solutions in solvents in which water is relatively soluble (ether, ethyl acetate) can be dried more economically by carrying out the drying process in two steps. A relatively inexpensive, high-capacity, low-intensity drying agent is used first, and the remaining water is then removed by treatment, in a second step, with a high-intensity drying agent. For example, the solution can be dried first with anhydrous sodium sulfate, which will combine with 127% of its weight of water to form a decahydrate, $Na_2SO_4 \cdot 10H_2O$. Then, the solution can then be further dried with anhydrous calcium sulfate, which has a very high intensity but a low capacity, since it combines with only 6% of its weight of water, to form a hemihydrate, $CaSO_4 \cdot \frac{1}{2}H_2O$.

Procedure. The procedure for drying a solution usually consists of adding a portion of the solid drying agent and allowing the mixture to stand for 5–15 minutes with occasional swirling. If a second liquid phase appears (a saturated aqueous solution of the drying agent), or if the drying agent clumps together, more should be added. It is easy to tell when an excess of anhydrous magnesium sulfate is present, since the suspension that forms when the mixture is swirled is quite cloudy and settles relatively slowly.

Table 15.3. Suitability of various drying agents for drying pure solvents

You must take care not to try to dry a liquid or solution with something that will react with it. Sodium, potassium, and hydrides must be kept away from acids, alcohols, phenols, and other acidic compounds; sodium and potassium must be kept away from alkyl or aryl halides.

	Drying Agent					
Solvent	P_2O_5	KOH, NaOH	BaO, CaO	K_2CO_3	$CaCl_2$	$MgSO_4$
Alcohols	−		+			
Aldehydes, ketones	−	−	−	+		
Alkanes	+				+	
Alkenes	+				+	
Alkyl halides	+				+	+
Amines	−	+	+	+	−	
Aromatic hydrocarbons	+				+	
Aryl halides	+				+	+
Ethers	−		+		+	+[a]
Nitriles	+			+		

(+) *Recommended for use* (−) *Advised against using*

[a] Magnesium sulfate is satisfactory for drying ethyl ether for use in the Grignard reaction [*J. Chem Educ.* **39,** 578 (1962)].

15.3 DRYING OF SOLVENTS AND LIQUID REAGENTS

Pure organic liquids (solvents and reagents) can be dried by treatment with various materials much in the way in which solutions are dried. The drying agent can act either by forming a relatively stable hydrate or by entering into an essentially irreversible chemical reaction with the water. Table 15.3 indicates which drying agents are useful for drying certain compounds or classes of compounds. If drying is the only thing to be done, you can simply add the drying agent to the liquid in the bottle in which it is stored, providing for venting if hydrogen gas will be evolved. Drying may be practically complete in less than half an hour, but it sometimes takes several days.

Drying by Azeotropic Distillation

In the special case of solvents in which water is sparingly soluble and which form minimum-boiling azeotropes with water, the solvent can be dried by distillation. The initial distillate will contain the water as the minimum-boiling azeotrope; the remainder of the distillate will be the dry solvent. Solvents that can be dried in this way include benzene, toluene, xylene, hexane, heptane, petroleum ether, carbon tetrachloride, and 1,2-dichloroethane. As long as the condensate appears cloudy, water is being removed; after about 10% of the volume of the solvent has been distilled, all the dissolved water and the water ad-

Table 15.4. Suitability of various drying agents for drying gases

	Drying Agent				
Gas	P_2O_5	$CaCl_2$	$CaSO_4$	Al_2O_3	*conc.* H_2SO_4
Air		+	+	+	+
Nitrogen		+	+	+	+
Hydrogen	+	+		+	+
Carbon dioxide	+	+		+	+

(+) *Recommended for use*

sorbed on the walls of the flask and condenser should be gone.

When any of these solvents must be dry for use as a solvent (or reagent) in a reaction, it is often most convenient to add a 10–15% excess to the reaction flask and then to remove the excess by distillation. If this is done, both the dissolved water and the water adsorbed on the inside of the apparatus will be removed as a minimum-boiling azeotrope. Sometimes it is possible to dry one or more of the reagents as well by adding it before the distillation.

The presence of water in a liquid can sometimes be detected by adding a small amount of anhydrous cobaltous chloride or anhydrous cobaltous bromide. If water is present, the blue color of the anhydrous salt gives way to the pink color of the hydrate. One form of anhydrous calcium sulfate commercially available (Drierite) comes as granules whose surface is impregnated with cobaltous chloride. A few granules of this material is useful for testing for the presence of water.

Figure 15.4. Gas-drying tube.

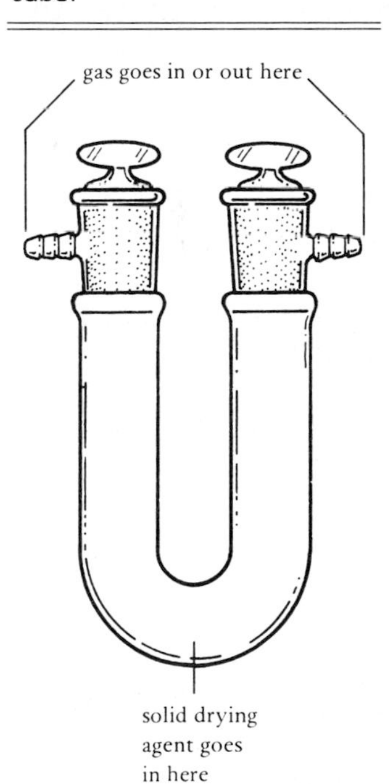

15.4 DRYING OF GASES

Gases are most conveniently dried by passing them through a tube packed with a granular adsorbent. Table 15.4 indicates the adsorbents that have been recommended for a number of different gases. The size of the column of adsorbent and the flow rate should be such that a contact time of at least five seconds is attained. Figure 15.4 illustrates a gas-drying tube.

Table 15.4 also indicates certain gases that can be dried by bubbling them through concentrated sulfuric acid.

Reference

1. D. R. Burfield, K.-H. Lee, and R. H. Smithers, "Desiccant Efficiency in Solvent Drying. A Reappraisal by Application of a Novel Method for Solvent Water Assay," *J. Org. Chem.* **42,** 3060 (1977). As the authors say in their abstract, the results range from the expected to the highly surprising.

Determination of Physical Properties

The physical and chemical properties of a substance are determined by the molecular structure of the substance. Different physical states of a substance differ not in molecular structure but only in the relationships between the molecules. The different intermolecular relationships are determined by the structure of the molecules.

Thus, *samples that have identical physical and chemical properties must be identical at the molecular level,* and substances that differ in chemical or physical properties must differ in molecular structure. Consequently, you can tell whether two samples are samples of the same substance by comparing their physical and chemical properties.

When a pure sample is available, its physical and chemical properties serve as the standards by which another sample of that substance can be identified; in a similar way, the properties of a mixture can serve to characterize a mixture. When a sample is not pure (when it is a mixture), the progress of its purification can be followed by comparison of its properties before and after the application of separation procedures.

In the preparative experiments in this book, the identity and purity of the products of the reactions can be established by comparison of certain of their physical and chemical properties with the properties of substances whose molecular structure and degree of purity are assumed to be known. The physical properties most often used in this way are the melting or boiling point and the infrared or nuclear magnetic resonance spectrum.

Since the physical and chemical properties of a substance are thought to be determined entirely and uniquely by the molecular structure of the substance, it follows that *we can infer the molecular structure of the substance from the physical and chemical properties of the substance.* After all, the properties of a substance are the only experimental knowledge we will ever have of that substance, and hence any theoretical description must be inferred from these observable properties. It is impossible to give a brief explanation of how empirical evidence has in the past been translated into structural theory and how, in general, theory is developed, but in the following discussions of physical properties and their interpretation by reference to molecular structure, many examples are presented that show how conclusions about the molecular structure of a substance can be inferred from its physical properties.

The physical properties most often used in the past to characterize substances were the boiling point, melting point, density, index of refraction, and optical rotation (Sections 16 through 20). For substances of unknown structure, determinations of molecular weight and solubility characteristics were often helpful (Sections 21 and 22). More recently, spectrometric methods have become very widely used, both for the characterization of substances and for structure determination. Of these, the most common are infrared, ultraviolet-visible, nuclear magnetic resonance, and mass spectrometry (Sections 23 through 26).

16 Boiling Point

The boiling point is one of the most often reported physical properties of a liquid, since the boiling point can usually be determined while the liquid is purified by distillation. The boiling point is often used, with the density and index of refraction (Sections 18 and 19), to establish the identity and to estimate the purity of a liquid.

16.1 EXPERIMENTAL DETERMINATION OF BOILING POINT

If sufficient material is available, the boiling point of a liquid can be determined by distillation (Section 9.6). The temperature of the vapor should be observed during the course of the distillation, and the temperature range over which most of the material distills should be taken as the boiling point. Since the boiling point is a function of the pressure, the barometric pressure or, in a distillation under reduced pressure, the pressure of the system should be recorded.

Micro boiling point determination

If less than a milliliter of liquid is available—as, for instance, when the sample has been isolated by vapor-phase chromatography—a *small-scale method* may be used. In this procedure, a 10- to 15-cm length of glass tubing 3–5 mm in inside diameter is sealed shut at one end by heating in a flame. Two or three drops of the liquid are added to this sample tube, and a length of melting point capillary, sealed about 5 mm from one end, is dropped in with the sealed end down. (This small tube is most conveniently made by melting shut a melting point capillary near the middle and cutting off all but 5 mm on one side of the seal.) The sample tube is fastened to a thermometer by a 2- to 3-mm slice of rubber tubing used as a small rubber band (see Figure 16.1). The thermometer is then supported in a melting point bath (Figure 17.4) and heated until a very rapid, steady stream of bubbles issues from the sealed capillary. The bath is then allowed to cool slowly, and the temperature at which a bubble just fails to come out of the capillary and the liquid starts to enter it is taken as the boiling point of the liquid.

Correction of boiling point

If the barometric pressure is not 760 Torr, the observed boiling point may be corrected to the temperature that would be expected at 760 Torr. This correction amounts to about 0.5°C for each 10-Torr deviation of atmospheric pressure from 760 Torr; the observed boiling point will be low if the atmospheric pressure is low.

Figure 16.1. Apparatus for small-scale determination of boiling point.

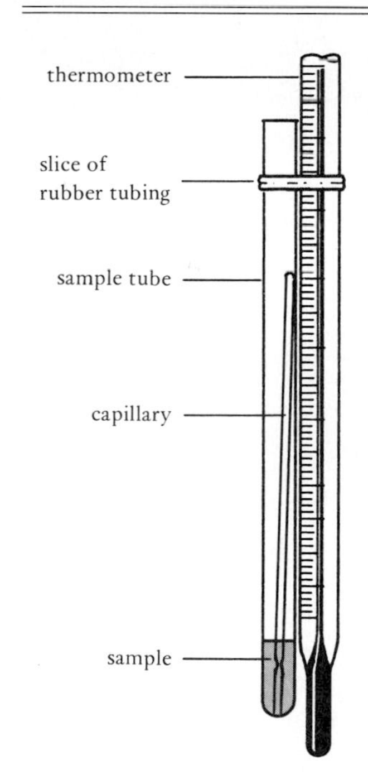

In addition to errors introduced by experimental variables such as the rate of heating, superheating, and the presence of impurities, the observed boiling point may be in error for two other reasons. The first is that the thermometer may not be correctly calibrated (the scale of degrees may not be in exactly the right place on the stem of the thermometer). The extent of any error of calibration can be determined by using the thermometer to experimentally determine the melting point of pure samples of solids of known melting point (Section 17.1). The second possible error depends on the design of the thermometer. Many thermometers are calibrated with the understanding that they will be immersed only to a certain line engraved upon the stem. Some thermometers, however, are calibrated with the understanding that the entire thermometer will be at the temperature that is being determined. If only part of the thermometer is at this temperature, the extent of this error, which is the result of unequal coefficients of thermal expansion of mercury and glass, can be calculated according to the following equation:

$$\text{Correction to be added to } t_1 = KN(t_1 - t_2)$$

where K = the apparent coefficient of thermal expansion of mercury in glass: 0.000159 for mercury in soft glass at 200°C; 0.000167 for borosilicate (Pyrex) glass at 200°C

N = The length, in degrees, of the part of the mercury column not heated to the temperature of the bulb: the length of the exposed column of mercury

t_1 = the observed temperature: the temperature of the mercury bulb

t_2 = the temperature at the middle of the exposed mercury column; this can be determined by a second thermometer

A typical emergent stem correction for an observed temperature reading of 200°C might be +5°C.

One of the best ways to estimate (and thus take into account) thermometer error and many experimental errors is to determine the boiling point of a pure sample of a substance of known boiling point under the same experimental conditions used to determine the unknown boiling point. Table 16.1 lists some liquids, and their boiling points, that have been recommended for use as standards for boiling point determinations.

16.2 BOILING POINT AND MOLECULAR STRUCTURE

The variation of boiling point with the structure of covalently bonded molecules can be resolved into three contributing factors: the molecular weight, the nature of the functional group, and the degree of branching of the molecule. In order to see the regularity in the effect of any one factor, the other two must be held constant.

Boiling Point and Molecular Weight

As an illustration of the effect of increasing molecular weight, Tables 16.2 and 16.3 list the boiling points of the first members of the homologous series of straight-chain alkanes and primary alkyl chlorides. From these data, it appears that *boiling points increase with increasing molecular weight* (the nature of the functional group and degree of branching being held constant). It is also apparent that the increase in boiling point per additional methylene group is not constant, but decreases with increasing molecular weight. For this reason, a graph of boiling point versus molecular weight (or number of carbon atoms) is a curved line.

Boiling Point and the Nature of the Functional Group

As an illustration of the effect of the nature of the functional group, Table 16.4 presents the boiling points of a number of straight-chain

Table 16.1. Reference liquids for boiling point determinations

Compound	*Normal b.p.* (°C)	ΔT/10 *Torr* (°C)[a]
Acetone	56.1	0.39
Water	100.0	0.37
Bromobenzene	156.2	0.53
Nitrobenzene	210.9	0.48
Quinoline	237.5	0.59
Benzophenone	305.9	0.6

[a] Variation of boiling point with pressure.

Table 16.2. Normal alkanes: boiling points and enthalpies and entropies of vaporization

	Boiling Point			*Increase*			
Compound	°C	*K*	*Molecular Weight*	*b.p.*	*m.w.*	ΔH_{vap} (*cal/mole*)	ΔS_{vap} (*cal/mole-degree*)
C_1	−161.49	111.66	16.042			1,955	17.51
C_2	−88.63	184.52	30.068	72.86	14	3,517	19.06
C_3	−42.07	231.08	44.094	46.56	14	4,487	19.42
C_4	−0.50	272.65	58.120	41.57	14	5,352	19.63
C_5	36.07	309.22	72.146	36.57	14	6,160	19.92
C_6	68.74	341.89	86.172	32.67	14	6,896	20.17
C_7	98.47	371.62	100.198	29.72	14	7,575	20.38
C_8	125.67	398.82	114.224	27.20	14	8,214	20.60
C_9	150.80	423.95	128.250	25.13	14	8,777	20.70
C_{10}	174.12	447.27	142.276	23.32	14	9,390	20.99

Table 16.3. Primary, straight-chain alkyl chlorides: boiling points and enthalpies and entropies of vaporization

	Boiling Point			*Increase*			
Compound	°C	*K*	*Molecular Weight*	*b.p.*	*m.w.*	ΔH_{vap} (*cal/mole*)	ΔS_{vap} (*cal/mole-degree*)
C_1	−24.22	248.93	50.5			5,126	20.59
C_2	12.27	285.42	64.5	36.49	14	5,832	20.43
C_3	46.60	319.75	78.5	34.33	14	6,512	20.37
C_4	78.44	351.59	92.6	31.84	14	7,174	20.40
C_5	107.76	380.91	106.6	29.32	14	7,824	20.54
C_6	134.50	407.65	120.6	26.74	14	8,458	20.75
C_7	159.1	432.3	134.6	24.65	14	9,091	20.98
C_8	182.0	455.2	148.7	22.9	14	9,673	21.25
C_9	203.4	476.6	162.7	21.4	14	10,250	21.51
C_{10}	223.4	496.6	176.7	20.0	14	10,800	21.75

Table 16.4. Boiling points and enthalpies and entropies of vaporization of straight-chain compounds with approximately the same molecular weight but with different functional groups

Compound	*Molecular Weight*	*Boiling Point* °C	*Boiling Point* K	ΔH_{vap} (*cal/mole*)	ΔS_{vap} (*cal/mole-degree*)
1-Pentene	70	30.0	303.2	6,021	19.86
1-Fluorobutane	76	32.5	305.7	6,264	20.49
Diethyl ether	74	34.3	307.5	6,355	20.67
Pentane	72	36.1	309.3	6,157	19.91
1-Chloropropane	78	45.7	318.9	6,594	20.68
Diethylamine	73	55.5	328.7	6,888	20.96
Methyl acetate	74	57.8	331.0	7,454	22.52
n-Butylamine	73	77.8	351.0	7,678	21.88
Butyraldehyde	72	74.6	347.8	7,880	22.66
2-Butanone	72	79.6	352.8	7,837	22.21
Propionitrile	69	117.4	390.6	7,937	20.32
1-Butanol	74	117.6	390.8	10,505	26.88
Propionic acid	74	139.3	412.5	7,318	17.74
Dimethylformamide	73	149.6	422.8	9,164	21.67
N-Methylacetamide	73	206	479		
Propionamide	73	213	486	14,860	30.56

compounds with approximately the same molecular weight but with different functional groups. It is apparent that *the more polar the functional group, the higher the boiling point.* (Actually, the degree of polarity of the functional group is inferred from the b.p.'s of the substances.)

Boiling Point and Degree of Branching

As an illustration of the effect of the degree of branching, Table 16.5 presents the boiling points and molecular structures of the eight isomers of molecular formula $C_5H_{11}Cl$. The general trend is that *the more highly branched the molecule* (the functional group and the molecular weight being the same), *the lower the boiling point.* A comparison of the last three entries in the table illustrates the general observation that if the branches are on the same carbon atom rather than on different ones, the boiling point will be a little lower.

16.3 BOILING POINT AND THE ENTHALPY AND ENTROPY OF VAPORIZATION

For processes that take place at a constant temperature, such as vaporization or freezing of a liquid, the entropy change (ΔS) is equal to the

Table 16.5. Boiling points of the isomeric alkyl chlorides of molecular formula $C_5H_{11}Cl$

Compound	*b.p.* (°C)	*Compound*	*b.p.* (°C)
C—C—C—C—C—Cl	108	Cl | C—C—C—C—C	97
C | C—C—C—C—Cl	97	C C | | C—C—C—Cl	91
C | C—C—C—C—Cl	98	C | C—C—C—Cl | C	86
C | C—C—C—C—Cl	100	C | C—C—C—Cl | C	84

enthalpy change (ΔH) divided by the temperature at which the process occurs:

$$\Delta S = \frac{\Delta H}{T}$$

At the normal boiling point, then,

$$T_{\text{vap}} = \frac{\Delta H_{\text{vap}}}{\Delta S_{\text{vap}}}$$

Thus, the high boiling points are the result of large heats of vaporization and small entropies of vaporization. It remains to see how the factors of molecular weight, nature of the functional group, and degree of branching contribute to these quantities.

Enthalpy of Vaporization and Molecular Structure

The heat of vaporization is interpreted as a measure of the amount of energy required to separate the molecules against an attractive intermolecular force in the change, at constant temperature, from the liquid state to the vapor state. The stronger the intermolecular forces, the higher the heat of vaporization. Four different kinds of intermolecular forces can be distinguished: (1) van der Waals forces, proportional to the square of the polarizability, (2) dipole–dipole forces, proportional to the fourth power of the dipole moments, (3) induced dipolar forces, proportional to the polarizability and the square of the dipole moment, and (4) hydrogen-bonding forces, usually due to the presence of O—H or N—H groups in the molecule (Reference 1).

Table 16.6. Boiling points of substituted butanes

Compound[a]	*Boiling Point* (°C)	*Molecular Weight*	*Intermolecular Forces*	ΔH_{vap} (*cal/mole*)	ΔS_{vap} (*cal/mole-degree*)
R—H	0	58	van der Waals only	5,342	19.63
R—F	33	76	van der Waals and dipolar	6,264	20.49
R—OCH_3	70	88	"		
R—Cl	78	93	"	7,174	20.40
R—Br	102	137	"	7,613	20.32
R—CHO	103	86	"	8,550	22.7
R—$COOCH_3$	127	116	"		
R—COCl	128	121	"		
R—I	131	184	"	7,983	19.77
R—C≡N	141	83	"	8,669	20.82
R—NO_2	152	103	"	10,000	23.5
R—NH_2	78	73	van der Waals, dipolar, and hydrogen bonds	7,678	21.88
R—SH	99	90	"	7,700	20.72
R—OH	118	74	"	10,505	26.88
R—COOH	187	102	"	11,891	25.89

[a] R = *n*-Butyl

The increase in boiling point with increasing molecular weight (the nature of the functional group and the degree of branching remaining constant) is due to larger van der Waals forces between the heavier molecules. The higher the molecular weight (the more CH_2 units), the larger the total intermolecular attraction by van der Waals forces. This results in an increase in heat of vaporization with increasing molecular weight. Data for the first members of the series of normal alkanes and alkyl chlorides are presented in Tables 16.2 and 16.3.

The decrease in boiling point with increasing branching (the nature of the functional group and the molecular weight remaining constant; see Table 16.5) can also be interpreted by looking at the influence of structure on the magnitude of the van der Waals forces. A more highly branched and thus more compact molecule will have less surface area and therefore a smaller total intermolecular attraction because of van der Waals forces.

When functional groups are present that cause the molecule to have a dipole moment, dipolar and induced dipolar forces can contribute to the intermolecular attraction in addition to the van der Waals forces. The higher boiling points of ethers, aldehydes, ketones, esters, nitriles, and nitro compounds (and, to a certain extent, alkyl halides) can be explained by the presence of these additional forces. The presence of

O—H or N—H (or, to a smaller extent, S—H) groups in a molecule adds a further particularly effective, localized type of dipole–dipole attraction (the "hydrogen bond"), which is due almost uniquely to these functional groups. The relatively high boiling points (for a given molecular weight and degree of branching) of alcohols, thiols, primary and secondary amines, phenols, carboxylic acids, and unsubstituted and monosubstituted amides are due to this additional relatively large force. Tables 16.4 and 16.6 present data that illustrate the relative effectiveness of different functional groups in contributing to a high boiling point.

Hydrogen bond

The boiling point increase that results from the introduction of a single halogen atom is due partly to the creation of a dipole moment (about the same for —F, —Cl, and —Br, and slightly less for —I) and partly to the introduction of a polarizable atom (the polarizability is least for fluorine and increases through chlorine to bromine to iodine; polarizability generally increases with atomic number or number of electrons). The effect of an iodine atom occurs mostly through its polarizability, and the effect of fluorine is almost entirely due to any dipole moment resulting from its presence. It is reasonable to account for the relatively low boiling points of fluorocarbons in terms of the low polarizability of fluorine.

Entropy of Vaporization and Molecular Structure

The entropy of vaporization is interpreted as a measure of the increase in disorder that results when a collection of molecules is changed from a relatively confined state to a relatively free state (or from a small volume to a large volume, or from a state with widely spaced energy levels to a state with closely spaced energy levels). In general, the entropy of vaporization for members of a homologous series increases only slowly with increasing molecular weight (Tables 16.2 and 16.3) and is independent of the degree of branching. Thus, the effect of molecular weight and degree of branching upon boiling point is almost entirely through the heat of vaporization and not through the entropy of vaporization. In fact, for most substances of moderate boiling point, the entropy of vaporization is equal to 20–22 calories/mole-degree (Trouton's rule).

There are, however, some interesting exceptions. Alcohols (and water) generally have an unusually large entropy of vaporization. This is interpreted by saying that the increase in disorder is unusually great when an alcohol or water is vaporized, and the reason for this is that the liquid state is relatively ordered or structured compared to other liquids. This extra degree of structure, presumably due to the formation of chains or networks of hydrogen bonds, is lost upon vaporization. Thus, from the point of view of the entropy change upon vaporization, we would expect alcohols and water to have unusually low boiling points. This is not the case, however; alcohols (and water) have relatively high boiling points compared to other compounds of the same molecular weight and degree of branching. The reason is that additional energy is required to break up the hydrogen bonds, which results in an unusually high heat of vaporization for alcohols and water. In fact, the greater heat of vaporization more than compensates for the increased entropy of vaporization. Acetic acid has an unusually low heat of vaporization, less than that for pentane. On the basis of heats of vaporization, acetic acid might be expected to boil below pentane. However, it also has an unusually low entropy of vaporization—about 14.5 cal/mole-degree. If acetic acid had a "normal" entropy of vaporization, it would boil at −3°C, rather than at 118°C. The low entropy of vaporization is interpreted by saying that acetic acid must retain some order or structure in the vapor phase. Vapor density measurements support this by indicating an average molecular weight considerably higher than 60, and both phenomena are interpreted in terms of partial dimerization through hydrogen bond formation in the vapor phase. Since not all the molecules are separated upon vaporization, the low heat of vaporization can thus be explained as well.

$CH_3-C(=O)-OH$
acetic acid

$CH_3CH_2CH_2CH_2CH_3$
pentane

Boiling points might be expected to vary much more widely than they do, except that differences in heat of vaporization and entropy of vaporization tend to balance each other. An increase in the intermolecular force would be expected to increase the order of the liquid, and thus lead to a larger entropy of vaporization, as well as to increase the heat of vaporization. Table 16.7 lists some compounds that have about the same boiling points (and therefore the same ratio of heat of vaporization to entropy of vaporization), but that show a considerable variation in these quantities.

Table 16.7. Enthalpies and entropies of vaporization of selected compounds boiling near 120°C

Compound	*Molecular Weight*	*Boiling Point* °C	*Boiling Point* K	ΔH_{vap} (*cal/mole*)	ΔS_{vap} (*cal/mole-degree*)
Acetic acid	60	118.5	391.7	5,662	14.46
Butyronitrile	71	117.4	390.6	7,937	20.32
n-Butanol	74	117.6	390.8	10,505	26.88
2-Hexanone	100	127.4	400.6	8,243	20.38
Ethyl butyrate	114	118.9	392.1	8,673	22.12
n-Octane	114	125.8	398.8	8,214	20.60
n-Butyl iodide	184	130.5	403.7	7,983	19.77

Problems

1. A liquid was observed to distill between 206 and 207.5°C when the atmospheric pressure was 743 Torr. The thermometer used was a total-immersion thermometer, and the exposed length was equivalent to 225°C. Room temperature was about 27°C. Calculate the expected normal boiling point of the liquid. What factors contribute to uncertainty in this value?

2. A liquid was observed to distill between 121.5 and 122.5°C under the conditions described in Problem 1. Under the same conditions, toluene distilled between 109 and 109.5°C. Calculate the expected normal boiling point of the liquid. What factors contribute to the uncertainty of this value?

3. What correction should be applied when a partial-immersion thermometer is used as a total-immersion thermometer?

4. The boiling points (in °C) of a series of analogous chlorine and fluorine compounds are presented in the following table. Rationalize the different trends shown by the different halogens.

	CH_4	CH_3X	CH_2X_3	CHX_3	CX_4
Fluorine	−161	−78	−51	−82	−129
Chlorine	−161	−24	40	62	76

Which trend would you expect the analogous bromine compounds to follow?

5. The following table presents values for the enthalpy and entropy of vaporization for several aliphatic carboxylic acids. How do you interpret these data?

Compound	ΔH_{vap} *(cal/mole)*	ΔS_{vap} *(cal/mole-degree)*	*b.p.* (°*C*)
Formic acid	5,318	14.26	101
Acetic acid	5,558	14.19	118
Propionic acid	9,998	24.14	141
n-Butyric acid	10,780	24.70	163
i-Butyric acid	10,630	24.92	153
Valeric acid	11,890	25.90	183

6. Rationalize the differences in boiling points of members of each group of compounds.

a. 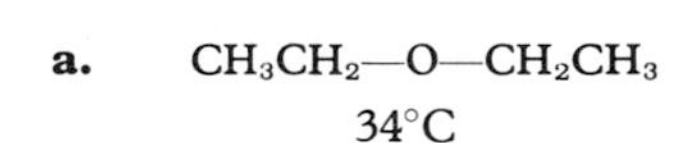

34°C

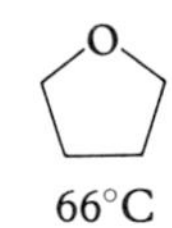

66°C

b. $CH_3-\overset{\overset{O}{\|}}{C}-O-CH_2CH_3$ 77°C; γ-butyrolactone 206°C

c. cyclohexylamine (NH_2) 134°C; cyclohexanol (OH) 160°C; aniline (NH_2) 184°C; phenol (OH) 182°C

d. $C_6H_5CH_2CH_2CH_3$ 158°C; $C_6H_5CH_2CH{=}CH_2$ 157°C; $C_6H_5CH{=}CHCH_3$ 177°C

7. Assign structures to the following substances on the basis of their boiling points.
   a. Isomers of molecular formula C_3H_9N. *Boiling points:* 49°, 35°, 35°, and 3°C.
   b. Isomers of molecular formula $C_4H_{10}O$. *Boiling points:* 118°, 108°, 100°, 83°, 39°, 34°, and 31°C.
   c. Methyl ethers of molecular formula $C_5H_{12}O$. *Boiling points:* 70°, 61°, 61°, and 55°C.

Reference

1. A discussion of the origin of these forces is presented by G. M. Barrow in *Physical Chemistry,* 4th edition, McGraw-Hill, New York, 1979, pp. 541–558.

17 Melting Point

The normal melting point of a solid is defined as the temperature at which the solid and liquid are in equilibrium at a total pressure of 1 atmosphere (the vapor pressure of the solid is usually very much less than 1 atmosphere at the melting point). In contrast to the volume change that accompanies the vaporization of a liquid, the change in volume that takes place upon the melting of a solid is very small. Hence, the melting point of a solid, unlike the boiling point of a liquid, is practically independent of any ordinary pressure change. Since the melting point of a solid can be easily and accurately determined with small amounts of material, it is the physical property that has most often been used for the identification and characterization of solids.

17.1 EXPERIMENTAL DETERMINATION OF THE MELTING POINT

There are several methods by which melting points can be determined, and the choice of method depends mainly upon how much material is available.

Melting Points from Cooling Curves

If large amounts of the solid are available (a gram or so), the most accurate method for the determination of the melting point is to heat the sample until it is melted (probably by means of an oil bath, Section 32.1) and then allow it to cool slowly for crystallization, keeping track of the temperature of the sample as a function of time by means of an immersed thermometer or thermistor. At first, the temperature will be observed to fall as the liquid loses heat to the surroundings. When crystallization begins, however, the heat evolved during this process (ΔH_f, the heat of fusion) will maintain the temperature at a constant value until crystallization is complete. At this point, the temperature will again fall as the solid loses heat to the surroundings. If the material is pure, the temperature of the sample remains constant during the entire process of solidification; this temperature is the melting point. Figure 17.1 illustrates this expected cooling curve for a pure substance. It is not unusual for the temperature to fall a little below the melting

Figure 17.1. Expected cooling curve for a pure substance.

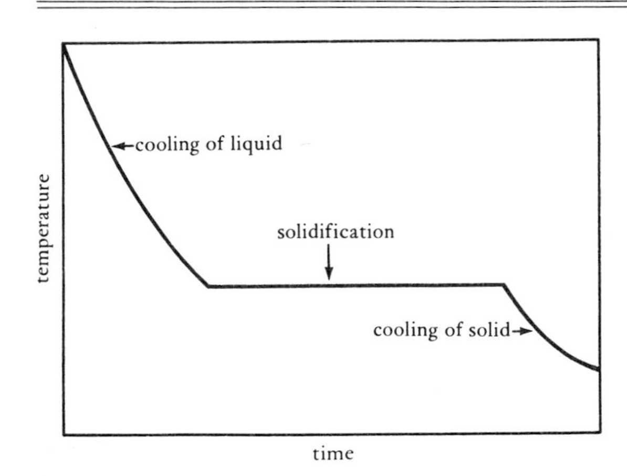

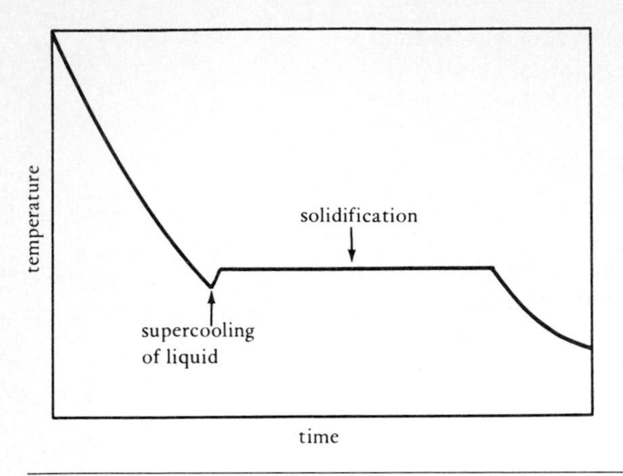

Figure 17.2. Expected cooling curve, with supercooling, for a pure substance.

Supercooling

point before crystallization begins. When this happens, the sample is said to have *supercooled*. The heat evolved as crystallization takes place then warms the sample to the melting point (see Figure 17.2).

This procedure can also be used for determining the freezing point of substances that are liquids at room temperature. The liquid is cooled in a cold or refrigerated bath. During the cooling time, the temperature of the bath should be only a little below the expected freezing point. (Several trials can be made with the same sample, and the temperature of the bath can be adjusted according to the results of the previous trial.)

Thermometer calibration

This procedure is the one that should be used for calibrating a thermometer or checking the calibration of a thermometer by means of solids of known melting point. In this case, any disagreement between the reading of the thermometer and the true melting point, which is assumed to be known, is attributed to an error in the calibration of the thermometer. Table 17.1 lists a number of substances of known melting point that have been recommended for use in this method of calibration.

Table 17.1. Reference substances for calibration of thermometers by melting point determination

Compound	*Melting Point* (°C)
Ice	0.0
m-Dinitrobenzene	89.7
Benzoic acid	121.7
Salicylic acid	160.4
3,5-Dinitrobenzoic acid	205
Sym.-di-*p*-tolyl urea	268

Capillary Melting Points

A much smaller sample is required for the method that is most often used for the determination of the melting point of a solid. A few crystals of the compound are placed in a thin-walled capillary tube 10–15 cm long and about 1 mm in inside diameter, which has been sealed at one end. The capillary, which contains the sample, and the thermometer are then suspended in an oil or air bath that can be heated slowly and evenly. The temperature *range* over which the sample is observed to melt is taken as the melting point. Obviously, if you are to know the temperature at which the crystals are melting, the thermometer and sample must be at the same temperature while the sample melts. This requires that the rate of heating of the bath be very low as the melting point is approached (about 1 degree per minute). Otherwise, the temperature of the mercury in the thermometer bulb and the temperature of the crystals in the capillary will not be the same as the temperature of the bath liquid, and probably not equal to each other. The transfer of heat energy by conduction takes place rather slowly.

A melting range

If the approximate temperature at which the sample will melt is not known, a preliminary melting point determination should be made in which the temperature of the bath is raised quickly. Then the more accurate determination should be carried out, with a low rate of heating near the melting point. A preliminary melting point can be determined within ten minutes, which is the time it would take to raise the temperature of the bath 10 degrees at 1 degree per minute. It should be obvious that if you cannot estimate the melting point within about 20 degrees, then two melting point determinations (one fast, one slow) will take less time than one determination that is slow over a wide range.

Figure 17.3. Arrangement of sample and thermometer for melting point determination.

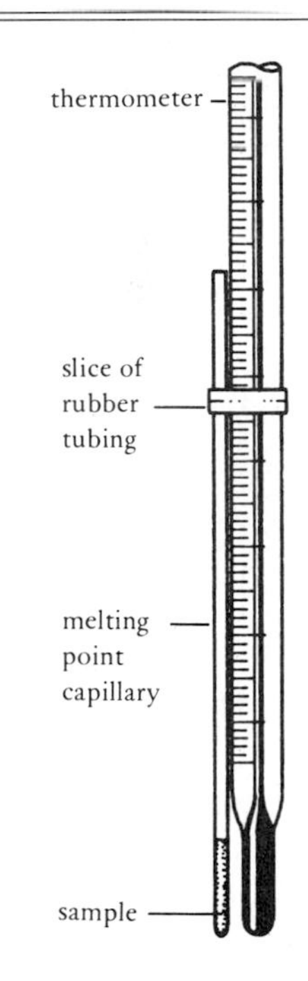

Usually, the melting point capillary can be filled by pressing the open end into a small heap of the crystals of the substance, turning the capillary open end up, and vibrating it by drawing a file across the side to rattle the crystals down into the bottom. If filing does not work, drop the tube, open end up, down a length of glass tubing about 1 cm in diameter (or a long condenser) onto a hard surface such as a porcelain sink, stone desk top, or the iron base of a ringstand. The solid should be tightly packed to a depth of 2–3 mm. If the sample sublimes rapidly at the melting point, it will be necessary to seal the capillary before attempting to determine the melting point; in an unsealed tube, the sample would sublime to the cooler part of the capillary above the bath. The capillary can be melted shut about 2 cm above the sample by briefly holding that part of the capillary in a small burner flame. The entire sealed portion should be immersed in the bath during heating. When an oil bath is used, the capillary can be fastened to the thermometer by means of a small slice of rubber tubing used as a rubber band (see Figure 17.3); if the capillary is straight, it may stick to the thermometer by the capillary action of the bath oil without the help of a rubber band. If a compound begins to decompose near the melting

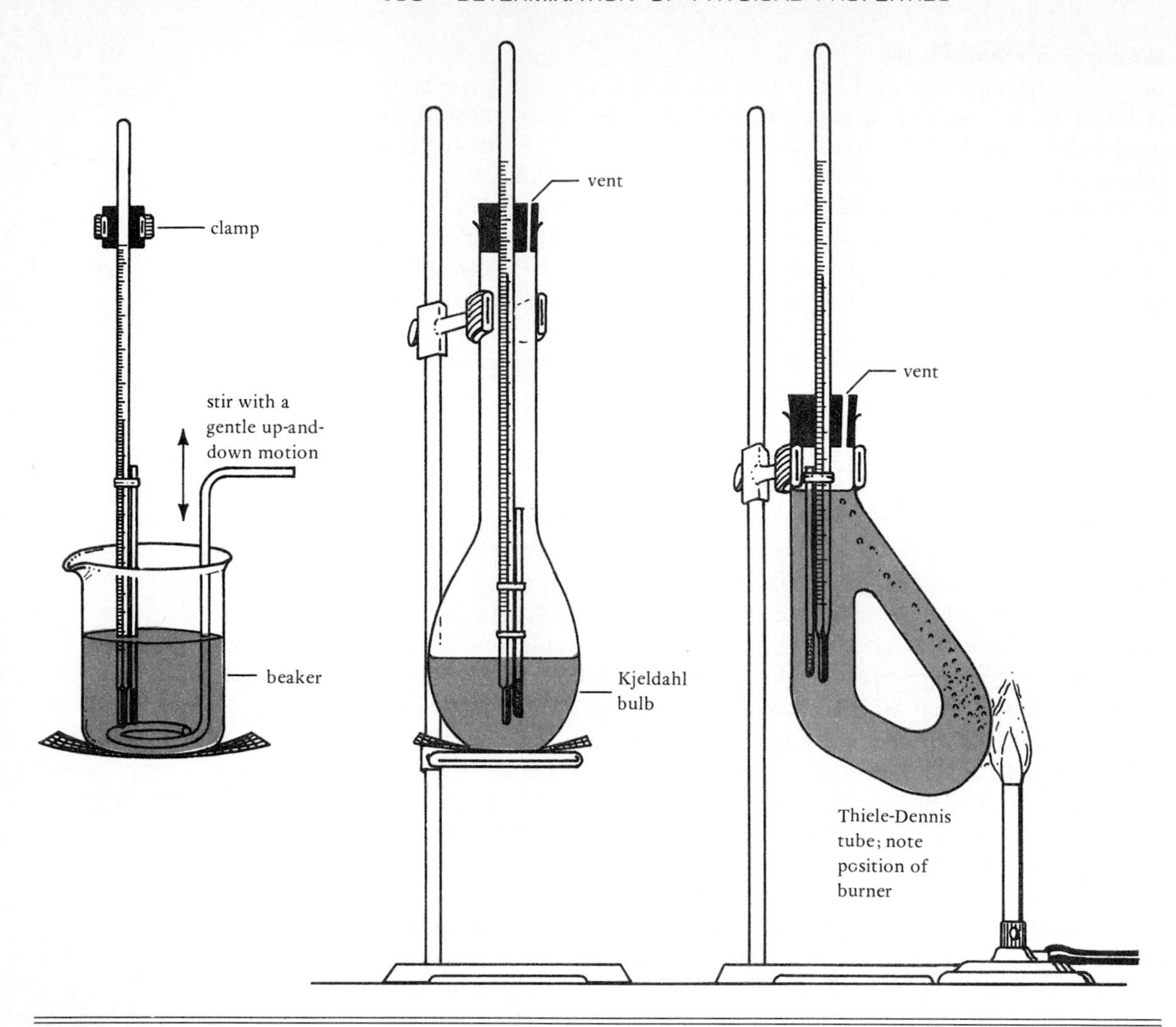

Figure 17.4. Heating baths for melting point or boiling point determination.

point, the capillary with the sample should be placed in the bath after the temperature has been raised to within 5 or 10 degrees of the expected melting point, so as to minimize the length of time that the sample is heated.

Melting point baths

There are many types of oil baths that can be used in a melting point determination, as well as in a boiling point determination. The simplest use a burner flame and depend upon convection for mixing; the more elaborate use an electric immersion heater and are stirred. For greater accuracy and convenience, the latter type is required. It is very easy to heat at a low and steady rate with an electric immersion heater, but almost impossible with a flame. Several heating baths are illustrated in Figure 17.4. It is dangerous to exceed 200°C with a bath oil such as mineral oil. Other liquids such as silicone oils, which have higher flash points, can be used at higher temperatures.

Thermometer correction

As in the determination of the boiling point, the emergent stem correction of the thermometer must be taken into account (see Section 16.1), as well as any correction for error in calibration of the thermometer.

Capillary melting points are properly compared with one another, but occasionally they are considerably different from melting points determined from cooling curves.

Micro Hot-Stage Melting Points

A fast and easy method for the determination of the melting point of a solid is to heat a few crystals of the sample between a pair of microscope cover glasses on an electrically heated metal block while observing the crystals with the aid of a magnifying glass. This method is very convenient and has the advantage of requiring as little as a single crystal, permitting good temperature control, and being very convenient. However, complete thermal equilibrium between the sample, block, and thermometer is not possible, since the thermometer is inside the block and the sample is on the surface, exposed to the cooler atmosphere. For this reason, observed block melting points often appear to be higher than capillary melting points; the higher the melting point, the greater the difference. However, a melting point quickly determined on a block can serve as an approximate melting point for the determination of a capillary melting point.

Melting range

Although a pure solid would be expected to have a single temperature at which solid and liquid are in equilibrium, most samples appear to melt over a small temperature *range*. This occurs because, with capillary or block melting points, the temperature of the bath or block rises a little during the time it takes the sample to melt. The presence of impurities in the sample can also cause the sample to melt over a range of temperatures, as explained in Section 17.2. Thus, the "melting point" will usually be reported as two temperatures between which the sample was observed to melt, a *melting range*.

17.2 THE MELTING POINT AS A CRITERION OF PURITY

A dilute solution of a liquid begins to freeze at a temperature somewhat lower than the freezing point of the pure liquid. The lowering of the freezing point (assuming that the material that separates out is the pure solvent) is given by

$$\Delta T = \frac{RT_f^2}{\Delta H_f} x_B$$

where ΔT is the difference between the freezing points of the dilute solution and the pure solvent, T_f is the freezing point of the pure solvent, ΔH_f is the heat of fusion of the pure solvent, and x_B is the mole fraction of the solute (impurity) (1). Thus, the presence of an impurity makes itself known by a reduction of the freezing point of the sample. As the pure solvent crystallizes from solution, the concentration of the impurity must increase (x_B increases), and the freezing point of the solution must fall. The result is that the cooling curve for such a solution will appear as in Figure 17.5: freezing not only starts at a lower temperature, but as Figure 17.5 shows, it also becomes complete only over a range of temperatures.

Thus, a "sharp" melting point (actually, a melting *range* of less than 1°C) is often taken as evidence that the sample is fairly pure, and a wide melting range is evidence that it is not pure. If the identity of the substance has been established by other experiments, the degree of purity can be estimated not only from the melting range, but from the difference between the actual temperature at which melting is complete (or freezing begins) and the known melting point of the substance, or ΔT in the equation. Since ΔT is inversely proportional to ΔH_f, as well

Figure 17.5. Expected cooling curve for an impure substance. Both solid and liquid are present between A and B.

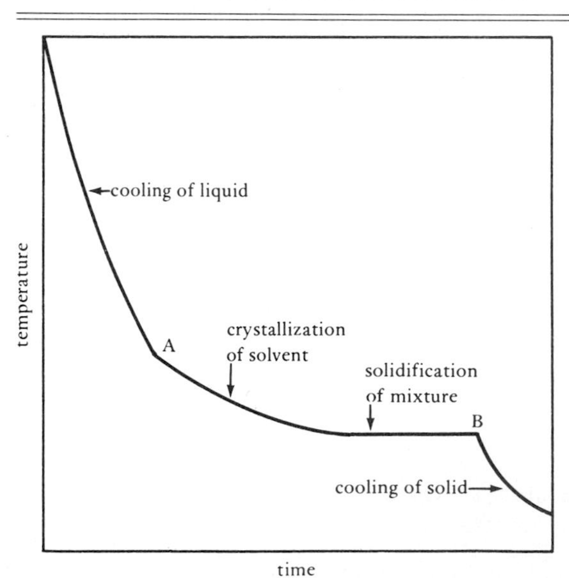

as proportional to x_B, a given level of impurity, x_B, can result in a large depression of the freezing point (when ΔH_f is small) or a small depression of the freezing point (when ΔH_f is large).

17.3 THE MELTING POINT AS A MEANS OF IDENTIFICATION AND CHARACTERIZATION

If two substances have different melting points (outside of experimental uncertainty), they must have different molecular structures or different configurations, as with *cis* and *trans* isomers or other diastereomers. If the melting points of two substances are the same (within experimental uncertainty), the molecular structures may be, but are not necessarily, the same; even with the same structure, the configurations may be enantiomeric. These statements hold, of course, only for pure substances. Recall that "sharp" melting points as determined by experiment are often taken as satisfactory evidence of purity.

For capillary melting points, experimental uncertainty is derived not only from any impurities in the sample and the procedure by which the melting point is determined, but also from the subjective interpretation of what is seen. The precise point at which melting starts is often difficult to distinguish from shrinking or sintering. Even with good technique, the uncertainty of the temperature at which melting begins may be several degrees; agreement between different determinations by different people on the same sample by the same method may not be within a degree.

17.4 MIXTURE MELTING POINTS

Mixtures of different substances generally melt over a range of temperatures, and melting is usually complete at a temperature that is below the melting point of at least one of the components. Thus, the nonidentity of two substances of the same melting point can often be established by determining the melting point of mixtures of the two. If each individual sample melts "sharply" (and at the same temperature, of course), and if an intimate mixture of the two, made by rubbing approximately equal amounts together, melts over a wide range, the two substances are not the same. Usually you wish to establish the identity rather than the nonidentity of two samples, so it is unfortunate that the converse is not always true: the absence of a depression of the melting point or of a wide melting range of the mixture is not certain evidence that the two substances are identical in molecular structure and configuration. Examples of the various types of behavior that can be observed for mixtures of different substances, as well as their interpretation in terms of phase diagrams, are given in References 2 and 3.

Double melting points and the behavior of polymorphic forms are also described in *Technique of Organic Chemistry* (3).

17.5 MELTING POINT AND MOLECULAR STRUCTURE

Systematic variations of melting point with changes in structure are not as obvious or predictable as are the variations of boiling point. Although melting points do generally increase with increasing molecular weight, the first members of homologous series often have melting points that are considerably different from what would be expected on the basis of the behavior of the higher homologs. In some homologous series of straight-chain aliphatic compounds, melting points alternate: the melting point of successive members of the series is higher or lower than that of the previous member, depending on whether the number of carbon atoms is even or odd. Sometimes, as with the normal alkanes, the melting points of successive members of the series always increase, but by a larger or smaller amount, depending upon whether the number of carbons is even or odd. As with boiling points, compounds with polar functional groups generally have higher melting points than compounds with nonpolar functional groups, but in contrast to the case with boiling points, highly branched or cyclic molecules (relatively symmetrical molecules) tend to have higher melting points than their straight-chain isomers. The combined effects of branching or the presence of rings, then, are to reduce the range of temperature over which the liquid can exist at a vapor pressure of less than 760 Torr. In extreme cases, a liquid range does not exist at a vapor pressure of less than 760 Torr; at atmospheric pressure, the substance will sublime without melting. Hexachloroethane and perfluorocyclohexane behave in this way.

Alternation of m.p. in homologous series

Polarity; Symmetry

```
    Cl  Cl
    |   |
Cl—C———C—Cl
    |   |
    Cl  Cl
```
hexachloroethane

Melting Point and the Enthalpy and Entropy of Fusion

Like the boiling point of a liquid, the absolute temperature at which a substance melts equals the heat of fusion, ΔH_f, in calories per mole, divided by the entropy of fusion ΔS_f, in calories per mole-degree:

$$T_f = \frac{\Delta H_f}{\Delta S_f} \qquad (17.5\text{-}1)$$

That this should be so can be seen most quickly by realizing that at the melting point the liquid and solid phases are in equilibrium, which means that at this temperature no change in free energy will accompany the conversion of liquid to solid (or solid to liquid), i.e., $\Delta G_f = 0$. Since

$$\Delta G_f = \Delta H_f - T_f \Delta S_f = 0$$

then

$$\Delta H_f = T_f \Delta S_f$$

and Equation 17.5-1 follows from this. Another way of looking at this is to say that, given definite values for ΔH_f and ΔS_f, there is only one temperature at which ΔH_f can equal $T\Delta S_f$ (that ΔG_f can equal 0; that the two phases can be in equilibrium). This temperature is called the melting point, T_f.

The Enthalpy and Entropy of Fusion and Molecular Structure

The increase in boiling point with molecular weight, and the influence of the nature of the functional group and degree of branching on the boiling point, are interpreted as the effect of structure on the heat of vaporization; to a first approximation, the entropy of vaporization has a constant value of 20–22 calories per mole-degree. With melting points, however, both the heat of fusion and the entropy of fusion can vary widely, even for apparently closely related compounds. Table 17.2 presents the melting points, heats of fusion, and entropies of fusion of several sets of related compounds.

Table 17.2. Melting points, heats of fusion, and entropies of fusion of selected compounds

	Melting Point		ΔH_f	ΔS_f
Compound	°C	*K*	(*cal/mole*)	(*cal/mole-degree*)
n-Butyl alcohol	−87.3	185.9	2130	11.48
t-Butyl alcohol	25.2	298.4	1620	5.44
Diethyl ether	−116.2	157.0	1649	10.50
o-Dichlorobenzene	−16.7	256.5	3080	12.02
m-Dichlorobenzene	−26.2	247.0	3010	12.19
p-Dichlorobenzene	53	326	4340	13.31
o-Dibromobenzene	1.8	275.0	2030	7.38
m-Dibromobenzene	−6.9	266.3	3150	11.84
p-Dibromobenzene	87	360	4780	13.27
o-Xylene	−25.2	248.0	3250	13.11
m-Xylene	−47.9	225.3	2765	12.27
p-Xylene	13.3	286.4	4090	14.28
Cyclohexane	6.6	279.7	673	2.28
Methylcyclohexane	−126.6	146.6	1613	11.01
Cyclohexanol	23.5	296.7	406	1.37
Phenol	41	314	2771	8.82
n-Octane	−56.8	216.4	4957	22.91
2-Methylheptane	−109.0	164.1	2451	14.94
2,3,3-Trimethylpentane	−100.7	172.5	370	2.12
2,2,3,3-Tetramethylbutane	100.7	373.9	1802	4.82

The other isomeric octanes (values for four isomers are not reported) have melting points between −126 and −91°C, heats of fusion between 3070 and 1700 cal/mole, and entropies of fusion between 17.8 and 10.69 cal/mole-degree.

The higher melting point of *n*-butyl alcohol compared to diethyl ether is a consequence of the larger heat of fusion of the more polar alcohol. Similarly, the higher melting points of the *p*-disubstituted benzenes compared to the *ortho* and *meta* isomers is due to the larger heat of fusion of the *para* isomer, since the differences in the entropy of fusion among the isomers are either small or in the wrong direction to account for the facts.

The unusually high melting point of cyclohexane compared to methylcyclohexane is primarily the result of an unusually low entropy of fusion for cyclohexane. Similarly, the relatively high melting point of *tert*-butyl alcohol compared to its isomers is the result of a relatively low entropy of fusion. The entropy of fusion is primarily the result of increased freedom of rotation of the molecule as a whole, and of parts of the molecule with respect to other parts, in the liquid phase compared to the solid phase. The increase in vibrational and translational entropy upon melting is relatively small; the large increase in translational entropy occurs upon vaporization, and there is considerable vibrational motion possible in the crystal. A small entropy of fusion, then, implies either that the molecule has relatively little rotational freedom in either the solid or liquid state (cyclic, polycyclic molecules), or that it can gain certain degrees of rotational freedom, with their associated increase in entropy, in the solid state without melting (highly branched, highly symmetrical molecules). Cyclohexane is known to have a transition at 185.9 K with an associated entropy change of 8.63 cal/mole-degree, and similar transitions have been observed for other highly symmetrical molecules.

The data for the isomeric octanes show how complex and variable the relationships can be between structure and melting point, heats of fusion, and entropies of fusion.

Problems

1. **a.** Calculate the freezing point depression, ΔT, for a sample of *tert*-butyl alcohol that contains 0.01 mole fraction of an impurity.
   b. Do the same for a similar solution of cyclohexanol.
2. **a.** A sample of a substance of known molecular structure starts to freeze at 57.00°C and is half frozen at 56.80°C. What would be the melting point of the pure substance?
   b. The sample in part **a** is remelted, and sufficient solute of known molecular weight is added to give a solution containing 1 mole percent of this solute (you must assume that the solvent is pure). The sample now starts to freeze at 56.90°C. What mole fraction of impurity was present in the original sample?

References

1. G. M. Barrow, *Physical Chemistry*, 4th edition, McGraw-Hill, New York, 1979, p. 297.
2. A. I. Vogel, *A Textbook of Practical Organic Chemistry*, 3rd edition, Wiley, New York, 1957, p. 21.
3. E. L. Skau, J. C. Arthur, and H. Wakeham, *Technique of Organic Chemistry*, Vol. I, Part I, 3rd edition, A. Weissberger, editor, Interscience, New York, 1959, p. 287.

18 Density; Specific Gravity

Before the advent of the spectroscopic methods, density was one of the most important physical properties by which liquids were characterized. Since the density is often reported and tabulated, especially in the older literature, and is fairly easy to determine, it is still often used to characterize and identify liquids.

18.1 EXPERIMENTAL DETERMINATION OF THE DENSITY

The *density* is defined as the *mass per unit volume;* the units usually used by chemists are grams/mL. The determination of density, then, requires the determination of both the weight and the volume of a sample of the substance. Since density is a function of temperature, the temperature at which the density is determined should also be recorded. For most organic liquids, the density decreases by about 0.001 g/mL per degree increase in temperature.

Temperature dependence

The weight of the sample is determined by weighing the container when it is empty and then when it is filled (completely or to the mark) with the liquid. The difference gives the weight of the sample at the temperature T, or w^T_{sample}. The volume of the sample is determined by cleaning and refilling the container with water and reweighing. Subtracting the weight of the empty container gives the weight of an equal volume of water at the temperature T, or $w^T_{H_2O}$. From the known density of water at the temperature of the samples, $d^T_{H_2O}$, the volume of the two samples can be calculated:

Weight

Volume

$$\text{Density of water at } T\,°\text{C} = \frac{w^T_{H_2O}}{\text{volume of sample}}$$

or

$$\text{Volume of sample} = \frac{w^T_{H_2O}}{d^T_{H_2O}}$$

Table 18.1. Density of water between 20° and 30°C

Temperature	*Density*
20	0.9982
21	0.9980
22	0.9978
23	0.9975
24	0.9973
25	0.9970
26	0.9968
27	0.9965
28	0.9962
29	0.9959
30	0.9956

The density d of the sample at T°C can now be calculated:

$$d^T_{\text{sample}} = \frac{\text{wgt of sample}}{\text{vol of sample}} = \frac{w^T_{\text{sample}}}{w^T_{H_2O}/d^T_{H_2O}} = \frac{w^T_{\text{sample}}}{w^T_{H_2O}} \times d^T_{H_2O} \qquad (18.1\text{-}1)$$

Table 18.1 gives the density of water at several temperatures.

The container used for density determinations can be as simple as a volumetric flask. A pycnometer, which is essentially a U-shaped tube, is often used with large samples, although small pycnometers may be constructed from a piece of glass tubing. The pycnometer is most easily filled by drawing in the liquid with gentle suction until it is filled past the mark, and adjusting to the mark by touching a piece of filter paper to the capillary tip; the excess will be drawn out into the filter paper by capillary action. A 1- or 2-mL specific gravity bottle can be used for small samples. The stopper for such a bottle has a capillary that allows it to be filled (to the top of the capillary) without trapping any air bubbles inside the bottle. (See Figure 18.1). It is possible to use a 0.05-mL syringe as a container if the sample is very small. In this case, the weighings must be done to 0.1 milligram.

Sometimes it is desirable to determine the density at a particular temperature, rather than at the temperature of the laboratory. If so, the container and sample must be brought to the desired temperature by storage in a bath maintained at this temperature. The final small adjustment to the desired volume must be done after thermal equilibrium is attained.

Specific gravity, which is related to density and sometimes confused with density, is defined as the ratio of the weight of a certain volume of liquid to the weight of an equal volume of water:

$$\text{Specific gravity} = \frac{w^T_{\text{sample}}}{w^T_{H_2O}} \text{ (dimensionless)}$$

Comparison of this expression with Equation 18.1-1 will show that the density equals the specific gravity times the density of water. Since the density of water is practically equal to 1, the density and specific gravity have almost the same value.

The density of a 0.5-mL sample can conveniently be determined by means of the *Fisher-Davidson gravitometer.* In effect, the density of the unknown liquid is determined by comparing the height to which a column of unknown liquid can be raised by a slight vacuum with the height to which a liquid of known density (ethylbenzene) can be raised by the same slight vacuum. Since the heights are inversely proportional to the densities, and the density of ethylbenzene is known, the unknown density can be calculated. The instrument is actually calibrated in units of grams/mL.

Figure 18.1. A specific gravity bottle. As the stopper is put in, the air escapes through the capillary. "Full" means filled to the top of the capillary.

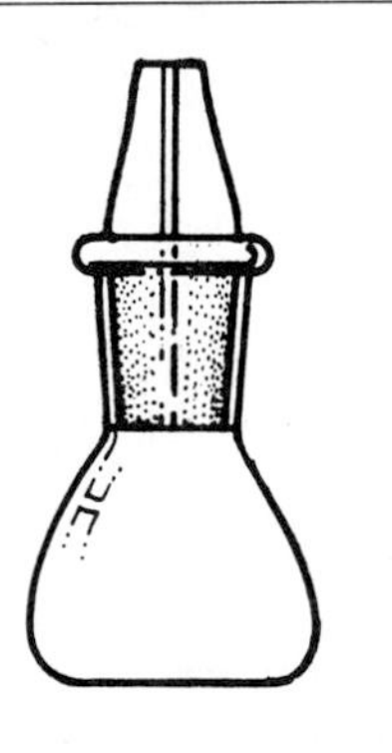

18.2 DENSITY AND MOLECULAR STRUCTURE

The density of a substance is determined mainly by the atomic weight of its constituent atoms: high atomic weight results in high density.

Most organic liquids have densities between 0.8 and 1.1 at room temperature. The structural possibilities for liquids that are more or less dense than this are quite limited. Aliphatic acyclic hydrocarbons, saturated and unsaturated, and aliphatic acyclic ethers and amines generally have densities less than 0.8. Aromatic hydrocarbons usually have densities between 0.86 and 0.9. Iodides and bromides have densities greater than 1.1, while alkyl chlorides have densities less than 1.

Compounds of density greater than 1.2 usually have at least one iodine or bromine atom, or two or more chlorine atoms. Compounds with two or more functional groups, especially compounds with relatively large intermolecular forces (relatively high boiling for their molecular weight), generally have densities greater than 1.

As the ratio of $—CH_2—$ units per functional group increases, the densities tend toward a value between 0.8 and 0.9.

Problems

1. The weight of a certain volume of a liquid at 25°C was determined to be 2.339 grams, and the weight of an equal volume of water at the same temperature was found to be 2.013 grams.
   a. Calculate the density (d^{25}) of the liquid.
   b. Calculate the specific gravity (d^{25}_{25} and d^{25}_{4}) of the liquid. *Note:* This notation means, in the first case, that the temperature of both the sample and the equal volume of water is assumed to be 25°C, and, in the second case, that the temperature of the sample is 25°C, but that of the water is 4°C.
2. How would densities determined on the moon differ from those determined on earth?

19 Index of Refraction

Along with the boiling point and density, the *index of refraction* is one of the physical properties most often used to identify and characterize liquids.

19.1 EXPERIMENTAL DETERMINATION OF THE INDEX OF REFRACTION

The index of refraction of a substance is the ratio of the speed of light in a vacuum to the speed of light in the substance. It is often measured by making use of the fact that the critical angle of refraction of light passing from one medium to another is a function of the refractive indexes of the two media. The Abbe refractometer in effect measures the critical angle for refraction of light passing from a liquid of unknown refractive index to a glass prism of known refractive index. It is calibrated in units of index of refraction (a dimensionless ratio). Since the index of refraction is a function of both the wavelength of light and the temperature, these must both be specified along with the measured refraction. The D line of sodium, $\lambda = 5890$ nm, is the wavelength for which the index of refraction is often reported, and at the temperature T, the index of refraction determined at this wavelength would be reported as n_D^T. The index of refraction of most organic liquids decreases between 3.5 and 5.5×10^{-4} per degree increase in temperature.

Wavelength dependence and temperature dependence

Abbe refractometer

With the Abbe refractometer, two drops of sample are required in the space between the prisms, and the instrument is adjusted until the field seen through the eyepiece appears as shown in Figure 19.1. The intersection of the cross-hairs should be on the border between the light and dark sections of the field, and the compensator should be set to sharpen and achromatize the border between the light and dark sections until the difference is as sharp and as near black and white as possible. When the cross-hairs and border are lined up as shown in Figure 19.1, the index of refraction is read from the scale. The value is reliable to ± 0.0002, provided that the instrument is properly calibrated. The calibration may be checked by determining the index of refraction of water, which is 1.3330 at 20°C and which decreases 0.0001 unit per degree increase in temperature between 20° and 30°C. With the compensator, the index of refraction determined with ordinary (white) light is very close to that which would be obtained using the light from a sodium lamp.

If no sharp boundary can be observed, insufficient sample was used, or the sample evaporated before the adjustment was completed. In making determinations on very volatile liquids, you can introduce the sample with a fine dropper through a little channel that leads to the space between the prisms, without having to open them up.

After use, the sample should be removed by means of a soft tissue and a little acetone or ethanol.

The thermometer indicates the temperature of the sample space. It can be maintained at a particular value by circulating water from a constant-temperature bath through the prism housings, using the hose connections provided.

Figure 19.1. View through the eyepiece of the Abbe refractometer. When the refractometer is set to be read, the intersection of the cross-hairs falls on the border between the light and dark fields; the border should be as sharp and free from color as possible.

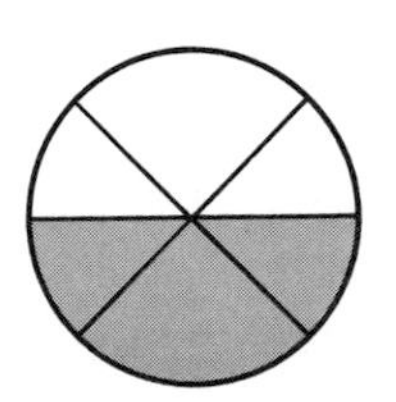

19.2 INDEX OF REFRACTION AND MOLECULAR STRUCTURE

The index of refraction of a substance is determined mainly by the polarizability of its constituent atoms and functional groups: the presence of atoms of higher atomic number and of conjugated unsaturation results in a higher index of refraction. Alkyl iodides and aromatic compounds usually have an index of refraction greater than 1.5000; most other compounds give a value lower than this.

As with the boiling point and density, the main use of experimental values for the index of refraction is for comparison with values reported in the literature for compounds whose structure is known. It is possible, however, to calculate the expected index of refraction for a substance from its molecular structure through the following relationship:

Molecular refraction

$$\text{Molecular refraction} = Mn_{\mathrm{D}}^{T} = \text{molecular weight} \times n_{\mathrm{D}}^{T}$$

Solving for n_{D}^{T} shows that the index of refraction equals the molecular refraction divided by the molecular weight. Contributions of individual structural features (atoms or bonds) to the molecular refraction can be found in Reference 1.

In a similar way, a related property, the molecular refractivity, can be estimated from independent contributions of structural features (1). The estimated value can then be compared with the experimental value, which depends upon the index of refraction and the density in this way:

Molecular refractivity

$$\text{Molecular refractivity} = R_{\mathrm{D}} = \frac{n^2 - 1}{n^2 + 2} \cdot \frac{mw}{d}$$

where n is the index of refraction at the wavelength of the D line of sodium, d is the density of the liquid at the same temperature at which the index of refraction was determined (preferably near 20°C), and mw is the molecular weight of the substance.

The fact that the molecular refraction and molecular refractivity can be estimated from the structural formula of a compound makes it possible to use the index of refraction (or index of refraction and density) as criteria of identity even though literature values may not be available.

Reference

1. A. I. Vogel et al., *J. Chem. Soc.* **1952** (514).

20 Optical Activity

An additional physical property that is of use in the identification and characterization of optically active solids or liquids is the *optical rotation*.

20.1 EXPERIMENTAL DETERMINATION OF OPTICAL ROTATION

The polarimeter

The optical rotation is measured by preparing a solution of the substance and then determining, by the use of a *polarimeter,* the direction and degree to which the plane of polarization of a beam of plane-polarized light is rotated upon passage through the solution. The polarimeter consists essentially of a polarizing prism (polarizer), which transmits from the monochromatic light source a beam of plane-polarized light, a trough to hold the sample tube, and a second polarizing prism (analyzer) that can be rotated so as to exactly compensate for the rotation that the solution induces in the polarized light beam. The correct position of the analyzer is determined by looking through the central eyepiece at the light transmitted by the instrument. Figure 20.1 indicates the appearance of the field for possible positions of the analyzer with respect to the plane of polarization of the beam. The angular displacement of the analyzer is read in degrees from the graduated circle with the aid of an auxiliary eyepiece and vernier, and the difference between the angular position of the analyzer with the sample in and out of the trough is the observed rotation, α.

Observed rotation

Concentration

The sample is prepared by making up a solution of known concentration by dissolving a weighed amount of the substance in a volumetric flask. A concentration of 0.5–2 grams per 100 mL is usually recommended. Solvents most commonly used are water, methyl and ethyl alcohols, and chloroform. A dilute solution is most desirable, but if the rotatory power of the substance is low, a higher concentration may be necessary. If the solution is not free from dust, it should be filtered. The volume of solution required varies with the length and diameter of the sample tube. Twenty-five milliliters should be enough in any case, and tubes are available that require 1 mL or less. With a liquid, the rotation can be determined using the neat liquid.

The sample tube is usually a 1- or 2-decimeter-long glass tube with a fitting for a brass screw cap at each end. It is filled by placing one end glass against one end of the tube (glass-to-glass) and then screwing on the cap, using a rubber washer between the end glass and the cap. With the tube standing vertically on this cap, the solution is added through the other end (possibly by means of a dropper or pipet) until the rounded meniscus stands above the end of the tube. The other end glass is then slid across the end of the tube, without leaving an air bubble in the tube; if air is trapped in the tube, remove the glass, add more solution if necessary, and try sliding the glass on again. The other brass cap is then screwed on, using a rubber washer as with the first cap. Screw the ends on firmly but not too tight, as strain in the glass of the end caps may cause them to rotate the beam slightly. Some polarimeter tubes are constructed so that they can be filled through a port in the middle with both ends capped. After placing the tube in the trough of the instrument, close the cover.

Figure 20.1. View through the eyepiece of the polarimeter. The analyzer should be set so that the intensity of all parts of the field is the same (b). When the analyzer is displaced to one side or the other, the field will appear as in (a) or (c).

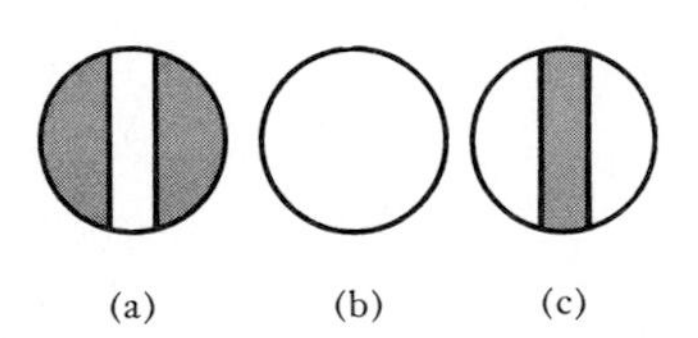

The image can be brought into focus, if necessary, by moving the eyepiece in or out. Distortion of the image may be due to an air bubble in the light path, or to the presence of density gradients in the sample, which are caused by variations either in concentration or in temperature.

The magnitude of the observed rotation, α, depends upon several variables that must therefore be recorded along with the observed rotation.

- *Concentration:* It is reasonable that the greater the concentration of molecules in the path of the beam, the greater should be the rotation. However, in many cases, the increase in rotation is not exactly proportional to the concentration. This is explained by postulating that solute–solute interactions, which are more prevalent at higher concentrations, affect the rotation differently from solute–solvent interactions, which are predominant at low concentrations.
- *Length of sample tube:* The rotation of the beam will be proportional to the number of molecules in the path; for a given concentration, the number of molecules in the path will be proportional to the length of the path. The path length is conventionally expressed in decimeters.
- *Solvent:* The observed rotation of the same substance at the same concentration in different solvents can be very different, even of opposite sign!
- *Temperature:* A change in concentration accompanies the expansion and contraction that a solution undergoes with change in temperature. This experimental variable can be eliminated by making up the solution at the temperature of the room that contains the polarimeter or the temperature at which the optical rotation will be determined. (Polarimeter tubes are available that can be thermostatted by passing water from a constant-temperature bath through a jacket surrounding the tube.) It can also be calculated from the coefficient of thermal expansion of the solvent. The most popular explanations of the rather large changes in the observed rotation that occur with changes in temperature are (1) the change with temperature of the relative populations of conformational isomers, which have different rotatory powers, and (2) the change with temperature of the solvation of the optically active molecules.
- *Wavelength of light:* The magnitude (and also the sign, in many cases) of the rotation depends on the wavelength of light used. Since the D line of sodium was used early, most rotations are reported at this wavelength, 5890 nm. Generally, the magnitude of rotation is larger at shorter wavelengths, and for this reason and the fact that the human eye (especially the dark-adapted eye) is

more sensitive to green light, the green line of the mercury arc (5461 Å) has been recommended for use.

Specific rotation

The optical rotation is usually reported as the specific rotation, $[\alpha]$:

for solutions, $$[\alpha]_D^T = \frac{100\ \alpha_D^T}{lc}$$

for pure liquids, $$[\alpha]_D^T = \frac{\alpha_D^T}{ld}$$

where $[\alpha]_D^T$ = specific rotation, at the D line of sodium at the temperature T

α_D^T = observed rotation—positive if clockwise; negative if counterclockwise—at the D line of sodium at the temperature T

l = length of sample tube, in decimeters

c = concentration of solution, in grams per 100 mL

d = density of the pure liquid at the temperature T, in grams/mL

The specific rotation $[\alpha]$ is a measure of the rotary power per unit path length per unit weight concentration. In taking your experimental data, you must report the solvent and the concentration as well, since the variation of rotation with concentration is not necessarily linear and is not usually independent of the nature of the solvent. And when comparing an experimental value of the specific rotation with a literature value, you must compare solvent and concentration as well as temperature and wavelength.

20.2 OPTICAL ACTIVITY AND MOLECULAR STRUCTURE

A compound that is optically active must be composed of molecules for which no conformation that can be attained by the molecules is the same as the mirror image of that conformation; *all* conformations must be chiral; *all* conformations must belong to point groups $\mathbf{C}_n$ or $\mathbf{D}_n$. Thus, if the molecules of a substance possess, in any conformation, a symmetry element whose corresponding symmetry operation involves reflection (a plane of symmetry, a center of symmetry, or a higher, even improper, axis of symmetry), the substance *cannot* exist in chiral forms, and no samples can be optically active.

The presence of a chiral center in the molecules of a substance permits the existence of chiral molecules and optically active samples. The most common chiral center is a carbon atom to which are bonded four different atoms or groups of atoms (a *chiral carbon*). In some rare cases, the expected optical activity will be too small to be observed, even though the molecules of the sample are known to be chiral. In other cases, the presence of a second chiral center similar to the first may give at least one conformation of the molecule, a plane or center of symmetry (a *meso form*), and samples of the meso isomer will be optically inactive. Of course, if the sample is a *racemic mixture* (an equimolar mixture of enantiomers), no optical activity will be observed.

The presence of a chiral center in the molecules of a substance is not required for molecular chirality, as illustrated by the existence of certain optically active allenes and biphenyls. Excellent discussions of the details of the relationships between molecular structure, molecular symmetry, and optical activity are presented in References 1 through 3.

If a sample is observed to be optically active, then, it is not a racemic mixture and it must be composed of chiral molecules. The absence of optical activity indicates that the sample is a racemic mixture or that its molecules possess a plane of symmetry, a center of symmetry, or a higher even improper axis of rotation in at least one conformation.*

Optical rotatory dispersion

The sign and magnitude of the rotation are a function of both the configuration of the molecule and the wavelength of the light, and a graph of the specific rotation of a compound versus wavelength is called its *optical rotatory dispersion curve* (ORD curve). Since enantiomers have equal but opposite optical rotations at all wavelengths, their ORD curves are exactly similar in shape but opposite in sign. A more useful observation is that similar or homologous compounds of the same relative configuration have similar ORD curves of the same sign. This makes it possible in certain cases to establish the relative configurations of optically active molecules without interconversion. If, in addition, the absolute configuration of one has been established, the absolute configuration of all is thereby established. A more complete discussion of the interpretation of ORD curves can be found in Reference 1.

Problems

1. The observed rotation at 25°C of a sample of α-phenylethylamine (Section 48) in a 10-cm tube, using a sodium lamp, is $-35.6°$. The density, d^{25}, of the amine is reported to be 0.953 g/mL. Calculate the specific rotation, $[\alpha]_D^{25}$.
2. The observed rotation α_D^{20} of a solution of 232 mg of cholesterol (Section 39) in 10 mL of chloroform in a 1-dm tube was $-0.73°$. Calculate the specific rotation $[\alpha]_D^{20}$.
3. How can a clockwise rotation of 10° be distinguished from a counterclockwise rotation of 350°?

(S)-(−)-α-phenylethylamine

* An exceptional case is discussed in Reference 2, p. 26.

4. For each molecular structure, determine (1) the total number of stereoisomers possible, and (2) which of the stereoisomers should be optically active.

a. 1-bromohexane
b. 2-bromohexane
c. 1,2-dibromohexane
d. 2,3-dibromohexane
e. 1,2-dibromocyclohexane
f. 1,3-dibromocyclohexane
g. 1,4-dibromocyclohexane
h. chloroallene ($CHCl{=}C{=}CH_2$)
i. 1,1-dichloroallene
j. 1,3-dichloroallene
k. 3-hexene
l. 2-chloro-3-hexene
m. 2,5-dichloro-3-hexene
n. 2,5-dichloro-3-hexyne

References

1. K. Mislow, *Introduction to Stereochemistry*, W. A. Benjamin, New York, 1965.
2. H. H. Jaffe and M. Orchin, *Symmetry in Chemistry*, Wiley, New York, 1965, Chapters 1 and 2.
3. M. Orchin and H. H. Jaffe, *J. Chem. Educ.* **47**, 246 (1970).
4. B. Testa, *Principles of Organic Stereochemistry*, Marcel Dekker, New York, 1979.
5. H. Kagan, *Organic Stereochemistry*, Halsted Press (John Wiley and Sons), London (New York), 1979.

21 Molecular Weight

One of the bits of information most useful in establishing the molecular structure of a new substance is the molecular weight. If the molecular weight is known with confidence to 1 atomic mass unit, all but a small fraction of the total structural possibilities can be eliminated by this fact alone.

If the problem is simply to establish the identity of a substance with a known compound, the determination of the molecular weight of the unknown is much less useful, since it provides only a single data point for comparison. The infrared spectrum, for example (Section 23), would provide many more points for comparison and would usually be easier to determine than the molecular weight. Of course, the molecular weight of any substance of known structure is also automatically known and thus can be compared to that of an unknown substance, whereas this is not true of the infrared spectrum and other physical and chemical properties.

21.1 MOLECULAR WEIGHT DETERMINATION BY MEANS OF MASS SPECTROMETRY

Mass spectrometry is currently the most commonly used method for the determination of molecular weight. It has the great advantages of being applicable to most substances that are at least slightly volatile (a vapor pressure of 10^{-2} Torr at a temperature of 100°C), of requiring a very small amount of sample, and of routinely giving the molecular weight to 1 atomic mass unit. The disadvantages of the mass spectrometric method are that mass spectrometers are expensive and difficult to maintain in top operating condition. Section 26.2 describes how the mass spectrum of a substance can be determined and how it can provide information about the molecular structure, as well as the molecular weight, of a substance.

21.2 MOLECULAR WEIGHT DETERMINATION BY OTHER METHODS

Other methods of molecular weight determination include measurement of freezing point depression, boiling point elevation, or osmotic pressure. In these methods, the concentration of a solution of the unknown substance is determined. Knowing the weight of the solute added, you can then calculate its molecular weight.

The equivalent weight of acidic or basic substances can be determined by titration. From the equivalents of titrant consumed, the equivalents of sample are known. Knowing the weight of sample used, you can calculate the weight per equivalent.

Methods for the determination of molecular weights by freezing point depression, using camphor and tertiary butyl alcohol as solvents, are given in References 1 and 2.

References

1. E. J. Cowles and M. T. Pike, *J. Chem. Educ.* **40,** 422 (1963).
2. M. J. Bigelow, *J. Chem. Educ.* **45,** 108 (1968).

22 Solubility

The solubility of a substance is significant for two reasons. First, before it is possible to start working with a compound in the laboratory, it is necessary to know or to be able to predict whether the substance will be soluble in water, aqueous acid, aqueous base, or organic solvents such as ether. Second, if the molecular structure of a substance is unknown, knowledge of its solubility characteristics can be the basis for certain deductions about its structure.

It must be emphasized that *solubility is a matter of degree;* that is, some compounds are more soluble than others, and the decision of where you wish to draw the line that says that substances less soluble than this will be called "insoluble" (in a particular solvent) and substances more soluble than this will be called "soluble" (in a particular solvent) is arbitrary. Very often, the line is drawn at 30 mg of substance per mL of solvent (3 g per 100 mL of solvent). Obviously there will be borderline cases, and one must look out for the use of different definitions of solubility and insolubility in the chemical literature.

22.1 SOLUBILITY OF LIQUIDS IN LIQUIDS

Mixing is normal

Although the mutual insolubility of "oil" and water is probably the most familiar example of the solubility behavior of two liquids, it is much more satisfactory to approach the phenomenon of solubility from the point of view that mutual solubility is the expected or normal behavior for two liquids, and that insolubility is exceptional or unusual behavior.

The two liquids benzene and toluene are mutually soluble in all proportions (miscible) for the same reason that red and white marbles will mix if you shake them together: the free energy of the mixed state is less than the free energy of the unmixed state because of the increase in entropy associated with mixing. The mixed state has the same potential energy as the unmixed state, but it is more probable than the unmixed state. If the intermolecular forces between unlike molecules and like molecules are the same, the heat of mixing, ΔH_{mix}, will be zero. Since the entropy of mixing, ΔS_{mix}, is always positive, the free-energy change upon mixing will, in this case, be negative.

$$\Delta G_{mix} = \Delta H_{mix} - T\Delta S_{mix} \tag{22.1-1}$$

In the special case when ΔH_{mix} equals zero, this equation becomes

$$\Delta G_{mix} = -T\Delta S_{mix} \tag{22.1-2}$$

The phenomenon of mixing is one of a very few in which the free-energy change can be considered to be primarily a function of the change in entropy rather than primarily a function of the change in enthalpy (energy). Thus, the expected or normal miscibility of liquids is observed when the heat of mixing, ΔH_{mix}, is negative (mixing is exothermic), zero, or only slightly positive (mixing is only slightly endothermic).

Insolubility

If ΔH_{mix} is large and positive (highly endothermic), it can more than compensate for $T\Delta S_{mix}$ and will cause ΔG_{mix} to be positive (Equation 22.1-1). If this is the case, the unmixed state will have a lower free energy than the mixed state, and mixing will not occur. The enthalpy of mixing, ΔH_{mix}, can be positive if the forces between like molecules are larger than the forces between unlike molecules, because it will then take more energy (work) to separate the like molecules (of both solvent and solute) to the larger average intermolecular distance of the mixed state than will be regained upon moving the unlike molecules together in the mixed state:

$$\Delta H_{mix} = E_{separation} + E_{association} \tag{22.1-3}$$

where $E_{separation}$ is the energy required to separate like molecules (always positive), and $E_{association}$ is the energy released upon moving unlike molecules together (always negative). The larger the intermolecular attractions, the greater the magnitude of each term of this equation.

Thus, the nonmiscibility of water with most organic liquids is the result of the large intermolecular attraction between water molecules (hydrogen bonding), which must be lost and remain uncompensated for in the mixed state when the organic liquid cannot form hydrogen bonds with water. It is in this way that the immiscibility of "oil" (or hexane, benzene, carbon tetrachloride, etc.) and water can be rationalized. On the other hand, methanol, ethanol, acetone, and acetic acid are miscible with water because the intermolecular attractions between the molecules of water and solvent are comparable in magnitude to the intermolecular attractions between water and water, and solvent and solvent. This results in a relatively small heat of mixing and therefore, by Equation 22.1-1, a negative free-energy change upon mixing. The often-quoted generalization that "like dissolves like" can be understood by realizing that substances whose molecules have similar intermolecular forces (type and magnitude) tend to mix.

O
‖
C
H_3C CH_3
acetone

"Like dissolves like"

Partial Solubility

Complete immiscibility, however, is an extreme behavior actually approached by few pairs of liquids; mercury and water appear to approach this extreme very closely. It is more usual for pairs of liquids that are not completely miscible to be slightly soluble in one another. For example, 100 mL of water will dissolve 7.5 g of diethyl ether, and 100 mL of ether will dissolve 1.3 g of water at 25°C; 100 mL of water will dissolve 8.5 g of *n*-butyl alcohol, and 100 g of *n*-butyl alcohol will dissolve 25 g of water at 20°C.

Partial solubility can be understood by realizing that the increase of entropy due to mixing is largest for the first material that dissolves. That is, ΔS_{mix} for going from mole fraction solvent = 1.0 to mole fraction solvent = 0.9 (or $x_{solute} = 0$ to $x_{solute} = 0.1$) is greater than ΔS_{mix} for going from $x_{solvent} = 0.9$ to $x_{solvent} = 0.8$ (or $x_{solute} = 0.1$ to

$x_{solute} = 0.2$). Since ΔH_{mix} might be expected to be approximately the same in both cases, ΔG_{mix}, which equals $\Delta H_{mix} - T\Delta S_{mix}$, can be negative for the formation of a dilute solution (when the average ΔS_{mix} per increase in concentration of solution is larger) and positive for the formation of a concentrated solution (when the average ΔS_{mix} per increase in concentration of solute is smaller). The larger and more positive the enthalpy of mixing ΔH_{mix}, the sooner ΔH_{mix} will equal $T\Delta S_{mix}$; the larger and more positive the enthalpy of mixing, the sooner the solution should be expected to become saturated with respect to the solute, the point at which ΔG_{mix} is equal to zero. Table 22.1 presents a qualitative summary of the solubility in water of organic compounds containing different functional groups.

Solubility and b.p.

Since both the heat of vaporization of a liquid (and thus the boiling point of a liquid) and the heat of mixing are functions of intermolecular forces, it is understandable that there is a correspondence between the solubility of certain organic compounds in water and their boiling points. For the four isomeric butyl alcohols, for example, the order of decreasing boiling point is the order of increasing solubility in water. In Section 16.2, lower boiling point was interpreted in terms of smaller intermolecular forces. Similarly, ΔH_{mix} would be expected to be more negative (or less positive) when less energy is required to separate the molecules from their average distance in the unmixed state to their

Table 22.1. Solubility in water of straight-chain organic compounds with different functional groups

Branching of the alkyl group increases the solubility (see Table 22.2).

	Number of Carbon Atoms						
Type of Compound	1	2	3	4	5	6	7
Alcohol	......	......	......	......	------	———	———
Aldehyde	......	......	......	------	———	———	———
Alkane	———	———	———	———	———	———	———
Alkene	———	———	———	———	———	———	———
Alkyl halide	———	———	———	———	———	———	———
Amide	......	......	......	......	------	———	———
Amine	......	......	......	......	------	———	———
Carboxylic acid	......	......	......	......	------	------	———
Methyl ester	......	......	......	......	------	———	———
Methyl ether	......	......	......	------	———	———	———
Methyl ketone	......	......	......	......	......	------	———
Nitrile	......	......	......	------	———	———	———
Thiol	......	———	———	———	———	———	———

Solubility is greater than 5 grams per 100 mL of water: ......
Solubility is between 1 and 5 grams per 100 mL of water: ------
Solubility is less than 1 gram per 100 mL of water: ———

average distance in the mixed state. Table 22.2 gives the boiling points and solubilities of the isomeric butyl alcohols and several other compounds.

It is generally true that the lower homologs corresponding to a given functional group are more soluble in water than are the higher homologs (see Table 22.1). This can be understood by realizing that the larger the hydrocarbon part of the molecule, compared to the functional group that can hydrogen-bond to water, the smaller the net intermolecular attraction between water molecules and the molecules of the organic liquid. This will result, according to Equation 22.1-3, in a more positive heat of mixing. Another way of looking at this is to say that the larger the alkyl group, the more the molecule will behave like the water-insoluble hydrocarbons.

Solubility increases with increasing T

The general trend of greater solubility at a higher temperature can be explained by Equation 22.1-1. Greater solubility means that the same amount of material can be dissolved in a smaller volume of solvent to form a more concentrated solution. Although the formation of a more concentrated solution involves a smaller ΔS_{mix} than does the formation of a dilute solution, the term $T\Delta S_{mix}$ can maintain the same value despite a decrease in ΔS_{mix} if T is greater.

"Salting out"

The fact that most organic compounds are less soluble in aqueous salt solutions than in water can be understood by realizing that ΔH_{mix} for a salt solution should be more positive than for water; mixing with salt solution requires additional energy to separate the charged ions. Thus, relatively soluble organic substances can often be *salted out* of aqueous solution by saturating the solution with sodium chloride or sodium sulfate.

Since ΔS_{mix} must always be positive, Equation 22.1-1 predicts that anything ought to be miscible with anything else if the temperature is

Table 22.2. Solubility in water and boiling point of isomeric compounds

Compound	*Boiling Point (°C)*	*Solubility in Water*
n-Butyl alcohol	118	8.3[a]
Isobutyl alcohol	108	9.6[a]
Sec-butyl alcohol	100	13.0[a]
Tert-butyl alcohol	83	Miscible in all proportions
Diethyl ether	35	7.5[a]
Methyl *n*-butyl ether	70	1.00[b]
Methyl isobutyl ether	58	1.24[b]
Methyl *sec*-butyl ether	59	1.79[b]
Methyl *tert*-butyl ether	54	5.89[b]

[a] Grams per 100 mL water at 20°C.
[b] Weight percent of ether in saturated aqueous solution at 20°C.

high enough. This will certainly be true when all the substances are above their critical temperatures, because they will then behave as gases. As vapors, mercury and water are completely miscible.

22.2 SOLUBILITY OF SOLIDS IN LIQUIDS

The process of dissolution of a solid in a solvent can be considered to be the sum of two consecutive processes: the melting of the solid (to give the supercooled liquid) followed by the mixing of the liquid with the solvent. Thus, the free-energy change for the dissolving of a solid can be considered to be the sum of the free-energy changes for these two processes:

$$\Delta G_{\text{dis}} = \Delta G_{\text{f}} + \Delta G_{\text{mix}} \tag{22.2-1}$$

Since

$$\Delta G_{\text{f}} = \Delta H_{\text{f}} - T\Delta S_{\text{f}} \qquad \text{and} \qquad \Delta G_{\text{mix}} = \Delta H_{\text{mix}} - T\Delta S_{\text{mix}}$$

it follows that:

$$\Delta G_{\text{dis}} = \Delta G_{\text{f}} + \Delta H_{\text{mix}} - T\Delta S_{\text{mix}} \tag{22.2-2}$$

$$\Delta G_{\text{dis}} = \Delta H_{\text{f}} - T\Delta S_{\text{f}} + \Delta H_{\text{mix}} - T\Delta S_{\text{mix}} \tag{22.2-3}$$

Solids are generally more soluble in a given solvent at higher temperatures than at lower temperatures (this is what makes recrystallization possible). At a higher temperature, less solvent will be needed to dissolve a given amount of material to give a saturated solution ($\Delta G_{\text{dis}} = 0$) because, although ΔS_{mix} will be smaller since a more concentrated solution is being formed, a larger value for T will compensate partially through the $T\Delta S_{\text{mix}}$ term, just as in the mixing of liquids that was discussed in the previous section, and additionally through the $T\Delta S_{\text{f}}$ term.

At the melting point of a solid, T_{f},

$$\Delta G_{\text{f}} = \Delta H_{\text{f}} - T_{\text{f}}\Delta S_{\text{f}} = 0 \tag{22.2-4}$$

or:

$$\Delta H_{\text{f}} = T_{\text{f}}\Delta S_{\text{f}} \qquad \text{or} \qquad \Delta S_{\text{f}} = \frac{\Delta H_{\text{f}}}{T_{\text{f}}} \tag{22.2-5}$$

By means of these last two relationships, ΔG_{f} at any temperature T can be expressed as a function of ΔH_{f} alone or ΔS_{f} alone:

$$\Delta G_{\text{f}} = \Delta H_{\text{f}} - T\left(\frac{\Delta H_{\text{f}}}{T_{\text{f}}}\right) = \Delta H_{\text{f}}\left(1 - \frac{T}{T_{\text{f}}}\right) \tag{22.2-6}$$

$$\Delta G_{\text{f}} = T_{\text{f}}\Delta S_{\text{f}} - T\Delta S_{\text{f}} = \Delta S_{\text{f}}(T_{\text{f}} - T) \tag{22.2-7}$$

(Remember from Section 17.5 that ΔG_f will equal zero only when $T = T_f$; that is, the solid and liquid will be in equilibrium only at a unique temperature, the melting point. Below the melting point, $T_f > T$, and therefore ΔG_f will be positive.) These relationships can each be substituted into Equation 22.2-2 to give

$$\Delta G_{dis} = \Delta H_f \left(1 - \frac{T}{T_f}\right) + \Delta H_{mix} - T\Delta S_{mix} \qquad (22.2\text{-}8)$$

$$\Delta G_{dis} = \Delta S_f(T_f - T) + \Delta H_{mix} - T\Delta S_{mix} \qquad (22.2\text{-}9)$$

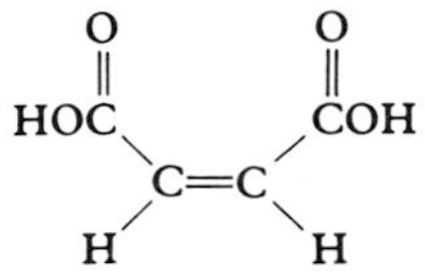

HOC(=O)–CH=CH–COH(=O) (cis)
maleic acid
m.p. 143°C
60 g/100 mL H_2O

Solubility and m.p.

HOC(=O)–CH=CH–COH(=O) (trans)
fumaric acid
m.p. 286°C
0.6 g/100 mL H_2O

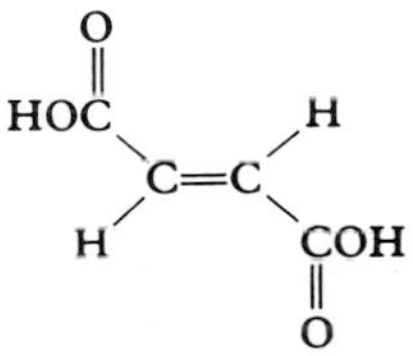

These last two relationships make it possible to understand why two (or more) substances that might be expected to have the same tendency to mix in a solvent can have different solubilities. For example, the *cis/trans* isomers, maleic and fumaric acid, might be expected to mix equally well with water, but the *cis* isomer, maleic acid, is 100 times more soluble than the *trans* isomer at room temperature (see Table 22.3). Similarly, the three position isomers of dinitrobenzene might be expected to mix equally well with benzene, but their solubilities range over a factor of 10 at 50°C (see Table 22.3). These phenomena can be understood by observing that *solubility decreases with increasing melting point of the compound.* If the difference in melting point is due entirely to a difference in the enthalpies of fusion ΔH_f, the entropies of fusion ΔS_f are thus assumed to be the same for all the isomers of the set and, according to Equation 22.2-9, the positive term $\Delta S_f(T_f - T)$ will be greater, the larger the T_f. Thus, the ΔS_{mix} term must remain larger for the higher-melting isomers, and saturation must occur at higher dilutions. Similarly, if the difference in melting point is due entirely to a difference in the entropies of fusion, the enthalpies of fusion are thus assumed to be the same for all the isomers of the set and, according to Equation 22.2-8, the positive term $\Delta H_f(1 - T/T_f)$ will be greater, the larger the T_f. Thus, again the ΔS_{mix} term must remain larger for the higher-melting isomers; that is, only relatively dilute solutions are possible for the higher-melting isomers. It seems reasonable that the result should be qualitatively the same if the differences in melting point are due to differences in both the enthalpy and entropy of fusion. More examples of this phenomenon are given in Table 22.3.

22.3 CLASSIFICATION OF COMPOUNDS BY SOLUBILITY: RELATIONSHIPS BETWEEN SOLUBILITY AND MOLECULAR STRUCTURE

In the following sections, structural features of molecules that lead to solubility under various conditions in different solvents will be de-

Table 22.3. Solubilities and melting points of isomeric compounds

Compound	*m.p.* (°C)	*Solubility in* *Water*	*Alcohol*	*Benzene*	*Hexane*
Maleic acid (*cis*)	143	60[a]	51[a]		
Fumaric acid (*trans*)	286	0.6[a]	5[a]		
d-Tartaric acid	170	139[a]	27[b]		
l-Tartaric acid	170	139[a]	27[b]		
dl-Tartaric acid	206	20[a]	2[b]		
o-Dinitrobenzene	116			17.5[c]	
m-Dinitrobenzene	90			37.6[c]	
p-Dinitrobenzene	170			3.1[c]	
Phenanthrene	100			18.6[b]	4.2[b]
Anthracene	217			0.6[b]	0.2[b]
o-Nitrobenzoic acid	148			0.35[d]	
m-Nitrobenzoic acid	142			0.80[d]	
p-Nitrobenzoic acid	240			0.08[d]	
o-Dihydroxybenzene	105	14[e]		1.15[e]	
m-Dihydroxybenzene	109	24[e]		0.40[e]	
p-Dihydroxybenzene	173	1.4[e]		0.04[e]	

[a] Grams per 100 mL solvent at 20°C.
[b] Grams per 100 mL solvent at 25°C.
[c] Grams per 100 mL solvent at 50°C.
[d] Grams per 100 mL solvent at 30°C.
[e] Mole percent of solute in saturated solution at 30°C.

scribed. This information makes it possible to predict the solubility behavior of a compound of known structure, and to exclude certain possibilities for the structure of an unknown substance on the basis of its solubility under different conditions. Again, a compound will be classified as "soluble" if it is soluble to the extent of 30 or more milligrams per milliliter of solvent (3 or more grams per 100 mL of solvent).

Solubility in Water

Covalent substances are soluble in water at room temperature only if the molecules have a functional group that can form hydrogen bonds with water—a functional group, in other words, that contains a nitrogen, oxygen, or sulfur atom. The higher homologs are decreasingly soluble in water, and with monofunctional compounds they become "insoluble" at about five carbon atoms (see Table 22.1). If two or more functional groups are present, especially if they include the amino or hydroxy group, compounds with more than five or six carbon atoms may be water soluble.

Ionic substances (salts) are often very soluble in water. Although these compounds are solids and may have very large heats of fusion, solvation of the ions is often sufficiently exothermic that they are soluble in water.

If an organic liquid is soluble in water, it is most likely to be a relatively low-molecular-weight mono- or difunctional compound. Higher-molecular-weight and polyfunctional compounds are much more likely to be solids.

If an organic solid is soluble in water, it is most likely to be either a salt (the alkali metal salt of an acid or the hydrogen halide or sulfate salt of an amine), a polyfunctional compound that happens to be a solid (polyhydric phenols, sugars), or an amino acid (which exists as an inner salt).

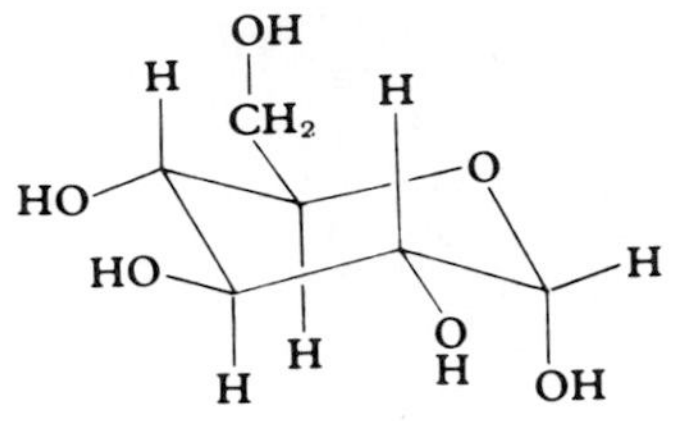

α-D-glucose
(a sugar)
very soluble
in water

Acidification of the aqueous solution of the salt of an organic acid will convert it to the free acid, which will separate from solution if it is not itself soluble in water or aqueous acid; similarly, making the solution of the salt of an amine basic will liberate the free base, which will separate from solution if it is not soluble in water or aqueous base. Salts of weak acids will hydrolyze in aqueous solution to form hydroxide ion, and salts of amines will hydrolyze to produce hydronium ion; the decrease or increase in acidity of the water in which the salt is dissolved can be detected by means of pH paper or a pH meter.

Solubility in Diethyl Ether

Almost all organic liquids are soluble in ether. A great many organic solids are soluble in ether. Those that are not are compounds with large heats of fusion and thus include salts and high-molecular-weight, high-melting covalent compounds.

Solubility in 5% Aqueous Sodium Hydroxide

Since the sodium salts of acids are generally very soluble in water, water-insoluble organic compounds that can be converted to their conjugate bases by 5% aqueous sodium hydroxide will dissolve in this solvent. Such compounds include carboxylic acids, sulfonic acids, sulfinic acids; phenols; sulfonamides of primary amines, imides; some β-diketones and β-keto esters; mercaptans and thiophenols; some primary and secondary nitro compounds; and oximes. Acidification of the basic solution of such compounds should precipitate the sample.

It should be obvious that substances that are soluble in water would be expected to be soluble in aqueous base, with the exception of the salts of water-insoluble amines.

Some acid halides and some easily hydrolyzed esters will react to give soluble products upon standing with 5% aqueous sodium hydroxide.

Solubility in 5% Aqueous Sodium Bicarbonate

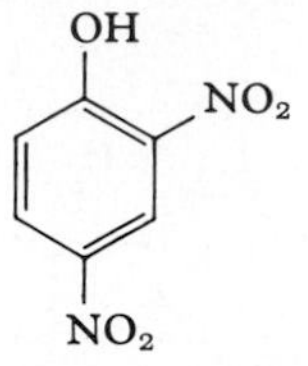

2,4-dinitrophenol

Of the substances listed as soluble in 5% aqueous sodium hydroxide, only carboxylic acids, sulfonic acids, sulfinic acids, and certain phenols with multiple electronegative substituents, such as 2,4,6-tribromophenol or 2,4-dinitrophenol, are sufficiently acidic to be converted to their conjugate bases by 5% aqueous sodium bicarbonate solution and thus to dissolve in this solvent. Dissolution of acids in sodium bicarbonate solution will result in the evolution of carbon dioxide gas.

Solubility in 5% Aqueous Hydrochloric Acid

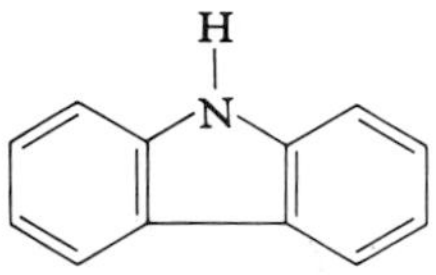

carbazole

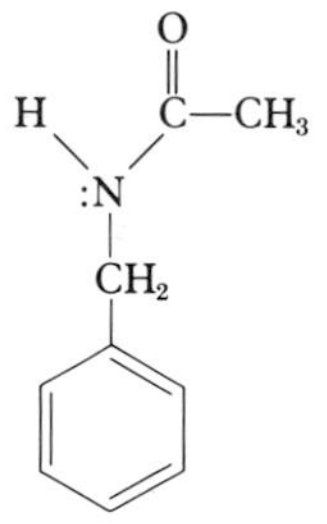

N-benzylacetamide

Five percent aqueous hydrochloric acid will convert many amines to their water-soluble hydrochloride salts and thus cause them to dissolve. Most aliphatic amines (primary, secondary, and tertiary) will be converted to their conjugate acids under these conditions, as well as most aromatic amines in which no more than one aromatic ring is attached directly to the nitrogen. Diphenylamine, carbazole, 2,4,6-tribromoaniline, and the nitroanilines are not sufficiently basic to be converted to their conjugate acids under these conditions.

N,N-disubstituted amides and N-benzylacetamide can be converted to their conjugate acids under these conditions, but not N-unsubstituted or most N-monosubstituted amides.

The hydrochloride salts of some basic substances are not soluble in 5% hydrochloric acid, although the salt may be soluble in water. Thus, if a solid remains after treatment with 5% hydrochloric acid, it should be separated from the supernatant liquid by suction filtration or centrifugation and its solubility in water determined. If doubt still remains as to whether the solid is the salt or the original compound, its melting point or infrared spectrum can be compared to that of the original material. Basification of the supernatant liquid should give a precipitate if the hydrochloride salt was partially dissolved.

It should be obvious that substances that are soluble in water would be expected to be soluble in aqueous acid, with the exception of the salts of water-insoluble acids.

Solubility in Concentrated Sulfuric Acid

Compounds containing an oxygen or nitrogen atom will dissolve in or react with cold concentrated sulfuric acid. Alkenes will react to form either a soluble alkyl hydrogen sulfate or an insoluble polymer, or will oxidize, usually with the evolution of heat. Polyalkylbenzenes will undergo sulfonation to give a product that is soluble in the mixture.

All saturated compounds containing only carbon, hydrogen, or halogen will be insoluble in concentrated sulfuric acid, as will those aromatic hydrocarbons or their halogenated derivatives that cannot be easily sulfonated.

Solubility in 85% Phosphoric Acid

Certain oxygen-containing compounds such as alcohols, aldehydes, methyl and cyclic ketones, and esters are soluble in 85% phosphoric acid, provided that they contain fewer than nine carbon atoms. Ethers are somewhat less soluble: di-*n*-propyl ether is soluble, whereas di-*n*-butyl ether and anisole are not.

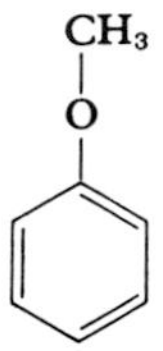

CH_3
O

anisole

22.4 TECHNIQUES FOR DETERMINATION OF SOLUBILITY

A sample size of 30 mg of a solid or one or two drops of a liquid to 1 mL of solvent may be satisfactory in most cases. The solid should be powdered. All solubility determinations should be made at room temperature, without heating.

In determining solubility in water or ether, add the solvent in portions to the sample in a small test tube (8-mm diameter by 50-mm length). Mix the contents of the tube well after each addition of solvent. Adding the solvent in portions will allow you to distinguish between substances that are freely soluble, barely soluble, and relatively insoluble. If the substance is insoluble in water, the resulting suspension can be used to determine solubility in hydrochloric acid or sodium hydroxide: add a drop or two of concentrated hydrochloric acid or 50% sodium hydroxide solution.

In determining solubility in 5% sodium hydroxide, 5% sodium bicarbonate (only for substances found to be soluble in 5% sodium hydroxide), or 5% hydrochloric acid (only for compounds that contain nitrogen), add the sample in portions to the solvent, mixing well after each addition. If the first portion does not dissolve completely, the fact that the material is of limited solubility is established, and the remainder need not be added. The solution should be neutralized after the removal of any insoluble residue.

When determining solubility in concentrated sulfuric acid, add the sample to the acid. Color changes, evolution of heat or a gas, or other evidence of reaction can be as significant as mere dissolving.

Solubility in 85% phosphoric acid should be determined only if the compound is soluble in concentrated sulfuric acid.

Problems

1. Perfluoroheptane (C_7F_{16}) dissolves in heptane only to the extent of 0.27 mole per mole of heptane. Account for the immiscibility of these two substances.
2. Account for the miscibility with water of polyethyleneglycol, a high-boiling substance whose formula is

$$HO{-}(CH_2CH_2O{-})_nCH_2CH_2OH$$

3. Account for the fact that, although diethyl ether boils lower than the isomeric butyl alcohols, it is less soluble in water (see Table 22.2).
4. Account for the lower solubility and higher melting point of the equimolar mixture of *d*- and *l*-tartaric acids compared to either pure enantiomer (see Table 22.3).
5. Evaluate the relative effectiveness of nitrogen, oxygen, and sulfur atoms in hydrogen bonding to water (see Table 22.1).
6. What is the significance of the way Equations 22.2-8 and 22.2-9 reduce when $T = T_f$?
7. Rationalize the trends in the solubilities of the following isomeric compounds:

Compound	*b.p.*	*m.p.*	*Solubility in Water (grams per 100 g)*
Valeric acid	187°C		3.4
Isovaleric acid	177°C		4.2
Trimethylacetic acid		36°C	2.2
n-Propyl acetate	101°C		1.9
Isopropyl acetate	88°C		3.2

8. The solubility of caffeine in several solvents is given; rationalize the high solubility in chloroform.

Solvent	*Solubility (grams per 100 g solvent; 18°C)*
Chloroform	11.2
Carbon tetrachloride	0.09
Ether	0.12
Ethyl acetate	0.73
Benzene	0.91

9. Explain the fact that some organic compounds, for instance many proteins, are more soluble in salt solutions than in pure water; that is, they can be "salted in" to aqueous solution.

References

Sources from which more information about the relationships between solubility and molecular structure can be obtained include

1. R. L. Shriner, R. C. Fuson, D. Y. Curtin, and T. C. Morrill, *The Systematic Identification of Organic Compounds,* 6th edition, Wiley, New York, 1980, p. 90.
2. N. D. Cheronis and J. B. Entrikin, *Identification of Organic Compounds,* Interscience, New York, 1963, p. 77.
3. A. I. Vogel, *A Textbook of Practical Organic Chemistry,* 3rd edition, Wiley, New York, 1957, p. 1045.

23 Infrared Absorption Spectrometry

Absorption spectrometric methods are relatively recent developments. Ultraviolet-visible spectrometry (Section 24) has been known the longest. It was followed by infrared spectrometry and, most recently, by nuclear magnetic resonance spectrometry (Section 25). Since equipment is widely available for the determination of IR, UV, and NMR spectra, these methods are now commonly and routinely used for the analysis and characterization of pure substances and mixtures.

Absorption spectrometry is the determination of the degree to which electromagnetic radiation (light; light energy) is absorbed by a substance over a range of wavelengths. The record of energy absorption versus wavelength is the absorption spectrum. The usual wavelength range for IR spectrometry is from 2 microns to 16 microns, although instruments are available that scan up to 200 microns. Table 23.1 indicates the relationship between the IR region and other regions of the electromagnetic spectrum, including the UV-visible region and the NMR region.

23.1 WAVELENGTH, FREQUENCY, AND ENERGY OF ELECTROMAGNETIC RADIATION

Electromagnetic radiation (light) can be specified by wavelength, frequency, or energy. The product of wavelength (λ; length) and frequency (v; time^{-1}) equals the speed of light (c; length per time):

$$\lambda v = c \qquad (23.1\text{-}1)$$

According to Equation 23.1-1, wavelength and frequency are inversely proportional to one another. That is, short wavelengths correspond to high frequencies, and vice versa.

Table 23.1. The electromagnetic spectrum: energy, frequency, and wavelength

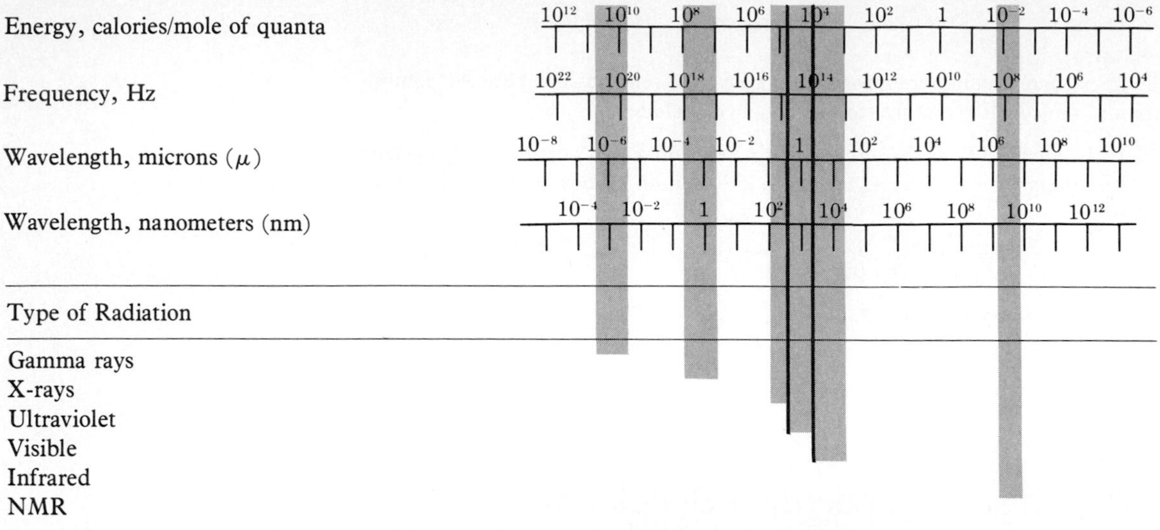

Scale										
Energy, calories/mole of quanta	10^{12}	10^{10}	10^{8}	10^{6}	10^{4}	10^{2}	1	10^{-2}	10^{-4}	10^{-6}
Frequency, Hz	10^{22}	10^{20}	10^{18}	10^{16}	10^{14}	10^{12}	10^{10}	10^{8}	10^{6}	10^{4}
Wavelength, microns (μ)	10^{-8}	10^{-6}	10^{-4}	10^{-2}	1	10^{2}	10^{4}	10^{6}	10^{8}	10^{10}
Wavelength, nanometers (nm)	10^{-4}	10^{-2}	1	10^{2}	10^{4}	10^{6}	10^{8}	10^{10}	10^{12}	

Type of Radiation
Gamma rays
X-rays
Ultraviolet
Visible
Infrared
NMR

Light as energy

Light energy is specified by the size of the *quantum,* or least increment of energy. The size of the quantum is directly proportional to the frequency of the light, as shown by Equation 23.1-2:

$$E = hv \qquad (23.1\text{-}2)$$

where E is the energy of the quantum, v is the frequency, and h is the proportionality constant, Planck's constant. Thus, large quanta are associated with high frequency and, by inverse proportionality, with short wavelength.

The relationships between wavelength, frequency, and energy are also illustrated in Table 23.1.

23.2 UNITS OF LIGHT ABSORPTION

Transmittance

The degree to which light is absorbed by a sample at a particular wavelength can be expressed in several ways. The first is as *transmittance, T,* which is defined in Equation 23.2-1 as the ratio of the intensity of light that passes through the sample to the intensity incident upon the sample (that is, the ratio that expresses the fraction of light that gets through the sample):

$$\text{Transmittance} = T = \frac{I_{\text{transmitted}}}{I_{\text{incident}}} = \frac{I_t}{I_i} \qquad (23.2\text{-}1)$$

Percent transmittance

The second is the *percent transmittance,* %*T,* which is defined as the transmittance × 100, Equation 23.2-2:

$$\text{Percent transmittance} = \%T = T \times 100 \qquad (23.2\text{-}2)$$

The third is the *absorbance, A*, which is defined as the logarithm of the reciprocal of the transmittance, Equation 23.2-3:

Absorbance

$$\text{Absorbance} = A = \log \frac{1}{T} = \log \frac{I_i}{I_t} \qquad (23.2\text{-}3)$$

The relationships between these units are illustrated in Table 23.2.

It is too bad to have to use "complicated" units like absorbance, but *only* absorbance is directly proportional to sample concentration and sample thickness (path length). The relationship between absorbance and concentration is given by Equation 23.2-4:

Absorbance and concentration

$$A = \epsilon c l \qquad (23.2\text{-}4)$$

where A is the absorbance; c is the concentration, in moles/liter; l is the path length, in cm; and ϵ, the extinction coefficient, is the proportionality constant. The practical consequences of this relationship between concentration and path length will be brought out in the discussions of sample preparation for infrared and ultraviolet spectroscopy.

Extinction coefficient

Most IR and UV spectrometers are calibrated in both percent transmittance and absorbance units.

23.3 INFRARED LIGHT ABSORPTION AND MOLECULAR STRUCTURE

In all forms of absorption spectrometry, an interpretation of a spectrum is a description of the ways by which light energy is absorbed by the molecules of the sample. In the infrared range, energy absorption is explained by an increase in the amplitude of various molecular vibrations. The frequency of the light absorbed equals the frequency of the

Mechanism of IR energy absorption

Table 23.2. Relationships between transmittance, percent transmittance, and absorbance units

%*T*	*T*	1/*T*	*Absorbance*
100	1	1	0
75	3/4	1.33	0.125
50	1/2	2	0.301
25	1/4	4	0.602
10	1/10	10	1.000
5	1/20	20	1.301
2	1/50	50	1.699
1	1/100	100	2.000
0	0	infinite	infinite

molecular vibration, and thus the higher-frequency vibrations will absorb the larger quanta, or the shorter-wavelength light.

For the purpose of interpreting infrared spectra, bonds between atoms in molecules can be thought of as springs, and the atoms as masses at the ends of the springs. According to this model, then, parts of the molecule can stretch and bend with respect to the rest of the molecule. The frequency of stretching and bending can be related to the atomic masses and the force constants of the springs by Equation 23.3-1:

$$\nu \propto \sqrt{\frac{k}{\mu}} \tag{23.3-1}$$

Here, ν is the frequency of stretching or bending, k is the force constant (a measure of the stiffness of the spring), and μ is the reduced mass. The reduced mass is defined by Equation 23.3-2:

$$\mu = \frac{M_1 M_2}{M_1 + M_2} \tag{23.3-2}$$

where M_1 and M_2 are the total masses at the two ends of the spring. If one part of the molecule is much lighter than the remainder ($M_1 \ll M_2$), then μ approximately equals the mass of the light part, M_1. In this case, Equation 23.3-1 reduces to Equation 23.3-3:

$$\nu \propto \sqrt{\frac{k}{M_1}} \tag{23.3-3}$$

and the frequency of bending or stretching of a small part of the molecule with respect to the remainder is proportional to the square root of the stiffness of the spring and inversely proportional to the square root of the mass of the small part. The fact that the absorption of energy by the increase in amplitude of the stretching vibration of C—H, C—F, C—Cl, C—Br, and C—I bonds occurs at lower and lower frequencies (longer wavelength; lower energy) can thus be interpreted by saying that the force constant k appears to remain about the same, while M_1 increases. Similarly, the fact that the stretching frequency of a multiple bond is higher than that of the corresponding single bond can be interpreted by saying that the force constant is larger for the multiple bond than for the single bond. It is harder to stretch a double or triple bond than a single bond.

23.4 INTERPRETATION OF INFRARED SPECTRA

Establishment of Identity of Samples

The simplest use of infrared spectra is to determine whether two samples are identical or not. If the samples are the same, their IR spectra (obtained under identical conditions) must be the same. If the samples are different, their spectra will be different. If the two samples are both pure substances very similar in structure, the differences in the spectra may be so small that it is not easy to see them; it may even be beyond the power of the instrument to detect them. The absorption peaks of the spectrum of an impure sample should be less intense than those of a pure sample, assuming equal amounts of sample, and the spectrum will contain additional peaks. The infrared spectra of enantiomers should be identical, and those of diastereomers should be different. Figures 47.1 and 47.2 show the IR spectra of a pair of enantiomers, the *R* and *S* isomers of carvone; the spectra are identical within experimental uncertainty. Figures 49.2 and 49.4 illustrate the IR spectra of a pair of diastereomers, the *trans* and *cis* isomers of 1,2-dibenzoylethylene. The spectra are obviously different.

Comparison of IR spectra is a convenient and relatively sensitive way to establish the identity of a substance with a sample of known structure and purity. Comparison of the IR spectra of different fractions obtained in a fractional distillation or of material before and after recrystallization is a good way to determine and follow the progress of a purification. Most IR spectra are obtained for use in these ways.

Monitoring a purification

Determination of Molecular Structure

A more sophisticated level of interpretation of an IR spectrum is to use it to establish the structure of an unknown material. Comparison of IR spectra of substances of known structure has led to the establishment of a great many correlations between wavelength (or frequency) of IR absorption and features of molecular structure. The presence or absence of absorption at certain wavelengths thus indicates the presence or absence of certain structural features. Table 23.3 presents a number of these correlations.

Detection of Functional Groups

Certain structural features can be established fairly easily. For example, if a substance contains only C, H, and O, the oxygen can be present only as C═O, O—H, or C—O—C (or a combination of these, such as the ester or carboxylic acid group). The presence or absence of absorption in the carbonyl region (~5.8–6.0 microns; ~1730–1670 cm^{-1}) or O—H region (~2.7–3.0 microns; ~3700–3300 cm^{-1}) can serve to eliminate or establish some of these possibilities. The nature of a functional group involving nitrogen can be inferred in a compound containing only C, H, and N in a similar way. If the compound contains both O and N and/or other atoms other than C and H, the problem is more difficult but can be approached in the same way.

O
‖
R—C—O—R
an ester

Table 23.3. Infrared absorption–structure correlations

		Range (microns)	*Intensity*	*Range* (cm^{-1})
C—H stretching vibrations				
Alkane		3.38–3.51	m–s	2962–2853
Alkene		3.23–3.32	m	3095–3010
Alkyne		3.03	s	3300
Aromatic		3.30	v	3030
Aldehyde		3.45–3.55	w	2900–2820
	and	3.60–3.70	w	2775–2700
C—H bending vibrations				
Alkane		6.74–7.33	v	1485–1365
Alkene				
monosubstituted (vinyl)		7.04–7.09	s	1420–1410
		7.69–7.75	w–s	1300–1290
		10.05–10.15	s	995–985
	and	10.93–11.05	s	915–905
disubstituted, *cis*		14.5	s	690
disubstituted, *trans*		7.64–7.72	m	1310–1295
	and	10.31–10.42	s	970–960
disubstituted, *gem*		7.04–7.09	s	1420–1410
	and	11.17–11.30	s	895–885
trisubstituted		11.90–12.66	s	840–790
Aromatic				
5 adjacent hydrogen atoms		13.3	v,s	750
	and	14.3	v,s	700
4 adjacent hydrogen atoms		13.3	v,s	750
3 adjacent hydrogen atoms		12.8	v,m	780
2 adjacent hydrogen atoms		12.0	v,m	830
1 isolated hydrogen atom		11.3	v,w	880
N—H stretching vibrations				
Amine, not hydrogen bonded		2.86–3.03	m	3500–3300
Amide		2.86–3.2	m	3500–3140
O—H stretching vibrations				
Alcohols and phenols				
not hydrogen bonded		2.74–2.79	v,sh	3650–3590
hydrogen bonded		2.80–3.13	v,b	3750–3200
Carboxylic acids				
hydrogen bonded		3.70–4.00	w	2700–2500
C—O stretching vibrations				
Esters				
formates		8.33–8.48	s	1200–1180
acetates		8.00–8.13	s	1250–1230
propionates and higher homologs		8.33–8.70	s	1200–1150
benzoates and phthalates		7.63–8.00	s	1310–1250
	and	8.69–9.09	s	1150–1100

Intensity: w = weak absorption, m = medium absorption, s = strong absorption, v = variable intensity of absorption, sh = sharp absorption, b = broad absorption.

		Range (microns)	Intensity	Range (cm^{-1})
Carbon–halogen stretching vibrations				
C—F		7.1–10.00	s	1400–1000
C—Cl		12.5–16.6	s	800–600
C—Br		16.6–20.0	s	600–500
C—I		~20	s	~500
C=C stretching vibrations				
Isolated alkene		5.99–6.08	v	1669–1645
Conjugated alkene				
C=C conjugated		6.25	m–s	1600
C=O conjugated		6.07–6.17	m–s	1647–1621
phenyl conjugated		6.15	m–s	1625
Aromatic		6.25	v	1600
		6.33	m	1580
		6.67	v	1500
	and	6.90	m	1450
C=O stretching vibrations				
Aldehydes				
saturated, aliphatic		5.75–5.81	s	1740–1720
α,β-unsaturated, aliphatic		5.87–5.95	s	1705–1680
Ketones				
saturated, acyclic		5.80–5.87	s	1725–1705
saturated, 6-membered ring and larger		5.80–5.87	s	1725–1705
saturated, 5-membered ring		5.71–5.75	s	1750–1740
α,β-unsaturated, acyclic		5.94–6.01	s	1685–1665
aryl, alkyl		5.88–5.95	s	1700–1680
diaryl		5.99–6.02	s	1670–1660
Carboxylic acids				
saturated, aliphatic		5.80–5.88	s	1725–1700
aromatic		5.88–5.95	s	1700–1680
Carboxylic acid anhydrides				
saturated, acyclic		5.41–5.56	s	1850–1800
	and	5.59–5.75	s	1790–1740
Acyl halides				
chlorides		5.57	s	1795
bromides		5.53	s	1810
Esters and lactones (cyclic esters)				
saturated, acyclic		5.71–5.76	s	1750–1735
saturated, 6-membered ring and larger		5.71–5.76	s	1750–1735
α,β-unsaturated and aryl		5.78–5.82	s	1730–1717
vinyl esters		5.56–5.65	s	1800–1770
Amides and lactams (cyclic amides)		5.88–6.14	s	1700–1630
Triple-bond stretching vibrations				
C≡N		4.42–4.51	m	2260–2215
C≡C		4.42–4.76	v,m	2260–2100

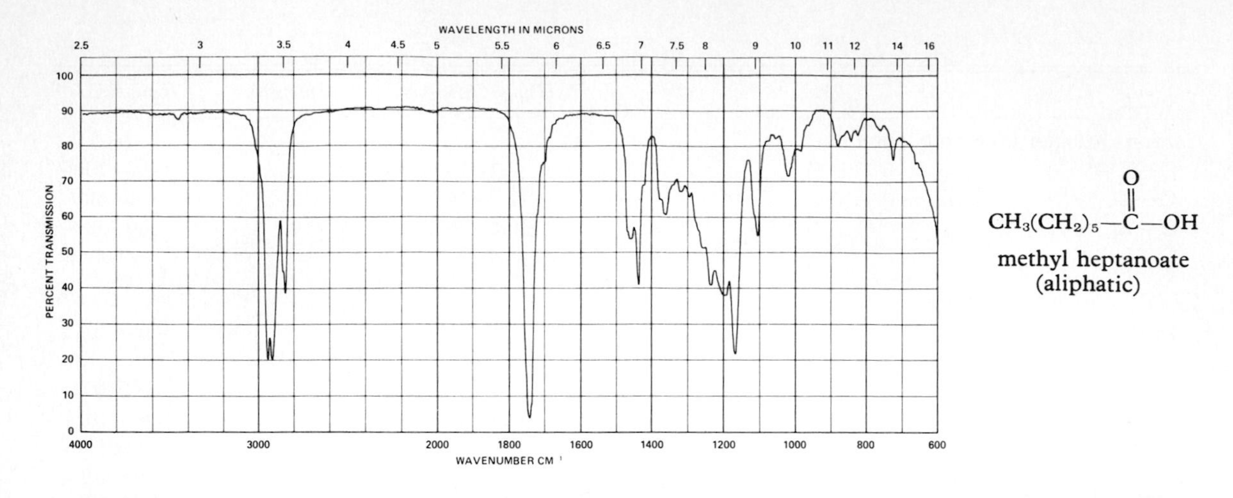

Figure 23.1. IR spectrum of the aliphatic compound methyl heptanoate; thin film.

Aromatic versus aliphatic

It is also fairly easy to establish whether a compound is primarily aromatic or primarily aliphatic. The IR spectrum of an aromatic compound (without a long aliphatic side chain) generally has only weak absorption in the C—H stretching region (~3.3 microns; ~3000 cm^{-1}), sharp bands in the 6–7-micron region (1660–1430 cm^{-1}), and strong absorption at wavelengths greater than 12 microns (less than 840 cm^{-1}), and gives the general impression of having sharp and symmetrical absorption bands. The IR spectrum of an aliphatic compound generally has relatively strong absorption in the C—H region (~3.4 microns; ~2950 cm^{-1}) and little or no absorption at wavelengths greater than 12 microns (less than 840 cm^{-1}) and gives the impression of having broad and not particularly symmetrical bands. Compare, for example, the spectra of methyl heptanoate and methyl benzoate, Figures 23.1 and 23.2. Cyclic and conformationally rigid compounds generally give spectra that contain a relatively small number of sharp, symmetrical bands; the spectrum of cyclohexanone, Figure 54.1, is an example.

$CH_3CH_2CH_2CH_2CH_2CH_2$—C(=O)—O—CH_3
methyl heptanoate
(aliphatic)

C_6H_5—C(=O)—O—CH_3
methyl benzoate
(aromatic)

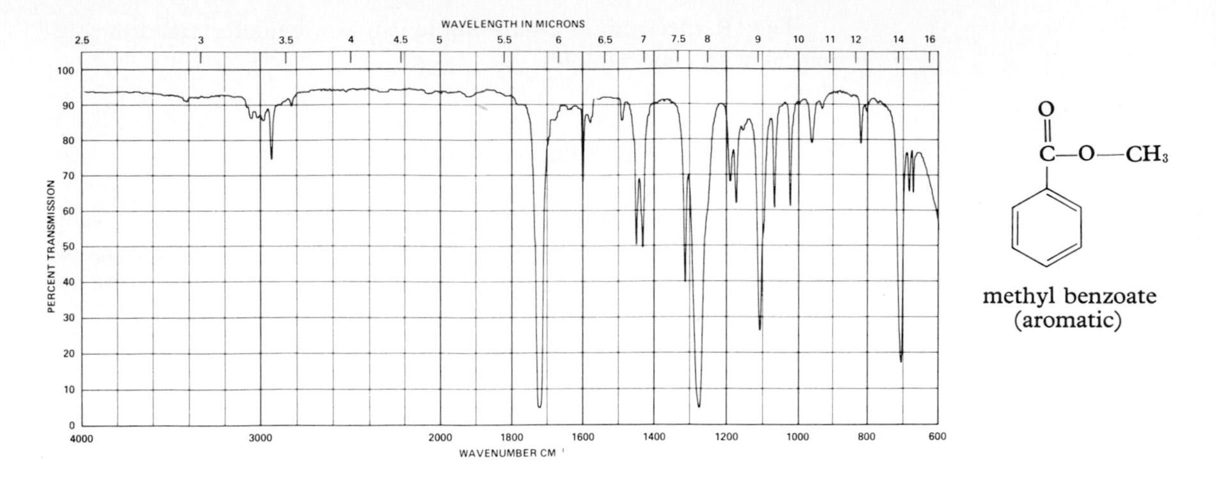

Figure 23.2. IR spectrum of the aromatic compound methyl benzoate; thin film.

23.5 SAMPLE PREPARATION

IR cell

For infrared absorption spectrometry, the sample whose absorption is to be determined must be placed in the beam of infrared radiation. The container (cell) or support for the sample must be transparent to infrared radiation and must therefore be made of one of a small number of materials, not including glass. The most commonly used material is sodium chloride, which is transparent between 2 and 16 microns (5000 and 600 cm^{-1}). Certain other metal halide salts are transparent over other wavelength ranges.

Liquid Samples

Thin film method

The spectrum of a pure liquid is most easily determined as a *film* between a pair of sodium chloride plates. One salt plate is positioned in the holder, a drop of the liquid is placed in the center, and the second plate is put on top. The top of the holder is then pressed or screwed on gently and the holder placed in the sample (near) beam of the instrument. Nothing need be placed in the reference (far) beam, since the only absorbing material in the sample beam is the liquid whose absorption spectrum is being determined. If the strongest peak in the spectrum absorbs more than about 98% of the light, tighten the top of the holder to make the film thinner and rerun the spectrum. The spectra shown in Figures 23.1 and 23.2, for example, were obtained from thin films of liquid.

The IR spectrum of a pure liquid can also be determined in solution, as with solid samples.

Solid Samples: The Solution Method

Solution method

The IR spectra of solids are most often determined *in solution.* Table 23.4 suggests appropriate concentrations for an average sample; the concentration should be adjusted, if necessary, so that the strongest peak will absorb between 90 and 98% of the light. As Equation 23.2-4 indicates, the absorbance of a sample is proportional to both its concentration and its thickness. Table 23.4 is based on the assumption that the sample thickness (the path length of the infrared cell) will be 0.2 mm. If a cell of 0.4-mm path length will be used instead, the amount of sample required will be only half that specified by the table. It is assumed, of course, that 0.5 mL of solution will be prepared in either case.

Double beam operation

The ideal solvent would be transparent over the entire infrared range of wavelength, but no liquids attain this ideal. For this reason, a so-called *double beam operation* must be used for obtaining the IR spectrum of solids in solution. The solution cell is placed in the sample beam. An identical cell containing pure solvent must be placed in the reference beam; its absorption is subtracted electronically by the instrument from the absorption of the sample beam, with the result that the spectrum recorded is that of the net absorption due to sample molecules in the sample beam. Solvents most often used for infrared spectrometry are carbon tetrachloride, chloroform, and carbon disulfide (Figures 23.3, 23.4, 23.5). Water is *never* used (except with special cells with water-resistant windows).

Solvents

Don't use water!

In those wavelength regions where the solvent absorbs strongly (more than about 95% of the light), the absorption recorded by the spectrophotometer is meaningless. This happens because, as the solvent absorption increases, the net absorption due to sample becomes a smaller fraction of the total absorption, and the difference in intensity of the two beams finally falls below the limit detectable by the instrument. You can verify this by placing your hand in the sample or refer-

Table 23.4. Milligrams of sample required for 0.50 mL of solution

A cell path length of 0.2 mm is assumed. For a different cell path length, adjust the amount of sample used so as to keep the number of molecules in the beam the same as for the examples given.

Molecular weight	100	200	300	400	500
Mg of sample	9	18	27	36	45

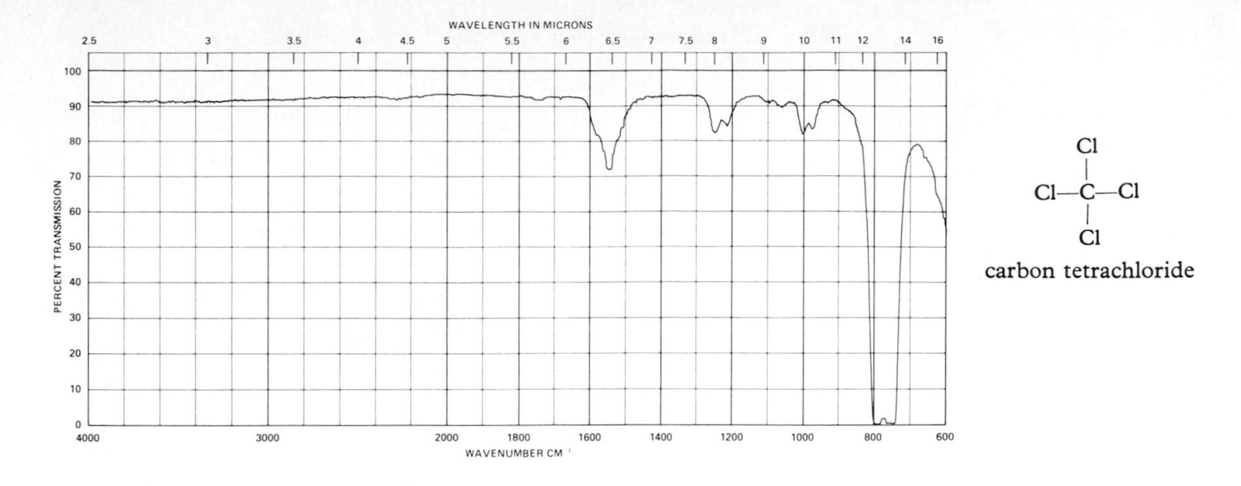

Figure 23.3. IR spectrum of carbon tetrachloride; 0.05-mm path length.

ence beam when the instrument is scanning a range of strong solvent absorption (for carbon tetrachloride, between 12.3 and 13.7 microns; 810 and 730 cm^{-1}; Figure 23.3) and seeing that the spectrum is unaffected. Between carbon tetrachloride or chloroform and carbon disulfide, sample absorption can be determined at all wavelengths between 2 and 16 microns (5000 and 600 cm^{-1}; see Figures 23.3, 23.4, and 23.5). If the recorder pen runs either upscale or downscale in a range of intense solvent absorption, the balance control is not adjusted correctly, and the peaks of the rest of the spectrum will be distorted.

Balance-control setting

Figure 23.4. IR spectrum of chloroform; 0.05-mm path length.

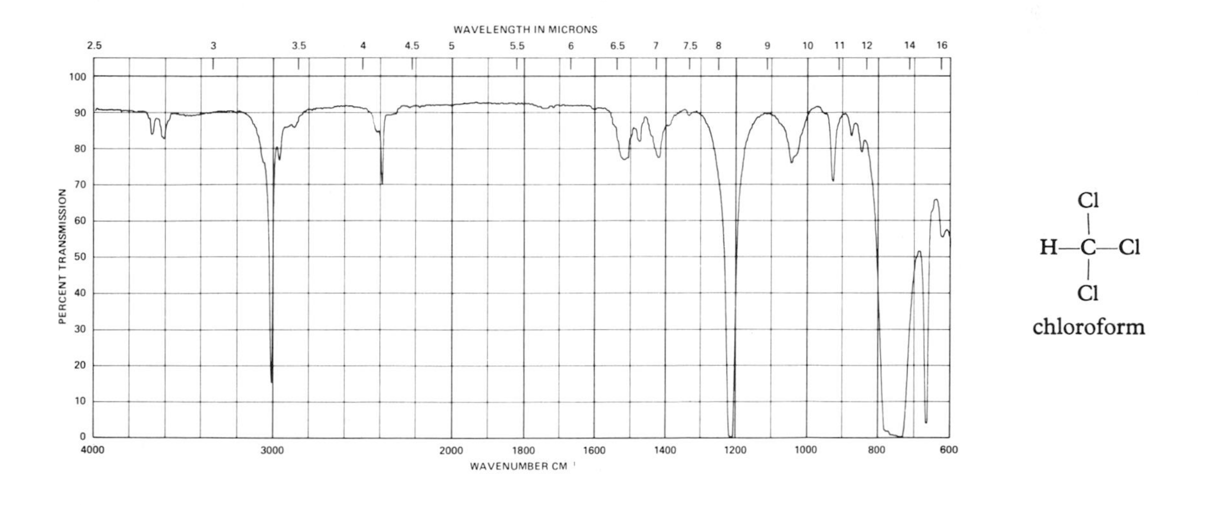

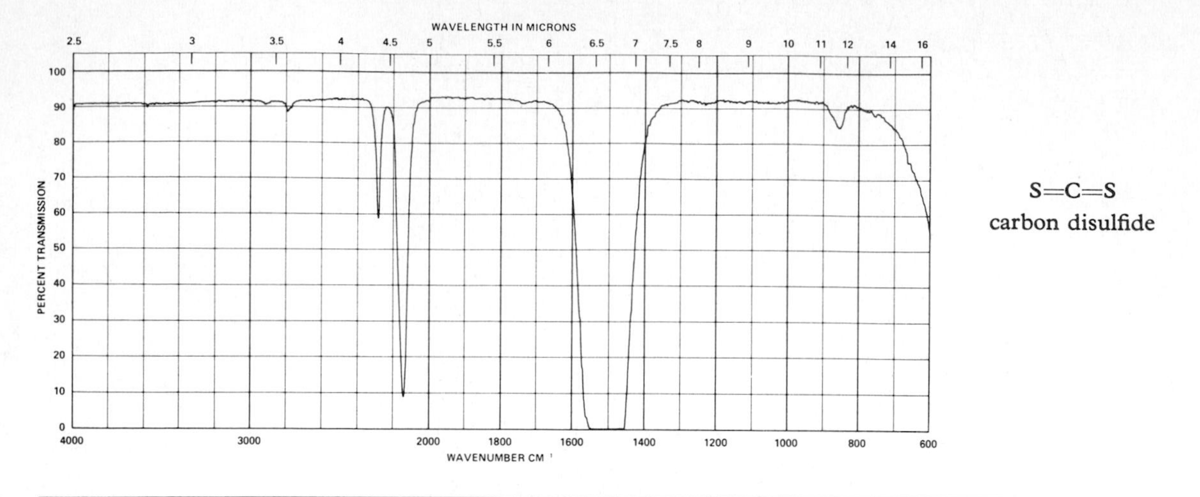

Figure 23.5. IR spectrum of carbon disulfide; 0.05-mm path length.

Solid Samples: The KBr Method

The infrared spectrum of a solid can also be determined in the solid state. One way is to grind 1–2 mg of sample with about 100–400 mg of anhydrous potassium bromide in a clean mortar and to press the resulting mixture into a translucent wafer, using a die. The wafer is then mounted in the sample beam, and an attenuator is placed in the reference beam. The function of the attenuator is to compensate for the loss of sample beam intensity due to scattering. The screen of the attenuator is set so that the recorder pen indicates 100% transmission at the wavelength at which the sample is most transparent. The appropriate wavelength can be found by a trial scan of the spectrum or by a shortcut appropriate to the individual instrument. Use of the attenuator gives a more normal-looking spectrum, since it in effect moves the spectrum to the part of the chart where a given amount of absorbance corresponds to the largest movement of the pen; compare, for example, the distance of pen travel from 0.0 to 0.1 and from 1.0 to 1.1 absorbance units.

Attenuator

O—H Absorption

Absorption in the O—H region in a spectrum obtained by the KBr pellet method must be interpreted with great caution, since it is hard to make sure that no water is in the KBr or gets into the sample during preparation.

Solid Samples: The Mull Method

A second method of determining the IR spectrum of a solid in the solid state is to grind (mull) 2–5 mg with a drop of mineral oil (paraffin oil) or hexachlorobutadiene. The spectrum of the mull is then determined as a liquid film. Since the recorded spectrum will be that of the compound

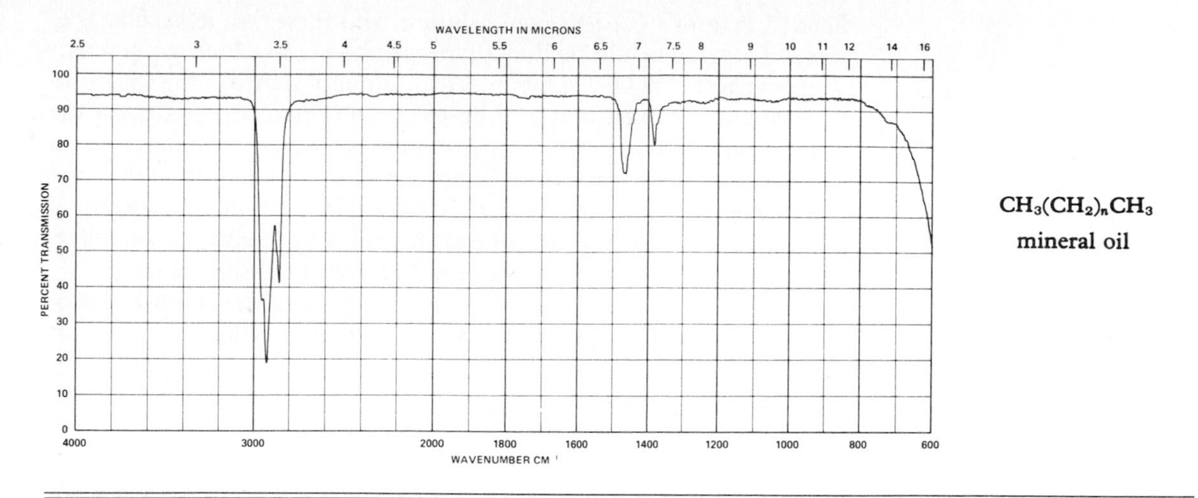

Figure 23.6. IR spectrum of mineral oil; thin film.

plus the mulling agent, the spectrum of the mulling agent must be mentally subtracted from the recorded spectrum. The spectra of mineral oil and hexachlorobutadiene are presented in Figures 23.6 and 23.7.

In Section 23.4, it was stated that the spectra of identical materials should be the same *if* the spectra are obtained under the same conditions. The reason for this qualification is that for some compounds the appearance of the spectrum depends upon the nature of the sample. For example, the appearance of the O—H part of the spectrum of an alcohol depends upon the concentration of the solution. The reason is that in dilute solution a smaller fraction of the molecules are hydrogen

Figure 23.7. IR spectrum of hexachlorobutadiene; thin film.

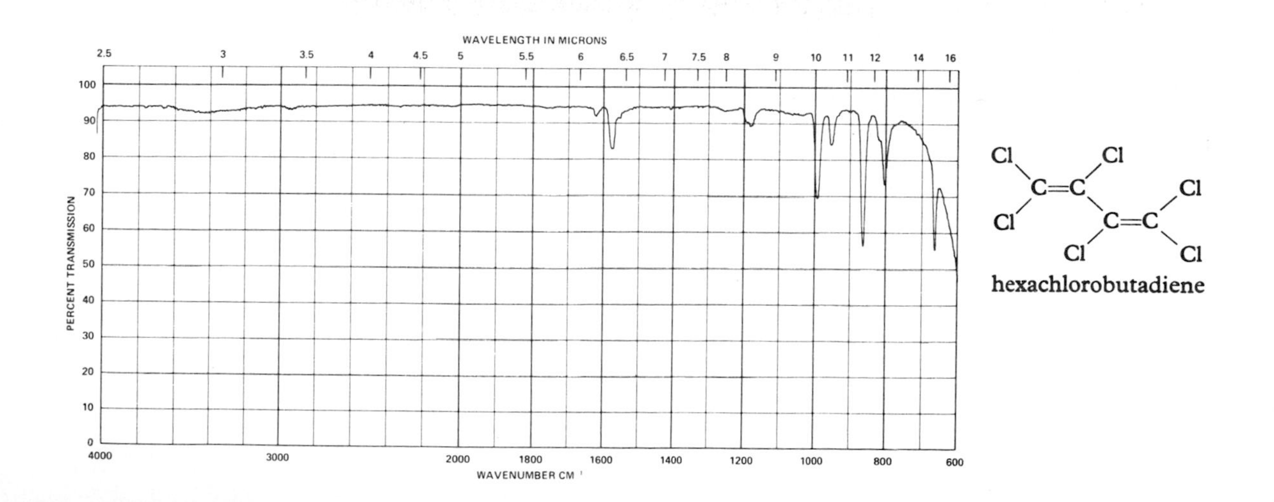

bonded than in a concentrated solution, and these two molecular species—hydrogen bonded and not hydrogen bonded—absorb at different wavelengths in the O—H region. Also, the appearance of the spectrum of a sample obtained by the KBr technique often depends upon the exact details of its preparation, presumably because of differences in particle size and distribution. The spectra of solids obtained in the solid state sometimes differ from spectra obtained in solution, and spectra of the same compound in different solvents will almost always look different because of different regions of strong solvent absorption.

Reference 11 presents an excellent introduction to the practice of infrared spectrometry, especially for sample preparation.

Problems

1. Saturated aliphatic esters (Structure **I**) absorb at about 5.71–5.76 microns. α,β-Unsaturated esters (**II**) absorb at longer wavelength, while vinyl esters (**III**) absorb at shorter wavelength. Explain.

R—C(=O)—O—R	C=C—C(=O)—O—R	R—C(=O)—O—C=C
I	**II**	**III**

2. Interpret the trend in the wavelength of the carbonyl absorption of these compounds.

R—C(=O)—Cl	R—C(=O)—O—R	R—C(=O)—H	R—C(=O)—R	R—C(=O)—O—H	R—C(=O)—NH_2
5.47	5.74	5.78	5.83	5.84	6.10 microns
1795	1742	1730	1715	1712	1665 cm^{-1}

3. Figures 23.8 and 23.9 show the infrared spectra of androstenedione and testosterone (Section 79). Which is the spectrum of androstenedione, and which is the spectrum of testosterone? Explain.

androstenedione

testosterone

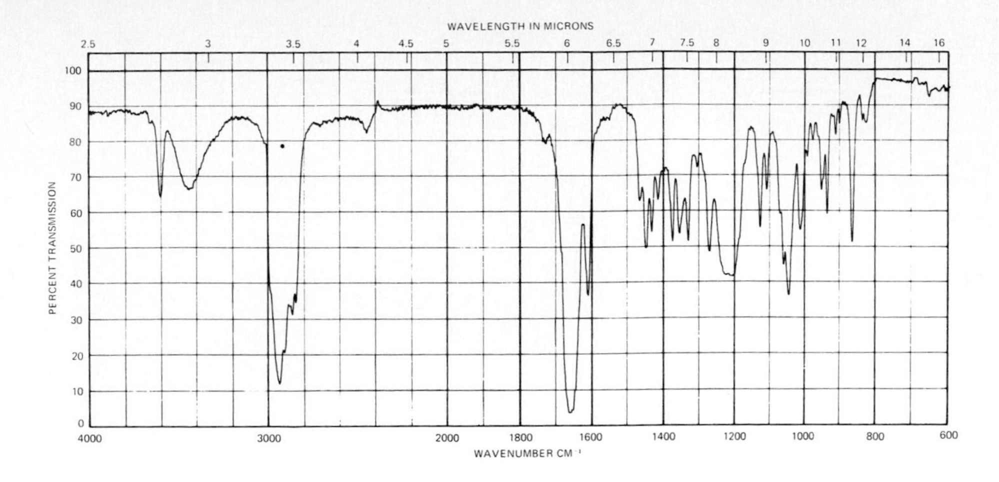

Figure 23.8

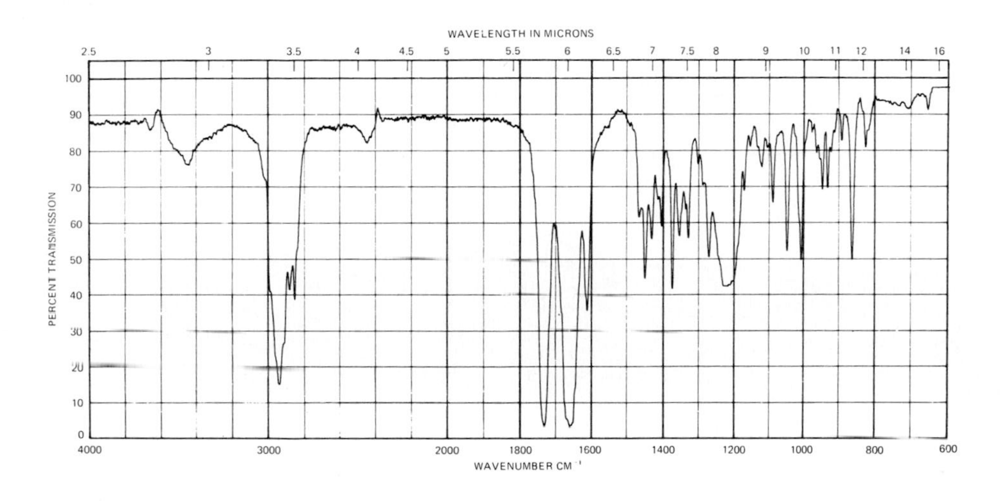

Figure 23.9

4. Figures 23.10 and 23.11 show the infrared spectra of pregnenolone acetate and progesterone (Section 79). Which is the spectrum of pregnenolone acetate, and which is the spectrum of progesterone? Explain.

CH_3, $C{=}O$, CH_3, CH_3, $CH_3{-}\overset{O}{\overset{\|}{C}}{-}O$

pregnenolone acetate

CH_3, $C{=}O$, CH_3, CH_3, O

progesterone

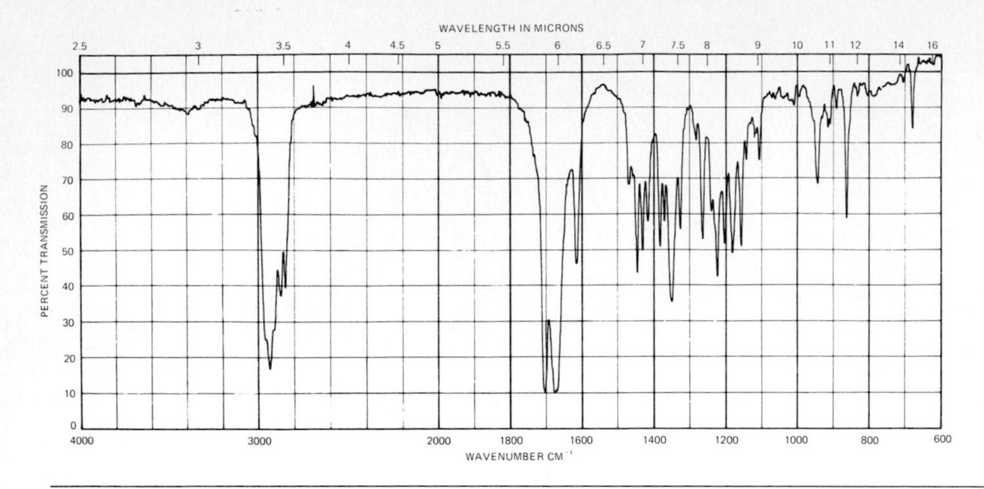

Figure 23.10

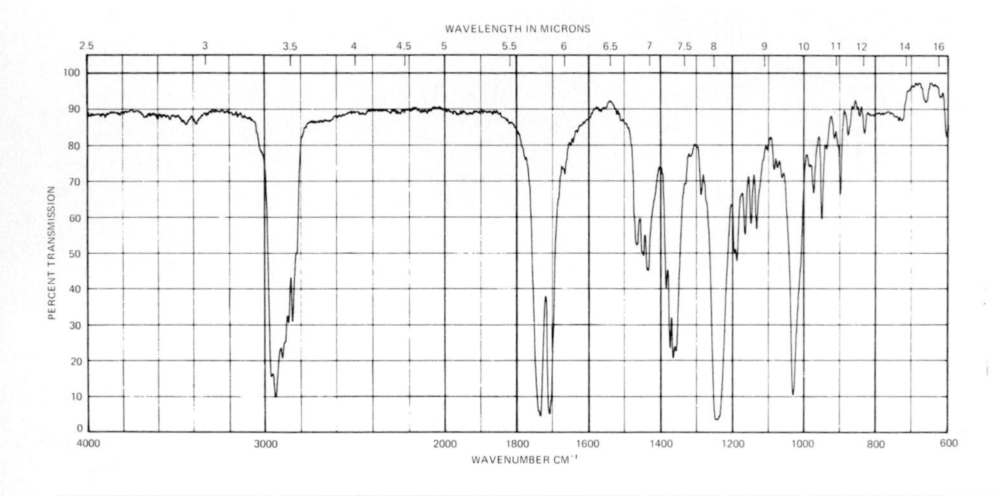

Figure 23.11

References

A brief introduction to infrared absorption spectrometry can be found in

1. P. R. Jones, *Chemistry* **38,** 5 (1965).

Introductory discussions that include spectra-structure correlations are

2. R. M. Silverstein, G. C. Bassler, and T. C. Morrill, *Spectrometric Identification of Organic Compounds*, 4th edition, Wiley, New York, 1981, pp. 95–180.
3. J. R. Dyer, *Applications of Absorption Spectroscopy of Organic Compounds*, Prentice-Hall, Englewood Cliffs, N.J., 1965, pp. 22–57.
4. D. J. Pasto and C. R. Johnson, *Organic Structure Determination*, Prentice-Hall, Englewood Cliffs, N.J., 1969, pp. 109–158.

5. A. D. Cross and R. A. Jones, *Introduction to Practical Infrared Spectroscopy*, 3rd edition, Plenum, New York, 1969.
6. R. T. Conley, *Infrared Spectroscopy*, 2nd edition, Allyn and Bacon, Boston, 1972.

An extensive presentation of spectra-structure correlations is made in:

7. L. J. Bellamy, *The Infrared Spectra of Complex Molecules*, 3rd edition, Halsted, New York, 1975.

Indexes of published infrared spectra include:

8. H. M. Hershenson, *Infrared Absorption Spectra: Index for 1945–1957*, Academic Press, New York, 1959.
9. H. M. Hershenson, *Infrared Absorption Spectra: Index for 1958–1962*, Academic Press, New York, 1964.

A catalog of about 8000 infrared spectra is

10. C. J. Pouchert, *The Aldrich Library of Infrared Spectra*, Aldrich Chemical Co., Inc., Milwaukee, Wisconsin, 1970 (2nd edition, 1975; 3rd edition, 1981).

Finally, an extensive discussion of methods of sample preparation is presented in

11. J. S. Swinehart, *Organic Chemistry*, Appleton-Century-Crofts, 1969, Appendix III.

24 Ultraviolet-Visible Absorption Spectrometry

Ultraviolet-visible spectrometry is the determination of the degree to which a substance absorbs ultraviolet and visible light; the output of a UV-visible spectrometer is a record of energy absorption versus wavelength, the UV-visible spectrum. Usually the wavelength ranges are from 200 nanometers to 360 nm (near ultraviolet range) and from 360 nm to 800 nm (visible range; see Table 23.1). Absorption at shorter wavelengths is more difficult to measure because air absorbs in this region, and absorption at longer wavelengths involves a different mechanism of energy absorption (IR spectrometry; Section 23.3). Since absorption of light energy in these wavelength ranges is the result of the promotion of electrons in lower energy levels to higher energy levels, UV-visible spectrometry is often called *electronic absorption spectrometry*.

Mechanism of UV energy absorption

24.1 ULTRAVIOLET-VISIBLE LIGHT ABSORPTION AND MOLECULAR STRUCTURE

Ultraviolet-visible spectrometry, the oldest of the spectrometric methods, found its greatest use in the determination of the structure of molecules that contain conjugated systems of double bonds. The presence of conjugated unsaturation was always found to result in one or more intense absorption maxima at wavelengths greater than 200 nm. Very useful correlations were deduced for the positions and intensities of these peaks as a function of structure for conjugated dienes (Table 24.1), α,β-unsaturated ketones (Table 24.2), and substituted aromatic compounds.

Table 24.3 summarizes the wavelength maximum and intensity of absorption to be expected for certain other types of compounds. From this, you can see that saturated compounds with O, N, S, or halogen atoms can absorb light in the ultraviolet range because of the presence of unshared pairs of electrons, *n* electrons, on these atoms. Nonconjugated aldehydes and ketones absorb weakly because of the possibility of excitation of a nonbonded pair of electrons into an antibonding π orbital, an $n \rightarrow \pi^{\star}$ transition. However, nonconjugated alkenes cannot absorb light by this mechanism. With the exception of the tail end of an intense absorption band, whose maximum falls at a wavelength shorter than 200 nm ("end absorption"), they are transparent in the ultraviolet region.

Table 24.1. Conjugated dienes: correlations between structure and wavelength of absorption maximum

Usually, the agreement between calculated values and observed values is within 5 nanometers.

	Nanometers
Base values	
Acyclic, conjugated dienes, and conjugated dienes contained in two nonfused 6-membered-ring systems	217
Conjugated dienes contained in two fused 6-membered-ring systems (heteroannular dienes)	214
Conjugated dienes contained in a single ring (homoannular dienes)	253
Increments	
C=C extending conjugation	+30
Each alkyl substituent, including ring residues	+5
Each ring, 6-membered or less, to which the diene double bonds are exocyclic	+5
Each —Cl or —Br substituent	+17
Alkoxy or acyloxy substituent	0
Solvent corrections	
No solvent corrections are necessary.	

From A. Ault, *Problems in Organic Structure Determination*, McGraw-Hill, New York, 1967, p. 16. Reprinted with permission.

Table 24.2. α,β-Unsaturated aldehydes and ketones: correlations between structure and wavelength of absorption maximum

Usually, the agreement between calculated values and observed values is within 5 nanometers.

	Nanometers
Base values	
α,β-Unsaturated ketones	215
α,β-Unsaturated aldehydes	210
Cyclopentenones	205
Increments	
C=C extending conjugation	+30
α-Alkyl substituent, including ring residue	+10
β-Alkyl substituents, including ring residues	+12
γ, δ, or further alkyl substituents, including ring residues	+18
Each ring, 6-membered or less, to which a C=C is exocyclic	+5
C=C contained in a 5-membered ring (except cyclopentenones)	+5
Enolic α or β—OH	+35
(Alkoxy and acyloxy substituents are treated as alkyl groups.)	
Solvent corrections	
Water	−8
Methanol and ethanol	0
Chloroform	+1
Dioxane	+5
Ether	+7
Hexane	+11

From A. Ault, *Problems in Organic Structure Determination*, McGraw-Hill, New York, 1967, p. 15. Reprinted with permission.

24.2 INTERPRETATION OF UV-VISIBLE SPECTRA

The two features of an ultraviolet-visible spectrum that can give information concerning molecular structure are (1) the wavelengths at which the maximum light is absorbed, and (2) the intensity of absorption at that wavelength. The way in which the degree of absorption is expressed and measured experimentally is explained in Section 23.2. Certain correlations of these two parameters with molecular structure are summarized in the three preceding tables.

Since the development of infrared and nuclear magnetic resonance spectrometry, ultraviolet spectrometry has become less important for the determination of structure of organic molecules. For this purpose, the newer methods are simply more convenient and more reliable. For example, the establishment of the presence of a carbonyl group is much

Table 24.3. Correlations between ultraviolet light absorption and molecular structure

Type of Compound	*Transition*	*Wavelength*[a]	ϵ[b]
Alkane	$\sigma \rightarrow \sigma^{\star}$	<200	
Nonconjugated alkene	$\pi \rightarrow \pi^{\star}$	<200	
Nonconjugated alkyne	$\pi \rightarrow \pi^{\star}$	<200	
Alcohols	$n \rightarrow \sigma^{\star}$	185	10^2
Ethers	$n \rightarrow \sigma^{\star}$	185	10^3
Amines	$n \rightarrow \sigma^{\star}$	200	10^3
Mercaptans	$n \rightarrow \sigma^{\star}$	200; 230	10^3; 10^2
Sulfides	$n \rightarrow \sigma^{\star}$	210; 240	10^3; 10^2
Alkyl chlorides	$n \rightarrow \sigma^{\star}$	175	10^2
Alkyl bromides	$n \rightarrow \sigma^{\star}$	210	10^2
Alkyl iodides	$n \rightarrow \sigma^{\star}$	260	10^2
Nonconjugated aldehydes	$n \rightarrow \pi^{\star}$	290	10
Nonconjugated ketones	$n \rightarrow \pi^{\star}$	280–290	10
Carboxylic acid derivatives	$n \rightarrow \pi^{\star}$	200–210	10–10^2
Conjugated alkenes	$\pi \rightarrow \pi^{\star}$	Table 24.1	
Conjugated aldehydes and ketones	$\pi \rightarrow \pi^{\star}$	Table 24.2	
Aromatic compounds	$\pi \rightarrow \pi^{\star}$	200–visible	

[a] Approximate wavelength maximum, in nanometers.
[b] Approximate extinction coefficient; see Section 23.2.

easier and more certain through infrared spectrometry. The principal application of UV spectrometry to determination of molecular structure lies at present in the study of more complex molecules containing conjugated systems.

Ultraviolet spectrometry is, however, still widely used as a quantitative method of analysis: to determine concentrations. This is because many compounds absorb strongly even at low concentration, and because UV spectrometers can determine the degree of absorption of a sample with great accuracy.

24.3 COLOR AND MOLECULAR STRUCTURE

Color is the result of a selective reflection or transmission of light in the visible range of the electromagnetic spectrum. A white or colorless substance reflects or transmits all incident light equally well. For any substance to appear colored, then, it must selectively absorb light of certain wavelengths in the visible part of the spectrum. For most organic compounds, the absorption of visible light requires the presence of a π electronic system that contains four or more conjugated double bonds. Thus, colored compounds are most often found to be aromatic rather than aliphatic substances.

A substance that absorbs light near the blue end of the visible spectrum appears yellow. As the wavelength of visible absorption moves toward the red end of the spectrum, the observed color becomes orange, red, and finally purple or blue. Colored organic compounds that are not yellow or orange are very unusual.

The most common types of colored compounds are nitro-substituted phenol or aniline derivatives. Less common types are quinones and compounds containing the azo (—N=N—) system. Many dyes and colorings combine these two structural features (see Section 68).

24.4 SAMPLE PREPARATION

Usually, a suitable sample for UV-visible spectrometry will be a solution of the substance in a solvent transparent to UV and visible radiation. Thus, choices of sample concentration and solvent must be made, in addition to a choice of cell type.

Concentration

As stated in Section 23.2, absorbance is proportional to the concentration of the sample (c, in moles per liter) and the sample thickness, or path length (l, in cm): $A = \epsilon c l$.

A knowledge of the value of the extinction coefficient, ϵ, at the absorption maximum is of use in the determination of the molecular structure of the sample. Therefore, you will want to determine the spectrum at a concentration that will give an absorbance of about 1 or a little less, because the value of ϵ can be estimated most accurately under these conditions. Since the standard sample cell has a path length of 1.00 cm, the concentration, c, in moles per liter, will equal $1/\epsilon$ when $A = 1$; and for A not to exceed 1, c should be equal to or less than $1/\epsilon$. Table 24.4 summarizes the concentration, in moles per liter, required for an absorbance of 1 for absorption bands of various extinction coeffi-

Table 24.4. Concentration c required for an absorbance of 1 for absorption bands of various extinction coefficients ϵ

	ϵ	c
Very weak band	10	0.1 M
Weak band	100	0.01 M
Medium band	1,000	0.001 M
Strong band	10,000	0.0001 M

cients. Table 24.5 presents some data that may be useful in preparing samples of various concentrations. Of course, if the molecular weight of a substance is not known with certainty, the calculated value of the extinction coefficient will contain a corresponding uncertainty.

There are two important practical consequences of the fact that values for ϵ for different functional groups can differ by several powers of 10. First, observing that a sample of 0.0001 *M* concentration appears to be completely transparent ($A \approx 0$) does not necessarily mean that absorption maxima with $\epsilon \approx 100$ or less are absent. This occurs because, at a concentration this low, they would show an absorbance of only 0.01 or less, which could easily be overlooked. Similarly, a strong absorption band can completely obscure a weak one. The second consequence is that for the accurate determination of extinction coefficients that differ by much more than a power of 10, spectra obtained at different concentrations will be needed. The UV spectra presented in Figures 70.3 and 70.4, obtained at concentrations that differ by a factor of 100, illustrate this very nicely.

Solvent

The ideal solvent for UV-visible spectrometry is completely transparent over the entire spectral range. In contrast to the case with IR spectrometry, many solvents do approach the ideal fairly closely. These include cyclohexane, ethanol, methanol, and water. In order to compensate for light absorption by the solvent and the cell (and any absorption by impurities in the solvent), UV-visible spectrometers are generally run in the double-beam mode wherein the absorption of pure solvent in a second cell is automatically subtracted from the absorption of the sample solution. The spectrum recorded is the difference spectrum, and it represents the net absorption due to the solute.

Beware of contamination

Since small concentrations of many compounds result in very intense absorption, a small amount of an impurity introduced during

Table 24.5. Milligrams of sample required for 5 mL of solution as a function of molecular weight and desired concentration[a]

	Molecular Weight				
Concentration	100	200	300	400	500
0.1 *M*	50	100	150	200	250
0.01 *M*	5	10	15	20	25
0.001 *M*	0.5	1	1.5	2	2.5
0.0001 *M*	0.05	0.1	0.15	0.2	0.25

[a] The most dilute solutions must be made by dilution of a more concentrated solution.

careless sample preparation may dominate the spectrum and completely obscure the spectrum of the compound you are interested in.

Cell Type

The standard UV-visible cell is a rectangular cell having a 1-cm path length and approximately a 3-mL volume. Circular cells and cells of different path lengths are also available.

For visible spectra, Pyrex glass cells are appropriate. For spectra in the UV range, the much more expensive quartz cells will be needed, since Pyrex absorbs strongly at wavelengths shorter than 320 nm.

Problems

1. Four derivatives of cholesterol—A, B, C, and D—have the following spectral properties:

   A IR: strong absorption near 5.95 microns
   UV: λ_{max} = 230 nm; ϵ = 10,700

   B IR: strong absorption near 5.95 microns
   UV: λ_{max} = 241 nm; ϵ = 16,600

   C IR: no absorption near 5.95 microns
   UV: λ_{max} = 234 nm; ϵ = 20,000

   D IR: no absorption near 5.95 microns
   UV: λ_{max} = 315 nm; ϵ = 19,800

   To each of the compounds, assign one of the following molecular structures—**I, II, III,** or **IV**—where R is

CH_3 CH_3
CH_3

R O **I**

R O **II**

R **III**

R **IV**

2. Absolute ethanol (anhydrous ethanol) can be prepared by adding benzene to 95% ethanol and distilling. The first distillate is a ternary azeotrope of benzene, water, and ethanol; the next is a binary azeotrope of benzene and ethanol. Benzene has a maximum extinction coefficient of 200 at 256 nm. What is the minimum concentration of benzene (mg/liter) that could be detected by UV spectroscopy, assuming that an absorbance as little as 0.02 could be determined?
3. LSD has a maximum extinction coefficient of less than 1000. What are the chances of detecting 10 micrograms of LSD by means of UV spectrometry?
4. Calculate the wavelength to be expected for the $\pi \rightarrow \pi^*$ transition for 2-methyl-2-pentenal. Compare your calculated value with the position of the peak in the spectrum shown in Figure 70.3.

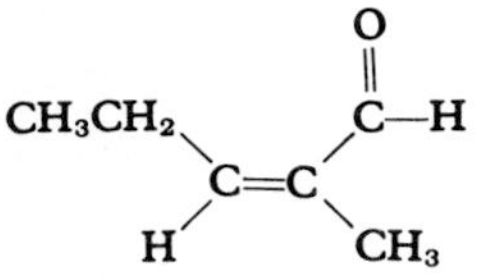

CH_3CH_2—CH=C(CH_3)—CHO (structure: CH_3CH_2 and H on one carbon, C(=O)—H and CH_3 on the other)

2-methyl-2-pentenal

References

1. R. M. Silverstein, G. C. Bassler, and T. C. Morrill, *Spectrometric Identification of Organic Compounds*, 4th edition, Wiley, New York, 1981, pp. 305–331.
2. J. R. Dyer, *Applications of Absorption Spectroscopy of Organic Compounds*, Prentice-Hall, Englewood Cliffs, N.J., 1965, pp. 4–21.

25 Nuclear Magnetic Resonance Spectrometry

Mechanism of energy absorption

Nuclear magnetic resonance spectrometry (NMR) measures the absorption of "light" energy in the radiofrequency portion of the electromagnetic spectrum. The mechanism of energy absorption is the reorientation of magnetic nuclei with respect to a large external magnetic field, H_{ext}, in which the sample is placed. Unlike a compass needle in the earth's magnetic field, which can have a great many different orientations, a nuclear magnet can have only a few. The proton, for example, can have only two orientations with respect to the field: one "with" the field (of lower energy) and one "against" the field (of higher energy). NMR spectrometry is based on the fact that the energy of the quantum of absorbed *radiofrequency* (rf) *radiation* goes into changing the orientation of the magnetic nucleus from being "with" the magnetic field to being "against" the field.

25.1 SHIELDING; CHEMICAL SHIFT

Since the difference in energy of the proton in the two states "with" and "against" depends upon the size of the magnetic field that the proton experiences (H_{nuc}), and since the energy for transitions between the

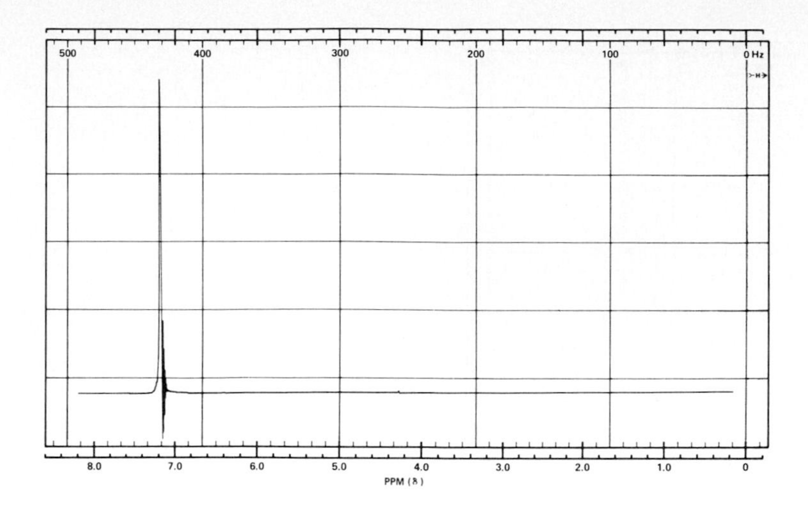

H H H H H H
benzene

Figure 25.1. NMR spectrum of benzene; neat.

two states ($h\nu$) is supplied by the rf field, the following relationship holds:

$$h\nu \propto H_{\text{nuc}}$$

where $h\nu$ is, according to Planck's relationship, the energy of the quanta of frequency ν. Protons experiencing different magnetic fields H_{nuc} will therefore absorb rf energy at different frequencies. For an NMR spectrometer operating at a frequency of 60 Megahertz (60 MHz), H_{nuc} must equal 14,092 gauss for energy to be absorbed (for *resonance* to occur).

Resonance condition

$H_{\text{nuc}} = 14{,}092$ gauss

The usefulness of NMR spectrometry to the organic chemist stems from the fact that *protons in different structural environments experience a different* H_{nuc} *for the same* H_{ext}. For a particular proton, the difference between H_{nuc} and H_{ext} is called the *shielding*:

$$H_{\text{nuc}} = H_{\text{ext}} - H_{\text{shielding}}$$

Thus, if the sample in the magnetic field H_{ext} is irradiated with a constant rf field of exactly 60 MHz and H_{ext} is slowly increased, energy will be absorbed by the sample each time H_{nuc} (which equals $H_{\text{ext}} - H_{\text{shielding}}$) becomes equal to 14,092 gauss. The graph of energy absorption versus H_{ext} is the NMR spectrum. The NMR spectrum of benzene shows only a single instance of energy absorption (a single resonance; see Figure 25.1), whereas the NMR spectrum of *p*-xylene shows two resonances, one for the methyl protons and one for the ring protons (Figure 25.2). It is important to remember that shielding is propor-

CH_3 H H H H CH_3
p-xylene

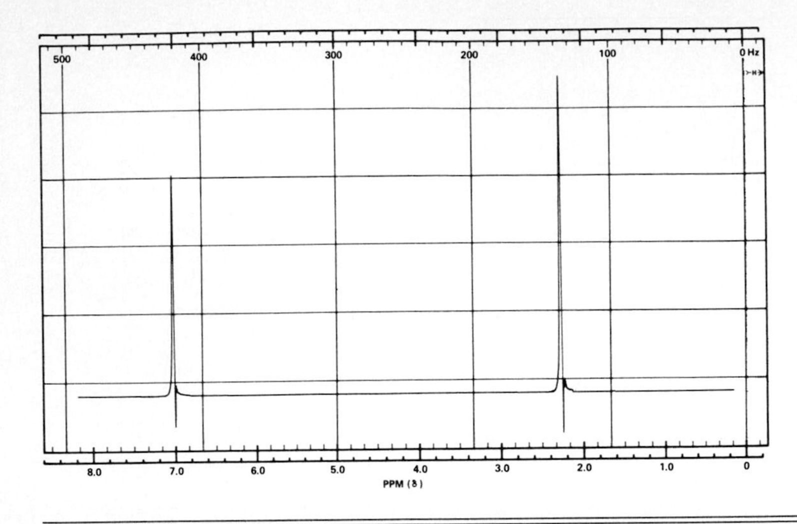

CH_3
H H
H H
CH_3
p-xylene

Figure 25.2. NMR spectrum of *p*-xylene; neat.

CH_3
CH_3—Si—CH_3
CH_3
tetramethylsilane (TMS)

Scales of chemical shift

tional to H_{ext} and that differences in shielding, or chemical shift differences, will also be proportional to H_{ext}.

Chemical shifts are usually reported as the amount by which the H_{ext} required for the resonance of the sample substance varies from the H_{ext} required for the resonance of a *reference substance, tetramethylsilane* (*TMS*). A unit of convenient size is one part per million (ppm) of H_{ext}, and the TMS resonance is taken sometimes as 0 (δ scale) and sometimes as 10 (τ scale). For comparison with coupling constants (see Section 25.2), chemical shift differences must be expressed in frequency units; 1 ppm of 60 MHz is 60 Hz. The relationships between the various units and scales are shown in Table 25.1.

Table 25.1. Units of chemical shift

Increasing H_{ext}: →
Increasing Frequency: ←
Increasing Shielding: →

11	10	9	8	7	6	5	4	3	2	1	0	−1	δ
−1	0	1	2	3	4	5	6	7	8	9	10	11	τ
330	300	270	240	210	180	150	120	90	60	30	0	−30	Hz[a]
660	600	540	480	420	360	300	240	180	120	60	0	−60	Hz[b]
1100	1000	900	800	700	600	500	400	300	200	100	0	−100	Hz[c]
											↑ TMS resonance		

[a] 30-MHz instrument.
[b] 60-MHz instrument.
[c] 100-MHz instrument.

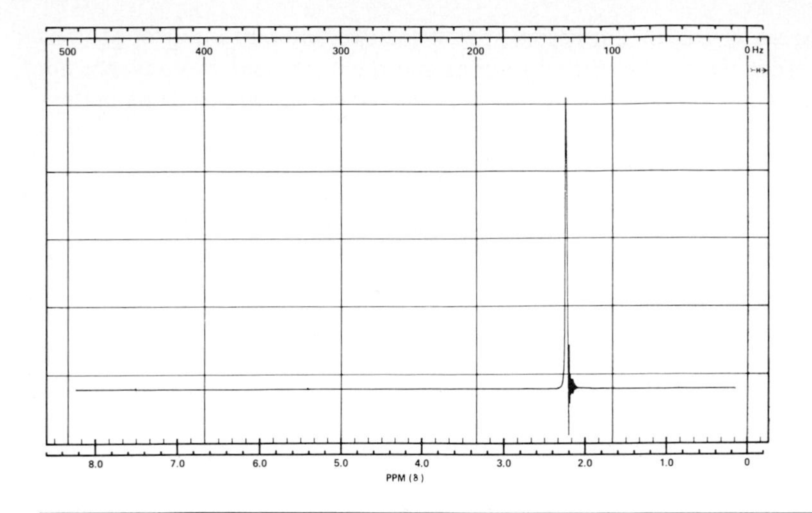

H
H—C—I
H
methyl iodide

Figure 25.3. NMR spectrum of methyl iodide; neat.

25.2 SPLITTING

Although the presence of resonances at different chemical shifts gives information about the numbers of different functional groups in which protons are present, the features of the NMR spectrum that provide the most detailed information about molecular structure are those that are the result of the *effect upon one another of protons in different groups at different chemical shifts* (coupling).

Coupling

While the NMR spectrum of methyl iodide (Figure 25.3) consists of a single peak, the NMR spectrum of ethyl iodide (Figure 25.4) shows

Figure 25.4. NMR spectrum of ethyl iodide; neat.

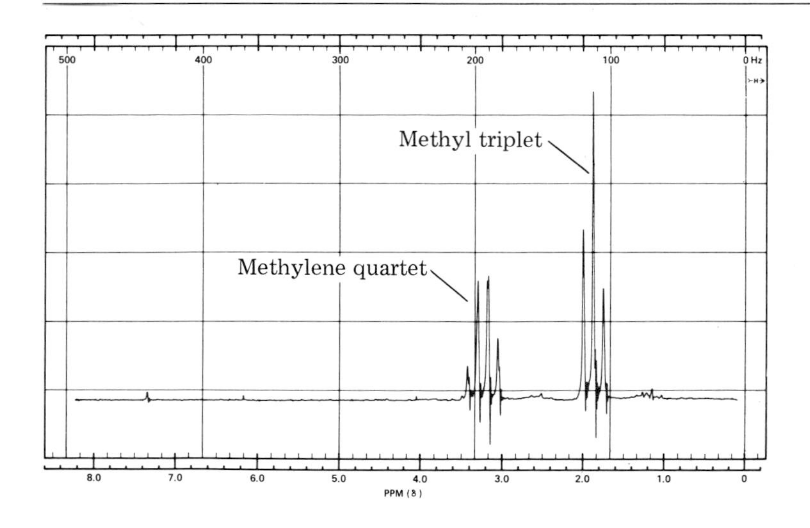

H H
I—C—C—H
H H
ethyl iodide

two *sets* of peaks. The resonances of the two different kinds of protons in ethyl iodide, the methyl protons and the methylene protons, are not each a single peak, but each is a set of peaks, or ***multiplet***. The methylene protons

Multiplets

```
   H            H  H
   |            |  |
H—C—I        H—C—C—I
   |            |  |
   H            H  H
methyl iodide   ethyl iodide
```

have caused the methyl resonance to appear as a triplet of approximate relative intensity 1 : 2 : 1, and the methyl protons have caused the resonance of the methylene protons to appear as a quartet of approximate relative intensity 1 : 3 : 3 : 1.

In general, it is necessary to use the methods of quantum mechanics to calculate the expected multiplicity of a resonance. It is fortunate that in many cases a very much simpler analysis is adequate.

Interpretation of Splitting of the CH_3 Resonance of CH_3CH_2I

1:2:1 Triplet

Consider first the appearance of the methyl resonance of ethyl iodide as a 1 : 2 : 1 triplet due to the presence of the two adjacent methylene protons. Because of the small energy difference for a proton in either of the two states, *a particular proton in a magnetic field of 14,092 gauss at room temperature has an almost equal probability of being in either the lower or the upper spin state*—of being oriented, that is, either "with" or "against" the external magnetic field. Therefore, for the purpose of this analysis, you can estimate that of the ethyl iodide molecules one-fourth will have both the methylene protons oriented "with" H_{ext} (↑ ↑), one-fourth will have both the methylene protons "against" H_{ext} (↓ ↓), and one-fourth will be in each of the two states possible where one proton is "with" and one "against" H_{ext} ($\frac{1}{4}$ as ↑ ↓ and $\frac{1}{4}$ as ↓ ↑).

The effect of having both of the two adjacent methylene protons "with" the external field is to bring the methyl protons of those molecules into resonance at a slightly lower value of H_{ext}, since the two protons "with" the field will augment H_{ext} for the neighboring methyl group. This effect of the adjacent protons can be expressed by Equation 25.2-1, where $H_{coupling}$ in this case is positive:

$$H_{nuc} = H_{ext} - H_{shielding} + H_{coupling} \tag{25.2-1}$$

Thus, the methyl resonance for one-fourth of the molecules will occur at slightly lower field (smaller H_{ext}) than would be expected in the absence of coupling.

The methyl resonance for half of the molecules will occur at the same chemical shift as would be expected in the absence of coupling. These are the molecules in which the two methylene protons are oriented in opposite directions, cancelling one another, so that in this case, in Equation 25.2-1, $H_{coupling}$ has a value of zero.

In the remaining fourth of the molecules, the adjacent methylene protons will have their spins oriented so that both oppose H_{ext}, diminishing the effect of H_{ext} on the methyl groups of these molecules. These molecules will thus have their methyl resonances at a slightly higher H_{ext} than would be expected in the absence of coupling; in this case, in Equation 25.2-1, $H_{coupling}$ is negative. In ethyl iodide, as in most ethyl groups, the magnitude of this effect, the coupling constant J, is about 7 Hz or 7/60 of 1 ppm in a 60-MHz machine.

It is important to notice in this analysis that the multiplicity of the methyl resonance is interpreted as an effect of the two *neighboring* protons. In this case, the protons of the methyl group have no apparent effect upon the multiplicity of their own resonance.

Splitting is due to neighbors

Interpretation of the Splitting of the CH_2 Resonance of CH_3CH_2I

The multiplicity of the resonance of the methylene group of ethyl iodide can be explained in a similar way. One-eighth of the molecules would be expected to have the protons of the adjacent methyl group all "with" H_{ext} ($\uparrow\uparrow\uparrow$), three-eighths would be expected to have two "with" and one "against" ($\frac{1}{8}$ $\uparrow\uparrow\downarrow$, $\frac{1}{8}$ $\uparrow\downarrow\uparrow$, and $\frac{1}{8}$ $\downarrow\uparrow\uparrow$), three-eighths would be expected to have one "with" and two "against" ($\frac{1}{8}$ $\uparrow\downarrow\downarrow$, $\frac{1}{8}$ $\downarrow\uparrow\downarrow$, and $\frac{1}{8}$ $\downarrow\downarrow\uparrow$), and one-eighth would be expected to have all three "against" ($\downarrow\downarrow\downarrow$). Thus, the molecules are divided into four groups of relative population 1 : 3 : 3 : 1, in which the methylene protons are expected to go into resonance at slightly different values of H_{ext} because of the effect of the adjacent methyl protons. Again, the multiplicity of the resonance of a group of protons is explained by the effect of the adjacent protons; the multiplicity of a resonance does not depend upon the number of protons in the group corresponding to that resonance.

1:3:3:1 Quartet

Splitting is due to neighbors

In general, when a simple analysis such as this applies, the number of peaks in a multiplet is one more than the number of adjacent protons; the "N + 1 Rule": a single adjacent proton will split the resonance of a group of neighboring protons into a 1 : 1 doublet; two adjacent protons will split the resonance of a group of neighboring protons into a 1 : 2 : 1 triplet; etc. (Table 25.2). The relative intensities of the peaks of a multiplet follow the coefficients of the binomial expansion: 1 : 1, 1 : 2 : 1, 1 : 3 : 3 : 1, 1 : 4 : 6 : 4 : 1, 1 : 5 : 10 : 10 : 5 : 1, etc.

N + 1 Rule

This analysis is appropriate *if two conditions are met.* First, the

Table 25.2. The N + 1 Rule

Number of Neighbors	*Number of Peaks in Multiplet*	*Name of Multiplet*	*Relative Intensities within Multiplet*
0	1	singlet	1
1	2	doublet	1 : 1
2	3	triplet	1 : 2 : 1
3	4	quartet	1 : 3 : 3 : 1
4	5	quintet	1 : 4 : 6 : 4 : 1
5	6	sextet	1 : 5 : 10 : 10 : 5 : 1
6	7	septet	1 : 6 : 15 : 20 : 15 : 6 : 1
.	.	.	.
.	.	.	.
.	.	.	.
N	N + 1		

average coupling constant, $\mathcal{J}$, between each proton in one group and each proton in the other must be the same, and second, the coupling constant, $\mathcal{J}$, must be small compared to the chemical shift difference between the two groups, $\Delta\delta$. Magnetic equivalence must prevail.

Magnetic equivalence

Figures 25.5 through 25.10 present the NMR spectra of some compounds that can be readily interpreted by this simple method of analysis. Figure 25.5 shows the spectrum of 1,1,2-trichloroethane, in which the methylene protons appear as an approximately 1 : 1 doublet and the single proton as an approximately 1 : 2 : 1 triplet. In Figure 25.6, the two equivalent end protons of 1,1,2,3,3-pentachloropropane

Figure 25.5. NMR spectrum of 1,1,2-trichloroethane; CCl_4 solution.

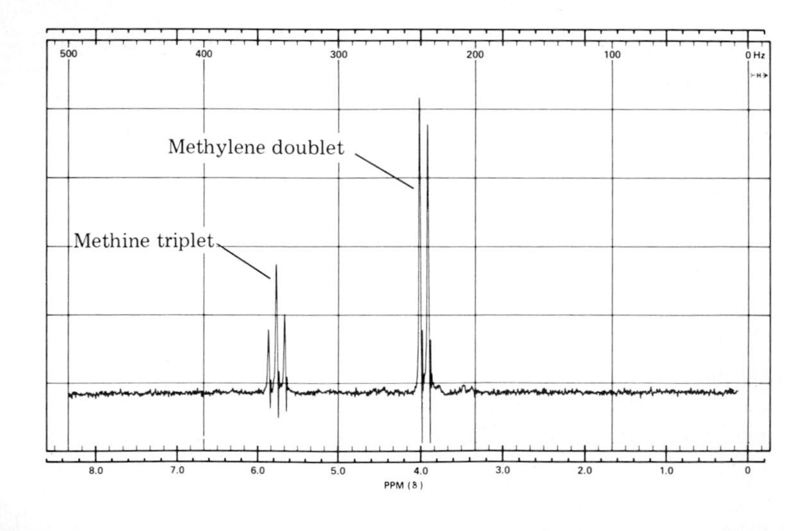

Cl H
| |
H—C—C—Cl
| |
Cl H

1,1,2-trichloroethane

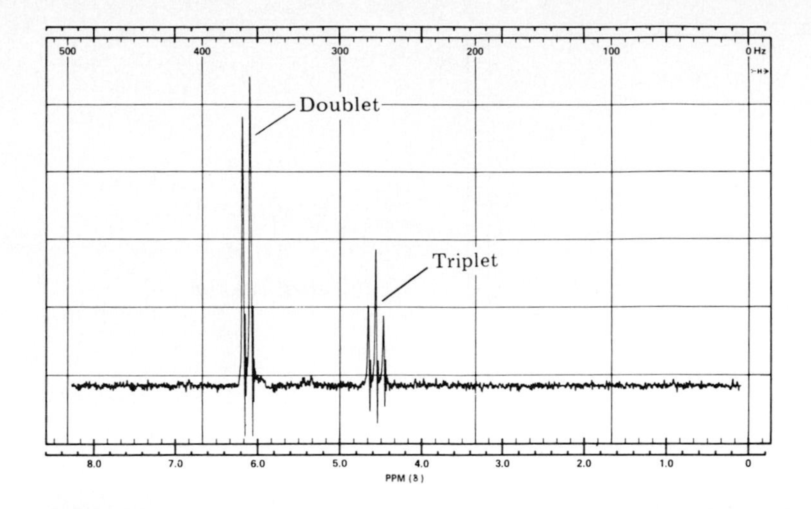

Cl H Cl
H—C—C—C—H
Cl Cl Cl

1,1,2,3,3-pentachloropropane

Figure 25.6. NMR spectrum of 1,1,2,3,3-pentachloropropane; CCl_4 solution.

appear as an approximately 1 : 1 doublet, and the middle proton appears as an approximately 1 : 2 : 1 triplet.

Cl H
H—C—C—Cl
Cl H

1,1,2-trichloroethane

Cl H Cl
H—C—C—C—H
Cl Cl Cl

1,1,2,3,3-pentachloropropane

Figure 25.7. NMR spectrum of 1,1-dichloroethane; neat.

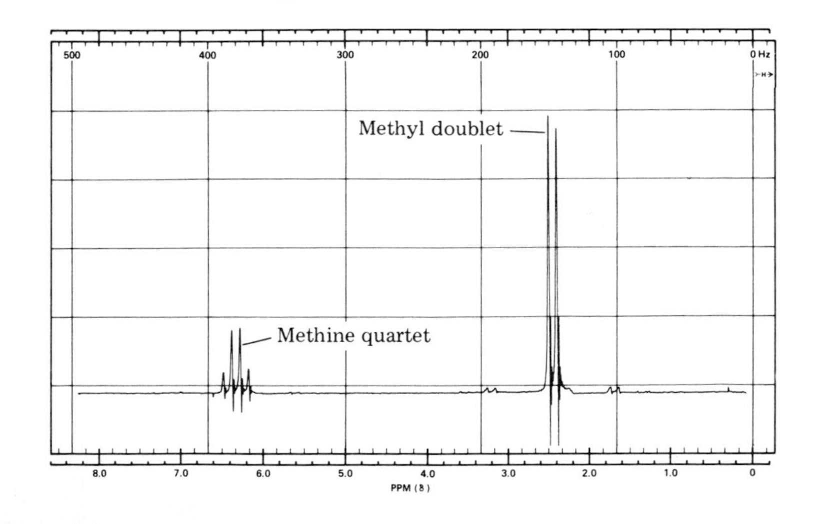

Cl H
H—C—C—H
Cl H

1,1-dichloroethane

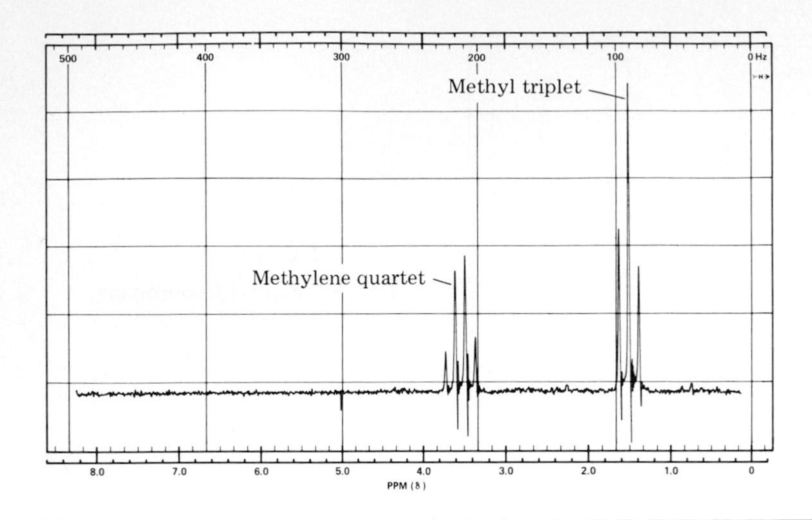

H H
Cl—C—C—H
H H
ethyl chloride

Figure 25.8. NMR spectrum of ethyl chloride; CCl_4 solution.

The ethyl resonance

Figure 25.7 shows the NMR spectrum of 1,1-dichloroethane, in which the methyl group appears as an approximately 1 : 1 doublet, and the single proton as an approximately 1 : 3 : 3 : 1 quartet. Figure 25.8 shows the NMR spectrum of ethyl chloride, which illustrates again the characteristic resonance of the ethyl group: an approximately 1 : 3 : 3 : 1 quartet for the methylene protons and an approximately 1 : 2 : 1 triplet for the methyl protons. Notice that the actual relative intensities of the peaks within a multiplet differ slightly from the predicted 1 : 1, 1 : 2 : 1,

Figure 25.9. NMR spectrum of 1,2-dichloroethane; neat.

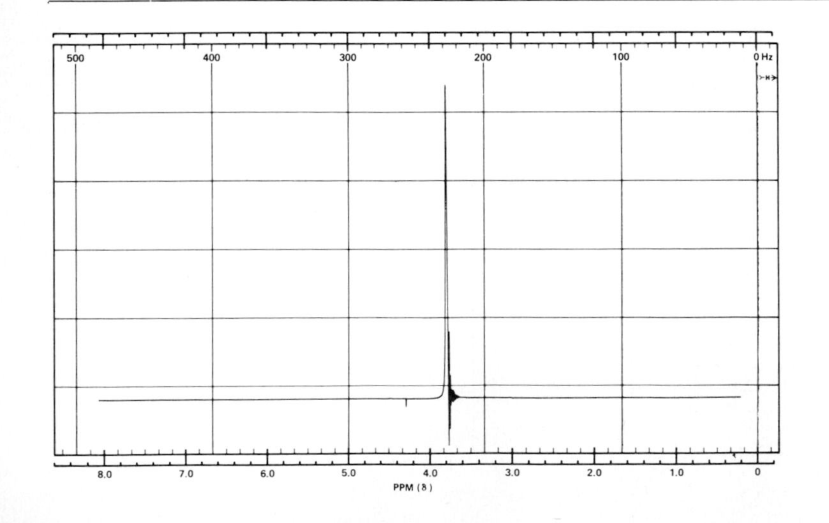

H H
Cl—C—C—Cl
H H
1,2-dichloroethane

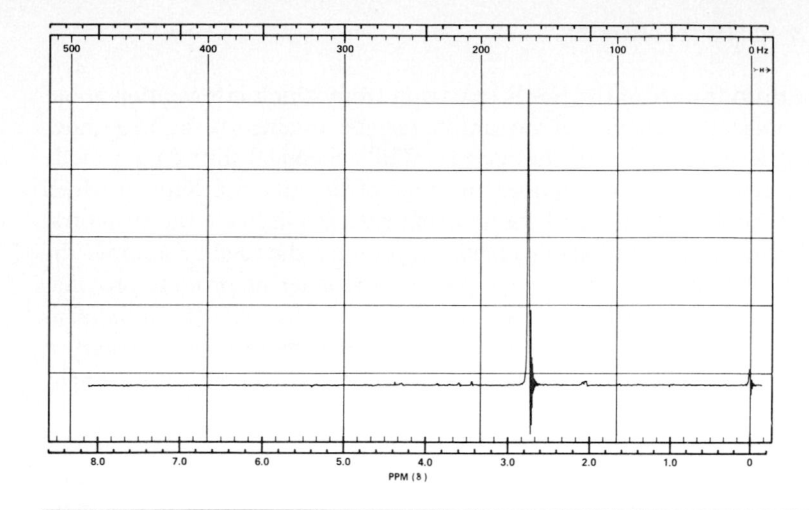

Cl H
Cl—C—C—H
Cl H
1,1,1-trichloroethane

Figure 25.10. NMR spectrum of 1,1,1-trichloroethane; CCl_4 solution.

etc., relative intensities. The distortion is such that the stronger peaks are on the side nearer the resonance of the nuclei to which they are coupled. In this sense, the peaks of a multiplet can be said to "point" toward the resonance of the nuclei with which they are coupled. Thus, in Figure 25.8, the methylene quartet is said to point toward the methyl resonance, and the methyl triplet is said to point toward the methylene quartet. By now you should expect that the NMR spectra of 1,2-dichloroethane and 1,1,1-trichloroethane should each consist of a single peak. They do; see Figures 25.9 and 25.10.

"Pointing"

Cl H
H—C—C—H
Cl H
1,1-dichloroethane

H H
Cl—C—C—H
H H
ethyl chloride

H H
Cl—C—C—Cl
H H
1,2-dichloroethane

Cl H
Cl—C—C—H
Cl H
1,1,1-trichloroethane

You should also notice from this series of spectra that protons bonded to a carbon bearing a chlorine atom are deshielded; two chlorine atoms appear to deshield more than one. The dependence of shielding upon molecular structure is discussed in more detail in Section 25.4.

25.3 THE INTEGRAL

The third feature of the NMR spectrum from which information about the molecular structure of the sample may be obtained is the area under the absorption peaks, or the *integral*. While chemical shift data provide information about the number and type of functional groups in which sets of protons occur, and the splitting patterns indicate the structural and geometrical relationships between protons, the total area under the peaks of a multiplet that corresponds to any set of protons provides information about the number of protons in that set: the integral is proportional to the number of protons. The degree to which energy is absorbed by a set of protons is independent of its structural environment.

The area under the peaks of an NMR spectrum is determined electronically by use of the NMR spectrometer in a separate operation after obtaining the spectrum. Figure 25.11 shows the NMR spectrum of ethyl iodide and its integral. The vertical displacement of the second (uppermost) trace is proportional to the area under the corresponding peaks. The total area of the methylene quartet is seen to be two-thirds that of the methyl triplet. You can also estimate in the spectra of Figures 25.4 through 25.8 that the relative area of each multiplet corresponds to the relative number of protons involved.

25.4 NUCLEAR MAGNETIC RESONANCE AND MOLECULAR STRUCTURE

The amount of energy absorbed by a proton is independent of its structural environment, and therefore the integral has the same value

Figure 25.11. NMR spectrum and integral of ethyl iodide; neat.

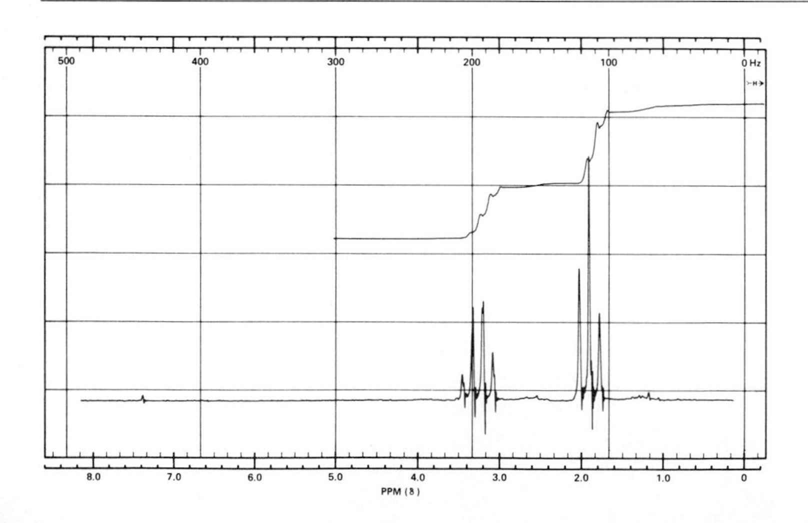


```
   H  H
   |  |
I—C—C—H
   |  |
   H  H
```

ethyl iodide

Table 25.3. Correlations between chemical shift and molecular structure

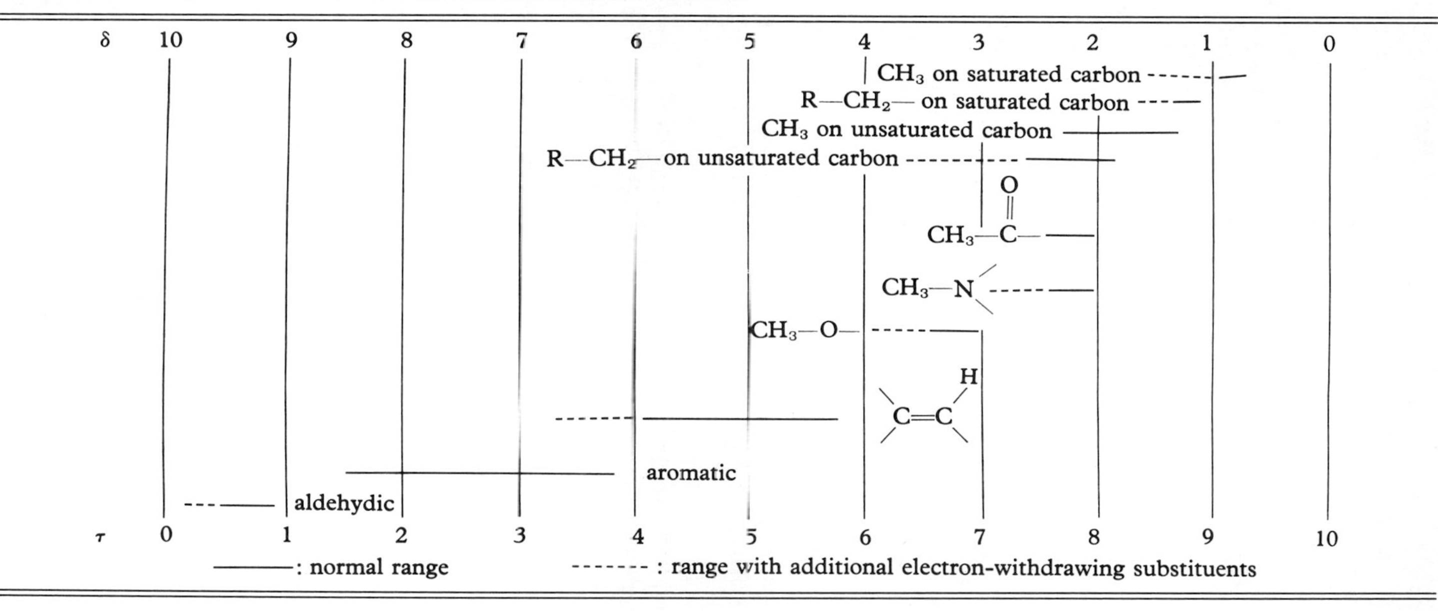

Table 25.4. Correlations between coupling constants and molecular structure

	J (Hz)		J (Hz)
H—C—C—H	~7	>C=C(H)—C—H	~7
H—C—C—C—H	~0	H(C=C)C—H (cis)	~2
>C=C(H)H	0–2	H(C=C)C—H (trans)	~1.5
H(C=C)H (cis)	~10	H—C—C(=O)—H	2–3
H(C=C)H (trans)	~17	H—C₆H₄—H (benzene ring)	o ~ 9 m ~ 3 p ~ 0

Structure and chemical shift

for every proton. In contrast, the shielding or chemical shift (shielding relative to a standard) does depend upon the structural environment of a proton. Table 25.3 presents some of the correlations between chemical shift and molecular structure. From the table, you can see two trends. (1) Methyl protons are more shielded than analogous methylene protons, which, in turn, are more shielded than methine protons. (2) Electron-withdrawing groups tend to deshield adjacent protons.

Structure and coupling constants

Coupling constants depend in a very sensitive way on the structural and geometrical relationships between nuclei. Examples of the ways in which coupling depends on relationships between nuclei are presented in Table 25.4. A study of the table suggests that, in general, coupling does not extend over more than three bonds, but that sp^2 hybridized atoms are better conductors of coupling than are sp^3 hybridized atoms.

25.5 INTERPRETATION OF NMR SPECTRA

If the NMR spectrum is that of a pure substance of unknown molecular structure, the goal of interpretation is to deduce the structure of the molecules of the sample. In contrast to the case with IR spectrometry, it is sometimes possible to deduce the structural formula of a substance from its molecular formula and NMR spectrum alone, without the use of reference spectra. The isomers of molecular formula $C_2H_4Cl_2$, for example, can be distinguished by their NMR spectra alone (Figures 25.7 and 25.9).

These are the steps of such an interpretation:

1. From the integral, determine the relative number of protons responsible for each multiplet. If the molecular formula is known, the absolute number of protons can be calculated.
2. Determine the chemical shift of the protons of each multiplet. The chemical shift is the difference between the shielding of the protons of the multiplet and the protons of TMS. When the spectrum is run, the instrument is normally adjusted so that the resonance of TMS falls on the zero of the δ scale (10 on the τ scale).
3. Determine the coupling constant between nuclei in different sets at different chemical shifts. In the examples illustrated by the figures up to this point, the spacing between peaks of a multiplet equals the coupling constant.
4. The final and most difficult step is to think of a molecular structure that accounts for the information determined in steps 1, 2, and 3. Correlation tables such as Tables 25.3 and 25.4 help out here.

Unfortunately, few spectra will be as easy to interpret as the examples that have been presented. The first reason is that in the examples the chemical shift difference was always large compared to the coupling

Figure 25.12. NMR spectrum of 1,2,3-trichlorobenzene; CCl_4 solution.

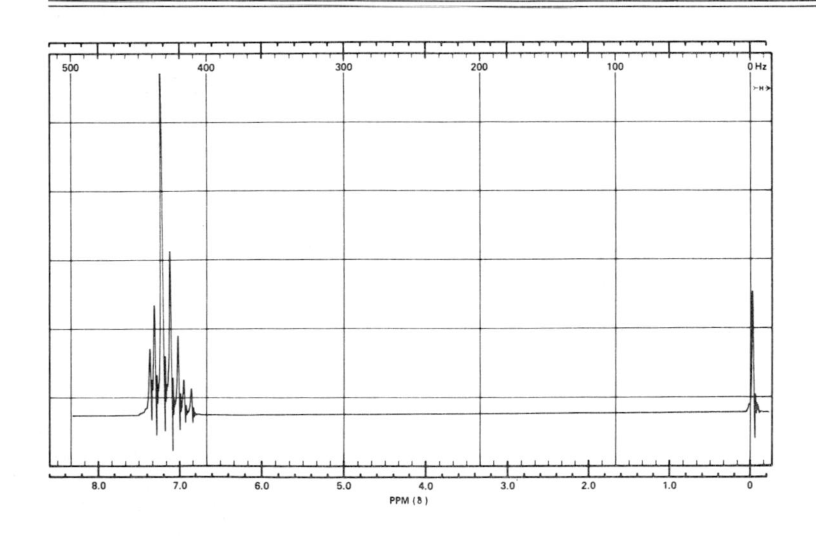

Cl
Cl Cl
H H
H
1,2,3-trichlorobenzene

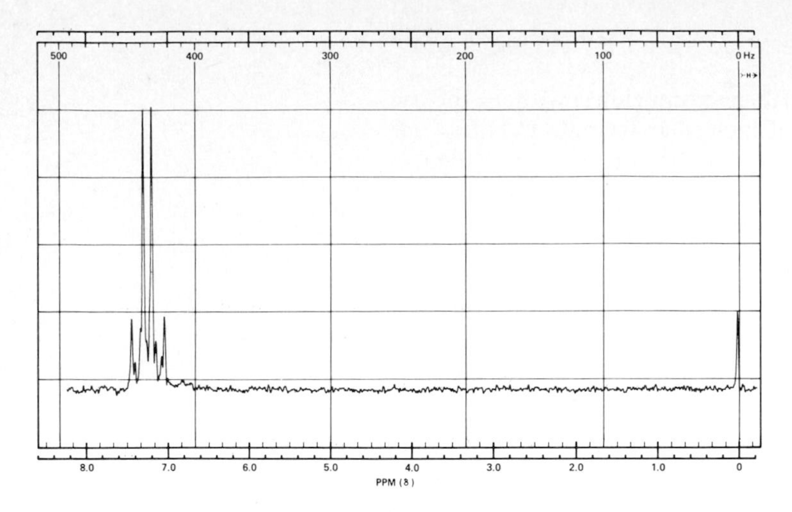

Cl
H H
H H
Br
1-chloro-4-bromobenzene

Figure 25.13. NMR spectrum of 1-chloro-4-bromobenzene; CCl_4 solution.

constant. Figure 25.12 shows the NMR spectrum of 1,2,3-trichlorobenzene, which, according to the simple analysis, should show a 1 : 1 doublet for the B protons, and a 1 : 2 : 1 triplet for the A proton. The NMR spectrum actually consists of seven lines.

Cl
Cl Cl
H_B H_B
H_A
1,2,3-trichlorobenzene

Cl
H_A H_A
H_B H_B
Br
1-chloro-4-bromobenzene

Figure 25.14. NMR spectrum of *o*-dichlorobenzene; neat.

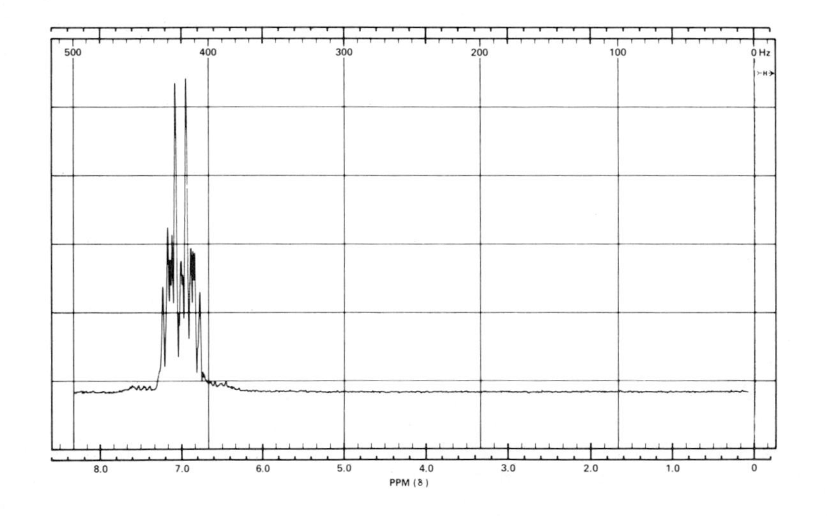

H
H Cl
H Cl
H
o-dichlorobenzene

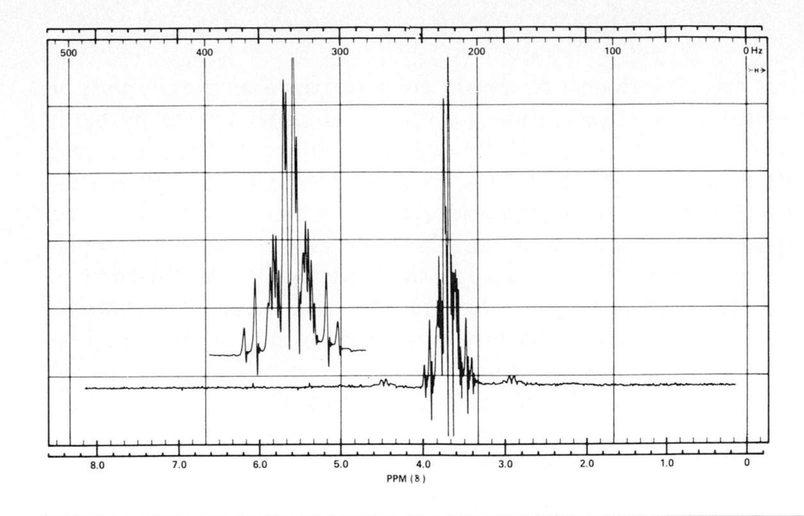

H H
Br—C—C—Cl
H H
1-bromo-2-chloroethane

Figure 25.15. NMR spectrum of 1-bromo-2-chloroethane; neat.

The second reason for difficulty in interpreting spectra is that in the examples described so far, the coupling constants between each proton in one set and every proton in the other set were exactly equal (the interset coupling constants were equal; the two sets of chemical shift equivalent nuclei were magnetically equivalent). Figure 25.13 shows the NMR spectrum of 1-chloro-4-bromobenzene. Here, the interset coupling constants are not equal: $J_{AB(ortho)}$ is not equal to $J_{AB(para)}$.

Similarly, as shown in Figures 25.14 and 25.15, the spectra of *o*-dichlorobenzene and 1-bromo-2-chloroethane are complex because of the inequality of the interset coupling constants. The analysis of the NMR spectra of compounds such as these in which the protons experience magnetic nonequivalence is discussed in References 1 and 2.

Magnetic nonequivalence

H_A, H_B, Cl, H_B, Cl, H_A
o-dichlorobenzene

H_A H_B
Br—C—C—Cl
H_A H_B
1-bromo-2-chloroethane

Reference 3 considers in more detail the concepts that have been presented in this section, and it discusses the interpretation of the proton NMR spectra of a number of additional compounds of known structure. Reference 4 is a collection of more than 300 proton NMR spectra of simple compounds that illustrate in a systematic way the various features of NMR spectra.

25.6 SAMPLE PREPARATION

Solvent

Deuterochloroform

Nuclear magnetic resonance spectra are determined on pure liquids or solutions. Carbon tetrachloride is a very good NMR solvent in that it contains no protons itself. However, some compounds are not very soluble in it. Chloroform ($CHCl_3$) is a better solvent for many substances than is carbon tetrachloride, but it contains a proton whose resonance can obscure some of those of interest in the sample. Deuterochloroform ($CDCl_3$) has the solvent properties of chloroform, and the deuterium resonance is far from the resonances of the protons. For routine work where solubility is not a problem, carbon tetrachloride should be used. If the compound is not very soluble in carbon tetrachloride, deuterochloroform should be tried. Other deuterated solvents, such as hexadeuteroacetone, are available but expensive. Water-soluble samples can be run in deuterium oxide (D_2O).

Concentration

Sample concentration should be about 10–30% by weight or volume. A total volume of 0.5 mL is all that is needed.

NMR tube

The spectrum is determined with the sample in an *NMR tube,* a thin-walled glass tube 5 mm in diameter and 180 mm long, sealed at the bottom. A tight-fitting plastic cap closes the top of the tube.

TMS

The reference standard, TMS, must be added if it is not already present (1%) in the NMR solvent as purchased. Since TMS boils at 27°C, it is best added by a microsyringe that has been stored in the freezer. Only 5–10 microliters is needed. Recap the NMR tube immediately after adding the TMS; invert the tube several times to mix. Before attempting to insert the sample tube into the probe of the instru-

Figure 25.16. NMR spectrum of a compound of molecular formula $C_6H_{12}O_2$; CCl_4 solution.

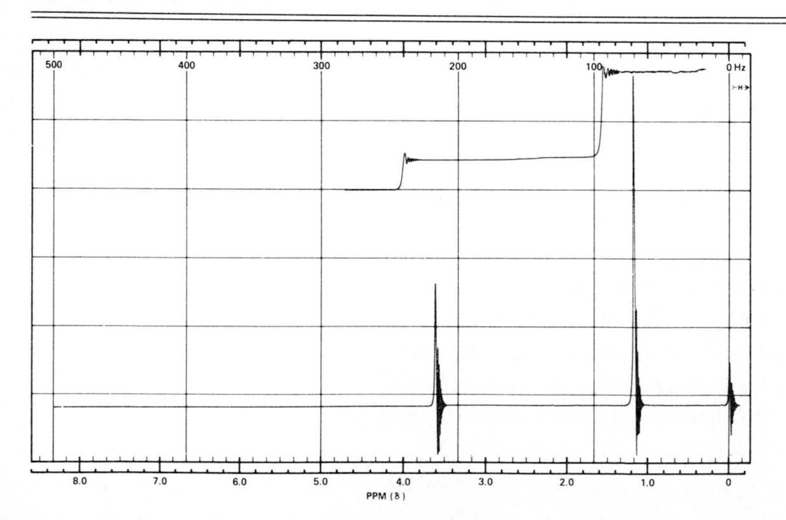

ment, wipe the outside clean and insert the tube into the white Teflon turbine ("spinner"), which allows the tube to be spun in the probe. Position the tube in the spinner with the depth gauge, and then place the tube and spinner in the probe.

Wipe the tube

Problems

1. Which isomer will give exactly one peak in its NMR spectrum?
   a. 1,1-Dichloroethane or 1,2-dichloroethane?
   b. Cyclobutane or methylcyclopropane?
   c. CH_3—O—CH_3 or CH_3CH_2—O—H?
   d. *ortho-*, *meta-* or *para-*dichlorobenzene?
   e. Isomers of molecular formula C_5H_{12}?
   f. Isomers of molecular formula C_4H_9Br?
2. Predict the appearance of the NMR spectra of the following compounds:

   a. CH_3—C(=O)—CH_3 (acetone)

   b. CH_3—O—C(=O)—O—CH_3 (dimethylcarbonate)

   c. CH_3—C(=O)—O—CH_3 (methyl acetate)

   d. CH_3CH_2—O—CH_2CH_3 (diethyl ether)

   e. CH_3—C(=O)—CH_2CH_3 (2-butanone)

   f. CH_3—C(=O)—O—CH_2CH_3 (ethyl acetate) and

   CH_3—O—C(=O)—CH_2CH_3 (methyl propionate)

3. Figure 25.16 shows the NMR spectrum of a compound of molecular formula $C_6H_{12}O_2$. What molecular structure is most consistent with the spectrum?
4. Figure 25.17 shows the NMR spectrum of a compound of molecular formula $C_7H_{14}O$. What molecular structure is most consistent with the spectrum?
5. Figure 25.18 shows the NMR spectrum of a compound of molecular formula C_5H_8. What molecular structure is most consistent with the spectrum?
6. Figure 25.19 shows the NMR spectrum of a compound of molecular formula $C_6H_{14}O_2$. What molecular structure is most consistent with the spectrum?
7. Figure 25.20 is the NMR spectrum of a compound of molecular formula $C_{10}H_{12}O$. The portion of spectrum at the left is a continuation of the rest of

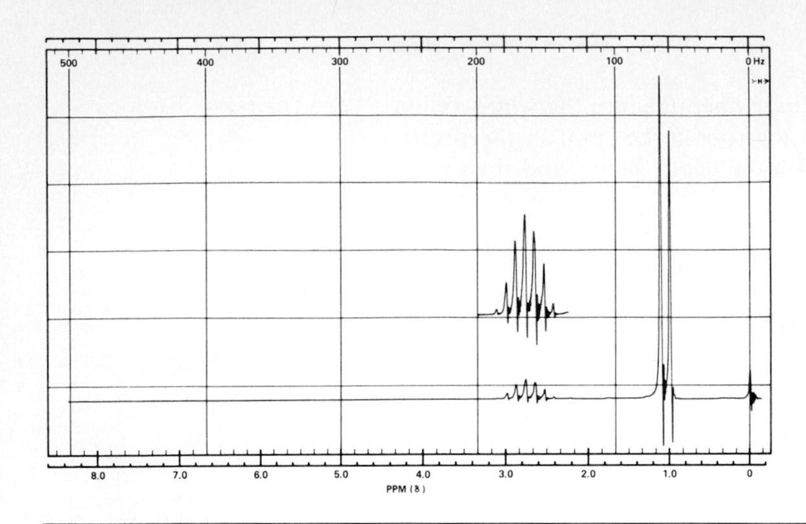

Figure 25.17. NMR spectrum of a compound of molecular formula $C_7H_{14}O$; CCl_4 solution.

the spectrum. The inset is an amplification of the resonance at $\delta = 3$. The relative areas under the multiplets are as follows:

δ	*Relative Area*
1.15	6
3.0	1
7.55	4
9.95	1

What molecular structure is most consistent with the spectrum?

Figure 25.18. NMR spectrum of a compound of molecular formula C_5H_8; CCl_4 solution.

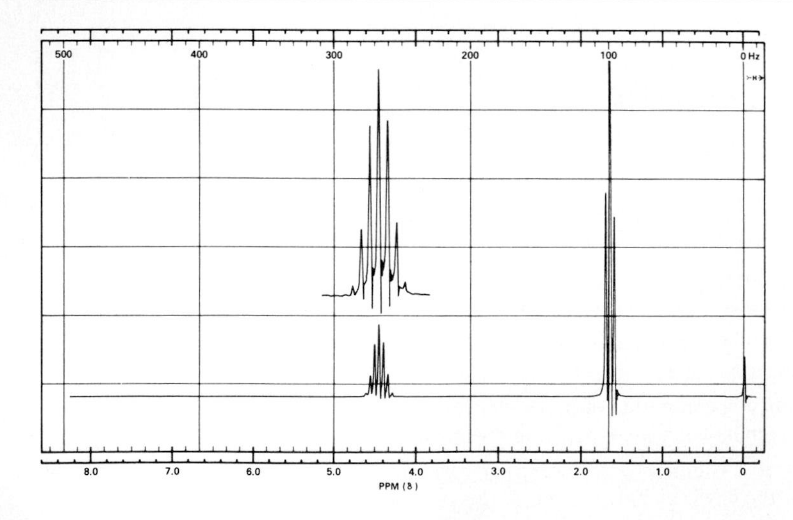

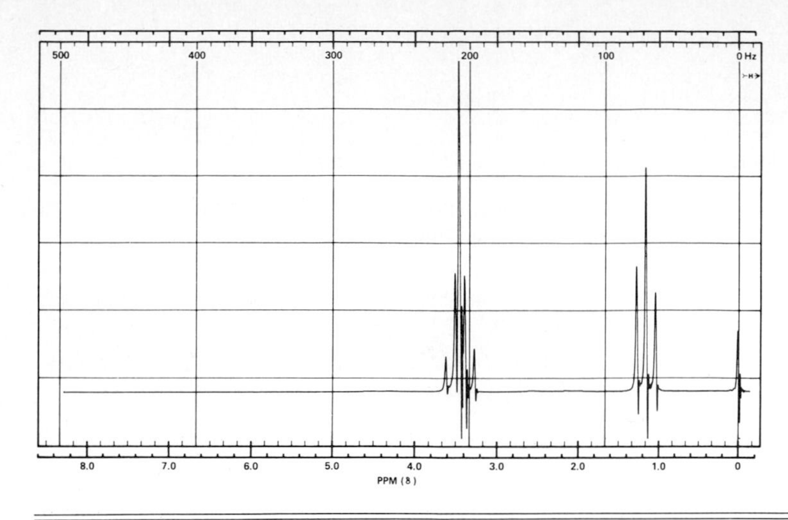

Figure 25.19. NMR spectrum of a compound of molecular formula $C_6H_{14}O_2$; CCl_4 solution.

8. Figure 64.4 shows the NMR spectrum of 2,4-dinitrobromobenzene. The doublet on the left is the resonance of one hydrogen; the pair of doublets in the middle is the resonance of a second hydrogen; and the doublet on the right is the resonance of the third hydrogen. Assuming that the spacings between lines of a multiplet are equal to coupling constants involving neighboring hydrogens, decide which hydrogen is responsible for each multiplet in the spectrum. Typical coupling constants between hydrogens *ortho, meta,* and *para* to each other are given in Table 25.4.

Figure 25.20. NMR spectrum of a compound of molecular formula $C_{10}H_{12}O$; CCl_4 solution.

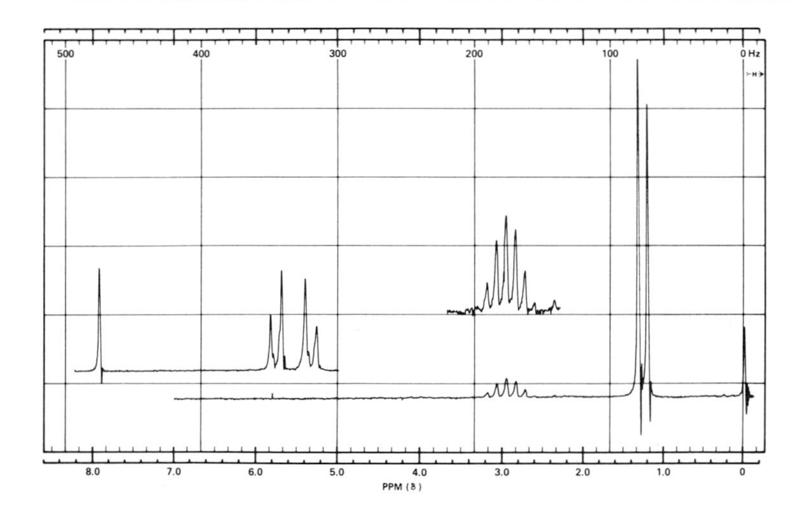

References

Introductory texts:

1. R. M. Silverstein, G. C. Bassler, and T. C. Morrill, *Spectrometric Identification of Organic Compounds,* 4th edition, Wiley, New York, 1981, pp. 181–303.
2. J. R. Dyer, *Applications of Absorption Spectroscopy of Organic Compounds,* Prentice-Hall, Englewood Cliffs, N.J., 1969, pp. 58–132.
3. A. Ault and G. O. Dudek, *An Introduction to Proton Nuclear Magnetic Resonance Spectroscopy,* Holden-Day, San Francisco, 1976.

Collections of spectra:

4. A. Ault and M. R. Ault, *A Handy and Systematic Catalog of NMR Spectra,* University Science Books, Mill Valley, California, 1980.
5. *High Resolution NMR Spectra Catalog,* Varian Associates, Palo Alto, California; Vol. 1, 1962; Vol. 2, 1963.
6. C. J. Pouchert and J. R. Campbell, *The Aldrich Library of NMR Spectra,* Aldrich Chemical Co., Inc., Milwaukee, Wisconsin; ten volumes plus index, 1974 (2nd edition, 1983; two volumes).

26 Mass Spectrometry

Mass spectrometers are not generally available in the undergraduate organic laboratory, but mass spectrometry is of sufficient importance that we present a brief introduction to the theory of mass spectrometry and the interpretation of mass spectra.

26.1 THEORY OF MASS SPECTROMETRY

In mass spectrometry, a vaporized sample of a substance is bombarded with a beam of electrons, and the relative abundance of the resulting positively charged molecular fragments is determined. The mass-to-charge ratio of each fragment is measured, but, since almost every fragment will bear a single positive charge, the mass-to-charge ratio will be numerically equal to the mass of the fragment. The record of relative abundance versus mass-to-charge ratio that is produced from the substance by the mass spectrometer is called the *mass spectrum* of the substance.

Mass spectrum

The production of positively charged ionic fragments of various masses occurs in this way. An electron from the electron beam knocks an electron from a molecule to give the parent ion, or molecular ion, $M\cdot^{+}$ (which must now have an unpaired electron):

Ionization

$$M + e^{-} \rightarrow M\cdot^{+} + 2e^{-} \qquad (26.1\text{-}1)$$

The molecular ion may fall apart very quickly (within 10^{-8} to 10^{-10} second) to give one charged and one uncharged fragment:

$$M\cdot^{+} \rightarrow A^{+} + B\cdot \quad \text{or} \quad A\cdot + B^{+} \tag{26.1-2}$$

$$M\cdot^{+} \rightarrow C^{+} + D\cdot \quad \text{or} \quad C\cdot + D^{+} \tag{26.1-3}$$

etc.

The charged fragmentation products may themselves undergo further fragmentation to give smaller charged and neutral pieces. The relative abundance of positively charged fragments of various mass-to-charge ratios is characteristic of the molecules of a substance and can serve to identify the substance.

Fragmentation

The relative abundance of the positively charged ions of various mass-to-charge ratios (m/e) is determined by accelerating the ions by an electrostatic field. The kinetic energy acquired by each ion in this process is equal to eV, where e is the charge on the particle and V is the potential difference through which the ion has fallen:

Acceleration

$$\text{Kinetic energy} = eV = \tfrac{1}{2}mv^2 \tag{26.1-4}$$

where m is the mass of the ion, v is its velocity, and "kinetic energy" is the kinetic energy gained upon acceleration in the electrostatic field. Solving for v^2 gives

$$v^2 = \frac{2eV}{m} \tag{26.1-5}$$

The accelerated ions then move into a magnetic field whose direction is perpendicular to their path. The action of the magnetic field is to force the ions to follow a circular path: the force of the magnetic field is a centripetal force. Since the magnitude of the force on an ion equals Hev (where H is the strength of the magnetic field, e is the charge of the ion, and v is its velocity) and the centripetal acceleration is given by v^2/r (where v is the tangential velocity and r is the radius of the path) we can write, according to Newton's second law, the equation

Sorting by mass

$$Hev = m\frac{v^2}{r} \tag{26.1-6}$$

Solving for v and squaring gives

$$v^2 = \frac{H^2e^2r^2}{m^2} \tag{26.1-7}$$

Substituting into this equation the value of v^2 that results from the acceleration by the electrostatic field (Equation 26.1-5) gives

$$\frac{2eV}{m} = \frac{H^2e^2r^2}{m^2} \qquad (26.1\text{-}8)$$

and solving for m/e gives

$$\frac{m}{e} = \frac{H^2r^2}{2V} \qquad (26.1\text{-}9)$$

Thus, for constant electrostatic and magnetic fields V and H, fragments of different m/e will be separated because they will follow paths of different radii in the magnetic field. A detector that can determine the relative numbers of ions that have followed paths of different radii can thus provide the desired record of relative abundance of fragments of different m/e. An alternative often used is to position a detector so that it can continuously record the intensity of the beam of particles that follows a path of a given, fixed, radius and vary either the magnetic field H or the electrostatic field V in such a way that ions of increasing m/e will fall in succession upon the detector. A record of the output of the detector versus H^2 or $1/V$ is equivalent to a record of relative abundance versus m/e. Reference 1 describes how a mass spectrometer is constructed so that it will perform these functions.

Detection

26.2 INTERPRETATION OF MASS SPECTRA

Three types of information can be obtained from a mass spectrum: the molecular weight of a substance, the molecular formula of a substance, and the molecular structure of a substance.

Determination of Molecular Weight

One might expect that the mass that corresponds to the largest m/e of the mass spectrum would be the molecular weight of the substance. This is essentially true, but with two important qualifications. The first is that if the initially formed molecular ions undergo rapid fragmentation, there may not be enough of these ions left to accelerate and detect; the peak for the molecular ion, the parent peak P, may not be strong enough to be seen. This is not often the case, but the possibility must be kept in mind.

Parent peak

The second qualification is best introduced by an example. In the mass spectrum of carbon monoxide, a peak at $m/e = 29$ appears with about 1.12% of the intensity of the parent peak P, which corresponds to the molecular ion at $m/e = 28$. The reason for this is that carbon from natural sources contains an amount of ^{13}C equal to 1.08% of the amount of naturally occurring ^{12}C, and that oxygen from natural sources contains an amount of ^{17}O equal to 0.04% of the amount of ^{16}O. The peak at *m/e* = 29 is due to the presence of $^{13}C^{16}O$ and $^{12}C^{17}O$ isotopes of carbon monoxide, which are 1.08% + 0.04% = 1.12% as abundant as $^{12}C^{16}O$. Thus, the molecular weight of a substance can be determined from its mass spectrum when the molecular ion is not too unstable, by identifying the peak of highest *m/e* (the parent ion P), not counting peaks due to molecules that contain small amounts of heavier isotopic atoms (P + 1, P + 2, etc.). The molecular weight obtained from a mass spectrum is the exact molecular weight, not an approximate weight as with molecular weights determined from freezing point depression or titration, for example.

Determination of Molecular Formula

The fact that substances from natural sources are composed of a certain calculable fraction of molecules that contain atoms of heavier isotopes often makes it possible to eliminate all but a few of the molecular formulas that correspond to a given molecular weight. This is done by comparing the predicted intensities of the P + 1 and P + 2 peaks that can be calculated for different molecular formulas with the experimental P + 1 and P + 2 intensities. For example, as we have seen, from carbon monoxide we would expect a P + 1 peak at *m/e* = 29 that is 1.12% as intense as the parent peak at *m/e* = 28. Also, since ^{18}O is 0.20% as abundant as ^{16}O, a P + 2 peak due to $^{12}C^{18}O$ that is 0.2% as intense as P would be expected. (The amounts of $^{12}C^{18}O$ and $^{14}C^{16}O$ are negligible.) On the other hand, molecular nitrogen, which also has a nominal molecular weight of 28, would be expected to have a P + 1 peak at *m/e* = 29 that would be 0.76% as intense as P, since ^{15}N occurs 0.38% as abundantly as ^{14}N (0.38% × 2 atoms of N = 0.76%). And the P + 2 peak at *m/e* = 30 would be expected to be 0.0038 × 0.0038 = 0.001444% as intense as P. If the experimental uncertainty is sufficiently small, the intensities of P + 1 and P + 2 relative to P will allow the assignment of CO or N_2 as the molecular formula of a sample of a gas whose parent peak appears at *m/e* = 28.

Tables have been prepared (Reference 1) of calculated intensities of P + 1 and P + 2 relative to P for many combinations of C, H, O, and N for integral values of P. With the help of these tables, you can greatly reduce the number of possible molecular formulas for a particular molecular weight, providing, however, that accurate intensities of P + 1 and P + 2 can be obtained from the mass spectrum. If the molecular ion, $M^{\cdot +}$, decomposes too quickly, the intensities of P + 1 and P + 2 may be too low to be measured with the necessary accuracy. The variations of this kind of analysis that must be made if the substance contains elements other than C, H, O, and N are described in Reference 1.

Determination of Molecular Structure

Since the relative intensities of the various positive ions produced upon electron bombardment in the mass spectrometer (fragmentation patterns) are different for substances of different molecular structure, it must be possible, at least in principle, to deduce the molecular structure of a substance from its mass spectral fragmentation pattern. Comparison of the mass spectra of many substances containing different functional groups has made it possible to rationalize many features of the fragmentation patterns as the result of the presence of various functional groups in the molecule. Conversely, then, it is possible to infer the presence of certain functional groups or structural features from the fragmentation patterns. The subject of the correlations between molecular structure and fragmentation patterns is very large and can only be hinted at in this brief discussion. Reference 1 is an excellent introduction.

26.3 HIGH-RESOLUTION MASS SPECTROMETRY

So far, we have assumed that mass spectrometers can distinguish between ions that differ in mass by one atomic mass unit (unit-resolution mass spectrometry). Instruments are available, at a high price, that can resolve masses to approximately 0.001 atomic mass unit (*high-resolution mass spectrometry*). With this kind of resolution, it would be possible to distinguish between $^{12}C^{16}O$ (molecular weight = 27.9949) and $^{14}N^{14}N$ (molecular weight = 28.0062) without reference to the relative intensities of P + 1 and P + 2. In fact, the molecular formula of every fragment in the mass spectrum can be determined, if its mass can be measured to about 0.001 atomic mass unit.

Problems

1. For each case, explain why the parent peak will always be either odd or even. If the parent peak can be either odd or even, explain how this can be so.
   a. The compound contains carbon and hydrogen.
   b. The compound contains carbon, hydrogen, and oxygen.
   c. The compound contains carbon, hydrogen, and fluorine.
   d. The compound contains carbon, hydrogen, and nitrogen.
2. Bromine is made up of almost equal parts of ^{79}Br and ^{81}Br.
   a. Taking the parent peak of CH_3Br to be 94 (or 12 + 3 + 79), estimate the intensity relative to the parent peak of P + 1, P + 2, etc.
   b. Taking the parent peak of CH_2Br_2 to be 172 (or 12 + 2 + 79 + 79), estimate the intensity relative to the parent peak of P + 1, P + 2, etc.
   c. Do the same for $CHBr_3$ and CBr_4.
   d. In what simple way might the number of bromine atoms per molecule in a compound containing bromine be determined from its mass spectrum?

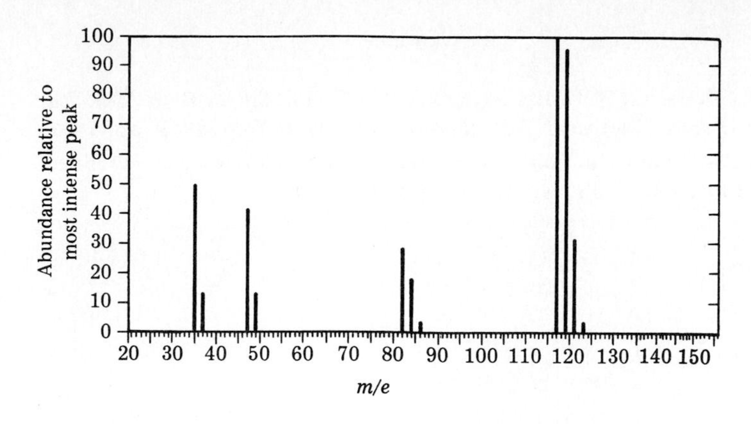

Cl
Cl—C—Cl
Cl

Figure 26.1. Mass spectrum of CCl_4.

3. Chlorine is made up of ^{35}Cl and ^{37}Cl in a ratio of about 3 to 1. Interpret the mass spectrum of carbon tetrachloride, shown in Figure 26.1. The parent peak of carbon tetrachloride fragments very rapidly.
4. The mass spectrum shown in Figure 26.2 is that of either methyl *p*-hydroxybenzoate or *p*-methoxybenzoic acid. Which substance can account for the spectrum? Explain.

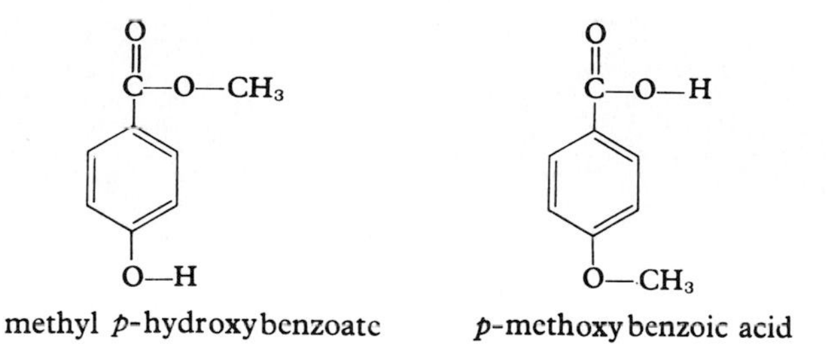

Figure 26.2

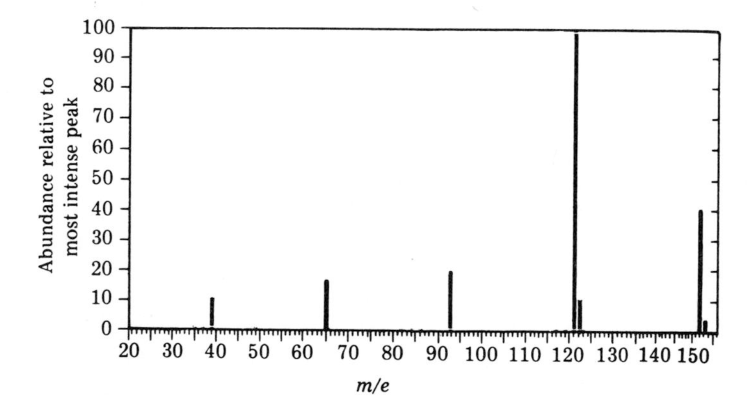

References

1. R. M. Silverstein, G. C. Bassler, and T. C. Morrill, *Spectrometric Identification of Organic Compounds*, 4th edition, Wiley, New York, 1981, pp. 3–93.
2. D. J. Pasto and C. R. Johnson, *Organic Structure Determination*, Prentice-Hall, Englewood Cliffs, N. J., 1969, pp. 243–294.

Determination of Chemical Properties; Qualitative Organic Analysis

The chemical properties of a substance, like its physical properties, can be used both for its characterization and for the determination of its molecular structure.

Usually the presence or the absence of a functional group is easily deduced, either by infrared spectrometry or by chemical methods. With the availability of infrared spectrometers, however (Section 23), chemical methods are less often used, and the information that can be obtained from the procedures of Section 28 will complement the information that can be obtained from a routine infrared spectrum.

The identity of a sample of a substance of unknown structure and a sample of a substance of known structure can often be confirmed by a comparison of the melting points of their solid derivatives. These derivatives, prepared by procedures described in Section 29, can be compared directly by the determination of melting point and mixed melting point, or indirectly by comparison of the melting point observed for the derivative of the unknown with melting points reported in the literature.

Part of an organic laboratory program is often devoted to *qualitative organic analysis*. At this time, you will be given a sample of a substance whose molecular structure is not known to you, an *unknown,* and you will be asked to identify the compound. While the best strategy will be different for each substance, there are several experiments that can be done routinely with any unknown. For liquids, one usually determines the boiling point (by distillation or by the micro method; Section 16), the density (Section 18), the index of refraction (Section 19), and the solubility behavior—and whether or not the compound is soluble in water, ether, 5% NaOH, 5% $NaHCO_3$, 5% HCl, conc. H_2SO_4, or 85% H_3PO_4 (Section 22). For solids, one normally determines the melting point (Section 17) and the solubility behavior. Other physical properties that can be determined routinely include the infrared spectrum (Section 23) and, if possible, the NMR spectrum (Section 25).

Fusion with sodium followed by qualitative tests for nitrogen (as cyanide), sulfur (as sulfide), and the halogens (Section 27.3) often provides important information, especially when positive indications are obtained.

Since time is always in short supply, qualitative tests for functional groups (Section 28) should be chosen with care so that the information obtained will not merely duplicate that which is already in hand. However, it is often a good idea to confirm tentative conclusions that have been based on other evidence.

If it is not possible to compare IR or NMR spectra of the unknown with those of an authentic sample, it is usually necessary to prepare one or more solid derivatives (Section 29) in order to confirm an identification. Preparing derivatives is also a very good way to develop chemical judgment and laboratory technique.

27 Qualitative Tests for the Elements

A common first step in the identification of an unknown substance is to determine what elements (other than carbon, hydrogen, and possibly oxygen) are present in the sample. This section describes qualitative tests for the detection of the presence of metals, nitrogen, sulfur, and the halogens.

27.1 IGNITION TEST; TEST FOR METALS

Procedure. Place 50–100 mg of the substance on a porcelain crucible cover and heat the cover with a flame. Heat gently at first, and then finally heat the cover to dull redness. If a metal was present in the sample, a nonvolatile residue will remain. Add a drop of water to the residue: the resulting solution should be basic; it should turn red litmus blue.

While the purpose of this test is to determine whether a solid is a metal salt or other metal-containing substance, other information can be obtained as well. If the substance burns with a fairly sooty flame, a high ratio of carbon to hydrogen is indicated (an aromatic rather than an aliphatic compound is indicated). If the substance burns with a blue rather than a yellow flame, a high percent of oxygen is indicated. Certain types of compounds can explode under these conditions (see Section 1.2).

Aliphatic versus aromatic

Warning

27.2 BEILSTEIN TEST; TEST FOR HALOGENS (EXCEPT FLUORINE)

Procedure. Form a small coil at one end of a 20-cm length of stiff copper wire by making a few turns around a glass rod 2 or 3 mm in diameter. Heat the coil in the oxidizing part of the flame of a Bunsen burner until it does not color the flame any longer. Allow the wire to cool (by waving it in the air) and then dip the coil into the material to be tested and heat it again in the same part of the flame. If the substance contains chlorine, bromine, or iodine, a transient green or blue-green color will be imparted to the flame; the color is due to copper that has been made volatile by combination with the halogen of the sample.

The Beilstein test is very sensitive. It should be confirmed by the sodium fusion test (Section 27.3) and the tests of Sections 28.11 and 28.13 to make sure that the effect was not due to the presence of a halogen-containing impurity—or to verify that the substance is not one that gives a positive test even though it contains no halogen. (Unfortunately, the test is not specific since some organic nitrogen compounds that do not contain halogen also give the effect.)

27.3 SODIUM FUSION TEST; TEST FOR NITROGEN, SULFUR, AND THE HALOGENS

In the sodium fusion test, a substance is heated with sodium under conditions that ensure the conversion of any nitrogen to cyanide ion, any sulfur to sulfide ion, and any halogens to the corresponding halide ions. The presence or absence of these ions in the solution that results from the fusion (the *fusion solution*) indicates the presence or absence of these elements in the sample.

Sodium Fusion

Procedure. Support a test tube 8 mm in diameter by 50 mm long by its lip in a hole in an asbestos board or sheet which is in turn supported on an iron ring. Add to the test tube about 40 mg of sodium (Note 1) and heat the test tube with a burner until the sodium has melted and the grey vapors rise about 2 cm in the tube. Adjust the rate of heating by adjusting the height of the burner flame or the height of the iron ring, so that this condition can be maintained. Then add, in two or three portions, 50 mg (or 2 or 3 small drops) of the compound over about a 30-second interval (Note 2). After the addition is complete, turn up the burner and heat the tube as strongly as possible until the lower 1 or 2 cm is red to orange-red hot; heat the tube at this rate for about 2 minutes (Note 3). Now, holding the asbestos sheet or board with a gloved hand or tongs, lower the red-hot tube into a small beaker (obviously no more than 50 mm high) that contains about 10 mL of distilled water (Note 4). The tube will shatter, and any residual sodium will react with the water. Boil the remains of the test tube with the water and filter by gravity. The clear, almost colorless filtrate (fusion solution) will be used in the tests that follow (Note 5).

Caution

Notes

1. For precautions to be taken in handling sodium, see Section 1.7. A piece the size of half a medium pea, or a 4-mm cube, will be satisfactory.
2. Safety glasses must be worn; nitro compounds, azo compounds, and polyhalogen compounds such as carbon tetrachloride or chloroform may cause a slight explosion. Try to drop the sample straight down, without letting it hit the side of the test tube, where it may simply evaporate.
3. It is not possible to heat too strongly. Some burners are too feeble; make sure that yours can give a strong, blue, nonluminous flame.
4. Safety glasses must be worn; hold the asbestos board at arm's length; don't aim the open top of the tube at anyone.
5. If the filtrate is more than just slightly yellow, it will usually be a waste of time to continue. Better do the procedure again and heat more strongly.

Tests for the Elements: Test for Nitrogen; Test for Cyanide Ion (Test a)

Procedure. Add 2 or 3 mL of the fusion solution to a test tube that contains 100–200 mg of powdered ferrous sulfate crystals. Heat the mixture cautiously and with shaking until it boils (Note 1). Then, without cooling, add just enough dilute sulfuric acid to dissolve the gelatinous hydroxides of iron. The appearance of a blue precipitate of ferric ferrocyanide, $Fe_4[Fe(CN)_6]_3$ (Prussian Blue), indicates that cya-

$Fe_4[Fe(CN)_6]_3$
blue

nide ion is present and, therefore, that the original compound contained nitrogen (Note 2). If no cyanide ion is present, the solution should be a pale yellow color (Note 3).

Notes

1. It is often recommended that 2 drops of a 30% aqueous solution of potassium fluoride be added before boiling. If sulfide ion is present, a precipitate of black ferrous sulfide may appear; it will dissolve upon acidification. Ferric ions are produced by air oxidation during boiling.
2. If a precipitate is not immediately apparent, allow the mixture to stand for 15 minutes and then filter it. After washing the filter paper with a little water to remove all the colored solution, any Prussian Blue present should be visible on the paper. If there is still doubt as to whether a precipitate of Prussian Blue was formed, another sodium fusion should be carried out and the test repeated.
3. A positive test for cyanide is good evidence for the presence of nitrogen in the original sample, but the presence of nitrogen should never be ruled out solely on the basis of a negative test here for cyanide ion. You should make sure that your technique is adequate by carrying out the sodium fusion and the test for nitrogen on a sample known to contain nitrogen. Nitrogen in a high oxidation state, for example as a nitro group, is especially difficult to detect.

Test for Sulfur; Test for Sulfide Ion (Test b)

Procedure. Acidify about 2 mL of the fusion solution with dilute acetic acid and then add a few drops of approximately 1% lead acetate solution. A black precipitate of lead sulfide indicates that sulfide ion is present and, therefore, that sulfur was present in the original sample.

PbS
black

Alternatively, 2 or 3 drops of a freshly prepared 0.1% solution of sodium nitroprusside (Note 1) may be added to 2 mL of the fusion solution. A purple coloration indicates the presence of sulfide ion.

Note

1. This may be prepared by dissolving a tiny crystal of sodium nitroprusside ($Na_2[Fe(CN_5)NO]$) in 1 or 2 mL of water.

Test for Halogen; Test for Halide Ion (Test c)

Procedure. Acidify 2 mL of the fusion solution with dilute nitric acid (Note 1) and add a few drops of 5% aqueous silver nitrate solution. A heavy precipitate indicates the presence of chloride, bromide, or iodide ion, or any combination of the three.

If only a single halide should be present, you may be able to distinguish among the possibilities since silver chloride is white (slightly purplish because of liberation of free silver by light) and is easily soluble in dilute ammonium hydroxide, silver bromide is slightly yellow and is only slightly soluble in ammonium hydroxide, and silver iodide is definitely yellow and is insoluble in ammonium hydroxide. To determine the solubility of the precipitate in dilute ammonium hydroxide, remove the supernatant liquid by centrifugation and decantation, add 2 mL of dilute ammonium hydroxide solution, and stir the mixture to see if the solid will dissolve.

AgCl: white
AgBr: slightly yellow
AgI: yellow

Note

1. If either cyanide or sulfide is present (Tests a and b above), the acidified solution must be boiled gently for a few minutes in order to expel hydrogen cyanide and hydrogen sulfide.

Test for Iodine; Test for Iodide Ion (Test d)

Procedure. Acidify 1 or 2 mL of the fusion solution with dilute sulfuric acid (Note 1). After cooling, add 1 mL of carbon tetrachloride; and then add drop by drop with good mixing a solution of chlorine water (Note 2). Iodide ion, if present, will be oxidized to free iodine under these conditions and will be extracted into the carbon tetrachloride to give a purple color (Note 3).

I_2
purple

Notes

1. If cyanide or sulfide is present (Tests a and b above), acidify the sample of fusion solution with dilute nitric acid, boil the mixture until its volume has been reduced by one-half in order to expel hydrogen cyanide and hydrogen sulfide, and dilute the resulting solution with an equal volume of distilled water.
2. A stabilized solution of sodium hypochlorite such as Clorox may be used; make sure that the test solution remains acidic to litmus by adding acid as necessary. If this type of commercial bleach solution is used, a blank should be run since often a yellow-brown color is produced just from the reagent. Chlorine water

may be prepared by acidifying 10% aqueous sodium hypochlorite solution with one-fifth of its volume of dilute hydrochloric acid.

3. The absence of iodide ion but the presence of bromide ion will result in the appearance of a brown color in the carbon tetrachloride layer, due to oxidation of the bromide ion to form bromine followed by extraction of the bromine into the carbon tetrachloride. If both iodide and bromide are present, the initial purple color will give way to a reddish-brown color.

Test for Bromine; Test for Bromide Ion (Test e)

In the absence of iodide ion, the conditions of Test d will result in a brown color being produced in the carbon tetrachloride layer, due to the oxidation of bromide to bromine. If iodide is present, a purple color will appear first. This will change to reddish-brown with the continued addition of chlorine water.

Br_2
reddish-brown

Test for Chlorine; Test for Chloride Ion (Test f)

If the presence of halide ion was indicated by Test c above and no color was produced in the carbon tetrachloride layer in Test d above, the halide ion present must be chloride.

If either iodide or bromide ion has been shown to be present by means of Test d, apply the following procedure.

Procedure. Acidify 1 or 2 mL of the fusion solution with glacial acetic acid, add 0.5 gram of lead dioxide, and boil the mixture gently until all iodine and bromine has been liberated and boiled off. When the mixture is allowed to stand so that the lead dioxide will settle to the bottom, the solution will be colorless when this point has been reached. Dilute the mixture with an equal volume of water, remove the excess lead dioxide either by filtration or centrifugation and decantation, and test for the presence of chloride ion in the filtrate by adding a little dilute nitric acid followed by a few drops of 5% aqueous silver nitrate solution. The formation of a white precipitate confirms the presence of chloride ion.

28 Qualitative Characterization Tests: Tests for the Functional Groups

Before the advent of spectrometric methods such as infrared, ultraviolet, nuclear magnetic resonance, and mass spectrometry (Sections 23 through 26), the determination of chemical properties was of paramount importance for the identification, characterization, and determination of structure of pure substances. Many reagents or reaction conditions were found to give moderately characteristic and specific results with substances that contain certain functional groups. These reagents or reaction conditions therefore serve to distinguish between substances with different functional groups, and to indicate the presence or absence of certain functional groups in substances of unknown structure.

Functional group tests

Sections 28.1 through 28.14 describe a number of tests that can distinguish between certain functional groups, or can indicate the presence or absence of certain functional groups. The comparison of the results of these tests on samples of substances of known and unknown structure will, in many cases, allow a tentative conclusion to be drawn about the nature of the functional group of the substance of unknown structure.

Different substances with the same functional group will, however, give slightly different results. The question is, then, how different can the results be before you must conclude that the substances have different functional groups? There will, obviously, be borderline cases. Also, certain individual substances can react atypically, to give either a "false positive test" (the unknown behaves as if a certain functional group were present, but it really isn't) or a "false negative test" (the unknown behaves as if the functional group were not there, but it really is). That is, most tests are not completely specific or characteristic. Finally, some tests are quite sensitive, and a small amount of a substance as an impurity might be sufficient to give a result that you could interpret as a "positive test."

The uncertainty in the interpretation of these tests can be decreased by testing the behavior of compounds of known structure under the same conditions. The substances of known structure should be either substances that are said to give typical or normal results, or, better, substances that are very similar in structure (especially if more than one functional group is present) to the suspected structure of the unknown substance.

Observations, not conclusions

When recording the results of one of these characterization tests, you must be careful to record *your observations* (a yellow-orange crystalline precipitate was formed within thirty seconds after adding two drops of the liquid of unknown structure; the mixture was not heated; no further changes occurred upon standing; similar behavior but slower than with two drops of acetone) and *not your tentative conclusion* (positive test for a ketone). It is impossible to reinterpret a conclusion; only data—observations—can be reinterpreted. If your conclusions are inconsistent, there is no way to reinterpret them, and you have wasted your time; data can be continually reinterpreted until the conclusions are consistent.

There are many hundreds of qualitative tests that can be used to characterize unknown substances and to distinguish between different functional groups. The procedures presented in these sections are only

Table 28.1. Summary of classification tests

Test	*Section*	*Application*
Ammonia	28.1	Ammonium salts; primary amides; nitriles
Baeyer's test: *see* Potassium permanganate test		
Benzenesulfonyl chloride (Hinsberg's test)	28.2	Amines (distinguishes among primary, secondary, or tertiary)
Bromine in carbon tetrachloride	28.3	Alkenes; alkynes
Chromic anhydride	28.4	Alcohols (distinguishes primary and secondary from tertiary)
2,4-Dinitrophenylhydrazine	28.5	Aldehydes and ketones
Ferric chloride	28.6	Phenols and enols
Ferric hydroxamate test	28.7	Acid halides and anhydrides; esters; amides; nitriles
Hinsberg's test: *see* Benzenesulfonyl chloride		
Hydrochloric acid/zinc chloride test (Lucas's test)	28.8	Alcohols (distinguishes among primary, secondary, or tertiary)
Iodoform test	28.9	Methyl ketones; secondary methyl carbinols
Lucas's test: *see* Hydrochloric acid/zinc chloride test		
Potassium permanganate test (Baeyer's test)	28.10	Alkenes; alkynes
Silver nitrate	28.11	Halides
Sodium hydroxide test	28.12	Esters; lactones; anhydrides
Sodium iodide	28.13	Halides
Tollens test	28.14	Aldehydes

a very small sample of such tests. They have been chosen because they involve some of the most familiar functional groups, and because the information they give complements the information that can be obtained from infrared and nuclear magnetic resonance spectra. Table 28.1 is a summary and index for the fourteen tests presented.

The references for each test are given at the end of this section. These references provide more information concerning the scope and limitations of certain procedures, variations in procedures, and exceptional behavior of certain compounds or types of compounds. You should consult these references in order to minimize the chances of being misled by the results of the tests.

28.1 DETECTION OF AMMONIA FROM AMMONIUM SALTS, PRIMARY AMIDES, AND NITRILES

Ammonium salts, primary amides, and nitriles that are relatively easily hydrolyzed will liberate ammonia under the conditions of the procedure that follows.

Materials Required.

20% aqueous sodium hydroxide solution

10% aqueous copper sulfate solution

$R{-}C({=}O){-}NH_2$

primary amide

$R{-}C{\equiv}N{:}$

nitrile

Procedure. Place 50 mg of the substance in a small test tube and add 2 mL of 20% sodium hydroxide solution. Fasten a small circle of filter paper tightly over the mouth of the test tube and wet it with a couple of drops of the copper sulfate solution. Heat the contents of the test tube to boiling for about one minute, and look to see if the filter paper has turned blue. The blue color is due to the formation of the copper-ammonia complex ion from ammonia either liberated from the ammonium salt or formed by hydrolysis of the primary amide or nitrile. (Ref. 1, p. 164; Ref. 2, pp. 109, 123; Ref. 3, pp. 404, 410, 798, 805.)

28.2 BENZENESULFONYL CHLORIDE (HINSBERG'S TEST)

Useful for distinguishing among primary, secondary, and tertiary amines.

Materials Required.

10% aqueous sodium hydroxide solution

Benzenesulfonyl chloride

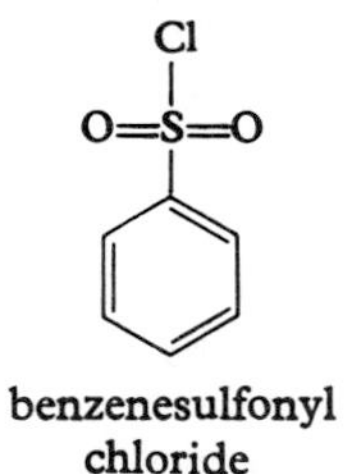

benzenesulfonyl chloride

Procedure. Add 5 mL of 10% sodium hydroxide to 0.2 mL of the liquid amine or 0.2 gram of the solid amine in a test tube, and then add 0.4 mL benzenesulfonyl chloride. Stopper the test tube and occasionally shake it vigorously over a period of 5 or 10 minutes, cooling the tube in water if it heats up a lot. By this time, all the benzenesulfonyl chloride should have reacted, either with the amine to form the sulfonamide or with the basic solution to form the water-soluble sodium benzenesulfonate; complete reaction is indicated by the disappearance of the distinctive odor of benzenesulfonyl chloride. Now make sure that the mixture is basic (if it is not, add 10% sodium hydroxide to make it strongly basic) and observe whether or not any insoluble material is present. Remove any insoluble material, by filtration if it is a solid and by decanting if it is a liquid.

Test the solubility of the insoluble material that was removed, both in water and in dilute hydrochloric acid.

Acidify the filtrate (or the basic solution that was separated by decantation, or the clear reaction mixture) and attempt to promote crystallization by scratching and by cooling. Benzenesulfonic acid is soluble in water.

Primary amines normally give a sulfonamide that is soluble in the basic reaction mixture; no insoluble material should be present at the end of the first part of the test. Acidification of the clear reaction mixture should then result in the precipitation or crystallization of the water-insoluble sulfonamide of the amine. However, the sodium salts of the sulfonamides of some primary amines—for example, cyclohexylamine through cyclodecylamine and certain high-molecular-weight amines—are not very soluble in 10% sodium hydroxide solution. If the unknown is one of these compounds, a precipitate will be present at the end of the first part of the test, but it will be found to be soluble in water.

Primary amines

Secondary amines normally give a sulfonamide that is insoluble in the basic reaction mixture; a solid residue will be present at the end of the first part of the test. This residue will be insoluble in both water and dilute hydrochloric acid. The filtrate should give no precipitate upon acidification.

Secondary

Tertiary amines normally give no reaction with benzenesulfonyl chloride; the amine itself should be present as the unchanged liquid or solid at the end of the first part of the test. The liquid or solid residue will be soluble in dilute hydrochloric acid.

Tertiary

It should be apparent that water-soluble tertiary amines and secondary amines with a carboxyl group on another part of the molecule will not show the typical behavior previously outlined. (Ref. 1, p. 119; Ref. 3, p. 650; Ref. 4, p. 230. See also Fanta and Wang, *J. Chem. Educ.* **41,** 280 (1964). For an alternative method, see Ritter, *J. Chem. Educ.* **29,** 506 (1952).)

28.3 BROMINE IN CARBON TETRACHLORIDE

This test is useful for indicating the presence of many olefinic or acetylenic functional groups. It should be used in conjunction with a similar test with aqueous potassium permanganate solution (Section 28.10).

$>C{=}C<$

olefin

$-C{\equiv}C-$

acetylene

Materials Required.

Bromine in carbon tetrachloride: a solution of 2 mL in 100 mL carbon tetrachloride

Carbon tetrachloride

Procedure. Dissolve 0.2 mL of a liquid or 0.1 gram of a solid in 2 mL of carbon tetrachloride, and add the bromine in carbon tetrachloride solution drop by drop with shaking. The presence of an olefinic or acetylenic linkage in the sample will cause more than 2 or 3 drops of the bromine solution to be required before the characteristic orange-brown color of bromine will persist for one minute. The presence of electron-withdrawing groups such as phenyl or substituted phenyl, carboxyl, or halogen will reduce the rate of addition and in some cases completely inhibit the reaction. The presence of phenols, enols, amines, aldehydes, or ketones can result in the consumption (as indicated by loss of color) of large amounts of bromine, accompanied by *the evolution of hydrogen bromide*, which can be detected by exhaling cautiously over the top of the test tube and noting the fog that is produced in the moist air. The test should be carried out in diffuse light, since light can catalyze a free-radical substitution reaction that will consume bromine and produce hydrogen bromide. (Ref. 1, p. 121; Ref. 2, p. 111; Ref. 3, p. 1058; Ref. 4, p. 190.)

28.4 CHROMIC ANHYDRIDE

Used to distinguish primary and secondary alcohols from tertiary alcohols.

Materials Required.

Reagent-grade acetone

Chromic anhydride reagent: prepared by pouring slowly and with good stirring a suspension of 25 grams chromic anhydride (CrO_3) in 25 mL conc. sulfuric acid into 75 mL of water and allowing the deep orange-red solution to cool to room temperature.

Procedure. Dissolve 1 drop of a liquid or 15–30 mg of a solid in 1 mL of reagent-grade acetone and then add 1 drop of the chromic anhydride reagent. Primary and secondary alcohols produce an opaque blue-green suspension within two seconds; tertiary alcohols give no visible reaction and the solution remains orange. Aldehydes give a positive test, but ketones, alkenes, acetylenes, amines, and ethers give negative tests for two seconds. Enols may give a positive test, and phenols cause the solution to turn a much darker color than the blue-green produced by primary and secondary alcohols. (See Bordwell and Wellman, *J. Chem. Educ.* **39,** 308 (1962). Ref. 1, p. 125; Ref. 2, p. 121; Ref. 4, p. 149. For a way to distinguish among primary, secondary, and tertiary alcohols according to ease of oxidation by potassium permanganate in acetic acid, see Ritter, *J. Chem. Educ.* **30,** 395 (1953).)

28.5 2,4-DINITROPHENYLHYDRAZINE

Useful for identifying aldehydes and ketones.

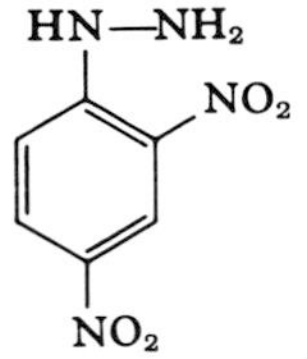

HN—NH_2, NO_2, NO_2

2,4-dinitro-phenylhydrazine

Materials Required.

2,4-Dinitrophenylhydrazine reagent: prepared by dissolving 3 grams of 2,4-dinitrophenylhydrazine in 15 mL conc. sulfuric acid and adding this solution, with good stirring, to a mixture of 20 mL of water and 70 mL of 95% ethanol. After thorough mixing, the reagent should be filtered.

95% ethanol

Procedure. Add a solution of 1 or 2 drops of the liquid, or 25–50 mg of the solid, in 2 mL of 95% ethanol to 3 mL of the 2,4-dinitrophenylhydrazine reagent. Mix the solution well and allow the mixture to stand for 15 minutes.

Most aldehydes and ketones give a solid 2,4-dinitrophenylhydrazone under these conditions. The derivative is yellow if the carbonyl group is not conjugated with a double bond or an aromatic ring. The 2,4-dinitrophenylhydrazones of most conjugated aldehydes and ketones are orange-red or red. The fact that the product is nonyellow should be interpreted with caution, as the color might be due to an impurity. The most obvious impurity is 2,4-dinitrophenylhydrazine itself, which is orange-red. (Ref. 1, p. 126; Ref. 2, p. 136; Ref. 3, p. 1060; Ref. 4, p. 162.)

28.6 FERRIC CHLORIDE SOLUTION

Useful for the recognition of phenols and enols.

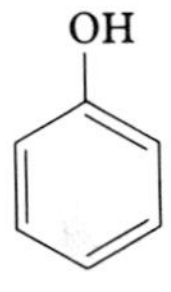

OH

phenol

OH

C=C

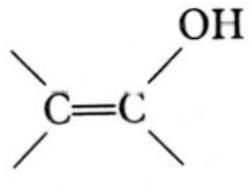

an enol

Materials Required.

Procedure a: 2.5% aqueous ferric chloride solution; Ethanol

Procedure b: Ferric chloride in chloroform; prepared by dissolving 1 gram ferric chloride in 100 mL chloroform; Pyridine

Procedure a. Dissolve 1 drop or 30–50 mg of the compound in 1 or 2 mL of water or a mixture of ethanol and water, and add up to 3 drops of the aqueous ferric chloride solution. Note color changes or formation of a precipitate; the color may not be permanent. Most phenols produce red, blue, purple, or green colorations; enols usually produce tan, red, or red-violet colorations. Some phenols, however, do not give a color.

Procedure b. Dissolve (or suspend) 1 drop of a liquid or 30 mg of a solid in 1 mL chloroform, and add 2 drops of the solution of ferric chloride in chloroform; then add 1 drop of pyridine and note color changes. This test appears to be more sensitive than the test described in Procedure a, and some phenols that do not give a color under the conditions of Procedure a will do so under these conditions. (Ref. 1, p. 127; Ref. 2, p. 147; Ref. 4, p. 348. Proc. a: Wesp and Brode, *J. Am. Chem. Soc.* **56,** 1037 (1934). Proc. b: Soloway and Wilen, *Anal. Chem.* **24,** 979 (1952).)

28.7 FERRIC HYDROXAMATE TEST

Useful for the identification of acid halides and anhydrides; esters; amides; and nitriles.

Test for Acid Halides and Anhydrides (Test a)

Materials Required.

1.0 *M* hydroxylamine hydrochloride in 95% ethanol

6 *M* hydrochloric acid

2 *M* hydrochloric acid

10% aqueous ferric chloride solution

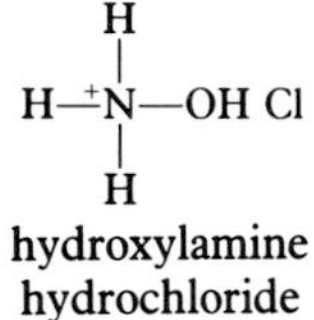
hydroxylamine hydrochloride

Procedure. Add 30–40 mg of the solid, or 1 drop of the liquid, to 0.5 mL of the hydroxylamine hydrochloride solution. Add 2 drops of the 6 *M* hydrochloric acid solution, warm the mixture slightly for 2 minutes, and then heat it to boiling for a few seconds. After cooling the solution, add 1 drop of the 10% ferric chloride solution.

A reddish-blue or bluish-red color will result if the original substance was an acid halide or anhydride. If the color is more red than blue, adjust the pH to 2 or 3 by dropwise addition of 2 *M* hydrochloric acid.

Tests for Esters of Carboxylic Acids (Test b)

Materials Required.

1.0 *M* hydroxylamine hydrochloride in 95% ethanol

2 *M* potassium hydroxide in methanol

2 *M* hydrochloric acid

10% aqueous ferric chloride solution

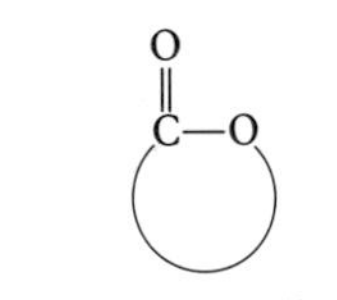
a lactone; a cyclic ester

Procedure. Add 30–40 mg of the solid, or 1 drop of the liquid, to 0.5 mL of the hydroxylamine hydrochloride solution. Now add the 2 *M* potassium hydroxide solution drop by drop until the mixture is basic to litmus, and then add 4 more drops. Heat the mixture just to boiling, cool it to room temperature, and add drop by drop (with good mixing) 2 *M* hydrochloric acid until the pH is approximately 3. Add 1 drop of the ferric chloride solution and note the color.

Esters of carboxylic acids, including lactones and polyesters, give definite magenta colors of varying degrees of intensity. Acid chlorides and anhydrides, and trihalo compounds such as benzotrichloride and chloroform give a positive test. Formic acid is the only acid that produces a color (red). Carbonic acid esters, sulfonic acid esters, urethanes, chloroformates, and esters of inorganic esters all give a negative result.

Test for Amides and Nitriles (Test c)

Materials Required.

1 *M* hydroxylamine hydrochloride in propylene glycol

1 *M* potassium hydroxide in propylene glycol

Propylene glycol
5% aqueous ferric chloride

Procedure. Add to 2 mL of the solution of hydroxylamine hydrochloride in propylene glycol a solution of 1 drop of the liquid, or 30–40 mg of the solid, dissolved in a minimum amount of propylene glycol. To this add 1 mL of the potassium hydroxide solution, and then boil the mixture for 2 minutes. Cool the solution to room temperature, and then add 0.5–1.0 mL of the ferric chloride solution. A red-to-violet color is a positive test; yellow colors should be interpreted as negative tests; brown colors or precipitates are indeterminate.

OH
|
CH_3—C—CH_2OH
|
H
propylene glycol

You should realize that esters, anhydrides, and acid halides will be converted to hydroxamic acids under these conditions as well as under the milder conditions of Tests a and b. Before a positive result in this test can be interpreted as indicating the presence of an amide or nitrile functionality, the possibility that the substance might be an ester, anhydride, or acid halide must be ruled out. (Ref. 1, p. 135; Ref. 2, pp. 119, 140, 144; Ref. 3, p. 1062. See also Davidson, *J. Chem. Educ.* **17,** 81 (1940).)

28.8 HYDROCHLORIC ACID/ZINC CHLORIDE TEST (LUCAS'S TEST)

Useful for distinguishing among lower-molecular-weight primary, secondary, and tertiary alcohols. Since the test requires that the alcohol initially be in solution, it is limited in its application to monohydroxylic alcohols more soluble than *n*-hexanol and to certain polyfunctional molecules.

Materials Required.
Lucas's reagent: prepared by dissolving 136 grams (1 mole) of zinc chloride in 89 mL (105 grams; 1 mole) conc. hydrochloric acid with cooling in ice.

Procedure. Add 2 mL of the reagent to 0.2 mL of the alcohol in a test tube; swirl to dissolve. Note the time required to form the insoluble alkyl chloride, which appears as a layer or emulsion.

The substitution of hydroxyl by chloride apparently takes place by an S_N1 mechanism, since tertiary alcohols, as well as allylic alcohols and benzyl alcohol, react very quickly. Secondary alcohols give indications of the formation of a second phase within 2 or 3 minutes, but primary alcohols react very much more slowly to form the insoluble alkyl chloride.

Tertiary alcohols will give a similar but slower reaction with concentrated hydrochloric acid alone. (Ref. 1, p. 131; Ref. 2, p. 121; Ref. 3, p. 261; Ref. 4, p. 150. See also Lucas, *J. Am. Chem. Soc.* **52,** 802 (1930).)

$$\mathrm{H{-}\overset{\displaystyle I}{\underset{\displaystyle I}{C}}{-}I}$$

iodoform

28.9 IODOFORM TEST

Useful for the identification of methyl ketones and secondary methyl carbinols.

Materials Required.

Iodine/potassium iodide reagent: prepared by adding 200 g potassium iodide and 100 g iodine to 800 mL distilled water, stirring until solution is complete

Dioxane

10% aqueous sodium hydroxide solution

Procedure. Dissolve 4 drops of a liquid or 100 mg of a solid in 5 mL of dioxane (use 1 mL of water in place of the dioxane if the compound is soluble in water to this extent). Add 1 mL of the sodium hydroxide solution and then the iodine/potassium iodide solution with shaking until the definite dark color of iodine persists. If less than 2 mL of the iodine/potassium iodide solution was consumed, place the test tube in a beaker of water at 60°C. If the dark color of iodine now disappears, continue to add the iodine/potassium iodide solution until the dark color that represents an excess of iodine is not discharged by heating at 60°C for 2 minutes. Now add 10% sodium hydroxide solution dropwise until the dark iodine color is gone, remove the tube from the heating bath, add 15 mL of water, and allow the mixture to cool to room temperature. Collect by suction filtration any solid that is formed and determine its melting point. Iodoform melts at 119–121°C. If the iodoform is reddish, dissolve it in 3 or 4 mL of dioxane, add 1 mL of 10% sodium hydroxide solution, and shake the mixture until the reddish color gives way to the lemon-yellow color of iodoform. Slowly dilute the mixture with water and collect the precipitated iodoform by suction filtration.

$$\mathrm{R{-}\overset{\displaystyle O}{\overset{\|}{C}}{-}CH_3}$$

a methyl ketone

Methyl ketones (and acetaldehyde) and methyl carbinols including ethanol (compounds that can be oxidized to methyl ketones by the reagent) give iodoform under these conditions; acetic acid does not. Compounds that can react with the reagent to generate one of these functional groups will also give iodoform; conversely, it is possible that the functionality that might be expected to result in the formation of iodoform can be removed by hydrolysis before iodoform formation is complete (see Fuson and Bull). (Ref. 1, p. 137; Ref. 2, p. 112; Ref. 3, p. 1068; Ref. 4, p. 167. See also Fuson and Bull, *Chem. Revs.* **15,** 275 (1934).)

28.10 AQUEOUS POTASSIUM PERMANGANATE SOLUTION (BAEYER'S TEST)

This test is useful for indicating the presence of most olefinic or acetylenic functional groups.

As a test for unsaturation, it should be used in conjunction with a similar test with a solution of bromine in carbon tetrachloride (Section 28.3).

Materials Required.

2% aqueous potassium permanganate solution

Reagent-grade acetone (free of alcohol)

Procedure. Dissolve 0.2 mL of a liquid or 0.1 gram of a solid in 2 mL of water or acetone. Add the potassium permanganate solution drop by drop with good shaking of the mixture. If more than 1 drop of the permanganate solution is consumed, as shown by the loss of the characteristic purple color, the presence of an olefin or acetylene or other functional group that can be oxidized by permanganate under these conditions is indicated. Such other groups include phenols and aryl amines, most aldehydes (but not benzaldehyde or formaldehyde) and formate esters, primary and secondary alcohols (see, however, Swinehart), mercaptans and sulfides, and thiophenols. Aryl-substituted alkenes are oxidized by permanganate under these conditions.

$$\mathrm{R{-}\overset{\overset{\displaystyle O}{\|}}{C}{-}H}$$
an aldehyde

$$\mathrm{R{-}O{-}\overset{\overset{\displaystyle O}{\|}}{C}{-}H}$$
a formate ester

R—SH
a mercaptan

Certain carefully purified alkenes are not oxidized under the conditions of this test (acetone solvent), but can be oxidized if ethanol is used instead; ethanol does not react with potassium permanganate within 5 minutes at room temperature.

In this test, you must take care not to be misled by the limited reaction of impurities. (Ref. 1, p. 149; Ref. 2, p. 112; Ref. 3, p. 1058; Ref. 4, p. 192. See also Swinehart, *F. Chem. Educ.* **41**, 392 (1964).)

28.11 ALCOHOLIC SILVER NITRATE SOLUTION

This test is useful for classifying compounds known to contain chlorine, bromine, or iodine. It should be used in conjunction with a similar test using sodium iodide in acetone (Section 28.13).

The rate of reaction of the halogen in this test gives an indication of its structural environment. Reference 1 includes a discussion of the relationship between structure and reactivity for organic halides. Table 28.2 summarizes the results that can be expected for the most common functional groups.

Materials Required.

2% silver nitrate in 95% ethanol

5% aqueous nitric acid

Procedure. Add 1 drop of the liquid (or 30–40 mg of the solid) to 2 mL of the silver nitrate solution at room temperature. If no precipitate is formed upon standing at room temperature for 5 minutes, heat the solution to boiling for 30 seconds and note whether or not a precipi-

Table 28.2. Alcoholic silver nitrate test

Water-soluble compounds that give an immediate precipitate at room temperature with *aqueous* silver nitrate:
- Salts of amines and halogen acids
- Low-molecular-weight acid halides (their water solubility results in their rapid hydrolysis to give halide ion)

Water-insoluble compounds that give an immediate precipitate at room temperature with *alcoholic* silver nitrate:
- Acid halides
- Tertiary alkyl halides
- Allylic halides
- Alkyl iodides
- α-Haloethers
- 1,2-Dibromides

Water-insoluble compounds that react slowly or not at all at room temperature, but give a precipitate readily at higher temperatures with *alcoholic* silver nitrate:
- Primary and secondary alkyl chlorides
- Geminal dibromides
- Activated aryl halides

Water-insoluble compounds that do not give a precipitate at higher temperatures with *alcoholic* silver nitrate:
- Unactivated aryl halides
- Vinyl halides
- Chloroform, carbon tetrachloride

tate is formed under these conditions. If a precipitate is formed, note the color and determine whether or not it dissolves upon addition of two drops of the nitric acid solution. Silver chloride is white, silver bromide is pale yellow, and silver iodide is yellow; the silver halides are insoluble in dilute nitric acid, but silver salts of organic acids will dissolve. (Ref. 1, p. 152; Ref. 2, p. 122; Ref. 3, p. 1059; Ref. 4, p. 202.)

28.12 SODIUM HYDROXIDE TEST

Useful for identifying esters, lactones, and anhydrides.

Materials Required.

0.1 *M* sodium hydroxide in 95% ethanol

Phenolphthalein indicator solution, in ethanol

95% ethanol

Procedure. Dissolve 0.1 gram of the substance in 3 mL ethanol. Add 3 drops of the phenolphthalein indicator solution and then add, dropwise, sufficient 0.1 *M* sodium hydroxide solution to turn the solution pink. Warm the mixture in a beaker of water at 40°C. Disappearance of the pink color indicates consumption of base.

Under these conditions, esters, lactones, and anhydrides are hydrolyzed, resulting in the consumption of base; additionally, some alkyl halides, amides, and nitriles will undergo solvolysis or hydrolysis and will likewise result in the use of base. In order to determine that the base was not consumed by an impurity, the cycle of adding base to turn the indicator pink and heating at 40°C to decolorize the solution should be repeated several times. (Ref. 1, p. 164.)

28.13 SODIUM IODIDE IN ACETONE

This test is useful for classifying compounds known to contain bromine or chlorine. It should be used in conjunction with a similar test using a solution of silver nitrate in 95% ethanol (Section 28.11).

Materials Required.

Acetone, reagent grade

Sodium iodide in acetone reagent: prepared by dissolving 15 g of sodium iodide in 100 mL reagent-grade acetone. The solution is colorless at first, but develops a pale lemon-yellow color on standing. When the color becomes definitely red-brown, the solution should be discarded.

Procedure. Add 2 drops of the liquid, or a solution of 100 mg of the solid dissolved in a minimum amount of acetone, to 1 mL of the sodium iodide in acetone solution. After mixing, allow the solution to

Table 28.3. Sodium iodide in acetone test

Formation of a precipitate within 3 minutes at 25°C:
Primary alkyl bromides
Benzylic and allylic chlorides and bromides
α-Halo ketones, esters, amides, and nitriles
Acid chlorides and bromides
Formation of a precipitate within 6 minutes at 50°C:
Primary and secondary alkyl chlorides
Secondary and tertiary alkyl bromides
Cyclopentyl chloride
Benzal chloride, benzotrichloride, bromoform, 1,1,2,2-tetrabromoethane
Unreactive under the conditions specified:
Tertiary alkyl chlorides
Cyclohexyl chloride and bromide
Vinyl halides; aryl halides
Chloroform; carbon tetrachloride, trichloroacetic acid

stand at room temperature for 3 minutes. Note whether or not a precipitate has formed (sodium chloride or sodium bromide) and whether or not the solution has become reddish-brown (liberation of iodine). If no change has occurred, heat the solution in a beaker of water at 50°C for 6 minutes and note any relevant changes.

This test depends upon two facts: (1) sodium chloride and sodium bromide, which can be formed in an S_N2-type displacement of the halide by iodide, are insoluble in acetone, and (2) iodine can be formed by the reaction of iodide with 1,2-dichloro and 1,2-dibromo compounds.

Table 28.3 summarizes the results that can be expected for some of the more common functional groups. (Ref. 1, pp. 169, 152; Ref. 3, p. 1059; Ref. 4, p. 204.)

28.14 TOLLENS REAGENT: SILVER-AMMONIA COMPLEX ION

Useful for distinguishing aldehydes from ketones and other carbonyl compounds.

Materials Required.

5% aqueous silver nitrate solution

10% aqueous sodium hydroxide solution

2 *M* aqueous ammonium hydroxide solution

Procedure. Clean a test tube thoroughly, preferably by boiling in it a 10% solution of sodium hydroxide and then discarding the sodium hydroxide solution and rinsing the test tube with distilled water. To the clean tube add 2 mL of the silver nitrate solution and 1 drop of the sodium hydroxide solution. Add the ammonium hydroxide solution drop by drop with good shaking until the dark precipitated silver oxide just dissolves.

Add 1 drop of the liquid or 30–50 mg of the solid to be tested, shake the tube to mix, and allow it to stand at room temperature for 20 minutes. If nothing happens, heat the tube in a beaker of water at 35°C for 5 minutes.

$$Ag[NH_3]_2^+ \longrightarrow Ag\downarrow$$

Aldehydes and other substances that can be oxidized by the silver-ammonia complex ion will reduce the silver ion to metallic silver, which will precipitate as a "mirror" on the test tube if it is sufficiently clean, or as a black colloidal suspension if the test tube is not clean.

The solution should not be allowed to stand after the test is completed, but should be discarded down the drain immediately, as the highly explosive silver fulminate may be formed on standing. Rinse the test tube with dilute nitric acid.

Substances that can be oxidized by means of the silver-ammonia complex ion include most aldehydes, "reducing sugars," hydroxylamines, acyloins, aminophenols, and polyhydroxyphenols. (Ref. 1, p. 173; Ref. 2, p. 138; Ref. 3, pp. 330, 1060; Ref. 4, p. 170.)

References

1. R. L. Shriner, R. C. Fuson, and D. Y. Curtin, *The Systematic Identification of Organic Compounds,* 5th edition, Wiley, New York, 1965.
2. N. D. Cheronis and J. B. Entrikin, *Identification of Organic Compounds,* Interscience, New York, 1963.
3. A. I. Vogel, *A Textbook of Practical Organic Chemistry,* 3rd edition, Wiley, New York, 1957.
4. R. L. Shriner, R. C. Fuson, D. Y. Curtin, and T. C. Morrill, *The Systematic Identification of Organic Compounds,* 6th edition, Wiley, New York, 1980.

29 Characterization Through Formation of Derivatives

Derivatives

Sections 29.1 through 29.31 describe a number of procedures whereby substances containing certain functional groups may be converted into reasonably well-behaved crystalline products (derivatives) through reaction at the functional group. As in the qualitative characterization tests of Section 28, the fact that the substance of unknown structure gives a solid product is consistent with the presence in its molecules of the suspected functional group, but the formation (or nonformation) of a solid product is by no means proof of the presence (or absence) of the suspected functional group. The reasons are the same as those discussed for the qualitative characterization tests: the derivative-forming reactions are not completely specific or characteristic (some substances may react atypically), and a small amount of an impurity, possibly water, may be sufficient to result in a solid product.

If a solid product is obtained, however, it can be purified—usually by recrystallization until successive recrystallizations do not raise the melting point significantly—and the melting point determined. Comparing the melting point of the derivative with the melting points recorded in the chemical literature for the corresponding derivatives of substances of known structure can often serve to eliminate all but one or two possibilities for the structure of the substance of unknown structure. A hypothetical case will show how this is done and will show some of the problems and pitfalls that must be considered and avoided.

Suppose the substance of unknown structure is a liquid that distilled between 96.5 and 98°C at 740 Torr. On the basis of various chemical and physical properties, you deduced that the liquid must be an alcohol. When you treated a sample with 3,5-dinitrobenzoyl chloride according to the procedure of Section 29.1, you obtained a solid that melted after two recrystallizations at 74.5–75.5°C. You presumed it to be the 3,5-dinitrobenzoate ester of the unknown alcohol.

Shriner, Fuson, and Curtin

At this point, alcohols of higher and lower boiling points have already been tentatively eliminated, because their boiling points are believed to be well outside the limits of the experimental error of the observed boiling range of the unknown. Shriner, Fuson, and Curtin (see References, end of this section) offer the following additional information:

		m.p. of Derivative		
Compound	*b.p.*	*α-Naphthylurethan*	*Phenylurethan*	*3,5-Dinitrobenzoate*
3-Butene-2-ol	96	—	—	—
Allyl alcohol	97	109	70	48
n-Propyl alcohol	97	80	51	74
sec-Butyl alcohol	99	97	65	75
tert-Amyl alcohol	102	71	42	117

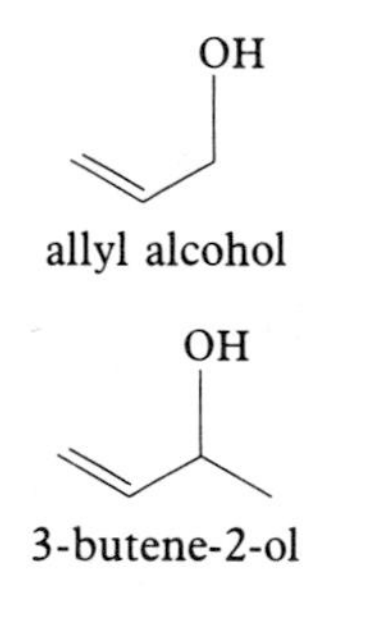

From these data, it appears that *tert*-amyl alcohol and allyl alcohol can also be eliminated, since the melting points reported for their 3,5-dinitrobenzoate esters are well outside the range of experimental error of the melting point observed for the 3,5-dinitrobenzoate of the unknown. If you knew the melting point of the 3,5-dinitrobenzoate of 3-butene-2-ol, you might be able to eliminate this as well. Now you can see that it might have been smarter to try to make the α-naphthylurethan of the unknown, since it might serve to distinguish among all possibilities.

Cheronis and Entrikin

Looking in Cheronis and Entrikin (see References) to try to find melting points for the derivatives of 3-butene-2-ol, you locate the following information:

		m.p. of Derivative		
Compound	*b.p.*	*α-Naphthylurethan*	*Phenylurethan*	*3,5-Dinitrobenzoate*
3-Butene-2-ol	—	—	—	—
Allyl alcohol	97.1	108	—	49
n-Propyl alcohol	97.15	105	—	74
sec-Butyl alcohol	99.5	97	—	76
tert-Amyl alcohol	102.35	75	—	116

You did not find what you wanted, but you did discover a remarkable disagreement between the value reported for the melting point of the α-naphthylurethan of *n*-propyl alcohol by Shriner, Fuson, and Curtin, and that reported by Cheronis and Entrikin. The disagreement between the other values reported for boiling points and melting points is typical and is an indication of the uncertainty of these values.

Looking in Vogel (see References) to find melting points for derivatives of 3-butene-2-ol and to get another opinion on the melting point of the α-naphthylurethan of *n*-propyl alcohol, you find the following information:

Vogel

		m.p. of Derivative			
Compound	*b.p.*	*α-Naphthyl-urethan*	*Phenyl-urethan*	*3,5-Dinitro-benzoate*	*Hydrogen 3-nitrophthalate*
3-Butene-2-ol	—	—	—	—	—
Allyl alcohol	97	109	70	50	124
n-Propyl alcohol	97	80	57	75	145
sec-Butyl alcohol	99.5	98	64	76	131
tert-Amyl alcohol	102	72	42	118	—

You did not find out anything here, either, about the melting points of derivatives of 3-butene-2-ol, but you did discover that Vogel and Shriner, Fuson, and Curtin agree on the melting point of the α-naphthylurethan of *n*-propyl alcohol. However, this indicates no more than that these authors are reporting a value from the same source in the original literature, and that Cheronis and Entrikin are reporting a value from a different source (assuming no typographical or transcription errors). You also see that Vogel and Shriner, Fuson, and Curtin report significantly different values for the melting point of the phenylurethan of *n*-propyl alcohol.

One more easy place to look for melting points of derivatives is the Chemical Rubber Company's *Handbook of Tables for Organic Compound Identification*, 3rd edition, which contains the following information:

CRC *Handbook of Tables*

		m.p. of Derivative			
Compound	*b.p.*	*α-Naphthyl-urethan*	*Phenyl-urethan*	*3,5-Dinitro-benzoate*	*Hydrogen 3-nitrophthalate*
3-Butene-2-ol	94–6	—	—	—	43–4
Allyl alcohol	97.1	108	70	49–50	124
n-Propyl alcohol	97.1	80; 76	57	74	145.5
sec-Butyl alcohol	99.5	97	64.5	76	131
tert-Amyl alcohol	102.3	72	42	116	—

From this research, you see that if the compound is an alcohol and if the solid formed from the unknown by treatment of a sample by the procedure of Section 28.1 is the corresponding 3,5-dinitrobenzoate ester, the boiling point of the unknown liquid and the melting point of the derivative serve to rule out all possibilities except *n*-propyl alcohol, *sec*-butyl alcohol, 3-butene-2-ol, and other alcohols of which we are not aware that boil near 96–98°C. You also see that the derivative that should have been made was not the 3,5-dinitrobenzoate but the hydrogen 3-nitrophthalate (by the procedure of Section 28.2), which would have allowed the elimination of at least two of the remaining possibilities. The moral of your search is that a few minutes of library research can save hours of laboratory work.

A minute in the lib. is worth an hour in the lab.

The next step would be to prepare the hydrogen 3-nitrophthalate of the unknown in order to reduce the number of possibilities for its structure. The α-naphthylurethan should not be prepared for two reasons: (1) on the basis of the present information, it would not allow elimination of 3-butene-2-ol, and (2) there is uncertainty about the melting point of this derivative of *n*-propanol.

Of course, the number of possibilities may be reduced by consideration of other physical and chemical properties. For example, the following values for density and index of refraction are reported in the *Handbook of Tables for Organic Compound Identification.*

Compound	d_4^{20}	n_D^{20}
3-Butene-2-ol	—	—
n-Propyl alcohol	0.80359	1.38499
sec-Butyl alcohol	0.80692	1.39495^{25}

It should be possible to distinguish between *n*-propyl alcohol and *sec*-butyl alcohol on the basis of the index of refraction, but of course 3-butene-2-ol cannot be ruled out by either the density or the index of refraction.

Assume now that you have prepared the hydrogen 3-nitrophthalate and found that it has a melting point of 144–145°C. All possibilities except *n*-propyl alcohol and alcohols of which we are not aware that boil near 96–98°C have now been eliminated. Of course, you would have to be very unlucky indeed to have an unknown sample that had the same boiling point as *n*-propyl alcohol and whose derivatives melted at the same temperature as those of *n*-propyl alcohol, but yet was not *n*-propyl alcohol.

The identity of the liquid of unknown structure and *n*-propyl alcohol could be established in several ways, if a sample of *n*-propyl alcohol were available. For example, the "identity" of the infrared or mass

Table 29.1. Procedures for the preparation of derivatives

The section numbers refer to the section in which the procedure for preparation will be found. The appendix numbers refer to the appendix in which the melting points of the derivatives can be found.

Type of Compound	*Type of Derivative*	*Section*	*Appendix*
Alcohol	Benzoate, *p*-nitrobenzoate, and 3,5-dinitrobenzoate esters	29.1	A.1
Alcohol	Hydrogen 3-nitrophthalate ester	29.2	A.1
Alcohol	Plenylurethan; α-naphthylurethan	29.3	A.1
Aldehyde	Methone derivative	29.4	A.2
Aldehyde	2,4-Dinitrophenylhydrazone	29.5	A.2
Aldehyde	Semicarbazone	29.6	A.2
Aldehyde	Oxime	29.7	A.2
Amide, unsubstituted	Hydrolysis	29.8	A.3
Amide, unsubstituted	9-Acylamidoxanthene	29.9	A.3
Amide, N-substituted	Hydrolysis	29.10	A.3
Amine, primary and secondary	N-Substituted acetamide	29.11	A.4
Amine, primary and secondary	N-Substituted benzamide	29.12	A.4
Amine, primary and secondary	N-Substituted *p*-toluenesulfonamide	29.13	A.4
Amine, primary and secondary	Phenylthiourea; α-naphthylthiourea	29.14	A.4
Amine, tertiary	Picrate	29.15	A.5
Amine, tertiary	Methiodide; *p*-toluenesulfonate	29.16	A.5
Carboxylic acid	Unsubstituted amide	29.17	A.6
Carboxylic acid	Anilide; *p*-toluidide; *p*-bromoanilide	29.18	A.6
Carboxylic acid	Phenacyl ester; *p*-substituted phenacyl esters	29.19	A.6
Carboxylic acid	*p*-Nitrobenzyl ester	29.20	A.6
Carboxylic acid anhydride	Anilide; *p*-toluidide; *p*-bromoanilide	29.18	A.6
Carboxylic acid ester	N-Benzylamide	29.21	A.7
Carboxylic acid ester	3,5-Dinitrobenzoate	29.22	A.7
Carboxylic acid ester	Hydrolysis	29.23	A.7
Carboxylic acid halide	Anilide; *p*-toluidide; *p*-bromoanilide	29.18	A.6
Ether, aromatic	Bromination product	29.24	A.8
Ether, aromatic	Picrate	29.15	A.8
Halide, aliphatic	S-Alkylthiuronium picrate	29.25	A.9
Halide, aromatic	*o*-Aroylbenzoic acid	29.26	A.10
Halid, aromatic	Oxidation	29.27	A.10
Halide	Anilide; *p*-toluidide; α-naphthalide	29.28	A.9
Hydrocarbon, aromatic	*o*-Aroylbenzoic acid	29.26	A.11
Hydrocarbon, aromatic	Oxidation	29.27	A.11
Hydrocarbon, aromatic	2,4,7-Trinitrofluorenone adduct	29.29	A.11
Hydrocarbon, aromatic	Picrate	29.15	A.11
Ketone	2,4-Dinitrophenylhydrazone	29.5	A.12
Ketone	Semicarbazone	29.6	A.12
Ketone	Oxime	29.7	A.12
Nitrile	Hydrolysis	29.8	A.13
Phenol	α-Naphthylurethan	29.3	A.14
Phenol	Bromination product	29.30	A.14
Phenol	Aryloxyacetic acid	29.31	A.14

spectra would be completely acceptable as evidence of the identity of the two substances. The comparison of spectra could also be made by way of published spectral data for *n*-propyl alcohol. Either the hydrogen 3-nitrophthalate or 3,5-dinitrobenzoate of *n*-propyl alcohol could be prepared, and the identity of the melting points and nondepression of the mixed melting point (or identity of the infrared spectra of the two derivatives) could establish the identity of the unknown liquid as *n*-propyl alcohol.

Finally, of course, it must be admitted that the NMR spectrum would immediately establish whether or not the unknown liquid was *n*-propyl alcohol, or, for that matter, any of the other four alcohols that were considered. The NMR spectrum would also reveal whether or not the unknown substance had a structure different from that of any of the five possibilities considered.

Table 29.1 is a summary and index for the 31 procedures for the formation of derivatives that are presented in the following sections. The references at the end of each procedure give more information concerning the scope and limitations of the procedures, possible procedural variations, and exceptional behavior of certain compounds or types of compounds; references that are given in abbreviated form are listed in full at the end of this section.

The book by Shriner, Fuson, and Curtin and the Cheronis and Entrikin book give many references to the original literature in which the procedures for preparations of derivatives are described. If you think that you know the identity of the substance for which you are attempting to prepare a derivative, you will often save time by looking up in these references the exact procedure used with that compound. You may find that the substance is best converted to the derivative by a modification of the general procedure, or that the derivative is formed in poor yield (by an experienced chemist, which might mean that you should consider another derivative), or that the product is best isolated in a certain way or recrystallized from a certain solvent, or that it has some interesting or characteristic property by which it may be recognized. This kind of information can sometimes save hours of lab time.

These two very useful books also give literature references to many more procedures for preparation of derivatives, both for substances with the functional groups for which procedures are given in this book and for compounds with other types of functional groups.

Questions

1. Consider the alcohols 3-butene-2-ol, allyl alcohol, *n*-propyl alcohol, *sec*-butyl alcohol, and *tert*-butyl alcohol. Which of these alcohols could be distinguished by the chromic anhydride test, Section 28.4?
2. Which of the alcohols of Problem 1 could be distinguished by Lucas's Test, Section 28.8?

29.1 BENZOATES, *p*-NITROBENZOATES, AND 3,5-DINITROBENZOATES OF ALCOHOLS

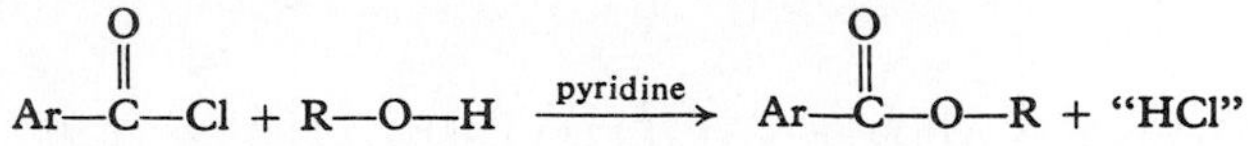

$$\underset{\text{acid chloride}}{\mathrm{Ar{-}\overset{\overset{\displaystyle O}{\|}}{C}{-}Cl}} + \mathrm{R{-}O{-}H} \xrightarrow{\text{pyridine}} \underset{\text{ester}}{\mathrm{Ar{-}\overset{\overset{\displaystyle O}{\|}}{C}{-}O{-}R}} + \text{“HCl”}$$

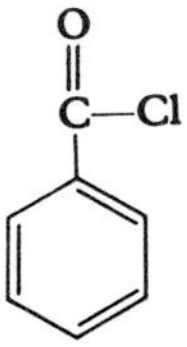

Procedure. Treat a solution of 0.5 gram of the alcohol in 3 mL of pyridine with about 2 grams of the appropriate acid chloride (benzoyl chloride, *p*-nitrobenzoyl chloride, or 3,5-dinitrobenzoyl chloride) with cooling in an ice bath. Heat the resulting mixture on the steam bath with exclusion of moisture for 10 minutes if the alcohol is primary or secondary, and for 30 minutes if it is tertiary (Note 1). Pour the mixture with stirring into 10 mL of ice water and cautiously acidify it with conc. hydrochloric acid. Thoroughly triturate the residue, which frequently is an oil, with 5 mL of 5% sodium carbonate solution. Finally, collect the solid by suction filtration and recrystallize it from aqueous alcohol, or methanol, or ethanol, or acetone/petroleum ether. (Ref. 1, pp. 246, 247; Ref. 2, p. 249; Ref. 3, p. 262; Ref. 4, pp. 156, 157.)

$C_6H_5{-}\overset{\overset{O}{\|}}{C}{-}Cl$

benzoyl chloride

Note

1. Tertiary alcohols are esterified poorly under these conditions.

29.2 HYDROGEN 3-NITROPHTHALATES OF ALCOHOLS

Derivative for primary and secondary alcohols (Note 1).

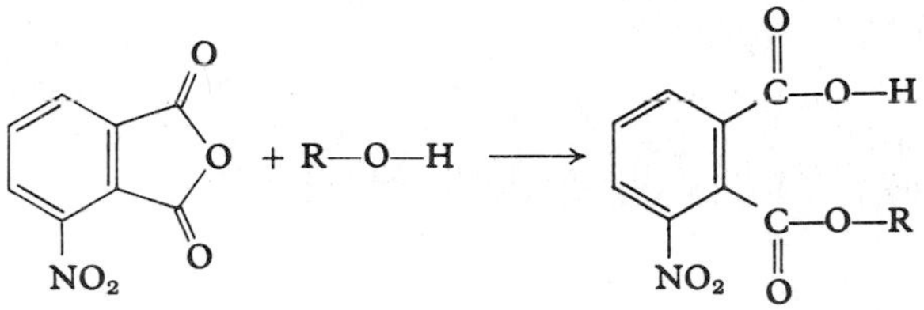

3-nitrophthalic anhydride (NO_2) + R—O—H ⟶ hydrogen 3-nitrophthalate (NO_2; C(=O)—O—H; C(=O)—O—R)

Procedure. Heat for 2 hours on the steam bath a mixture of 0.3 mL of the alcohol, 0.3 gram of 3-nitrophthalic anhydride, and 0.5 mL of pyridine. Pour the mixture onto ice, acidify with conc. hydrochloric acid, and isolate the solid ester either by filtration or by extraction with benzene or chloroform. Recover the acid phthalate by extraction into dilute sodium hydroxide followed by acidification of the extract and suction filtration. Recrystallize from water, alcohol/water, or toluene. (Ref. 1, p. 248; Ref. 4, p. 158.)

Note

1. Tertiary alcohols undergo elimination under these conditions.

29.3 PHENYL- AND α-NAPHTHYLURETHANS

Derivatives for primary and secondary alcohols (Note 1) and phenols.

$$\underset{\text{isocyanate}}{Ar{-}N{=}C{=}O} + R{-}O{-}H \longrightarrow \underset{\text{urethane}}{Ar{-}\underset{\displaystyle H}{\underset{|}{N}}{-}\overset{\displaystyle O}{\overset{\|}{C}}{-}O{-}R}$$

Procedure. Add a solution of 0.5 gram of the alcohol (Note 2) in 5 mL of ligroin (b.p. 80–100°C) to a solution of 0.5 gram of phenylisocyanate or α-naphthylisocyanate (Note 3) in 10 mL of the same solvent. Heat the mixture for 1–3 hours on the steam bath, filter it while hot (Note 2), and allow the filtrate to cool. Collect the product by suction filtration, and recrystallize it from ligroin or carbon tetrachloride (Ref. 1, p. 246; Ref. 2, p. 252; Ref. 3, pp. 264, 683; Ref. 4, p. 355.)

Notes

1. The urethans of tertiary alcohols are difficult to prepare by this method.
2. The alcohol must be dry; water leads to the formation of the very insoluble diphenylurea or di-α-naphthylurea. The hot filtration will remove most of this if it is formed, but some will crystallize with the product.
3. If a phenol is used, the α-naphthylurethan is the preferred derivative and the reaction should be catalyzed by the addition of a few drops of pyridine.

29.4 METHONE DERIVATIVES OF ALDEHYDES

Derivative for low-molecular-weight aldehydes.

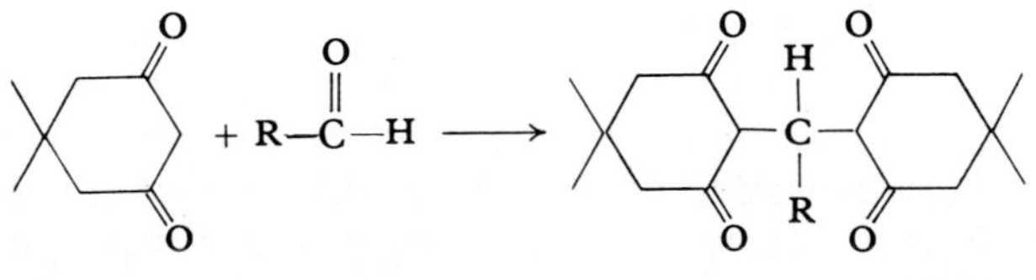

$$\text{Dimedon} + R{-}\overset{\displaystyle O}{\overset{\|}{C}}{-}H \longrightarrow (\text{dimedon})_2CHR$$

Dimedon

Procedure. Dissolve 300 mg of the reagent (Dimedon; dimethyldihydroresorcinol; 5,5-dimethyl-1,3-cyclohexanedione—Section 70.2; 2.1 mmole) in 4 mL of 50% aqueous ethanol. Add to this 1 mmole of the aldehyde and boil the mixture for about 30 seconds. Allow it to cool and stand for crystallization for at least 4 hours. Collect the product by suction filtration and recrystallize from a minimum amount of methanol and water. (Ref. 1, p. 254; Ref. 2, p. 262; Ref. 3, p. 332; Ref. 4, p. 180.)

29.5 2,4-DINITROPHENYLHYDRAZONES

Derivative for aldehydes and ketones.

$$O_2N-C_6H_3(NO_2)-NH-NH_2 + R-\overset{O}{\overset{\|}{C}}-R' \longrightarrow O_2N-C_6H_3(NO_2)-NH-N{=}C\begin{matrix}R\\R'\end{matrix}$$

2,4-dinitrophenylhydrazine — 2,4-dinitrophenylhydrazone

Preparation of a Solution of 2,4-Dinitrophenylhydrazine in 30% Perchloric Acid. Dissolve 1.2 grams (6 mmole) of 2,4-dinitrophenylhydrazine in a mixture of 16 mL of 60% perchloric acid plus 34 mL of water at room temperature.

Preparation of the Derivative. Take 4 mL of the perchloric acid solution of 2,4-dinitrophenylhydrazine and dilute it with 8 mL of water; stir well. Add quickly to this well-stirred mixture 0.5 mmole of the carbonyl compound as a 10–20% solution in ethanol. Collect the resulting 2,4-dinitrophenylhydrazone by suction filtration, and recrystallize it from ethanol, ethyl acetate, dioxane, dioxane/water, or ethanol/water. (Ref. 1, p. 253; Ref. 2, p. 260; Ref. 3, p. 344; Ref. 4, p. 179.)

29.6 SEMICARBAZONES

Derivative for aldehydes and ketones.

$$H_2N-\overset{O}{\overset{\|}{C}}-NH-NH_2 + R-\overset{O}{\overset{\|}{C}}-R' \longrightarrow H_2N-\overset{O}{\overset{\|}{C}}-NH-N{=}C\begin{matrix}R\\R'\end{matrix}$$

semicarbazide — semicarbazone

Preparation of an Alcoholic Solution of Semicarbazide Acetate. Grind 1 gram of semicarbazide hydrochloride with 1 gram of anhydrous sodium acetate in a mortar. Transfer the mixture to a flask, boil it with 10 mL of absolute ethanol, and filter the suspension while hot.

Preparation of the Derivative. Add to the freshly prepared solution of semicarbazide acetate about 0.2 gram of the carbonyl compound and heat the mixture under reflux on the steam bath for 30–60 minutes. Dilute the hot solution with water to incipient cloudiness, and allow it to cool slowly to room temperature. Collect the semicarbazone by suction filtration and recrystallize it from aqueous alcohol or alcohol. (Ref. 1, p. 253; Ref. 2, pp. 262, 320; Ref. 3, p. 344; Ref. 4, p. 179.)

29.7 OXIMES

Derivative for higher-molecular-weight aldehydes and ketones.

$$\underset{\text{hydroxylamine hydrochloride}}{H_3\overset{+}{N}{-}O{-}H\ Cl^-} + R{-}\overset{O}{\overset{\|}{C}}{-}R' \xrightarrow{\text{pyridine}} \underset{\text{oxime}}{H{-}O{-}N{=}C(R)(R')} + \text{“HCl”}$$

Procedure. Heat for 2 hours under reflux on the steam bath a mixture of 0.5 gram of the carbonyl compound, 0.5 gram of hydroxylamine hydrochloride, 3 mL of pyridine, and 3 mL of absolute ethanol. Remove the solvent by evaporation (by heating the mixture while drawing a current of air over it or with a rotary evaporator; Section 36) and recrystallize the residue from methanol or methanol/water. (Ref. 1, p. 289; Ref. 2, p. 319; Ref. 3, p. 721; Ref. 4, p. 181.)

29.8 CARBOXYLIC ACIDS BY HYDROLYSIS OF PRIMARY AMIDES AND NITRILES

Derivative for primary amides and nitriles.

$$\underset{\text{primary amide}}{R{-}\overset{O}{\overset{\|}{C}}{-}\underset{H}{\underset{|}{N}}{-}H} + H_2O \longrightarrow \underset{\text{carboxylic acid}}{R{-}\overset{O}{\overset{\|}{C}}{-}OH} + NH_3$$

$$\underset{\text{nitrile}}{R{-}C{\equiv}N} + 2H_2O \longrightarrow \underset{\text{carboxylic acid}}{R{-}\overset{O}{\overset{\|}{C}}{-}OH} + NH_3$$

Basic Hydrolysis

Procedure a (*easily hydrolyzed amides or nitriles*). Boil under reflux 10 mmole of the substance with 0.8 gram (20 mmoles) of sodium hydroxide in 3 mL of water until ammonia evolution can be detected no longer (4–10 hours; Notes 1 and 2).

Procedure b (*more difficultly hydrolyzed amides or nitriles*). Boil under reflux 10 mmole of the substance with 1.2 grams (20 mmoles) of potassium hydroxide in 4 mL of mono-, di-, or triethyleneglycol until ammonia evolution has ceased (about 5 hours; Note 1). At this time dilute the mixture with 10 mL of water.

Workup for Both Procedures a and b. Acidify the aqueous solution with 20% sulfuric acid. Collect the precipitated acid by suction filtration, wash it with water, and recrystallize it from water or aqueous methanol.

If the acid is a liquid or if it is relatively soluble in water, it should be converted to a phenacyl ester by the procedure of Section 29.19.

Acidic Hydrolysis: For Nitriles That Resist Basic Hydrolysis

Procedure. Heat in an oil bath at 160°C for 30 minutes, with stirring and using a reflux condenser, a mixture of 5 mL of 75% sulfuric acid, 200 mg of sodium chloride, and 10 mmole of the nitrile. Raise the temperature of the bath to 190°C and continue to heat and stir for another 30 minutes. Cool the mixture to room temperature and pour it onto 20 grams of ice. Collect the precipitate by suction filtration and dissolve it in 5 mL of 10% sodium hydroxide solution. Any insoluble amide should be removed by suction filtration and, if it is a major product, recrystallized from water or aqueous methanol. Acidify the filtrate to obtain the acid. Collect it by suction filtration and recrystallize from water, toluene, acetone/water, or alcohol/water. (Ref. 1, p. 291; Ref. 2, p. 321; Ref. 3, pp. 404, 410, 798, 805; Ref. 4, p. 282.)

Notes

1. Ammonia can be detected by the method described in Section 28.1.
2. If the amide or nitrile solidifies in the condenser, 0.5 mL of ethanol can be added to help to redissolve it; the alcohol should be removed by distillation at the end of the heating period.

29.9 9-ACYLAMIDOXANTHENES FROM AMIDES

Derivative for unsubstituted amides.

$$\text{xanthydrol} + H{-}NH{-}C(=O){-}R \longrightarrow \text{9-acylamidoxanthene}$$

xanthydrol amide 9-acylamidoxanthene

Procedure. Dissolve 2 mmoles (400 mg) of xanthydrol in 5 mL of glacial acetic acid (Note 1). Add to this 1–1.5 mmole of the unsubstituted amide (Note 2), heat the mixture in a beaker of water at 85°C for 20–30 minutes, and then allow it to cool for crystallization. Collect the product by suction filtration and recrystallize it from 2 : 1 dioxane : water or 2 : 1 ethanol : water. (Ref. 1, p. 256; Ref. 2, p. 269; Ref. 3, p. 405; Ref. 4, p. 284.)

Notes

1. If the solution is not clear, allow it to stand for a few minutes and decant or filter the clear liquid from the insoluble material.
2. If the amide is not soluble in glacial acetic acid, it can be added to the reaction mixture as a solution in 2 mL of ethanol; if this is done, add 1 mL of water to the mixture after heating and before cooling.

29.10 HYDROLYSIS OF N-SUBSTITUTED AMIDES

$$R{-}C(=O){-}N(R''){-}R' + H_3O^+ \xrightarrow[H_2O]{acid} R{-}C(=O){-}OH + H{-}\overset{+}{N}H(R''){-}R'$$

amide carboxylic acid amine hydrochloride

Procedure a. Boil under reflux for 2 hours a mixture of 0.5–1 gram of the amide and 10–20 mL of conc. hydrochloric acid. Cool the mixture and collect any solid by suction filtration. The solid may be the acid corresponding to the amide, the hydrochloride salt of the amine corresponding to the amide, or unchanged amide. If the solid is soluble in water, it is probably the salt of the amine. If it is insoluble in water but soluble in 5% aqueous sodium bicarbonate, it is probably the acid. The melting point as well as solubility properties will indicate unchanged starting material.

The amine can be recovered by making the solution alkaline (use care in neutralizing a strongly acidic solution) and then either filtering with suction or extracting with ether, depending upon the nature of the amine. It can then be purified by recrystallization or converted to a derivative by procedures suitable for primary and secondary amines.

The acid can be isolated from the solution by reacidification and extraction (if it were not soluble, it would have separated from the solution originally) and converted to a derivative.

Procedure b (*for amides that resist Procedure a*) (*Note 1*). Boil under reflux for 30 minutes to an hour a mixture of 1 gram of the amide in 10 mL of 70% sulfuric acid (Note 2). Cautiously pour 10 mL of water down the condenser and allow the solution to cool. Remove any solid by suction filtration and recover the amine as in Procedure a. (Ref. 1, p. 255; Ref. 2, p. 267; Ref. 3, p. 801.)

Notes

1. Benzanilide and related compounds, for example.
2. Prepared by pouring 8 mL of conc. sulfuric acid onto 6 grams of ice.

29.11 SUBSTITUTED ACETAMIDES FROM AMINES

Derivative for water-insoluble primary and secondary amines.

$$\underset{\text{amine}}{R{-}\ddot{N}(R'){-}H} + \underset{\text{acetic anhydride}}{CH_3{-}\overset{O}{\overset{\|}{C}}{-}O{-}\overset{O}{\overset{\|}{C}}{-}CH_3} \longrightarrow \underset{\text{acetamide}}{R{-}\ddot{N}(R'){-}\overset{O}{\overset{\|}{C}}{-}CH_3} + \underset{\text{acetic acid}}{CH_3{-}\overset{O}{\overset{\|}{C}}{-}O{-}H}$$

Procedure. Dissolve 0.5 gram of the amine in 25 mL of 5% hydrochloric acid. Add 5% sodium hydroxide solution in small portions until the mixture becomes cloudy (because of precipitation of the amine). Redissolve the precipitate by adding a little more 5% hydrochloric acid. Add about 10 grams of ice and then 5 mL of acetic anhydride. Next add with good stirring and all in one portion a previously prepared solution of 5 grams of sodium acetate trihydrate in 5 mL of water. Cool the mixture in an ice bath and allow it to stand (possibly overnight) for crystallization. (Compare the preparation of acetanilide, Section 63.1).

Collect the product by suction filtration and wash it well with water. Recrystallize from ethanol or ethanol/water or, if the crude product has been thoroughly dried, from toluene/cyclohexane. (Ref. 1, p. 259; Ref. 3, pp. 576, 652; Ref. 4, p. 232.)

29.12 SUBSTITUTED BENZAMIDES FROM AMINES

Derivative for primary and secondary amines.

$$\underset{\text{amine}}{R{-}\ddot{N}(R'){-}H} + \underset{\text{benzoyl chloride}}{Cl{-}\overset{\overset{O}{\|}}{C}{-}\phi} \xrightarrow{\text{pyridine}} \underset{\text{benzamide}}{R{-}\ddot{N}(R'){-}\overset{\overset{O}{\|}}{C}{-}\phi} + \text{"HCl"}$$

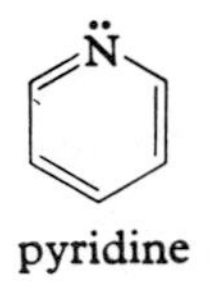
pyridine

Procedure. Add drop by drop 0.5 mL of benzoyl chloride to a solution of 0.5 gram of the amine in 5 mL of dry pyridine and 10 mL of dry toluene. Heat the resulting mixture in a beaker of hot water at 60–70°C for 30 minutes. Pour the mixture into 100 mL of water, separate the toluene layer, extract the aqueous layer with one 10-mL portion of toluene, and combine this with the toluene layer. Wash the combined toluene layers with water followed by 5% sodium carbonate solution, dry over anhydrous magnesium sulfate, and, after removing the magnesium sulfate by filtration, concentrate to a volume of 3 or 4 mL. Stir about 20 mL of hexane into the concentrated toluene solution in order to precipitate the derivative, collect the product by suction filtration, and wash it with hexane. Recrystallize the benzamide from cyclohexane/hexane or cyclohexane/ethyl acetate. (Ref. 1, p. 260; Ref. 2, p. 272; Ref. 3, p. 652; Ref. 4, p. 233.)

29.13 *p*-TOLUENESULFONAMIDES FROM AMINES

Derivative for primary and secondary amines (see Note 1).

$$\underset{\text{amine}}{R{-}\ddot{N}(R'){-}H} + \underset{p\text{-toluenesulfonyl chloride}}{Cl{-}SO_2{-}C_6H_4{-}CH_3} \longrightarrow \underset{p\text{-toluenesulfonamide}}{R{-}\ddot{N}(R'){-}SO_2{-}C_6H_4{-}CH_3} + HCl$$

Procedure. Boil for 30 minutes under reflux a mixture of 0.5 g of the amine, 1–1.5 g (5–8 mmoles) of *p*-toluenesulfonyl chloride (Note 2), and 3 mL of pyridine. Pour the reaction mixture into 5 mL of cold water and stir until the product crystallizes. Collect the sulfonamide by suction filtration, wash it well with water, and recrystallize it from alcohol or aqueous alcohol. (Ref. 1, p. 261; Ref. 2, p. 274; Ref. 3, p. 653; Ref. 4, p. 233.)

Notes

1. Compare with Hinsberg's test, Section 28.2.
2. An equivalent amount of benzenesulfonyl chloride can be used; the *p*-toluenesulfonamides are said to be more satisfactory derivatives.

29.14 PHENYLTHIOUREAS AND α-NAPHTHYLTHIOUREAS

Derivatives for primary and secondary amines.

$$\underset{\text{arylisothiocyanate}}{\text{Ar—N=C=S}} + \underset{\text{amine}}{\text{H—}\underset{\text{R}'}{\ddot{\text{N}}}\text{—R}} \longrightarrow \underset{\text{arylthiourea}}{\text{Ar—}\underset{\text{H}}{\ddot{\text{N}}}\text{—}\overset{\text{S}}{\overset{\|}{\text{C}}}\text{—}\underset{\text{R}'}{\ddot{\text{N}}}\text{—R}}$$

Procedure. Dissolve 0.2 gram of the amine in 5 mL of ethanol (Note 1) and add to this a solution of 0.2 g (0.18 mL; 1.5 mmole) of phenylisothiocyanate or 0.28 g (1.5 mmole) of α-naphthylisothiocyanate in 5 mL of ethanol. If no reaction takes place at room temperature, heat the mixture for one or two minutes. If no crystals separate upon cooling and scratching (as with aromatic amines), reheat the mixture for another 10 minutes and then cool again (Note 2). Collect the product by suction filtration and recrystallize it from ethanol. (Ref. 1, p. 261; Ref. 2, p. 275; Ref. 3, p. 422; Ref. 4, p. 234.)

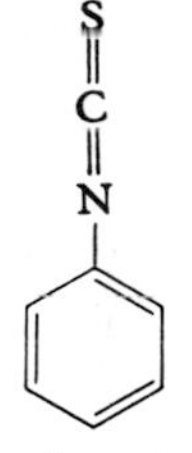

phenyl-isothiocycanate

Notes

1. These reagents are not reactive toward water (compare the analogous isocyanates, Section 28.3), and so an aqueous solution of the amine can be used if it is difficult to obtain the pure amine.
2. If no product can be obtained with an aromatic amine, the reaction can be tried without a solvent with heating for 10 minutes. The product can be isolated at the end of the reaction by the addition of 50% aqueous ethanol.

29.15 PICRATES

Derivative for tertiary amines, aromatic ethers, and aromatic hydrocarbons (Note 1).

$$R_3N: + \text{picric acid (2,4,6-trinitrophenol, OH)} \longrightarrow R_3\overset{+}{N}\text{—H} \quad \text{picrate salt (}O^-\text{)}$$

picric acid

picrate salt

$$Ar\text{—}H + \text{picric acid} \longrightarrow Ar\text{—}H \cdot \text{picric acid}$$

picric acid

picrate complex

Procedure a (*for tertiary amines; also for aromatic ethers and hydrocarbons that form relatively stable picrates*). Dissolve 2 mmoles of the substance in 5 mL of 95% ethanol, and add to this solution a solution of 2.2 mmoles of picric acid (500 mg) in 10 mL of 95% ethanol. Heat the mixture to boiling on the steam bath and allow it to cool slowly for crystallization. Collect the product by suction filtration and recrystallize it from ethanol (Note 2).

Procedure b (*for substances that form relatively unstable picrates*). Dissolve 5 mmoles of the substance in about 5 mL of boiling chloroform and add a solution of 5 mmoles (1.15 grams) of picric acid in 3 mL of boiling chloroform. Swirl the resulting mixture and allow it to cool for crystallization. Collect the product by suction filtration and recrystallize it from a minimum amount of chloroform (Notes 2 and 3). (For amine picrates see Ref. 1, p. 263; Ref. 2, p. 277; Ref. 3, p. 422; Ref. 4, p. 236. For ether and hydrocarbon picrates, see Ref. 1, p. 277; Ref. 2, pp. 298, 315; Ref. 3, pp. 518, 672; Ref. 4, p. 309.)

Notes

1. Certain other aromatic compounds such as primary and secondary amines and halides will form picrates.
2. The picrates of some substances cannot be recrystallized because they dissociate to too large a degree in solution.

3. The m.p. should be determined immediately because some picrates decompose.

29.16 QUATERNARY AMMONIUM SALTS: METHIODIDES AND *p*-TOLUENESULFONATES

Derivatives for tertiary amines.

$$R{-}\overset{\overset{\Large R'}{|}}{\underset{\underset{\Large R''}{|}}{N}}{:} + CH_3{-}I \longrightarrow R{-}\overset{\overset{\Large R'}{|+}}{\underset{\underset{\Large R''}{|}}{N}}{-}CH_3 + I^-$$

tertiary amine — quaternary methiodide

$$R{-}\overset{\overset{\Large R'}{|}}{\underset{\underset{\Large R''}{|}}{N}}{:} + CH_3{-}O{-}\overset{\overset{\Large O}{\|}}{\underset{\underset{\Large O}{\|}}{S}}{-}C_6H_4{-}CH_3 \longrightarrow R{-}\overset{\overset{\Large R'}{|+}}{\underset{\underset{\Large R''}{|}}{N}}{-}CH_3 \quad {}^{-}O_3S{-}C_6H_4{-}CH_3$$

tertiary amine — quaternary *p*-toluenesulfonate

Procedure. Dissolve 0.5 gram of the tertiary amine in twice its volume of nitromethane, acetonitrile, or alcohol (Note 1); do the same for 1 gram of the quaternizing agents (methyl iodide or methyl *p*-toluenesulfonate). Combine the solutions, allow the mixture to stand at room temperature for an hour, and then heat it for 30 minutes on the steam bath. If the quaternary salt crystallizes out, it should be collected by suction filtration; if not, the solvent should be removed under vacuum (Section 36). The crude product or residue should be recrystallized from ethyl acetate/ethanol. (Ref. 1, p. 262; Ref. 2, p. 277; Ref. 3, p. 660; Ref. 4, p. 235.)

Note

1. The solvents named are listed in decreasing order of suitability for the reaction.

29.17 CARBOXYLIC ACID AMIDES

A procedure for deriving amides from carboxylic acids. Lower-molecular-weight amides, which are relatively soluble in water, are difficult to isolate by this procedure.

$$R{-}\overset{O}{\overset{\|}{C}}{-}OH + Cl{-}\overset{O}{\overset{\|}{S}}{-}Cl \xrightarrow{\text{dimethylformamide}} R{-}\overset{O}{\overset{\|}{C}}{-}Cl + HCl + SO_2$$

thionyl chloride — acid chloride

$$R{-}\overset{O}{\overset{\|}{C}}{-}Cl + 2NH_3 \longrightarrow R{-}\overset{O}{\overset{\|}{C}}{-}NH_2 + \overset{+}{N}H_4 + Cl^-$$

acid chloride — amide

$$H{-}\overset{O}{\overset{\|}{C}}{-}N(CH_3)_2$$

dimethylformamide

Procedure. Boil under reflux on the steam bath, using a calcium chloride drying tube in the condenser, a mixture of 1 gram of the carboxylic acid, 5 mL of thionyl chloride, and 1 drop of dimethylformamide for 15–30 minutes. Pour the mixture *cautiously* into 15 mL of ice-cold conc. ammonium hydroxide. Collect the crude amide by suction filtration and recrystallize it from water or aqueous alcohol. (Ref. 1, p. 235; Ref. 3, p. 361; Ref. 4, p. 271.)

29.18 ANILIDES, *p*-TOLUIDIDES, AND *p*-BROMOANILIDES OF CARBOXYLIC ACIDS

Derivatives for carboxylic acids, acid halides, and anhydrides.

$$R{-}\overset{O}{\overset{\|}{C}}{-}OH + Cl{-}\overset{O}{\overset{\|}{S}}{-}Cl \xrightarrow{\text{dimethylformamide}} R{-}\overset{O}{\overset{\|}{C}}{-}Cl + HCl + SO_2$$

$$R{-}\overset{O}{\overset{\|}{C}}{-}X + 2\,Ar{-}\ddot{N}H_2 \longrightarrow R{-}\overset{O}{\overset{\|}{C}}{-}\underset{H}{\underset{|}{\ddot{N}}}{-}Ar + Ar{-}\overset{+}{N}H_3 + X^-$$

acyl halide or anhydride — aryl amine — aryl amide

Preparation of the Acid Chloride. Heat for 30 minutes 1 mmole of the acid with 0.8 mL (1.3 g; 11 mmole) of thionyl chloride, and 1 drop of dimethylformamide on the steam bath under reflux, using a calcium chloride drying tube on the condenser.

$:NH_2$ on benzene ring

aniline

Treatment of the Acid Halide or Anhydride with the Aromatic Amine. Add to the acid chloride just prepared, or to 1 mmole of the acid halide or anhydride, dissolved in dry toluene if it is a solid, a solution of 5 mmoles of the aromatic amine (aniline, *p*-toluidine, or *p*-bromoaniline) dissolved in 20 mL of toluene. Heat the mixture under reflux on the steam bath for about 15 minutes. Cool the mixture to room temperature, transfer it to a separatory funnel, and extract it with 2 mL of water, 5 mL of 5% hydrochloric acid, 5 mL of 5% sodium hydroxide, and 2 mL of water. Evaporate the toluene solution to dryness and recrystallize the residual amide from aqueous alcohol, methanol, or ethanol. (Ref. 1, p. 236; Ref. 2, p. 238; Ref. 3, p. 361; Ref. 4, p. 272.)

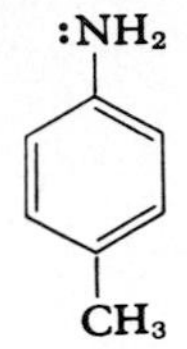

$:NH_2$

CH_3

p-toluidine

29.19 PHENACYL AND SUBSTITUTED PHENACYL ESTERS OF CARBOXYLIC ACIDS

Derivative for carboxylic acids.

$$Ar{-}\overset{\overset{\displaystyle O}{\|}}{C}{-}CH_2{-}Br + R{-}\overset{\overset{\displaystyle O}{\|}}{C}{-}O^- \longrightarrow Ar{-}\overset{\overset{\displaystyle O}{\|}}{C}{-}CH_2{-}O{-}\overset{\overset{\displaystyle O}{\|}}{C}{-}R + Br^-$$

phenacyl bromide phenacyl ester

Procedure a (*for pure acids*). Dissolve 100 mg (1 mmole) triethylamine in 2 mL of dry acetone and neutralize the solution by the addition of the carboxylic acid. To this solution, add a solution of 0.5 mmole of the phenacyl bromide (Note 1) (phenacyl bromide, *p*-chlorophenacyl bromide, *p*-bromophenacyl bromide, or *p*-phenylphenacyl bromide) in 3 mL of dry acetone. Allow the mixture to stand for 3 hours at room temperature (a precipitate of triethyl ammonium bromide will form after a short time); then dilute it with 10 mL of water and collect the precipitate by suction filtration. After thoroughly washing the crude phenacyl ester with 5% sodium bicarbonate solution and finally with water, recrystallize it from aqueous alcohol.

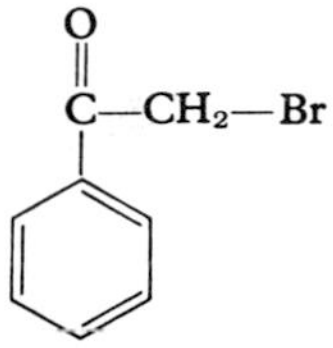

$\overset{\overset{\displaystyle O}{\|}}{C}{-}CH_2{-}Br$

phenacyl bromide

Procedure b (*for salts or aqueous solutions of the acid*). Dissolve 150 mg of the salt in 2 mL of water and add 2 drops of 5% hydrochloric acid; *or* add 5% hydrochloric acid to 2 mL of a basic aqueous or aqueous alcoholic solution of the acid containing about 100 mg of the acid until it is neutral to litmus, and then add 2 more drops of 5% hydrochloric acid. To this slightly acidic solution of the sodium salt of the acid (Note 2), add a solution of 200 mg of the phenacyl bromide (Notes 1 and 3) in ethanol and heat the mixture under reflux—(1 hour for a monocarboxylic acid; 2 hours for a dicarboxylic acid; 3 hours for a tricarboxylic acid). Occasionally, a solid separates from solution during the reflux period; it should be brought into solution by the addition of a few milliliters of ethanol. At the end of the heating period, allow the mixture to cool to room temperature and collect the crude phenacyl ester by suction filtration. It should be washed and recrystallized as in Procedure a. (Ref. 1, p. 235; Ref. 2, p. 243; Ref. 3, p. 362; Ref. 4, p. 270.)

Notes

1. The phenacyl halides are lachrymatory and tend to irritate the skin. They should be treated with respect.
2. The reaction mixture should not be alkaline; alkali causes the hydrolysis of the phenacyl bromide to the phenacyl alcohol.
3. *p*-Bromophenacyl bromide is converted to the relatively insoluble *p*-bromophenacyl chloride (m.p. 117°C) by chloride ion. If the mixture contains large amounts of chloride ion, another phenacyl halide should be used.

29.20 *p*-NITROBENZYL ESTERS OF CARBOXYLIC ACIDS

Derivative for carboxylic acids.

$$CH_2\text{—}Br\ (C_6H_4)\ NO_2 + R\text{—}\overset{O}{\overset{\|}{C}}\text{—}O^- \longrightarrow CH_2\text{—}O\text{—}\overset{O}{\overset{\|}{C}}\text{—}R\ (C_6H_4)\ NO_2 + Br^-$$

p-nitrobenzyl bromide *p*-nitrobenzyl ester

Procedure. The procedure for the preparation of phenacyl esters (Section 29.19) may be used with the following exceptions:

- *Procedure a: p*-Nitrobenzyl bromide is used in place of the phenacyl bromide; ethanol is used as solvent in place of acetone.
- *Procedure b: p*-Nitrobenzyl bromide is used in place of the phenacyl bromide.
- *Procedures a and b:* If *p*-nitrobenzyl chloride is substituted for *p*-nitrobenzyl bromide, 10 mg of sodium iodide should be added.

The benzyl halides are lachrymatory and tend to irritate the skin; they should be treated with respect. (Ref. 1, p. 235; Ref. 2, p. 244; Ref. 3, p. 362; Ref. 4, p. 270.)

29.21 N-BENZYLAMIDES FROM ESTERS

Derivative for esters.

$$\underset{\text{ester}}{R\text{—}\overset{\overset{\displaystyle O}{\|}}{C}\text{—}O\text{—}CH_3} + \underset{\text{benzylamine}}{\phi\text{—}CH_2\text{—}\ddot{N}H_2} \xrightarrow{NH_4Cl} \underset{\text{benzylamide}}{R\text{—}\overset{\overset{\displaystyle O}{\|}}{C}\text{—}\underset{\underset{\displaystyle H}{|}}{\ddot{N}}\text{—}CH_2\text{—}\phi} + CH_3OH$$

Procedure. Heat under reflux for about 1 hour a mixture of 0.5 g of the methyl or ethyl ester (Note 1), 1.5 mL of benzylamine, and 50 mg of ammonium chloride. Cool the mixture and stir it with a little water and finally a little dilute hydrochloric acid (Note 2). Collect the solid by suction filtration and recrystallize it from alcohol/water or acetone/water. (Ref. 1, p. 272; Ref. 2, p. 290; Ref. 3, p. 394; Ref. 4, p. 300.)

Notes

1. Higher esters must first be transesterified to the methyl ester by heating 0.6–1 gram under reflux for 30 minutes in 10 mL of anhydrous methanol in which 0.1 g of sodium metal or sodium methoxide has been dissolved. The residue obtained by evaporation of the excess methanol can then be used in the procedure described.
2. Avoid an excess of hydrochloric acid, as it may dissolve the product as well.

29.22 3,5-DINITROBENZOATES FROM ESTERS

Derivative for the alcohol fragment of an ester (Note 1).

$$\underset{\text{3,5-dinitrobenzoic acid}}{3,5\text{-}(NO_2)_2C_6H_3\text{—}\overset{\overset{\displaystyle O}{\|}}{C}\text{—}O\text{—}H} + \underset{\text{ester}}{R\text{—}\overset{\overset{\displaystyle O}{\|}}{C}\text{—}O\text{—}R'} \xrightarrow{H_2SO_4} \underset{\text{3,5-dinitrobenzoate ester}}{3,5\text{-}(NO_2)_2C_6H_3\text{—}\overset{\overset{\displaystyle O}{\|}}{C}\text{—}O\text{—}R'} + \underset{\text{acid}}{R\text{—}\overset{\overset{\displaystyle O}{\|}}{C}\text{—}O\text{—}H}$$

Procedure. Mix 0.5 gram of the ester with 0.5 gram of powdered 3,5-dinitrobenzoic acid, add a small drop of conc. sulfuric acid, and heat the mixture. Heat under reflux if the ester boils below 150°C, and at 150°C in an oil bath if it boils above 150°C; heat until the 3,5-dinitrobenzoic acid dissolves and then for an additional 30 minutes. Cool the mixture, dissolve it in 30 mL of ether, and extract it twice with 5% sodium bicarbonate solution to remove 3,5-dinitrobenzoic acid and sulfuric acid (*caution:* foaming from CO_2 evolution). Finally, after washing the ether extract with water, evaporate it to dryness and recrystallize the residue by dissolving it in a minimum of hot ethanol, adding water to incipient cloudiness, and allowing the solution to cool. It may be necessary to induce crystallization by scratching. (Ref. 1, p. 273; Ref. 2, p. 292; Ref. 3, p. 393; Ref. 4, p. 300.)

Caution

Note

1. This procedure is not satisfactory for esters of alcohols that are unstable to conc. sulfuric acid, such as tertiary alcohols or certain unsaturated alcohols. Higher-molecular-weight esters react slowly or not at all.

29.23 HYDROLYSIS OF ESTERS

$$\underset{\text{ester}}{\text{R—}\overset{\overset{\displaystyle \text{O}}{\|}}{\text{C}}\text{—O—R}'} + \text{HO}^- \xrightarrow{\text{H}_2\text{O}} \text{R—}\overset{\overset{\displaystyle \text{O}}{\|}}{\text{C}}\text{—O}^- + \text{H—O—R}'$$

Procedure. Boil under reflux a mixture of 1 gram of the ester and 10 mL of 1 *M* sodium hydroxide until a clear solution has been obtained (Note 1).

Some of the solution can be used to identify the acid corresponding to the ester by means of Procedure b in Section 29.19 (Note 2). The remainder of the solution can be saturated with potassium carbonate to salt out the alcohol, which can then be extracted and identified by one of the procedures described in Sections 29.1–29.3. (Ref. 2, p. 287; Ref. 3, p. 391; Ref. 4, p. 293.)

Notes

1. If the alcohol that corresponds to the ester is not soluble in water, it will be present as an oil after hydrolysis.
2. If the acid is relatively insoluble in water, it can be isolated by acidification of the hydrolysate, followed by suction filtration.

29.24 BROMINATION OF AROMATIC ETHERS

Derivative for aromatic ethers.

$$\text{Ar—H} + Br_2 \longrightarrow \text{Ar—Br} + \text{HBr}$$

Procedure. Add 1 mmole of the aromatic ether to the calculated amount of a 1% (by volume) solution of bromine in acetic acid (Note 1) that has been cooled in an ice bath; continue cooling during the addition. After a few more minutes, remove the mixture from the ice bath and allow it to stand at room temperature for 15 minutes. Isolate the product by diluting the mixture with water and collecting the precipitate by suction filtration. Remove traces of bromine from the product by washing it with dilute sodium bisulfite solution and then water. It can be recrystallized from alcohol or alcohol/water. (Ref. 1, p. 276; Ref. 2, p. 297; Ref. 4, p. 308.)

Note

1. A solution of 1 mL of bromine in 100 mL of acetic acid will contain 1 millimole of bromine per 5.2 mL of solution. Depending upon whether a mono-, di-, or tribromo derivative is expected, a 10–20% excess over 1, 2, or 3 millimoles of bromine should be used.

29.25 S-ALKYLTHIURONIUM PICRATES

Derivative for primary and secondary alkyl chlorides, bromides, and iodides.

$$\text{R—X} + H_2\ddot{N}\text{—}\underset{\text{thiourea}}{\overset{:S:}{\overset{\|}{C}}}\text{—}\ddot{N}H_2 \longrightarrow H_2\ddot{N}\text{—}\underset{+}{\overset{\overset{\overset{R}{|}}{:S:}}{\overset{|}{C}}}\text{—}\ddot{N}H_2 \quad X^-$$

$$H_2\ddot{N}\text{—}\underset{+}{\overset{\overset{\overset{R}{|}}{:S:}}{\overset{|}{C}}}\text{—}\ddot{N}H_2 \quad X^- + \underset{\text{picric acid}}{\text{2,4,6-}(NO_2)_3C_6H_2OH} \longrightarrow \text{Picrate salt} + \text{HX}$$

$HO—CH_2CH_2—OH$
ethylene glycol

Procedure (Reference 4). Place in a 10-mL standard-taper Erlenmeyer flask 5 mL of ethylene glycol, 300 mg of thiourea, 6 drops of the alkyl halide, and two boiling stones. Fit the flask with a reflux condenser and clamp the flask in position in an oil bath that has been preheated to 117°C. After the flask has been held at this temperature for 30 minutes (Note 1), add down the condenser 1 mL of a saturated solution of picric acid in ethanol (Note 2). Keep the flask in the oil bath for another 15 minutes; then remove it, cool it to room temperature, and then add 5 mL of water. Finally, cool the mixture in an ice bath for 15 minutes, with occasional swirling. Collect the S-alkylthiuronium picrate by suction filtration. The derivative can be recrystallized from methanol. (Ref. 1, p. 280; Ref. 2, p. 302; Ref. 3, p. 291; Ref. 4, p. 217. See also H. M. Crosby and J. B. Entrikin, *J. Chem. Educ.* **41,** 360 (1964).)

Notes

1. If decomposition occurs (odor of mercaptan; red color), repeat the experiment with the oil bath held at 65°C.
2. The solubility of picric acid is reported to be 1 gram per 12 mL of ethanol.

29.26 *o*-AROYLBENZOIC ACIDS FROM AROMATIC HYDROCARBONS

Derivative for aromatic hydrocarbons (Note 1).

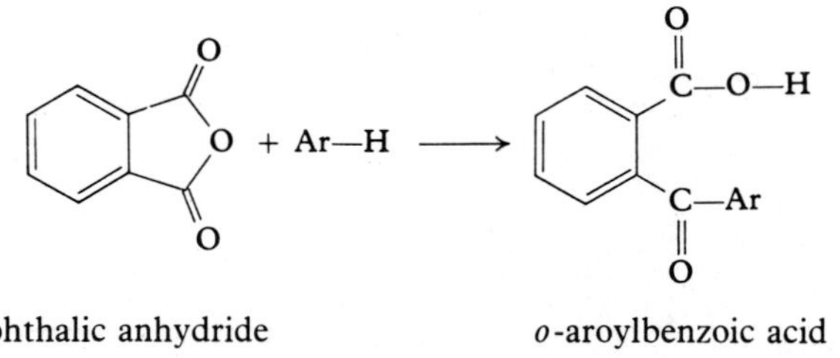

phthalic anhydride + Ar—H ⟶ *o*-aroylbenzoic acid

Procedure. Add 2.5 grams of powdered anhydrous aluminum chloride to a mixture of 0.5 gram of the hydrocarbon and 0.6 gram of phthalic anhydride in 2 or 3 mL of dry dichloromethane while the mixture is cooled in an ice bath. After the initial reaction has subsided, either allow the mixture to stand at room temperature or heat it under reflux until the evolution of hydrogen chloride has occurred (about half an hour); if the mixture has been heated, allow it to cool. Then add 5 grams of ice and 5 mL of conc. hydrochloric acid. When the addition product has been hydrolyzed, collect the product by suction filtration and wash it well with water (Note 2). The aroylbenzoic acid should be purified by dissolving it by heating in 5 mL of conc. sodium carbonate solution, boiling the solution with activated carbon for 5 minutes, filtering, and acidifying to a pH of 3 by the addition of 20% hydrochloric acid. The precipitated acid should then be collected by suction filtration, washed with water, and recrystallized from aqueous alcohol or (after drying) from toluene/petroleum ether. (Ref. 1, p. 285; Ref. 2, p. 519; Ref. 4, p. 235.)

Notes

1. Certain aryl halides can be characterized in this manner also.
2. If the product does not crystallize immediately, it should be allowed to stand overnight.

29.27 AROMATIC ACIDS BY OXIDATION BY PERMANGANATE

Derivative for aromatic hydrocarbons and aryl bromides and chlorides.

$$\mathrm{Ar{-}R} \xrightarrow{\mathrm{KMnO_4}} \mathrm{Ar{-}\overset{\overset{\displaystyle O}{\|}}{C}{-}O{-}H}$$

Procedure. Add 1 gram of the hydrocarbon to a solution of 3 grams of potassium permanganate and 1 gram of sodium carbonate in 75 mL of water and heat the mixture under reflux until the permanganate color has disappeared (in 15 minutes to 4 hours; Note 1). Cool the solution to room temperature and acidify it cautiously by the addition of 50% sulfuric acid (Note 2). Remove the manganese dioxide by the addition of a concentrated solution of sodium bisulfite (with good stirring and possibly heating on the steam bath) and, after cooling the mixture thoroughly in an ice bath, collect the acid by suction filtration. It can be recrystallized from water. (Ref. 1, p. 285; Ref. 2, pp. 315, 326; Ref. 3, p. 520; Ref. 4, p. 253.)

Notes

1. Potassium permanganate can be detected by dipping a stirring rod into the mixture and touching the rod to a piece of filter paper; a pink color in the ring around the dark spot of manganese dioxide indicates the presence of permanganate.
2. 50% sulfuric acid can be prepared by cautiously pouring 5 mL conc. sulfuric acid over 10 grams ice.

29.28 ANILIDES, *p*-TOLUIDIDES, AND α-NAPHTHALIDES FROM ALKYL HALIDES

Derivatives for alkyl halides and aryl bromides and iodides.

$$\text{R—X} + \text{Mg} \xrightarrow{\text{dry ether}} \underset{\text{Grignard reagent}}{\text{R—Mg—X}}$$

$$\text{R—Mg—X} + \underset{\text{aryl isocyanate}}{\text{Ar—N=C=O}} \longrightarrow \underset{\text{aryl amide}}{\text{Ar—NH—C(=O)—R}}$$

Preparation of the Grignard Reagent. Add a solution of 1 mL of the halide in 5 mL of absolute ether to 0.4 gram of magnesium turnings and a crystal of iodine in a dry flask.

Preparation of the Derivative. To the freshly prepared Grignard reagent, add a solution of 0.5 mL of the isocyanate (phenyl-, *p*-tolyl-, or α-naphthylisocyanate) in 10 mL of absolute ether. After swirling the mixture and allowing it to stand for 10 minutes, add 20 grams of crushed ice and 1 mL of conc. hydrochloric acid. After hydrolysis of the solid, separate the ether layer; then, after drying over anhydrous magnesium sulfate, evaporate the ether to give the derivative as a solid residue. The residual amide can be recrystallized from methanol, ethanol, or aqueous alcohol. (Ref. 1, p. 279; Ref. 3, p. 290; Ref. 4, p. 214.)

29.29 2,4,7-TRINITROFLUORENONE ADDUCTS OF AROMATIC HYDROCARBONS

Derivative for aromatic hydrocarbons.

$$\text{Ar—H} + \underset{\text{2,4,7-trinitrofluorenone}}{\text{[2,4,7-}(NO_2)_3\text{-fluoren-9-one]}} \longrightarrow \underset{\text{adduct}}{\text{Ar—H} \cdot \text{[2,4,7-}(NO_2)_3\text{-fluoren-9-one]}}$$

Procedure. Dissolve 100 mg of 2,4,7-trinitrofluorenone (0.3 mmole) in a mixture of 10 mL of absolute methanol or ethanol and 2 mL of benzene. To this hot solution add a solution of an equivalent amount (0.3 mmole) of the aromatic hydrocarbon in the minimum amount of

2 : 1 alcohol : benzene. Heat the mixture for 30 seconds and then allow it to cool for crystallization. Collect the product by suction filtration. It can be recrystallized from absolute ethanol. (Ref. 2, p. 315.)

29.30 BROMINATION OF PHENOLS

Derivative for phenols.

$$\underset{\text{phenol}}{\text{Ar—O—H}} + \text{Br}_2 \longrightarrow \text{Brominated phenol} + \text{HBr}$$

Procedure. Prepare a solution of 7.5 grams of potassium bromide and 5 grams (1.6 mL; 31 mmoles) of bromine in 50 mL of water. Add this solution, drop by drop, to a well-stirred solution of 0.5 gram of the phenol dissolved in water, dioxane, ethanol, or acetone until a weak yellow color persists. Precipitate the brominated phenol by stirring in 25 mL of cold water and collect the crude product by suction filtration. Wash the product with dilute sodium bisulfite solution to remove free bromine and then recrystallize the solid from ethanol or aqueous alcohol. (Ref. 1, p. 298; Ref. 4, p. 354.)

29.31 ARYLOXYACETIC ACIDS FROM PHENOLS

Derivative for phenols.

$$\underset{\text{phenol}}{\text{Ar—O—H}} + \underset{\text{chloroacetic acid}}{\text{Cl—CH}_2\text{—}\overset{\overset{\text{O}}{\|}}{\text{C}}\text{—O—H}} \longrightarrow \underset{\text{phenoxyacetic acid}}{\text{Ar—O—CH}_2\text{—}\overset{\overset{\text{O}}{\|}}{\text{C}}\text{—O—H}}$$

Procedure. Add 1.25 grams of monochloroacetic acid to a mixture of 1 gram of the phenol and 4 mL of 10 *M* sodium hydroxide solution. Add 1 or 2 mL of water as necessary to give a homogeneous solution. Heat the mixture for 1 hour on the steam bath, cool it to room temperature, dilute it with 10 or 15 mL of water, and acidify it to a pH of 3 with dilute hydrochloric acid. Extract the product with 50 mL of ether, wash the extract with 10 mL of water, and then extract the product from the ethereal solution by means of 25 mL of 5% aqueous sodium carbonate solution. Precipitate the product by acidifying the sodium carbonate solution with dilute hydrochloric acid (CO_2 evolution) and collect it by suction filtration. It can then be recrystallized from water. (Ref. 1, p. 298; Ref. 2, p. 331; Ref. 3, p. 682; Ref. 4, p. 354.)

References

1. R. L. Shriner, R. C. Fuson, and D. Y. Curtin, *The Systematic Identification of Organic Compounds,* 5th edition, Wiley, New York, 1965.
2. N. D. Cheronis and J. B. Entrikin, *Identification of Organic Compounds,* Interscience, New York, 1963.
3. A. I. Vogel, *A Textbook of Practical Organic Chemistry,* 3rd edition, Wiley, New York, 1957.
4. R. L. Shriner, R. C. Fuson, D. Y. Curtin, and T. C. Morrill, *The Systematic Identification of Organic Compounds,* 6th edition, Wiley, New York, 1980. This revision of the book cited in Reference 1 places a greater emphasis on physical methods of analysis. Its Tables of Derivatives are reproduced from Reference 1.
5. A. I. Vogel, *Vogel's Textbook of Practical Organic Chemistry,* 4th edition, revised by Furniss et al., Longman, London and New York, 1978. This revision of *Vogel* is less comprehensive than its predecessor, cited in Reference 3.
6. Z. Rappoport, *Handbook of Tables for Organic Compound Identification,* 3rd edition, The Chemical Rubber Co., Cleveland, Ohio, 1967.

Apparatus and Techniques for Chemical Reactions

A chemical reaction, in which a substance of one structure is converted to a substance of a different structure, typically includes (1) combining the materials that are to react, usually with a solvent or a catalyst, (2) maintaining the mixture at a specified temperature for a certain length of time, with stirring if necessary, and then (3) isolating the desired product by the separation procedures described in Sections 7–15. The next sections describe some of the ways by which these various phases of the operation can be carried out in the laboratory. Industrial-scale and microscale reactions usually require different procedures.

30 Assembling the Apparatus

The apparatus in which a reaction is to be carried out may be as simple as a single test tube or flask. More often, it is made up of several glass components, such as a boiling flask and condenser, plus sometimes a drying tube to exclude moisture, a separatory funnel by which a liquid reagent or solution can be added, a trap to adsorb harmful gases that

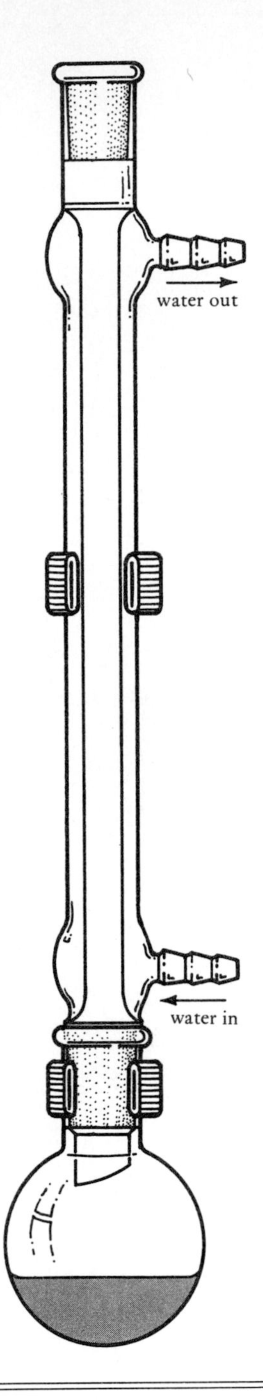

Figure 30.1. Apparatus for carrying out a reaction under reflux.

may be evolved, or any of a number of other parts. Figures 30.1 and 30.2 illustrate two of the assemblies that are most often used for carrying out reactions.

There is no doubt that the nicest and most convenient components from which the apparatus can be assembled are those that can be connected by ground-glass joints. The joints should normally be put together *without* a lubricant. A lubricant may be used on ground-glass joints for either of two reasons: (1) to make them vacuum-tight, and (2) to keep them from sticking together. If the joints are well ground, and most of them are, lubrication will be needed for vacuum tightness only in vacuum distillation at oil pump pressures (less than about 5 Torr). If the apparatus is taken apart *immediately* after use, a lubricant will be needed to prevent sticking only if strongly basic solutions (NaOH, KOH) have been used—as, for example, in the steam distillation of aniline (Section 60). If a lubricant must be employed, use only a very small amount spread out in a thin, even layer on the inner member. Avoid silicone-based lubricants, as they are practically impossible to remove and will prevent the glass from ever being wet by water again.

If the glass components do not have ground-glass joints, they must be joined by corks or rubber stoppers: one member is placed through a hole in the cork or stopper, and this is then plugged into the other.

As illustrated in Figures 30.1 and 30.2, the weight of the apparatus should be supported by clamps so that the tendency to twist is minimized.

31 Temperature Control

For two reasons, a reaction should be carried out at a definite, constant temperature. The first reason is that *the rate of a reaction increases sharply with increasing temperature, and a reaction will proceed at a convenient or desirable rate only within a relatively narrow temperature range.* The rate increases with temperature according to the relationship

$$\text{Rate} \propto e^{-\Delta G^{\ddagger}RT}$$

where $\Delta G^{\ddagger}$ is the free energy of activation of the reaction, R is the gas constant, and T is the absolute temperature. The second reason is that *the rates of competing reactions, or side reactions, increase faster with an increase in temperature than the rate of the desired reaction.* This happens because the free energies of activation of the side reactions are greater than the free energy of reaction of the desired reaction. Thus, the fraction of side products will be greater at higher temperatures. The choice of the optimum temperature at which to carry out a reaction, therefore, always represents a compromise between a desire to carry out

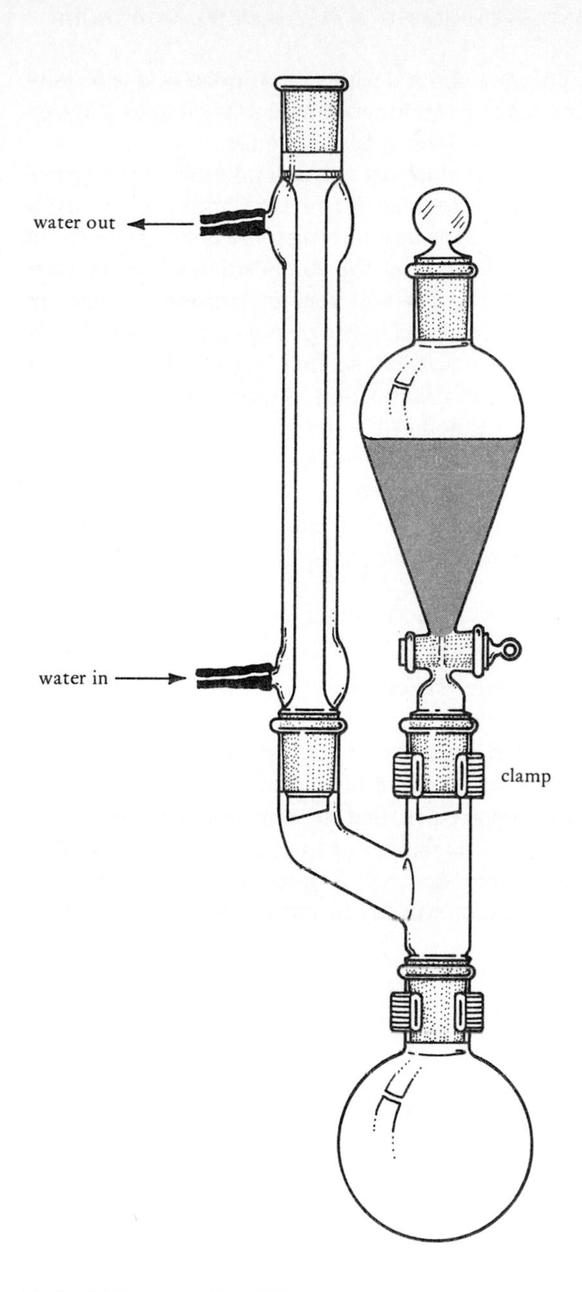

Figure 30.2. Apparatus that allows heating under reflux and addition of liquid reagent.

the reaction as fast as possible and a desire to minimize side reactions. The exponential dependence of rate upon temperature makes it necessary to control the temperature over a narrow range.

Heat under reflux

One of the easiest ways to maintain a reaction mixture at an approximately constant temperature is to heat the mixture until it boils, using a condenser to condense the vapor and to return the condensate to the flask (to heat under reflux). As long as the composition of the mixture does not change, the boiling point will remain constant. Since the composition must change as the reaction progresses, an excess of one of the reactants can be used as a solvent so that the boiling point will change less during the course of the reaction. Sometimes the solvent is chosen so that the mixture will boil at the desired temperature. This is the usual procedure if the reactants are solids.

The reaction mixture can be boiled by heating the flask with a flame, an oil or steam bath, or a heating mantle. These methods of heating are described in Section 32.1.

Control of Highly Exothermic Reactions

A highly exothermic reaction presents an additional problem since, as the rate of a reaction increases, the rate of heat evolution will increase. If the rate of heat evolution is greater than the rate at which heat can be removed (by condensation of the vapors of the boiling mixture and all other cooling methods), the reaction will proceed at an ever increasing rate as the temperature of the mixture rises until it boils over or blows out the condenser.

By rate of addition of reactant

One way to control a highly exothermic reaction is to add one of the reactants to the mixture at such a rate that the reaction can be maintained at the desired temperature (the boiling point or below) by the heat evolved in the reaction. The heat of the reaction may be the only source of heat, or, if the reaction is only moderately exothermic or if the temperature is relatively high, an external source of heat can be used as well. More often it is necessary to remove heat by external cooling. This will always be necessary if the temperature is to be kept below the boiling point of the mixture. When the mixture is not to be allowed to boil, a cooling bath must be used, and the rate of addition of the reactant will be determined by how efficiently heat can be removed from the mixture. Different types of cooling baths are described in Section 32.2.

Warning

Warning: The danger of this method of controlling exothermic reactions is that the reactant may unintentionally be added faster than it reacts. It is especially easy to add too much reactant at the beginning, when the temperature may be too low for the reaction to proceed at a reasonable rate. If too much reactant accumulates in the mixture, the reaction may accelerate out of control when it finally starts. The only way to be sure the reaction is under way and under control is to determine that the temperature rises when the reactant is added and starts to fall when the addition is stopped or slowed down.

The second method of controlling an exothermic reaction is to run the reaction in small batches, or on a small scale, in a large flask so that a large surface area is available for efficient cooling. When the temperature of the reaction starts to climb, the flask can be plunged into a cold bath and swirled for a short time; remove it when the temperature rise has been checked. The use of a large amount of solvent (and a still larger flask so that the mixture can be cooled efficiently in a cold bath) can serve to make the reaction more easily controlled, both by decreasing the rate of bimolecular reactions by dilution, and by increasing the total heat capacity of the mixture. This second method requires a little more judgment and skill on the part of the experimenter, and it is dangerous if large amounts are used.

From this discussion of exothermic reactions, it should be obvious that new or unfamiliar reactions should be carried out on a small scale until their exothermicity has been determined. Another, less obvious consequence is that, as a reaction is scaled up, more attention must be paid to the problem of heat transfer. If the volume of the reaction, and thus of the apparatus, is increased by a factor of 8, the surface area of the apparatus (through which the heat must flow) is increased by a factor of only $8^{2/3} = 4$. Whereas a cooling bath might not have been needed for the small-scale reaction, it might be required for the large-scale reaction.

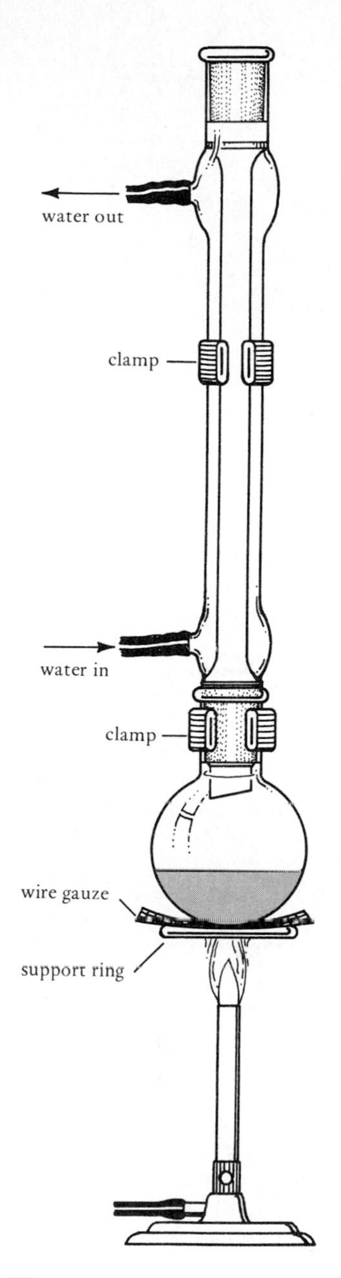

Figure 32.1. Heating with a burner flame. Rate of heating is determined by the size of the flame and the height of the flask above the flame. The wire gauze moderates and spreads the heat of the flame.

32 Methods of Heating and Cooling

The typical chemical reaction is not carried out at room temperature. The next two sections describe methods by which reaction mixtures can be heated or cooled.

32.1 HEATING

Heating with a Flame

One very common source of heat is a Bunsen burner flame. The height of the flame and the distance of the flask from the flame control the rate of heating (see Figure 32.1).

The advantages of using a burner as a source of heat are that it can be set up very quickly and easily, and that the mixture can be brought up to temperature very rapidly. However, a burner has three important disadvantages. The first is that it must *not* be left unattended. It might

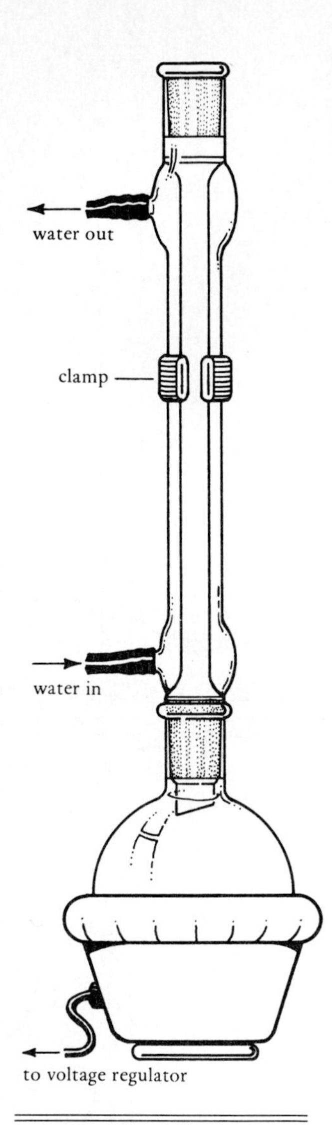

Figure 32.2. Use of a heating mantle. The mantle should be supported on an iron ring or other removable support.

go out, creating a gas leak that could lead to an explosion; or the height of the flame might change, providing either too little heat to maintain boiling or so much heat that the mixture boils over and catches fire. The second disadvantage is that if the flask should crack or break, or if the reaction mixture should boil over, the burner could also set the mixture on fire. The third disadvantage is that a burner flame is an exceedingly uneven source of heat. The glass at the bottom of the flask will be very much hotter than the rest of the flask, and undesired reactions may occur. If there is a solid in the reaction mixture, boiling may be uneven with lots of bumping. However, if relatively small amounts are involved, if the heating period is short, and if the mixture will be watched, heating with a flame will often be quite satisfactory.

Electrical Heating

Three methods of electrical heating are often used. In the first, an electrically heated jacket (*heating mantle*) is placed around the lower half of the flask (sometimes a second part is used around the upper half also). Figure 32.2 illustrates the use of such a mantle. The rate of heating is controlled by the voltage that is supplied to the mantle from the variable voltage transformer. A similar device, the Thermowell heater, does not require a variable voltage transformer. The rate of heating of the Thermowell device is set by a proportional controller, which determines the fraction of the time that current is allowed to flow through the heater.

The advantages of this method are that the apparatus can more safely be left unattended (overnight, for example); it is a very constant and very easily controlled source of heat; and it is not nearly as likely to result in ignition of flammable materials as is a flame. There are several disadvantages, however. One is that the glass of the flask is likely to be at a considerably higher temperature than the contents of the flask. As in heating with a flame, this can cause local superheating and possible decomposition, as well as uneven boiling. The temperature of the mantle must be determined by a thermocouple, and this is somewhat inconvenient. As a result, most chemists use a heating mantle without bothering to determine its temperature. Also, the level of the liquid in the flask should not fall below the top of the mantle. If it does, the glass between the top of the liquid and the top of the mantle will become much hotter than the rest of the flask. Material that splashes onto this part of the flask will be superheated. For this reason, heating with a mantle is not entirely satisfactory for a distillation in which the pot will end up nearly empty. Another disadvantage is that a heating mantle takes quite a long time to warm up and reach thermal equilibrium, perhaps an hour or more. Also, if you wish to turn off the heat in a hurry, it is not enough to turn off the electricity, because the mantle is hotter than the flask and it has a high heat capacity. The mantle itself must be removed. For this reason, you should never set up an apparatus using a heating mantle with the mantle resting on the desk top. An additional disadvantage is that the mantle must approximately fit the flask, and a different one may be required for each size of flask. You can buy at least a dozen burners for the price of a mantle and transformer. However, for a large-scale reaction, a heating mantle is more satisfactory than any other apparatus. A heating mantle will also be most useful when you will have to run the same reaction a number of times, and a relatively high temperature or long reaction time is required.

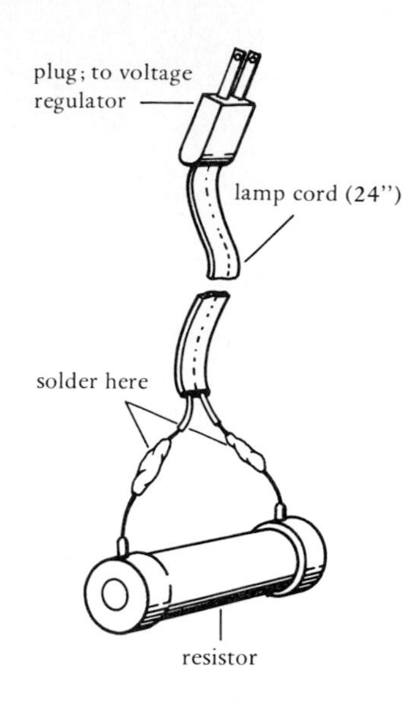

Figure 32.3. An inexpensive immersion heater, made from 125-ohm, 5-watt resistor (Ohmite "Brown Devil"), lamp cord, and plug.

A second method of electrical heating uses an immersion heater, supplied by a variable autotransformer, in a bath of a liquid heat-transfer medium. The immersion heater can be of the commercially available Calrod type, or can be easily made from a power resistor, a piece of lamp cord, and a plug (see Figure 32.3). The bath liquid can be a mixture of hydrocarbons (mineral oil), a silicone oil, or a polyethylene glycol fraction that freezes just below room temperature (Carbowax 600). The latter is most useful since it is cheap, can be used at temperatures up to about 175–180°C without appreciable decomposition, and has the great advantage of being completely soluble in water, which makes cleaning up much easier. The disadvantage of polyethylene glycol is that it is somewhat hygroscopic and, if exposed for a long time to high humidity, it can boil below the desired temperature; the water will boil out with continued heating.

The advantages of an electrically heated oil bath are that the temperature of the bath (and thus of the contents of the flask) can be determined by a thermometer in the bath; no superheating or local decomposition is possible, and, when a power resistor is used as the heating element, the heat capacity of the heater is so small that the rate of heating can be changed almost instantly by adjusting the voltage from the variable transformer. The electrically heated oil bath has the advantage over the heating mantle that it can be rapidly brought up to temperature, the rate of heating is more quickly and easily adjusted, the contents of the flask are more visible, and a magnetic stirrer can easily be used (Section 33) with it. The disadvantages of the oil bath are that it is a nuisance to store, it starts to smoke at elevated temperatures (about 150°C for mineral oil, acrid smell; about 180° for polyethylene glycol, burnt sugar smell), and it presents the danger of spilling hot oil if the container is tipped or broken during use. For small flasks (up to about 500 mL), at temperatures below 150–180°C, and for short reaction times, an electrically heated oil bath is most useful. For the constant and exact heat control required for a distillation, the electrically heated oil bath is unsurpassed. Of course, the immersion heater, like the heating mantle, requires a variable autotransformer.

Electric hotplate

A third method of electrical heating makes use of a hotplate. The hotplate, which may also contain a magnetic stirrer (Section 33), can be

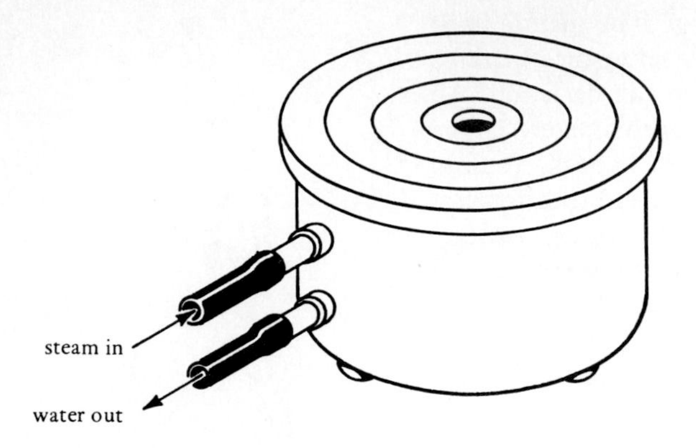

Figure 32.4. A steam bath. Rings can be removed to accommodate flasks of different sizes.

used either to heat a flask directly or to heat an oil bath in which the flask is supported. Since the surface of the hotplate will be much hotter than the contents of the flask, local superheating is possible, just as in heating with a flame. Indirect heating via an oil bath avoids the problem of superheating and generally provides a much more even and constant source of heat for the flask. The temperature of the bath can be determined with a thermometer.

Heating with a Steam Bath

A steam bath, illustrated in Figure 32.4, is often used to heat solutions that boil below about 90°C, or to heat a mixture to approximately 100°C. Most laboratories are supplied with steam from a central boiler; when this is the case, use of a steam bath eliminates the hazards of a flame. The steam bath heats the contents of a flask rapidly, but it has the disadvantage of being good for only one temperature, approximately 100°C.

Heating with Hot Water

A useful device for heating at temperatures below 100°C is a hot water bath—simply a beaker or steam bath full of hot water. Sometimes hot water from the faucet will be hot enough; sometimes it may be necessary to heat (and occasionally reheat) the water in the beaker or steam bath with a flame. A steam bath or hot water bath should be used with flammable substances whenever possible.

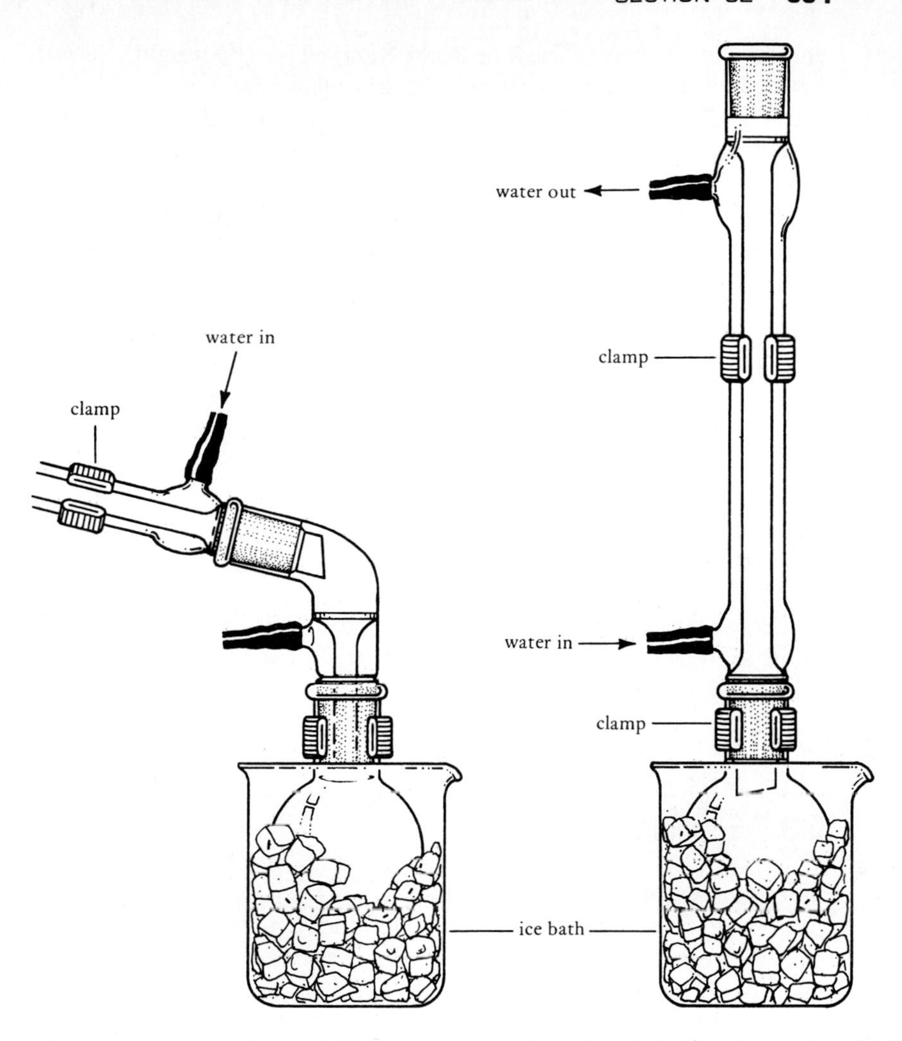

Figure 32.5. Use of an ice bath.

32.2 COOLING

Depending upon the temperature of the flask to be cooled, a cooling bath may be simply a beaker of water or a specially prepared cold bath. For cooling, exposure to the air is very inefficient compared to immersion in a cold liquid, since a liquid has a much higher heat capacity. For cooling below room temperature and down to about 0°C, a mixture of ice and water may be used (Figure 32.5). It is much more satisfactory for the ice to be in small pieces than as ice cubes. For cooling to a few degrees below zero, or for more efficient cooling to 0°C, a freezing mixture may be used. This is made by thoroughly mixing 1 part sodium

Cold water; freezing mixture

chloride and 3 parts of snow or finely chopped ice (by weight). Ice/salt baths, however, are fairly short-lived and inefficient.

For efficient cooling for extended periods, a Dry Ice/acetone bath may be used. This is prepared by adding lumps of Dry Ice (do not handle with bare hands) to acetone until the desired temperature is reached. The temperature can be maintained at any given level by adding Dry Ice as needed. Do not add the Dry Ice too rapidly, or the carbon dioxide that is evolved will cause the bath to overflow. The ultimate temperature that can be reached with such a bath is about −78°C; it must be measured with an alcohol or toluene thermometer, as mercury freezes at −40°C.

Dry Ice bath

A crystallizing dish is the most conveniently shaped container for hot or cold baths, since it has a flat bottom and straight sides, is wider than it is tall, and comes in several sizes. Don't forget that it is glass and can be broken. The life of a low-temperature bath can be greatly extended by using a Dewar flask (like a Thermos bottle) as a container, or by making the bath in a beaker that is nested inside a large beaker with a folded towel in between for insulation.

Efficient cooling

For most efficient cooling, the reaction mixture or solution should be in a large flask, to provide a large area of cold surface, and the mixture should be swirled or stirred continuously.

33 Stirring

A reaction mixture usually must be stirred. Stirring serves to mix two phases, or to mix a reagent as it is gradually added, or to promote efficient heat transfer to a cooling bath or from a heating bath, thus providing for a constant temperature throughout the mixture. If the reaction time is short, swirling the flask by hand may be an acceptable method of mixing. If the mixture is homogeneous and the reaction is neither very endothermic nor very exothermic, so that the transfer of heat does not have to be very efficient, occasional swirling in the heating or cooling bath may be adequate. If constant mixing or stirring is required for more than a short time, however, a stirring mechanism is almost a necessity. The boiling action of a vigorously boiling mixture can often provide adequate mixing, but boiling will be smoother and superheating can be minimized if the mixture is stirred.

Swirling by hand

Two stirring mechanisms are commonly used. The first makes use of a paddle or wire agitator, which extends on a shaft into the mixture from above the reaction vessel. The shaft is usually a glass sleeve with a ground joint that fits into the center ground-glass-jointed neck of the flask. The vigor of the stirring is determined by the speed of the motor, which is governed by a rheostat. Figure 37.1 illustrates this type of stirring apparatus.

Mechanical stirring

The second method makes use of a glass- or Teflon-covered bar magnet that is placed in the reaction flask. The flask is positioned over a motor, which turns a magnet fastened to its shaft (magnetic stirrer); as the motor-driven magnet turns, the magnet inside the flask turns and stirs the mixture. Simultaneous heating and stirring are often carried out by use of a combination magnetic stirrer-hotplate. If the flask is heated by the hotplate via an oil bath, the magnetic stirrer can be used to stir both the bath and the mixture. Figure 33.1 shows another arrangement in which a magnetic stirrer is used to stir both an electrically heated oil bath and the contents of a reaction flask. The magnet in the bath should be larger than the one in the flask.

Magnetic stirring

Direct mechanical stirring has an advantage over magnetic stirring in that mechanical stirring can provide the greater torque necessary to stir suspended solids or viscous mixtures or very large volumes. The mechanical stirrer can also be used when heating with a flame or with a

Figure 33.1. Magnetic stirrer arranged to stir electrically heated oil bath and reaction flask.

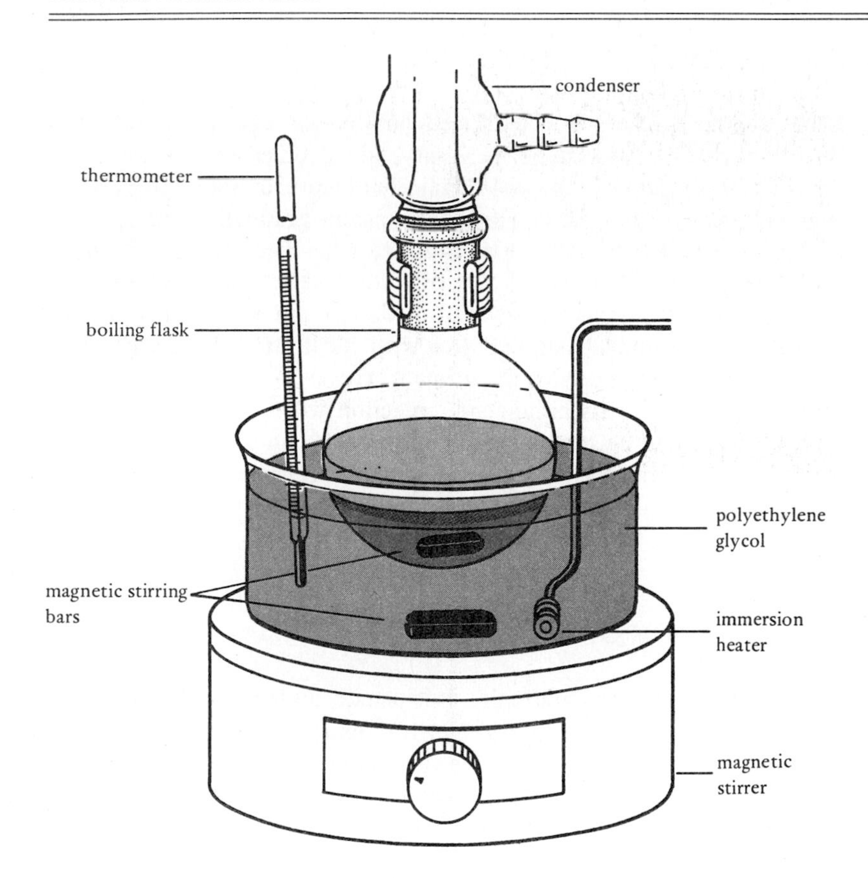

heating mantle, since the stirrer enters from the top of the flask. On the other hand, a magnetic stirrer is much easier to set up, and is generally much more convenient if the mixture needs only to be stirred and not heated or cooled at the same time. It is most useful when used in conjunction with an electrically heated oil bath, as the stirring motor can stir both the bath and the reaction mixture. This combination is especially suited for heating and stirring during a distillation, particularly a vacuum distillation. Use of a small magnetic stirrer in the pot will essentially eliminate superheating and bumping. Magnetic stirring is somewhat more convenient when the apparatus is not at atmospheric pressure, or when an inert atmosphere is being used, as magnetic stirring does not require a flask opening. Both types of stirring can be used when a heating or cooling bath is being employed, but with the magnetic stirrer, the flask must be positioned near the bottom of the bath, and the bath container must be made of nonferrous material.

Small air or water turbine-driven magnetic stirring motors are also available. These are useful for stirring "under water" in a constant-temperature bath.

34 Addition of Reagents

Many chemical reactions are carried out by combining the reagents (and solvent and catalyst, if necessary) all at once and keeping the mixture for a length of time at a certain temperature. Sometimes, however, it is necessary to add one or more reagents gradually over a period of time. This would be done, for example, if the reaction were strongly exothermic and the rate could be easily controlled only by adding one reagent a little at a time. Or it would be done if one reagent underwent an undesirable bimolecular reaction with itself: if this reagent were added slowly to an excess of the other materials to keep its concentration low, the rate of its bimolecular reaction with itself would be decreased relative to its bimolecular reaction with another substance. A gaseous reagent must usually be added slowly because of limited solubility; after the solution is saturated, it can be added no faster than it can be consumed.

34.1 ADDITION OF SOLIDS

If a solid must be added gradually, it is usually added as a solution. If it is necessary or desirable to add it as a solid, portions can often be dropped right into the flask or down the condenser; sometimes it will be necessary to poke a length of glass rod down the condenser to dislodge material that has stuck to the walls. If larger amounts need to be added, or if air or water vapor must be excluded from the reaction, an Erlen-

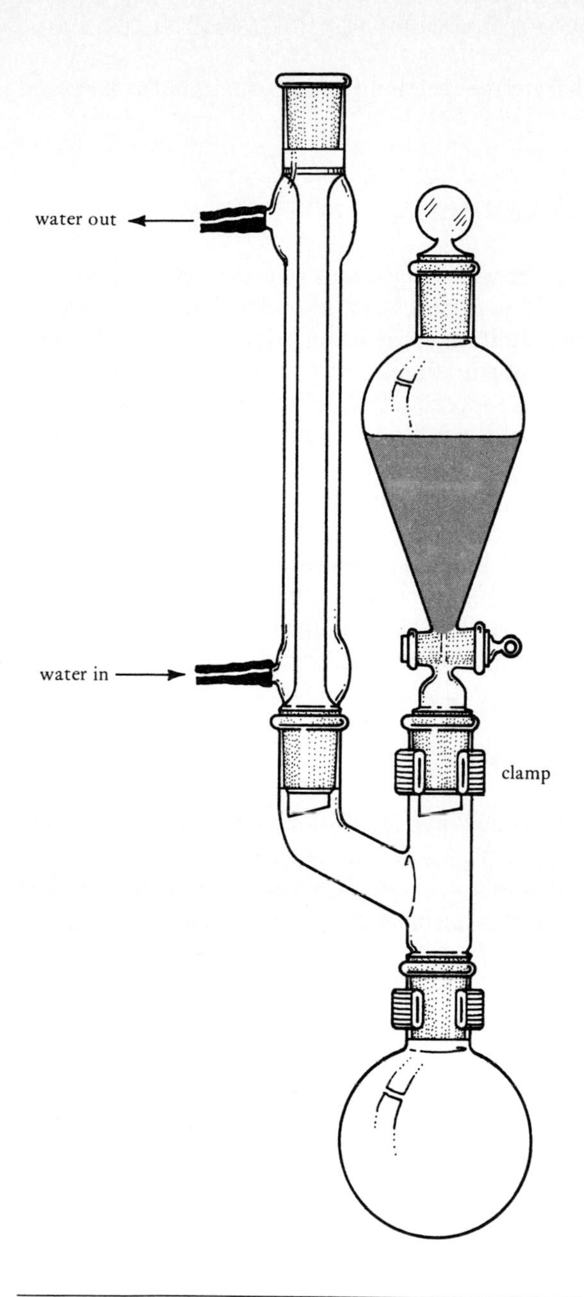

Figure 34.1. Reaction apparatus that allows addition of a liquid reagent or solution.

meyer flask containing the solid can be attached to a neck of the flask with a length of wide, thin-walled tubing (Gooch tubing). Portions can be added by shaking the solid in through the tubing. When the Erlenmeyer flask is hanging down between additions, the tubing is held shut, preventing the vapors from the reaction from getting into the flask and caking the solid.

34.2 ADDITION OF LIQUIDS AND SOLUTIONS

Liquids or solutions can be added by means of a separatory funnel, as shown in Figure 34.1. If it is necessary to add a liquid to a closed system, as when an inert atmosphere is being used, a pressure-equalizing addition funnel is required (Figure 37.1). The rate of addition is controlled by means of the stopcock. If the addition must be made very slowly over a long period of time, funnels with stopcocks are not very satisfactory since the flow rate is variable, increasing as the stopcock grease is dissolved away (unless a Teflon stopcock is used), and decreasing as dust or other bits of solid collect at the opening. In these cases an addition funnel can be used that has, instead of a stopcock, a narrow capillary tube fitted with a tungsten wire that may be moved in and out of the capillary (a Hershberg funnel). The flow rate is determined by the length of wire extending into the capillary.

Figure 34.2. A type of gas-addition tube.

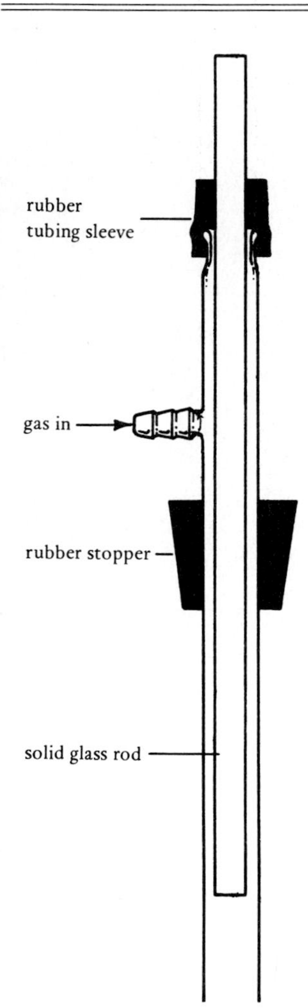

34.3 ADDITION OF GASES

A gas can be added to a reaction mixture through a tube that dips below the surface of the liquid. If the gas is only slightly soluble, it must be passed in only as quickly as it is consumed, in order not to waste it. The solubility of a gas decreases with increasing temperature, and stirring promotes more rapid solution. The gas should not be so soluble or reactive that the solution is sucked up the tube, and it should not react to produce a solid that will clog the addition tube.

If a gas is very soluble or very reactive, the addition tube must be open-ended and of wide bore so that the solution cannot be drawn up into it very far before the level falls below the end of the tube. Alternatively, the gas can be led just to the surface of the mixture in the vortex formed by the stirrer.

If the product of the reaction with the gas is a solid that tends to clog the tube, as in the treatment of a Grignard reagent with carbon dioxide, a solid rod can be incorporated into the wide-bore addition tube, as shown in Figure 34.2, so that the solid can be dislodged from time to time by pushing the rod down the addition tube.

Measuring Gases

If the weight of the gas added is large enough, compared to the weight of the cylinder, the amount added can be determined by the weight loss of the cylinder. A lecture bottle is a good source of gas, from this point of view. Sometimes it may be possible to pass the gas in at a steady rate. If so, the use of a flowmeter in the line will allow the amount of gas added to be determined from the length of time of addition. If the gas is not too low-boiling, it can be passed into a tube cooled in a Dry Ice/acetone bath and measured by volume (if the density of the liquid at the bath temperature can be estimated). When the desired volume has been condensed, the gas can be passed into the reaction mixture as it vaporizes. Section 83.1 describes this procedure for a chlorination reaction. Of course, if an excess of an inexpensive gas is not objectionable, the amount added need be estimated only very roughly.

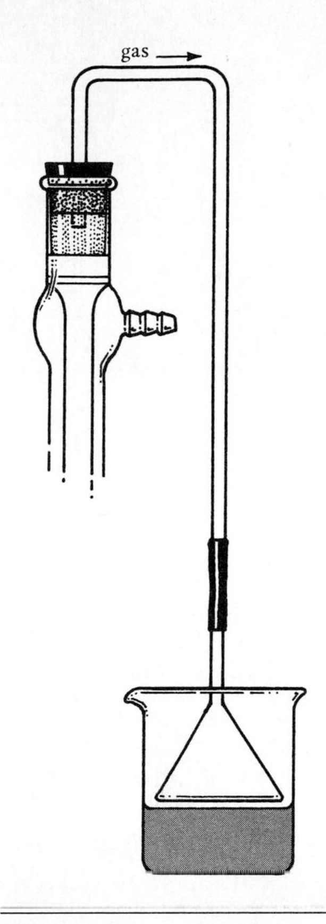

Figure 35.1. A way of absorbing soluble noxious gases. The funnel should not touch the solution in the beaker.

35 Control of Evolved Gases

Reactions in which a poisonous or irritating gas is formed are best carried out in a ventilated enclosure, or hood. If hood space is limited, it is possible, when the gas is readily soluble in water and not too poisonous, to set up or modify the apparatus in such a manner that the poisonous or irritating gas can be prevented from escaping into the atmosphere (to trap the gas). One way to do this is to lead the gas from the apparatus to a beaker of water, or an appropriate aqueous solution, where the gas will dissolve rather than escape into the room, as illustrated in Figure 35.1. This method is satisfactory if the rate of gas evolution is not too great and if small amounts are involved. The capacity and efficiency of this system can be increased by using a dilute sodium hydroxide solution (for acidic gases such as HCl or HBr) or a dilute solution of sulfuric acid (for basic gases such as ammonia) in the beaker.

Use the aspirator

Use of the *aspirator* provides another way to dispose of a water-soluble gas; an arrangement is illustrated in Figure 35.2. While it is slightly easier to set up, it is less reliable than the first method and it uses a lot of water.

Of course, gases that are not soluble in water, such as carbon monoxide, cannot be controlled by either of these methods.

36 Concentration; Evaporation

It is often necessary to either concentrate a solution or completely remove the solvent, in order to leave a nonvolatile solid or liquid residue. This is most often done to recover material after an extraction, to evaporate chromatographic fractions, or to concentrate the filtrate after recrystallization.

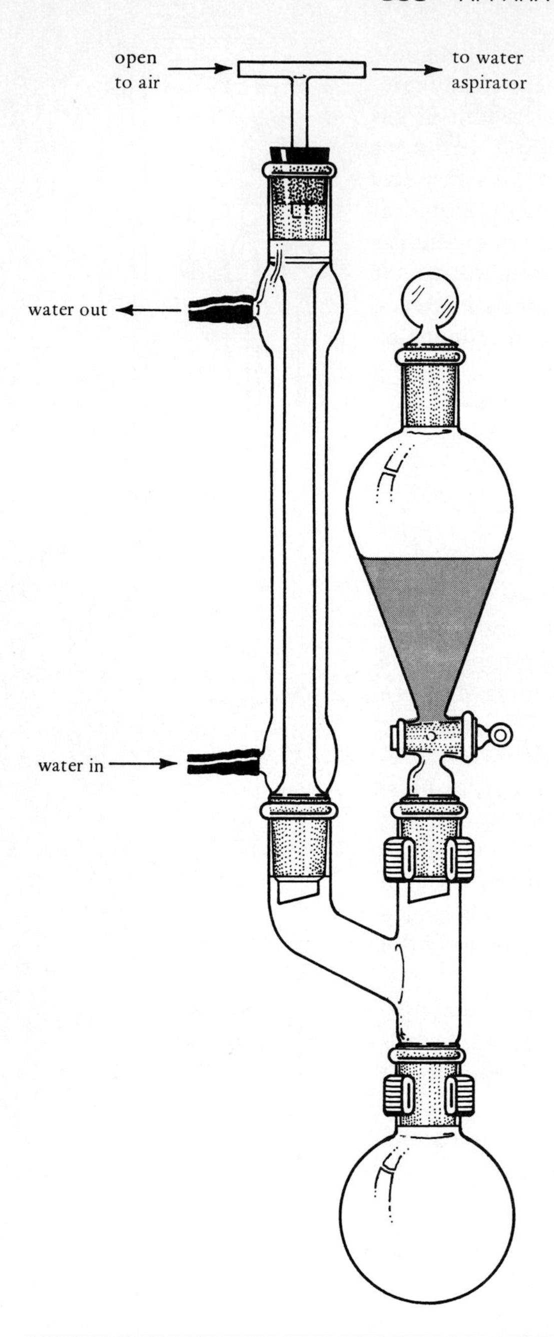

Figure 35.2. A method of control of noxious water-soluble gases. A vacuum adapter can be used in place of the cork and *T* tube.

Removal of Solvent by Distillation

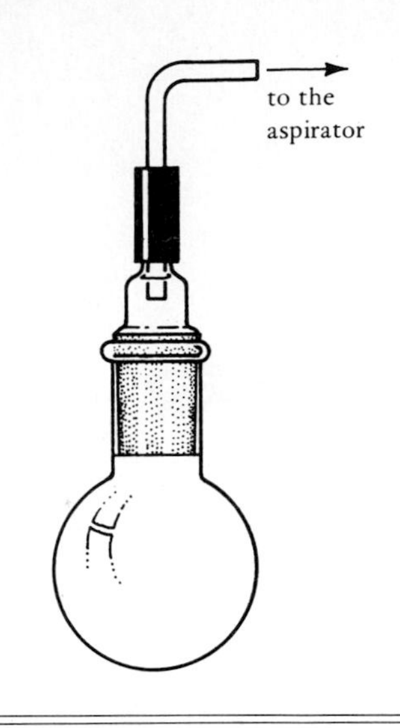

Possibly the most often used method of concentration or evaporation is the removal of solvent by distillation. If the source of heat is a steam bath or an oil bath, this method is entirely satisfactory for either partial or complete removal of the solvent.

The complete removal of solvent by distillation requires two steps. First, the solution is boiled at a bath temperature 25–50°C above the boiling point of the solvent until distillation has slowed almost to a complete stop. In the second step, *the residue is heated under vacuum*, usually at the same temperature at which the bulk of the solvent was removed, until the remainder of the solvent is gone, as illustrated in Figure 36.1. Evaporation can be judged to be complete when the residue shows no appreciable loss in weight between successive periods of heating under vacuum. If only a small amount of a volatile solvent is present, it can be removed by distillation, using an aspirator to remove the vapors, as illustrated in Figure 36.2.

Figure 36.1. Removal of solvent under vacuum. Heat the flask in a beaker of hot water or on the steam bath.

Figure 36.2. A way to concentrate a solution rapidly. The flask may be heated on the steam bath.

While you can concentrate a solution by distillation using a flame or heating mantle as a source of heat, you should not attempt to completely evaporate a solution by heating by either of these methods. If you attempt to remove the last traces of solvent by heating with a flame, you run the risk of decomposing the residue when it becomes viscous, or when a solid separates after the solution becomes more concentrated.

If a large volume of solvent is to be evaporated to give a small residue, it may be appropriate to fit a small flask with a separatory funnel so that solution can be added as the distillation progresses. In this way, the small residue will end up in a small flask, and mechanical losses can be minimized. Or the bulk of the solvent can be removed using a large flask and the remainder of the solution can be transferred (with rinsing) to a small flask for completion of the evaporation.

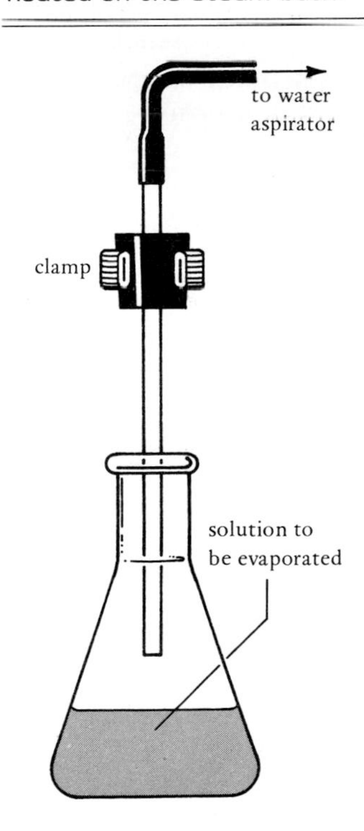

Other Techniques

If large amounts of a volatile solvent must be evaporated, it may be convenient to place the solution in a large shallow dish (evaporating dish) and keep the dish in the hood with the draft on until evaporation is complete. This method is slower, but it requires no attention.

It is possible to evaporate solutions very rapidly by using a *rotary evaporator*, a device that can apply an aspirator vacuum to a flask while rotating it in a heating bath (see Figure 36.3). Rotation of the flask minimizes superheating and bumping by keeping the contents well mixed, and provides a large surface area for evaporation by constantly rewetting the walls of the flask.

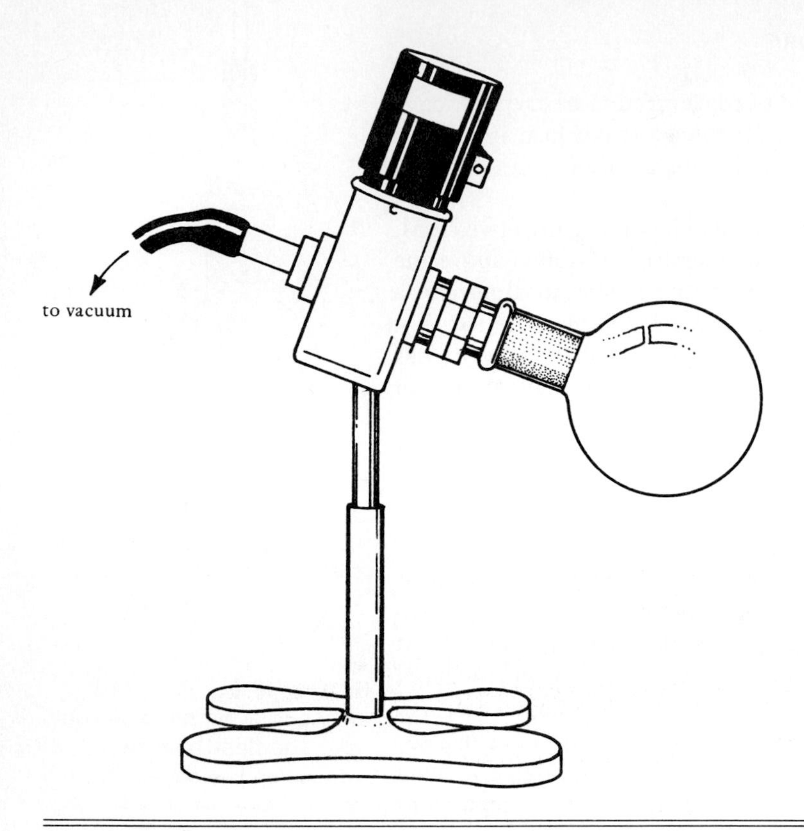

Figure 36.3. A rotary evaporator. The flask may be heated with warm water or steam.

37 Use of an Inert Atmosphere

Keeps out water and oxygen

Certain reactions, for instance the Grignard reaction, are better carried out in the absence of oxygen or water vapor because these substances either undergo or catalyze undesired side reactions. If the acceptable levels of oxygen and water vapor are not too low, simply flushing out the air of the reaction flask with dry nitrogen and maintaining a slight positive pressure of nitrogen will be adequate. Figure 37.1 illustrates the essential features of an arrangement to do this: the tank of nitrogen connected to the apparatus through a wash bottle for removal of traces of water, and the bubbler in the exit line to maintain the slight positive pressure and to indicate the flow rate. The apparatus is flushed initially at a moderate flow rate and is then maintained at a slight positive pressure throughout the reaction at a minimum flow rate.

If the acceptable levels of water or oxygen are rather low, usually

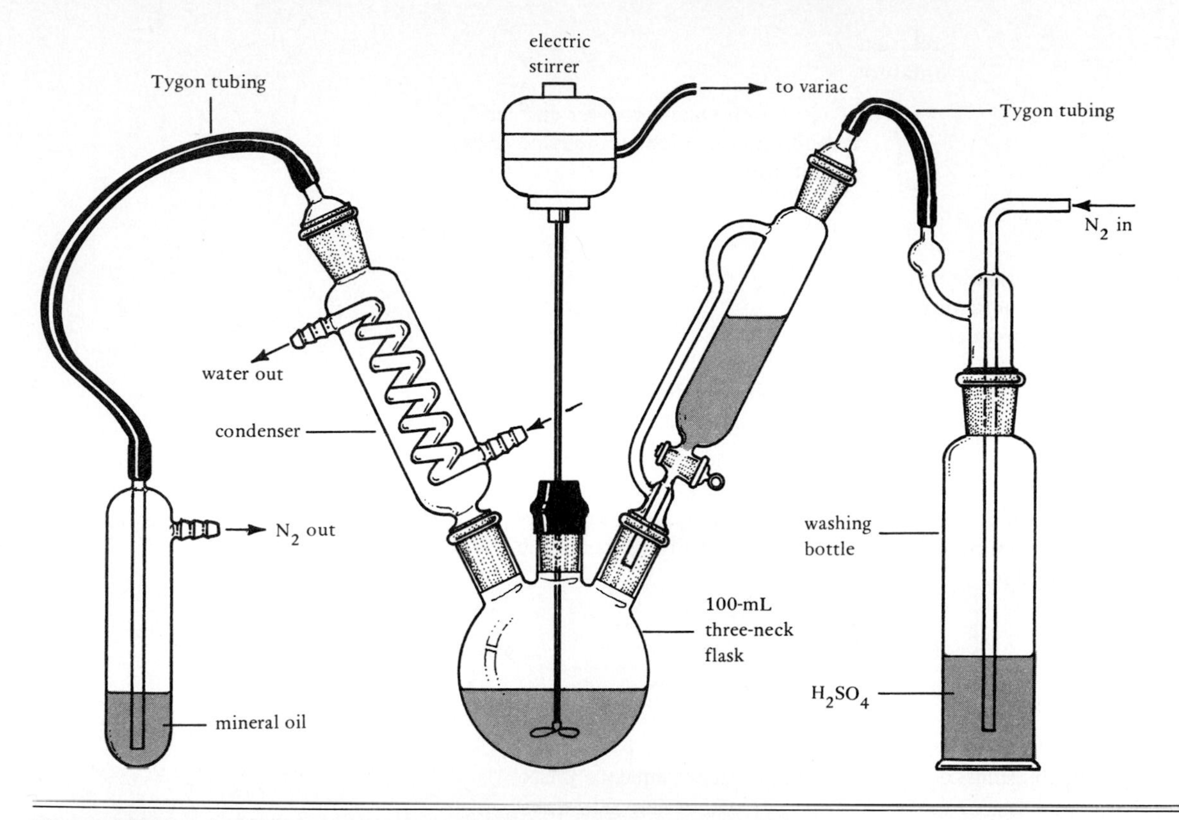

Figure 37.1. An apparatus for carrying out a reaction in an atmosphere of nitrogen. If magnetic stirring is used rather than the mechanical stirring shown, a three-neck flask is not needed. Instead, a single-neck flask fitted with a Claisen adapter may be used. The condenser will go in one neck of the adapter, and the pressure-equalizing addition funnel in the other.

because their action is catalytic, much more elaborate procedures must be used; these might include working on a vacuum line, recrystallization and redistillation of the reactants and solvents under nitrogen, and heating and pumping out of the reaction vessels.

38 Working Up the Reaction; Isolation of the Product

After the reaction has taken place, the product must be separated from the reaction mixture by one or more of the separation procedures described in Sections 7–15. Although each reaction mixture presents a

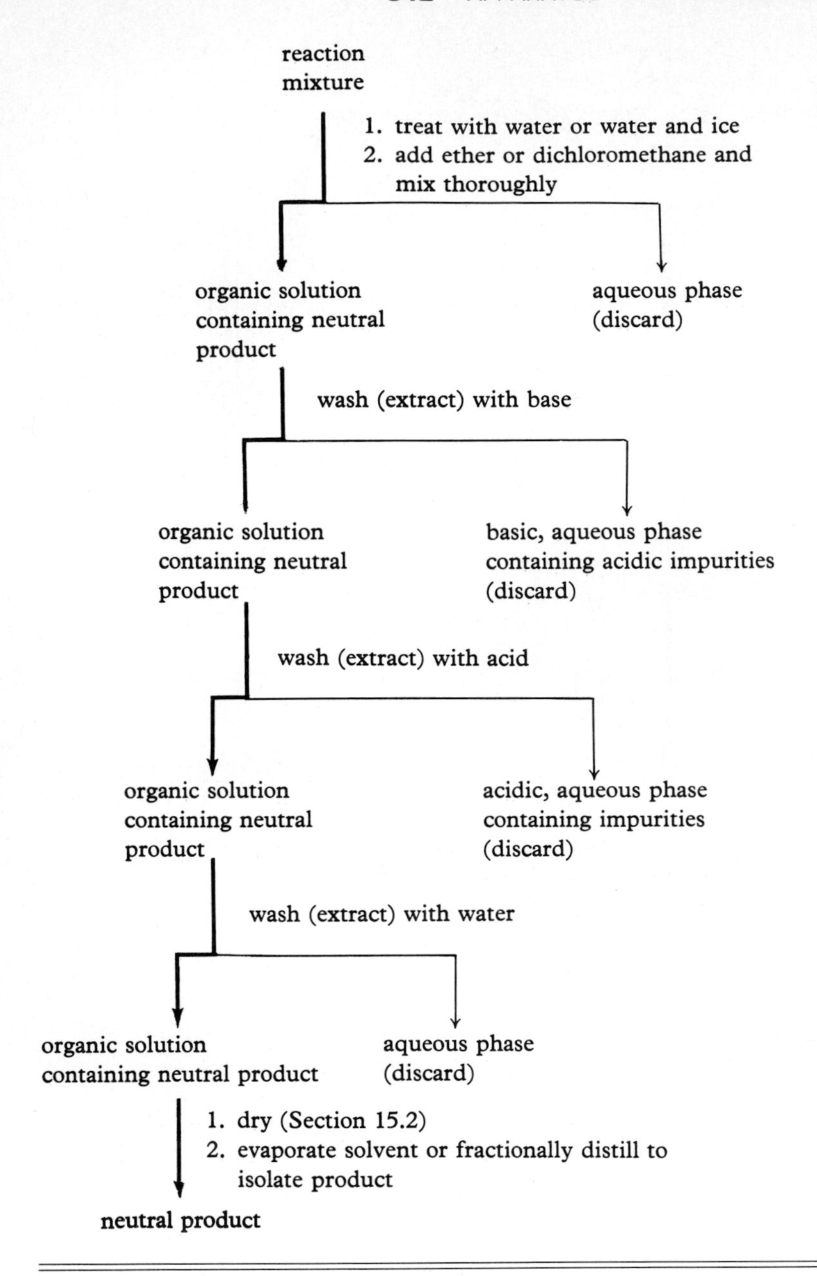

Figure 38.1. Flow diagram for isolation and purification of a neutral product.

unique problem in separation, most reactions can be worked up (the product isolated) according to one of a few general schemes.

If the product is a solid, it may crystallize or separate spontaneously from the reaction mixture. When it does this, the solid need only be collected by suction filtration and washed with the appropriate solvents. Depending on the degree of purity that is required and on the nature of any impurities, the solid can be further purified by recrystallization, sublimation, or some other method, perhaps column chromatography.

Suction filtration

If the product is a liquid, it can occasionally be isolated simply by fractional distillation of the entire reaction mixture. The initial product fraction can then be further purified, if necessary, by redistillation.

Fractional distillation

Usually, the workup will call for the addition of water, or water and ice, to the reaction mixture, or vice versa. If the product is a water-insoluble solid, it can then be collected by suction filtration, washed with appropriate solvents, dried, and recrystallized. If the product is a liquid—or often even if it is a solid—an organic solvent such as ether or dichloromethane can be added to dissolve it. The resulting ethereal or dichloromethane solution is first separated from the aqueous phase, using the separatory funnel (Section 13.3). The solution of the neutral product in the organic solvent is then washed with portions of either aqueous base (such as dilute sodium hydroxide or dilute sodium bicarbonate) or aqueous acid (such as dilute hydrochloric acid or dilute sulfuric acid). If the organic solution is washed with both acid and base, the washings must be done in succession. Usually, the order of washing with both acid and base is not important. Washing with base will remove acidic impurities, and washing with acid will eliminate basic contaminants, as described in Section 13.2. A final wash with water will then remove traces of the previous aqueous solution. The organic solution of the neutral compound is then dried over a suitable drying agent (Section 15.2), filtered by gravity to remove the hydrated drying agent (Section 7.1), and finally evaporated to remove the solvent if the product is a solid, or fractionally distilled if the product is a liquid. These operations are summarized in the flow diagram in Figure 38.1.

Aqueous workup

Isolation of a water-insoluble, neutral product: Figure 38.1

If the product is a water-insoluble acid, a typical aqueous workup will involve extraction from the organic solvent into a basic aqueous solution, such as dilute sodium hydroxide. After the organic phase has been removed by using the separatory funnel (Section 13.3), the basic aqueous solution of the salt of the acid is acidified and the free acid is then extracted into a fresh portion of the organic solvent. After the organic solution of the free acid has been separated from the aqueous layer, the organic solution is washed with a small portion of water (to remove traces of the previous aqueous solution) and dried over a suitable drying agent (Section 15.2). The hydrated drying agent is then removed by gravity filtration and the organic solution is evaporated to remove the solvent if the product is a solid, or fractionally distilled if the product is a liquid. This procedure is summarized in the flow diagram of Figure 38.2.

Isolation of a water-insoluble, acidic product: Figure 38.2

If the product to be purified is a water-insoluble base, an aqueous workup may involve extraction of the base from the organic solvent into

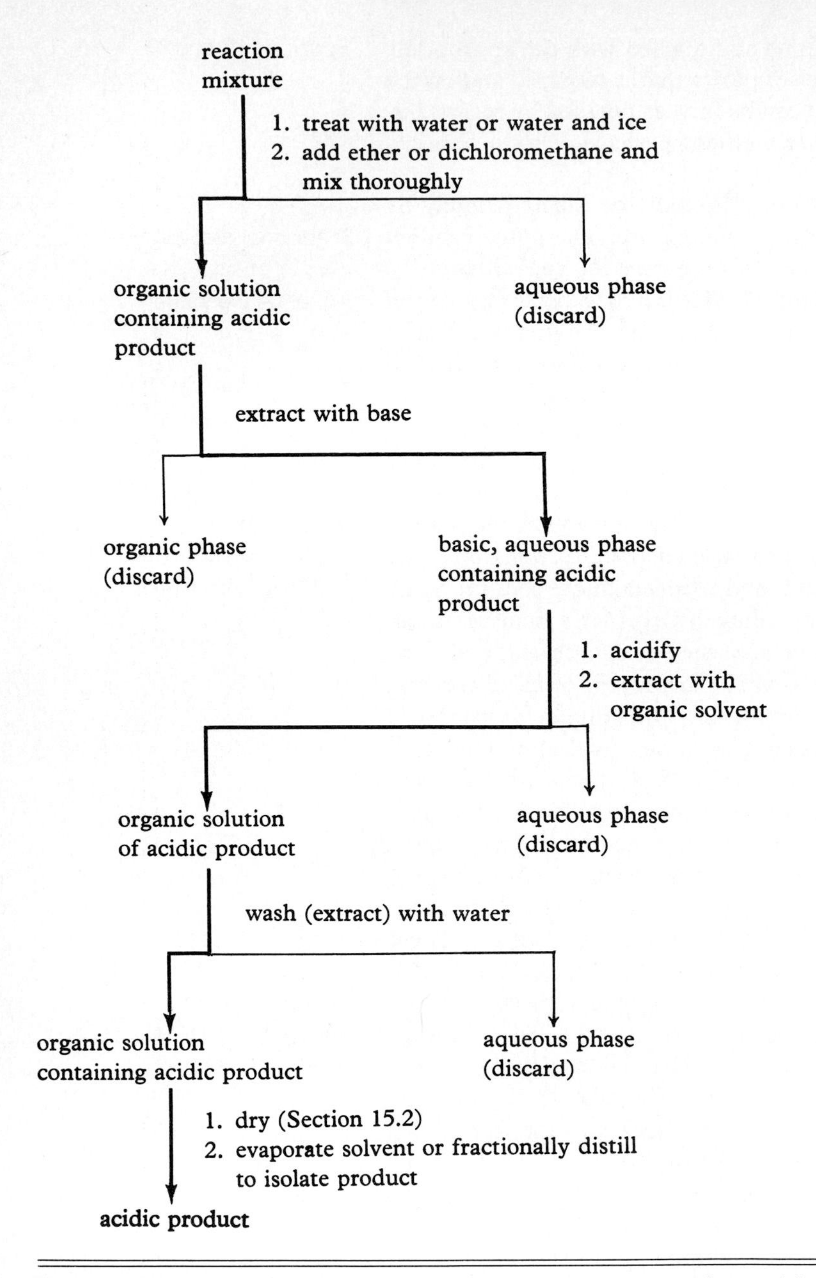

Figure 38.2. Flow diagram for isolation and purification of an acidic product.

Isolation of a water-insoluble, basic product: Figure 38.3

aqueous acid, basification of the aqueous solution, and reextraction by a fresh portion of the organic solvent. After the organic solution is dried, the product can be isolated either by evaporation or fractional distillation. These operations are summarized in the flow diagram of Figure 38.3.

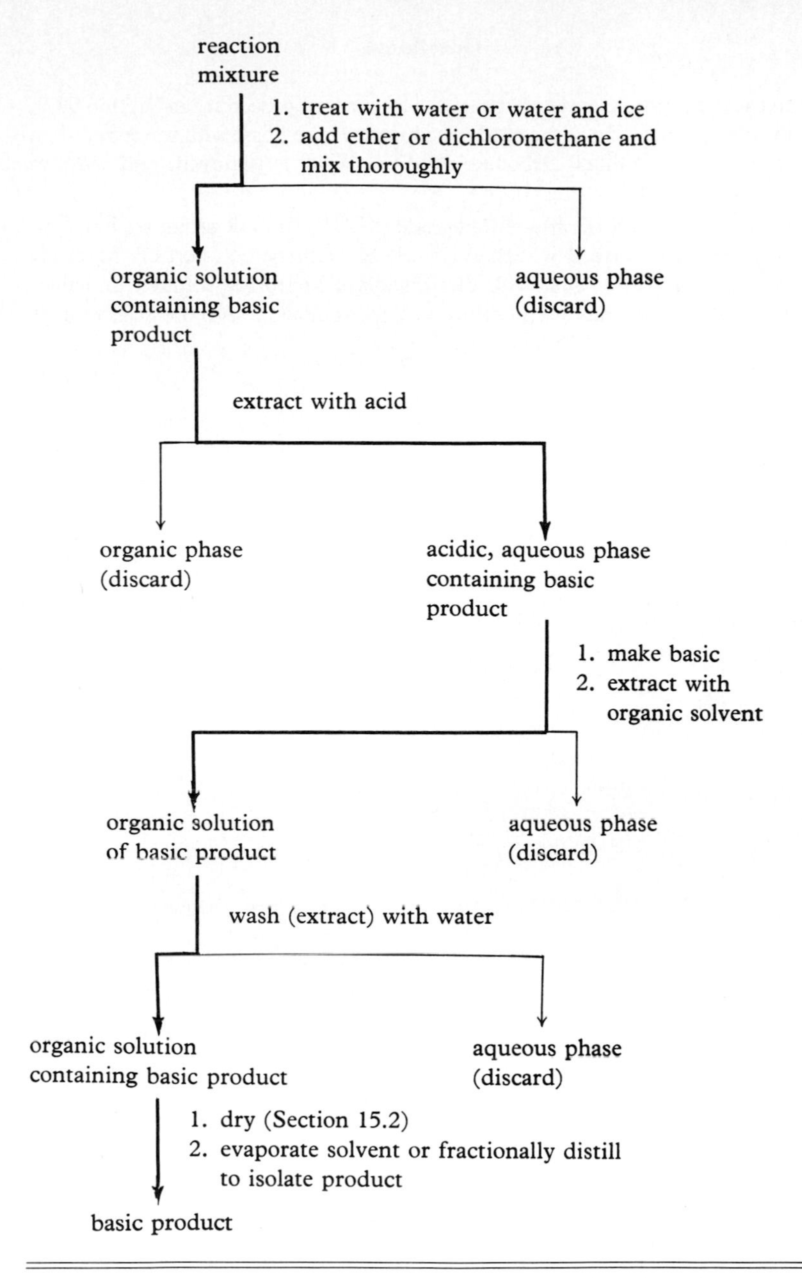

Figure 38.3. Flow diagram for isolation and purification of a basic product.

Questions

1. In the preparation of methyl benzoate by the first procedure in Section 59.2, the ethereal solution of the crude product is washed first with water and then with aqueous sodium carbonate solution. What is removed, and why, in each extraction?
2. In preparing N,N-diethyl-*m*-toluamide ("Off") from *m*-toluic acid and diethylamine, described in Section 63.3, the toluene solution of the crude product is extracted first with dilute sodium hydroxide solution and then with dilute hydrochloric acid solution. Explain how these extractions remove unreacted starting materials.

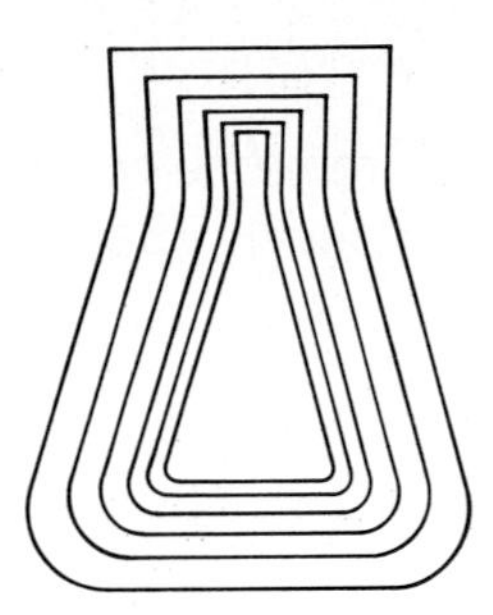

2 Experiments

- Separations
- Transformations
- Synthetic Sequences: Synthesis Experiments That Use a Sequence of Reactions

Three types of experiments are presented in the three main divisions of Part 2: separations, one-step transformations, and synthetic sequences.

Most of the *separations* are straightforward applications of the techniques that were described in Sections 7 through 15, in which the substance to be isolated is already present as a component of a mixture. The isolation of acetylsalicylic acid from aspirin tablets is an example.

Several simple experiments that illustrate a single technique are presented as exercises in the sections on recrystallization, distillation, extraction, and chromatography. These experiments include:

- Recrystallization of a dicarboxylic acid from water; Exercise 1, Section 8, page 61.
- Recrystallization of acetanilide, *m*-nitroaniline, or *p*-nitroaniline; Exercises 2, 3, and 4, Section 8, page 61.
- Distillation of water or methanol; Exercises 1 and 2, Section 9, page 85.

- Distillation of an azeotropic mixture of water and *n*-propyl alcohol; Exercise 3, Section 9, page 85.

- Distillation of a mixture of water and methanol; Exercise 4, Section 9, page 86.

- Separation by extraction of a mixture of an acid, a base, and a neutral compound; Exercise 3, Section 13, pages 115–117.

- Separation by extraction of a mixture of aspirin, phenacetin, and caffeine; Exercise 4, Section 13, pages 117–119.

- Paper chromatography of spinach leaves, carrots, food coloring, or ink; Exercises 1, 2, 3, and 4, Section 14.6, page 132.

Most of the *transformations* are a one-step synthesis of a compound, followed by its isolation and purification. Sometimes two procedures are given, and you must choose between them or combine the best parts of each.

The *synthetic sequences* are preparations that must be carried out by a series of reactions. Usually, the crude product of the preceding reaction can be used without purification.

The synthesis projects of the previous editions have been replaced in this edition by *variations*, alternative experiments in which the detailed procedure of the accompanying experiment must be adapted for use with an alternative starting material for preparation of a related product. For example, isoamyl bromide (Section 56.2), which can be converted to isoamyl acetate by treatment with potassium acetate in dimethylformamide (Section 58.2), can be prepared from isoamyl alcohol by the same method that is used for the preparation of *n*-butyl bromide from *n*-butyl alcohol (Section 56.1), after adjustments have been made that take into account the different molecular weights of the two alcohols and the different boiling points of the two alkyl bromides. A variation requires that the student *actively* determine the experimental details, rather than *passively* follow a set of directions.

Sometimes it will take more than one laboratory period to complete an experiment. Or there may be some other reason for setting an experiment aside for a while. With experience it is possible to recognize good stopping points. For example, one of the best is when a solution is being set aside to stand for crystallization. Other good times to stop are when a mixture is ready to be distilled or when a solution is being kept over a drying agent. Sometimes a good time to stop is at the end of a heating period. In many of the procedures, some of the possible stopping points are indicated by the symbol ■. All the experiments have been designed for use in a three-hour laboratory period. Some can be done easily in less time, but a few will require more than three hours for many workers. The estimates of the time required, which are given at the end of each procedure, are of the time needed by students who are taking their first course in organic chemistry. The estimates do not include the time needed for any optional recrystallizations.

Separations

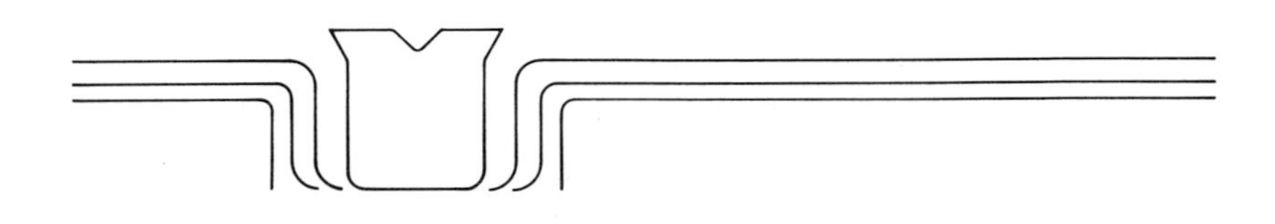

The separation procedures described in this section include the isolation of cholesterol from gallstones (Section 39), lactose from dry milk (Section 40), acetylsalicylic acid from aspirin tablets (Section 41), caffeine from tea or NoDoz (Section 42), piperine from black pepper (Section 43), clove oil from cloves (Section 44), eugenol from clove oil (Section 45), and (R)-(+)-limonene from grapefruit or orange peel (Section 46). The two remaining experiments involve the isolation of enantiomeric molecules. In the first of these, the R and S forms of carvone are isolated, one from oil of caraway and the other from oil of spearmint, and their odors, which are different, can be compared. In the second, the R and S forms of α-phenylethylamine are separated from the racemic mixture by a chiral reagent, (R),(R)-(+)-tartaric acid.

39 Isolation of Cholesterol from Gallstones

In this experiment, cholesterol is isolated from human gallstones by a simple crystallization process. On the average, the cholesterol content of human gallstones is about 75%.

The infrared spectrum of cholesterol is shown in Figure 39.1.

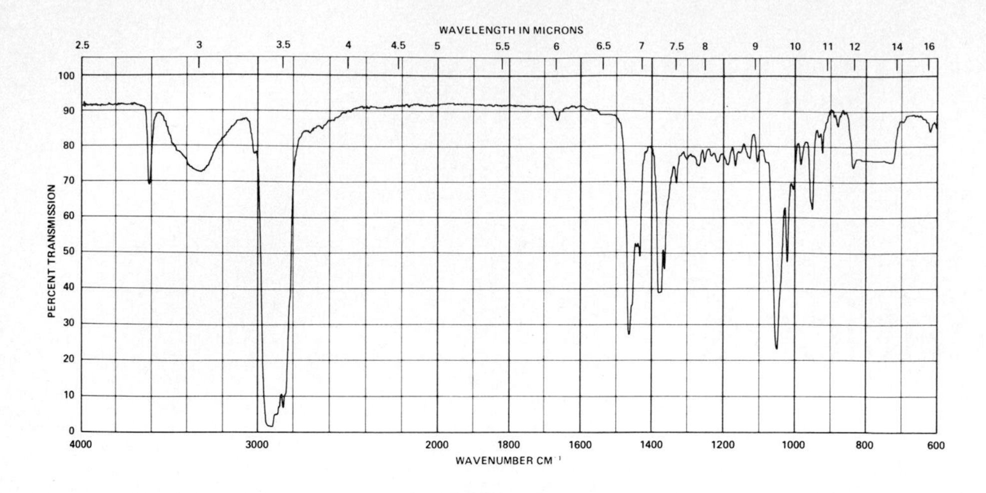

Figure 39.1. Infrared spectrum of cholesterol; CCl_4 solution.

cholesterol

Filtration: Section 7

Decolorization: Section 14.10

Recrystallization: Section 8

Procedure (Reference 1). Place a 2-gram sample of crushed gallstones in a 50-mL Erlenmeyer flask. To this, add 15 mL of 2-butanone (Note 1) and heat the mixture on the steam bath with occasional swirling for about 5 minutes. During this time most of the solid in the flask should dissolve. While the mixture is still hot, remove the brown residue of bile pigments by gravity filtration, collecting the filtrate in a 50-mL Erlenmeyer flask. Dilute the filtrate with 10 mL of methanol, add some decolorizing carbon, and warm the mixture on the steam bath for a minute or two. While the mixture is still hot, remove the carbon by gravity filtration and collect the filtrate in a 50-mL Erlenmeyer flask. Reheat the filtrate on the steam bath, add 1 mL of hot water, and swirl to dissolve any precipitated solid. (If it does not all dissolve, add a little methanol and heat with swirling until it does.) Allow the solution to come to room temperature, during which time cholesterol will separate as colorless plates. ■* Collect the crystals of cholesterol by suction filtra-

* The symbol ■ indicates a stopping point in the procedure.

tion, and wash them with a few milliliters of cold methanol. Leave the vacuum on for a few minutes to suck the crystals as dry as possible. Cholesterol can be recrystallized from methanol, using 35 mL per gram. Pure cholesterol melts at 149°C.

Melting point: Section 17.1

Cholesteryl benzoate, the ester of cholesterol and benzoic acid, is coverted to a "liquid crystal" when heated to a temperature above 147°C but below 180°C. Section 59.1 tells how to prepare cholesteryl benzoate and how to observe the liquid crystal phenomenon.

Note

1. The 10 mL of dioxane called for in previous editions has been replaced by 15 mL of 2-butanone, so as to avoid exposure to dioxane. Other alternatives are 15 mL of isopropyl alcohol or 10 mL of diethyl ether.

Time: less than 3 hours.

Reference

1. L. F. Fieser, *Organic Experiments,* Heath, Boston, 1964, p. 70.

40 Isolation of Lactose from Powdered Milk

In this experiment, lactose is recovered from the whey that remains after precipitation and removal of the milk proteins. Lactose is a disaccharide that, upon hydrolysis, yields a molecule of glucose and a molecule of galactose. Dry milk is approximately one-half lactose.

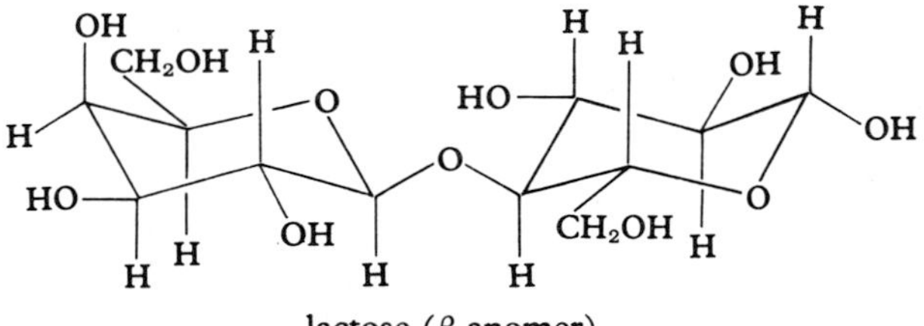

lactose (β anomer)

Procedure. Place 25 grams of nonfat dry milk powder in a 250-mL beaker (Note 1). Add 75 mL of warm water and stir to mix. Adjust the temperature of the mixture to between 40 and 50°C by heating or cooling. Add about 10 mL of 10% acetic acid solution (Note 2) and stir the mixture to coagulate the casein. Precipitation can be judged to be complete when the liquid changes from milky to clear. Remove the precipitated casein by filtering the mixture by gravity through cheesecloth (Note 3). Collect the filtrate in a 250-mL beaker. Add about 2 grams of calcium carbonate powder to the filtrate, stir it well, and boil the suspension for about 10 minutes (Note 4). Add to the hot mixture about as much decolorizing carbon as would cover a nickel, stir the mixture thoroughly, and filter it by suction through a layer of wet filter aid on a Buchner funnel (Note 5). ■ Transfer the filtrate to a 250-mL beaker and concentrate it to a volume of about 30 mL by boiling over a low flame with a wire gauze between the flame and the beaker (Note 6). When the volume has been reduced to 30 mL, turn off the burner and add 125 mL of 95% ethanol and about the same amount of decolorizing carbon as used before. Stir the mixture well and filter it through a layer of wet filter aid on a Buchner funnel (Note 5). Allow the clear filtrate to stand for crystallization for at least 24 hours in a stoppered Erlenmeyer flask. ■ Collect the crystals of lactose by suction filtration. They may be washed with a small amount of 95% ethanol. Yield: between 2.5 and 4.5 grams.

Filtration: Section 7

Decolorization: Section 14.10

Recrystallization: Section 8

Notes

1. The powdered milk is most easily measured by volume using a beaker; 25 grams is about 100 mL.
2. Ten percent acetic acid is prepared by diluting 10 mL of glacial acetic acid to 100 mL. More than 10 mL of 10% acetic acid seems to be required if the dry milk is not fresh.
3. Use two pieces of cheesecloth about 12 in. square. Lay them over a conical funnel large enough to contain all the curds. First decant as much as possible of the liquid from the coagulated casein into the funnel and then carefully transfer the wet protein mass onto the cheesecloth. After most of the liquid has drained through, a bit more can be recovered by wrapping the cheesecloth around the curds and squeezing.
4. The mixture will foam occasionally.
5. Add about 5 grams of filter aid (we have used Celite) to about 25 mL of water in a small beaker. Wet the filter paper on the Buchner funnel with water, fit the funnel to the suction flask, and apply a gentle suction. Swirl the mixture of filter aid and water to suspend the solid and pour the suspension all at once into the Buchner funnel. When the water has been drawn through the funnel, a damp pad of filter aid should be in place on top of the filter paper. Remove the funnel momentarily and pour the water out of the filter flask. Be careful not to disturb the layer of filter aid when pouring the mixture to be filtered into the funnel.
6. Some bumping and foaming occur.

Time: 3 hours plus ½ hour to isolate the product after crystallization is complete.

Reference

1. J. Cason and H. Rapoport, *Laboratory Text in Organic Chemistry*, Prentice-Hall, New York, 1950, p. 113.

41 Isolation of Acetylsalicylic Acid from Aspirin Tablets

Aspirin tablets consist of acetylsalicylic acid plus a binder that helps to keep the tablet from breaking up. In this experiment, the acetylsalicylic acid is recovered by a crystallization process. A procedure for the preparation of acetylsalicylic acid from salicylic acid is described in Section 59.4. The infrared and NMR spectra of acetylsalicylic acid are shown in Figures 59.6 and 59.7.

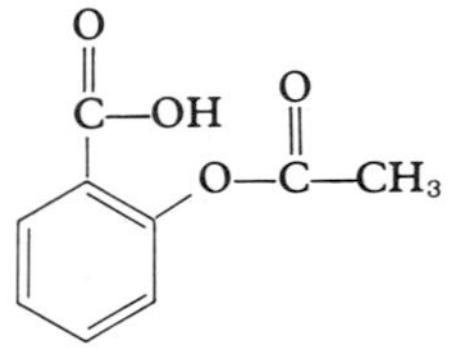

acetylsalicylic acid
(aspirin)

Procedure. Place ten 5-grain aspirin tablets (3.24 grams acetylsalicylic acid) in a 125-mL Erlenmeyer flask. Add to the flask 10 mL of 95% ethanol and heat the mixture to boiling on the steam bath. Swirl the hot mixture until the tablets disintegrate. The acetylsalicylic acid will dissolve, leaving a white residue (the binder). Filter the hot solution by gravity to remove the insoluble material and add to the filtrate 50 mL of cool water. Acetylsalicylic acid will begin to crystallize almost immediately. Allow crystallization to take place for about 10 minutes at room temperature and then for about 5 minutes in an ice bath. Collect the acetylsalicylic acid by suction filtration and wash the product with a little cold water. Recovery: about 2.4 grams (75%). The melting point of acetylsalicylic acid is reported to be 135–136°C.

Filtration: Section 7

Recrystallization: Section 8

Melting point: Section 17.1

Time: about 2 hours.

42 Isolation of Caffeine from Tea and NoDoz

In these experiments, caffeine is isolated from tea or NoDoz tablets by extraction into water followed by extraction into chloroform, evaporation, and recrystallization from ethanol. It may also be purified by sublimation (Section 12).

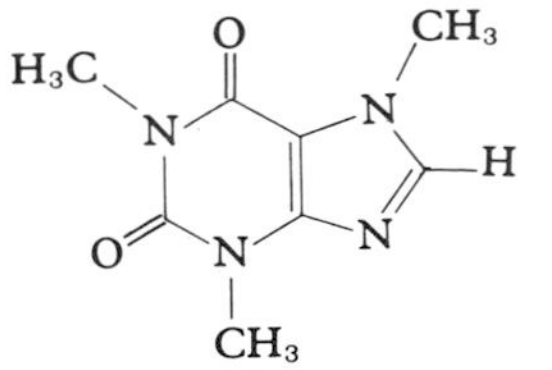

Caffeine is described as a cardiac, respiratory, and psychic stimulant, and as a diuretic. It is reported to have a melting range of 235–

caffeine

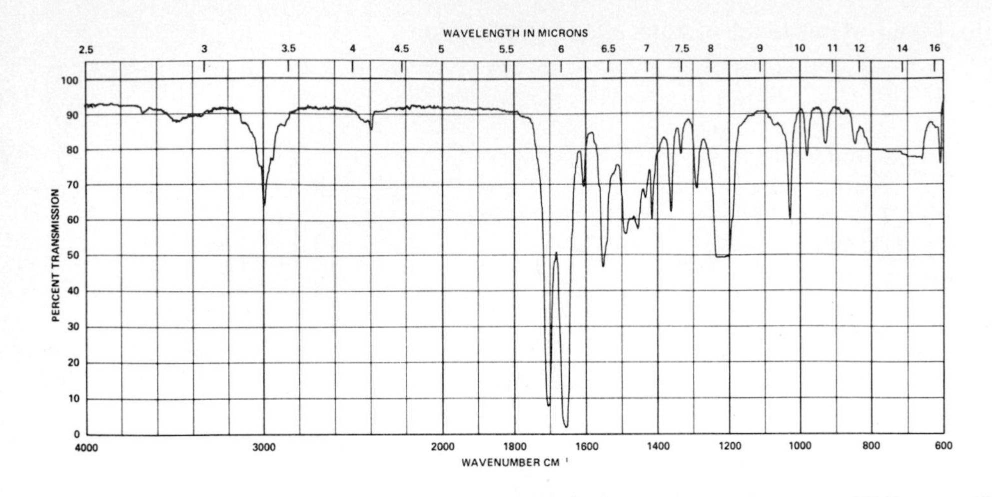

Figure 42.1. Infrared spectrum of caffeine; $CHCl_3$ solution.

236°C. The infrared spectrum is shown in Figure 42.1 and the NMR spectrum in Figure 42.2.

Caffeine from Tea

Procedure (Reference 1). Place 125 mL of water in a 500-mL Erlenmeyer flask. Add 12.5 grams of tea and 12.5 grams of powdered calcium carbonate (Note 1). Boil the contents of the flask with constant stirring for about 20 minutes. At this time, filter the hot mixture with suction, and press out the liquid from the tea leaves with a large cork. Transfer the filtrate to a 250-mL Erlenmeyer flask and add 100 mL of chloroform. Swirl or gently stir the two layers together for about 10 minutes (Note 2) and then separate them, using a separatory funnel. Distill the chloroform layer until only about 10 mL remain, transfer the remaining solution to a tared (weighed) 25-mL Erlenmeyer flask (filtering if necessary), and evaporate to dryness on the steam bath. Recrystallize the solid residue of caffeine from 95% ethanol, using 5 mL per gram (Note 3).

Filtration: Section 7

Extraction: Section 13

Concentration: Sections 9, 36

Recrystallization: Section 8

Caffeine from NoDoz

Procedure. Place 100 mL of water in a 250-mL Erlenmeyer flask. Add two NoDoz tablets that have been crushed and weighed. Heat the suspension to boiling to ensure solution of the caffeine in the NoDoz tablets; the binder will remain in suspension. Cool the mixture, add 100 mL of chloroform, and gently swirl or stir the two layers together for

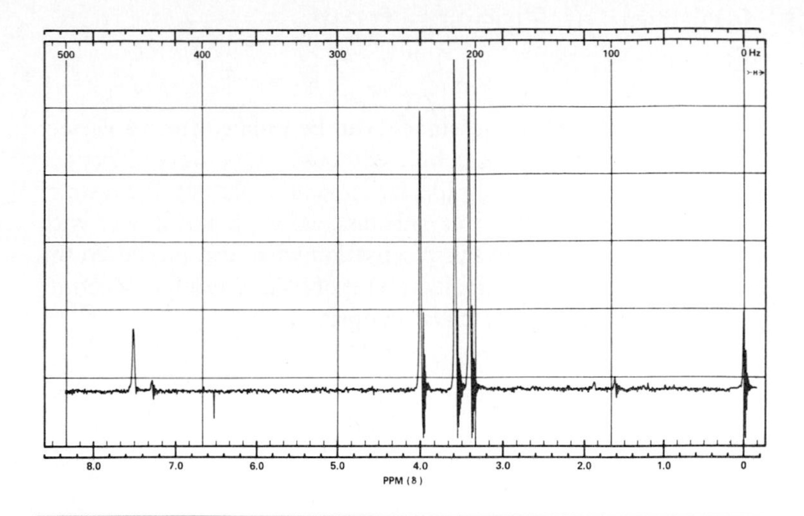

caffeine

Figure 42.2. NMR spectrum of caffeine; $CDCl_3$ solution.

about 10 minutes (Note 2). Continue as in the isolation of caffeine from tea.

Notes

1. R. H. Mitchell, W. A. Scott, and P. R. West suggest using 12.5 grams of tea *bags* and 9 grams of *sodium* carbonate [*J. Chem. Educ.* **51**, 69 (1974)]. This variation makes it possible to omit the suction filtration.
2. More vigorous extraction procedures often result in very troublesome emulsions. Gentle stirring of the mixture with a magnetic stirrer is quite satisfactory. The NoDoz extraction is less likely to form an emulsion.
3. One recrystallization from ethanol is sufficient and yields caffeine in the form of small needles. A reasonable recovery of recrystallized caffeine is about 25–50% of the total present, depending upon the thoroughness of the extraction. The greenish color that is present in the product of the extraction of tea can be completely removed by careful washing of the crystals with ethanol.

Time: about 3 hours.

Reference

1. G. K. Helmkamp and H. W. Johnson, Jr., *Selected Experiments in Organic Chemistry*, 2nd edition, W. H. Freeman, San Francisco, 1968, p. 157.

43 Isolation of Piperine from Black Pepper

Piperine, a very weakly basic substance, can be isolated from a variety of peppers by extraction with alcohol. Although the aroma of pepper cannot be due to the presence of piperine (it is not volatile), the taste of pepper may be due in part to this substance since piperine wet with alcohol produces an immediate sharp sensation when it is placed on the tongue. Piperine, along with a small amount of its Z,E isomer, accounts for about 10% of the weight of black pepper.

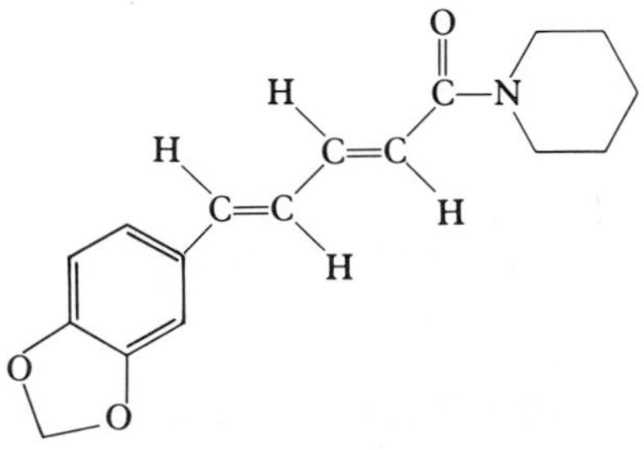

Sharp taste!

piperine

Procedure. Place 15 grams of ground black pepper and 1 gram of powdered calcium carbonate (Note 1) in a 250-mL boiling flask, add 100 mL of isopropyl alcohol, and, after fitting the flask with a reflux condenser, boil the mixture for about 1 hour on the steam bath.

Apparatus: Fig. 30.1

Steam bath: Section 32

At the end of the heating period, filter the mixture by gravity into a 125-mL Erlenmeyer flask, clean the 250-mL boiling flask, and return the filtrate to the boiling flask. Fit the boiling flask with a distillation adapter and condenser (see Figure 9.3), and boil off all but about 10 mL of the isopropyl alcohol (Note 2). Transfer the residual solution from the boiling flask to a 25-mL Erlenmeyer flask, and set the flask aside to cool for crystallization of piperine (Notes 3 and 4). Collect the product by suction filtration, using small portions of methyl alcohol to rinse the flask and wash the product. Yield: about 0.5 gram.

Filtration: Section 7

Concentration: Section 36

Crystallization: Section 8

Notes

1. The addition of calcium carbonate should prevent the extraction of acidic components of pepper.
2. Isopropyl alcohol boils at 80°C, only a little below the maximum temperature attainable on the steam bath. To make the distillation proceed quickly, clamp the boiling flask so that it is well down in the rings of the steam bath, and drape a towel over the

flask and the steam bath to make a tent that will hold steam around the top of the flask.

3. Crystallization occurs slowly, and the flask must be allowed to stand for at least 24 hours.
4. Alternatively, add 25 mL of water to the isopropyl alcohol solution of piperine, allow the mixture to stand for at least 24 hours so that precipitation will be complete, collect the solid by suction filtration, and recrystallize it from either isopropyl alcohol or acetone.

Question

1. Piperine is one member of a set of stereoisomers. What is the total number of isomers in this set, and what are the relationships among these stereoisomers?

44 Isolation of Clove Oil from Cloves

In this experiment, oil of cloves is isolated from cloves by steam distillation followed by extraction of the distillate. The extraction solution can be further treated to yield oil of cloves, or it can be used for the isolation of eugenol, as described in the procedure of Section 45.

Procedure (Reference 1). Place 75 grams of cloves (Note 1) and 250 mL of water in a 500-mL boiling flask. Clamp the flask so that it can be heated with a burner flame, and then fit it with a Claisen adapter. Fit the center neck of the Claisen adapter with a separatory funnel, and the side neck with a distillation adapter carrying a thermometer and a condenser set for downward distillation (Figure 44.1). Distill the mixture rapidly by heating the flask with a burner until no more oil can be observed to form in the condensate (Note 2). ■

Steam distillation: Section 11

Apparatus: Fig. 44.1

Extract the oil of cloves from the distillate with 50 mL of dichloromethane or petroleum ether, batchwise if necessary, as described in Section 13.3.

Extraction: Section 13

The oil of cloves can be isolated by drying the extract over anhydrous magnesium sulfate (Section 15) and removing the dichloromethane or petroleum ether by distillation on the steam bath (Section 36).

Drying: Section 15

Concentration: Sections 9, 36

Notes

1. Whole cloves are much more satisfactory than ground cloves.
2. Steam distillation for 1 hour will remove most of the oil of cloves.

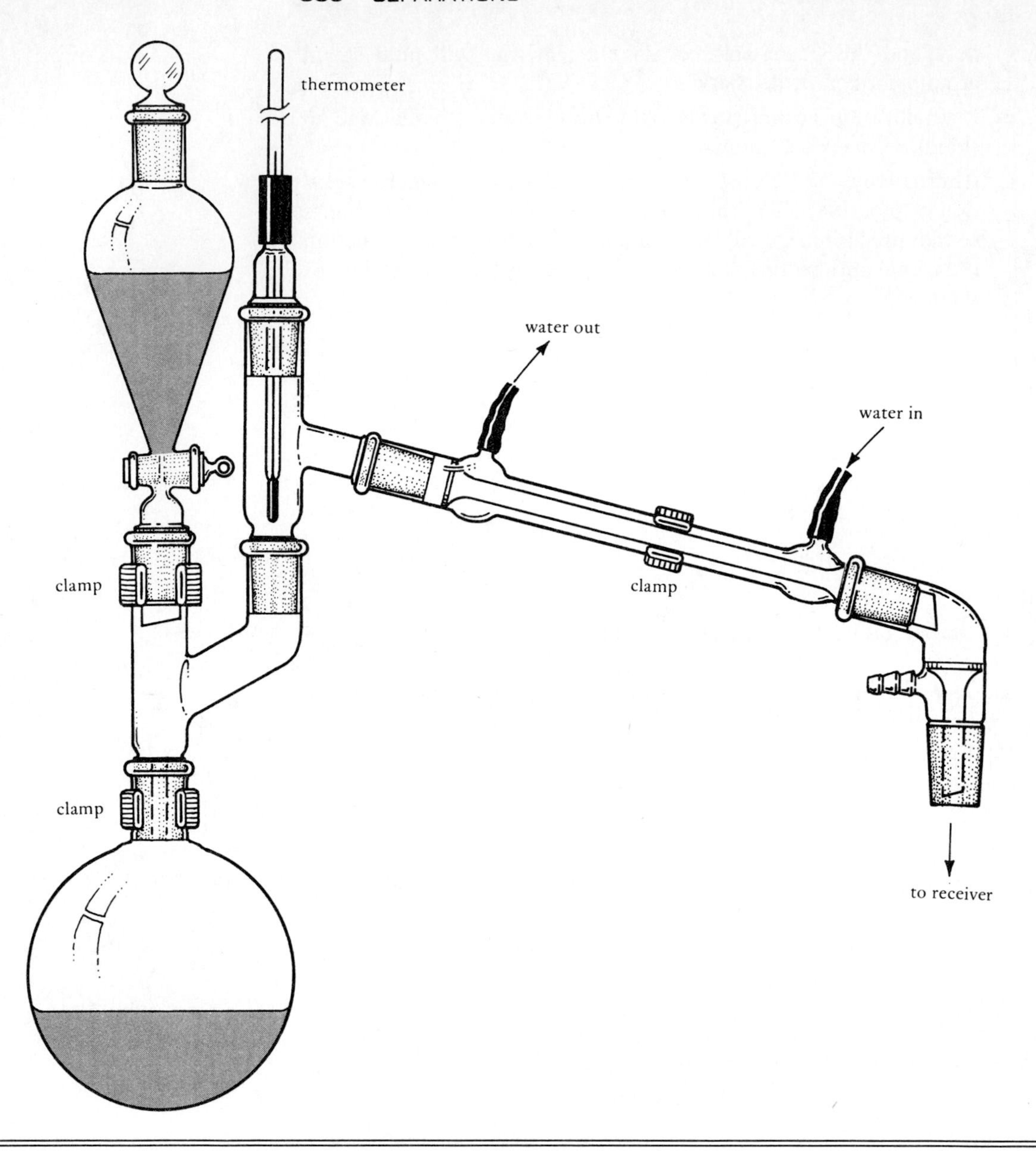

Figure 44.1. Apparatus for the steam distillation of cloves.

Time: 3 hours.

Reference

1. Information Division, Unilever, Ltd., Unilever House, Blackfriars, London.

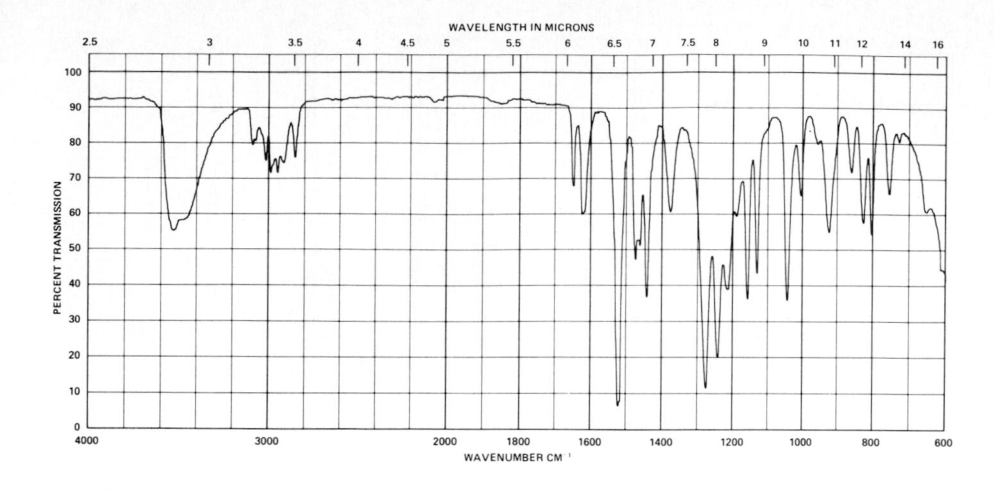

Figure 45.1. Infrared spectrum of eugenol; thin film.

45 Isolation of Eugenol from Clove Oil

In this experiment, eugenol is isolated from oil of cloves by extraction with base. Eugenol has the characteristic odor of cloves. Its infrared spectrum is shown in Figurc 45.1.

CH_2 CH CH_2

HO OCH_3

eugenol

Procedure (Reference 1). Dissolve 5 mL of oil of cloves in 50 mL of carbon tetrachloride and extract the solution with three 50-mL portions of 5% aqueous potassium hydroxide solution. Combine the basic extracts and acidify them to litmus with 5% aqueous hydrochloric acid; about 100 mL will be needed.

Extraction: Section 13

Extract the free eugenol from the aqueous mixture with 50 mL of carbon tetrachloride, dry the organic extract over anhydrous magnesium sulfate, and remove the carbon tetrachloride to give a residue of eugenol by distillation using a steam bath.

Drying: Section 15

Concentration: Sections 9, 36

Time: about 2 hours.

Reference

1. Information Division, Unilever Ltd., Unilever House, Blackfriars, London.

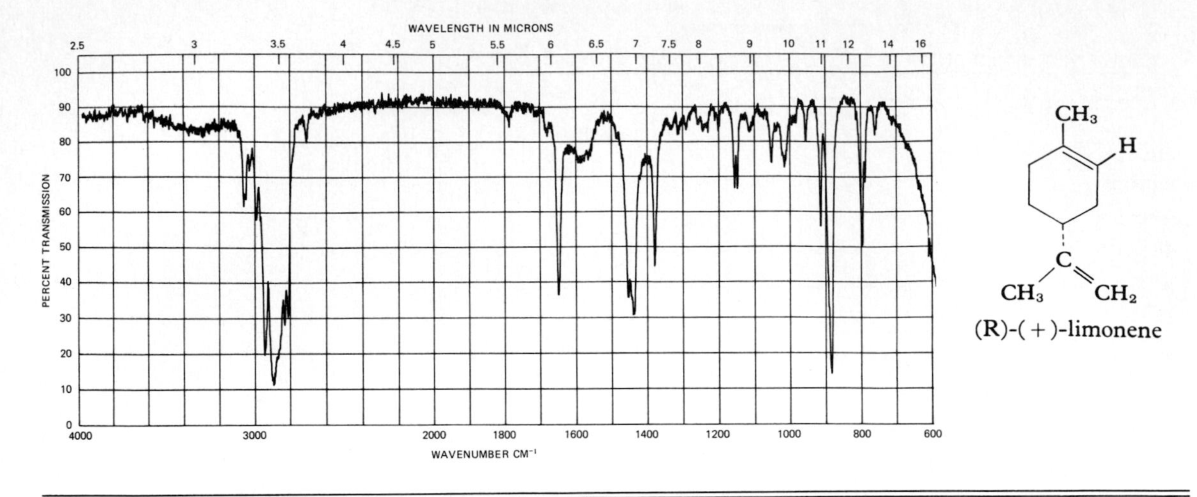

Figure 46.1. Infrared spectrum of (R)-(+)-limonene; thin film.

46 Isolation of (R)-(+)-Limonene from Grapefruit or Orange Peel

The major constituent of the steam-volatile oil of grapefruit or orange peel is (R)-(+)-limonene:

(R)-(+)-limonene

It can be isolated as a material of about 97% purity by a simple steam distillation of citrus fruit peels. It is responsible for the characteristic smell of citrus peel. The infrared spectrum of a commercial sample of (R)-(+)-limonene is shown in Figure 46.1.

Steam distillation: Section 11

Apparatus: Fig. 46.2

Extraction: Section 13

Procedure. Obtain the peel from a grapefruit or from two oranges. Cut the peel into pieces about 5–10 mm square, and put the peel into a 500-mL boiling flask along with about 250 mL of water. Clamp the flask so that it can be heated with a burner flame, and then fit it with a distillation adapter carrying a condenser set for downward distillation (Figure 46.2). Distill the mixture rapidly by heating the flask with a burner until you have collected about 50 mL of distillate.

Extract the limonene from the distillate with 20 mL of either pentane or dichloromethane, separate the extract by means of a separatory funnel, and dry the extract over anhydrous magnesium sulfate for a few minutes (Note 1). Isolate the limonene by filtering the extract by gravity from the magnesium sulfate into a tared (previously weighed) flask and removing the solvent by distillation on the steam bath. Yield: between $\frac{1}{2}$ and $1\frac{1}{2}$ mL. Physical properties reported for purified (R)-(+)-limonene include b.p.^{763} = 175.5–176°C, d_4^{21} = 0.8403, n_D^{21} = 1.4743, and $[\alpha]_D^{19.5}$ = +124.2°.

Drying: Section 15

Concentration: Sections 9, 36

Note

1. As an alternative, you can collect the distillate in two 20 × 150 mm test tubes. The limonene will separate to form a clear layer above the water and most of the limonene can be easily removed using a medicine dropper.

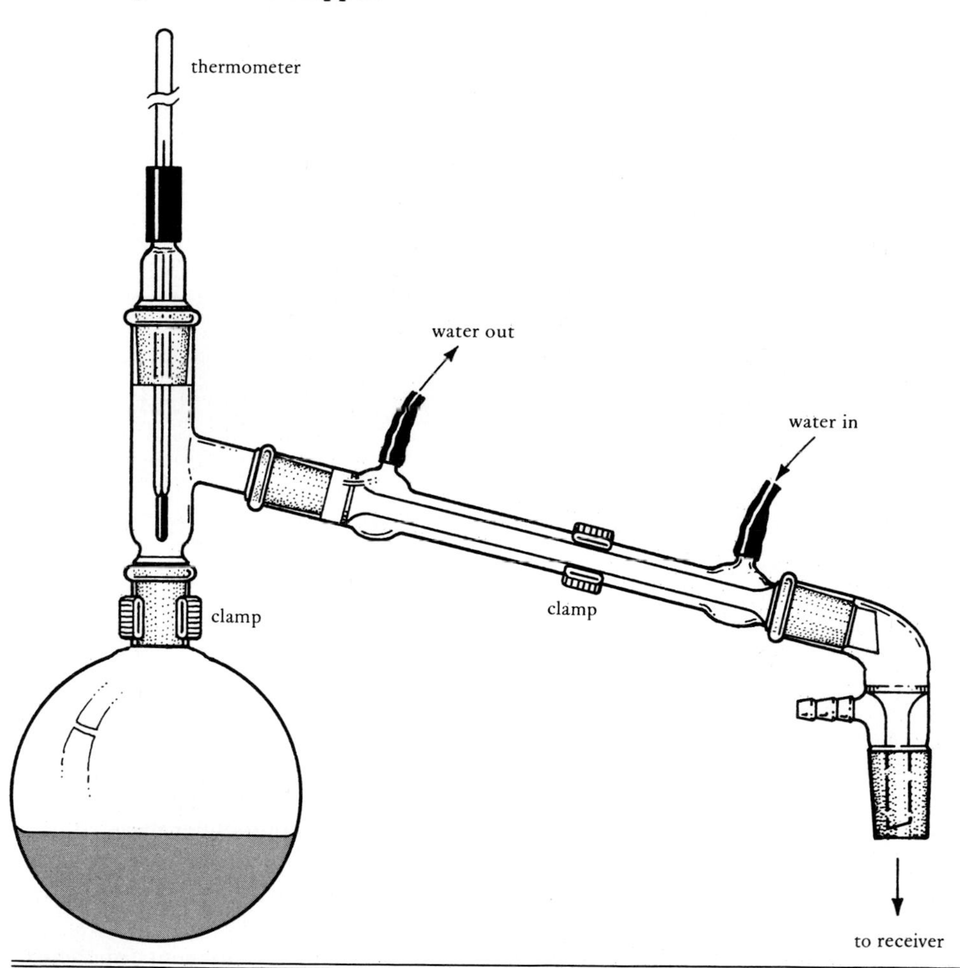

Figure 46.2. Apparatus for the steam distillation of grapefruit or orange peel.

Time: less than 3 hours.

Reference

1. F. H. Greenberg, *J. Chem. Educ.* **45**, 537 (1968).

47 Isolation of (R)-(−)- or (S)-(+)-Carvone from Oil of Spearmint or Oil of Caraway

Enantiomeric molecules have identical properties, with two exceptions. Enantiomers rotate the plane of polarization of plane-polarized light in opposite directions, and they interact differently with chiral molecules. The latter difference is like the difference in the ways that enantiomeric members of a pair of gloves will interact with a right hand. Enzymes, which are chiral molecules present in every living cell, generally react exclusively with one member of a pair of enantiomers. It is therefore not surprising that enantiomers have often been reported to differ in taste or smell. One report, in which the evidence for a difference in smell is especially convincing (Reference 1), describes the enantiomers of carvone. (R)-(−)-Carvone, isolable from oil of spearmint, smells like spearmint, while (S)-(+)-carvone, isolable from oil of caraway, smells like caraway. About nine persons out of ten can distinguish these enantiomers by smell (Reference 2).

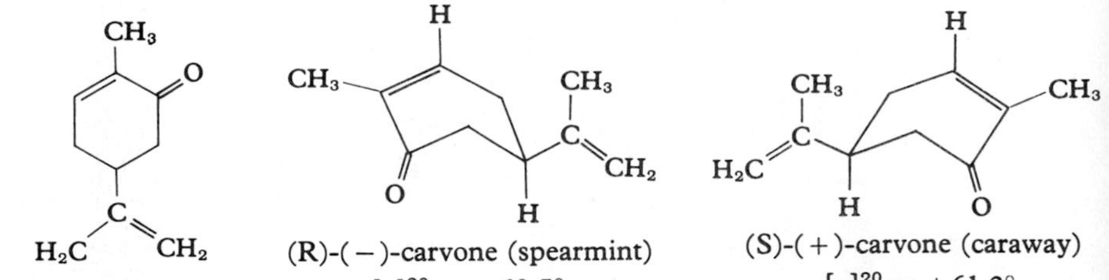

carvone

(R)-(−)-carvone (spearmint) $[\alpha]_D^{20} = -62.5°$

(S)-(+)-carvone (caraway) $[\alpha]_D^{20} = +61.2°$

The experiment in this section involves the isolation of the enantiomeric forms of carvone, one from each oil. The isolation can be effected by distillation at atmospheric pressure, distillation under reduced pressure, or column chromatography on silica gel. As you can see from Figures 47.1 and 47.2, the infrared spectra of the enantiomeric forms are identical.

The lower-boiling and more easily eluted components of the oils are mostly isomers of molecular formula $C_{10}H_{16}$:

(S)-(+)-α-phellandrene (S)-(+)-β-phellandrene (R)-(+)-limonene

Distillation at Atmospheric Pressure

Procedure. Put 10 mL of one of the oils and one or two boiling stones in a 25-mL round-bottom flask; place a Claisen adapter in the neck of the flask (Note 1). Clamp the flask in position on a wire gauze supported by an iron ring. Put a stopper in the center neck of the Claisen adapter. If you want, you may poke a small amount of stainless steel sponge into the curved part of the Claisen adapter in order to increase the efficiency of the separation (Section 9.4). Then place a distillation adapter in the side neck of the Claisen adapter. Fit a thermometer in the top neck of the distilling adapter and connect the side neck directly to a vacuum adapter (Figure 47.3). Heat the flask with a burner until the oil boils and continue to heat it strongly until material begins to distill. Then adjust the heat so that the distillate accumulates at a rate of 1 drop every 3–5 seconds (Note 2). Collect the distillate in at least 3 fractions: $C_{10}H_{16}$ isomers, intermediate fraction, and carvone fraction.

Distillation: Section 9

Apparatus: Fig. 47.3

Atmospheric-pressure boiling points for carvone and the isomers of molecular formula $C_{10}H_{16}$ have been reported as follows. Carvone: vari-

Figure 47.1. Infrared spectrum of (R)-(−)-carvone; thin film.

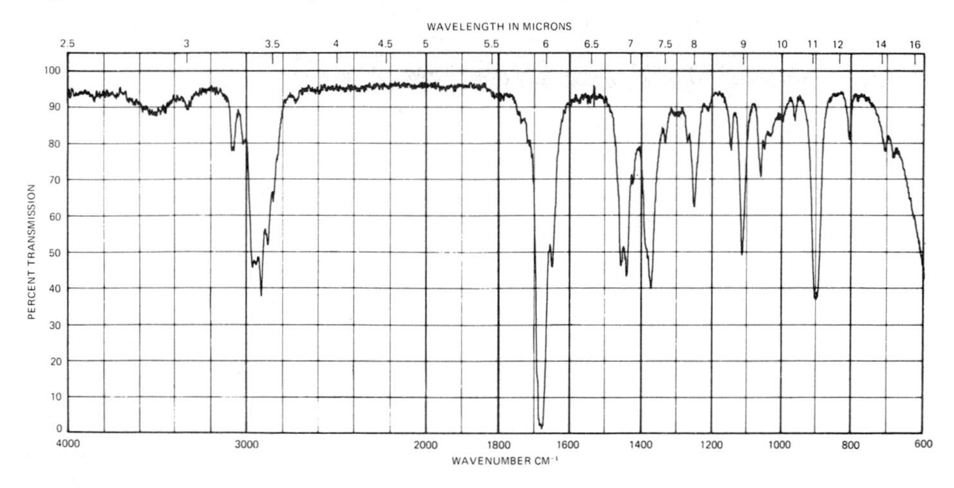

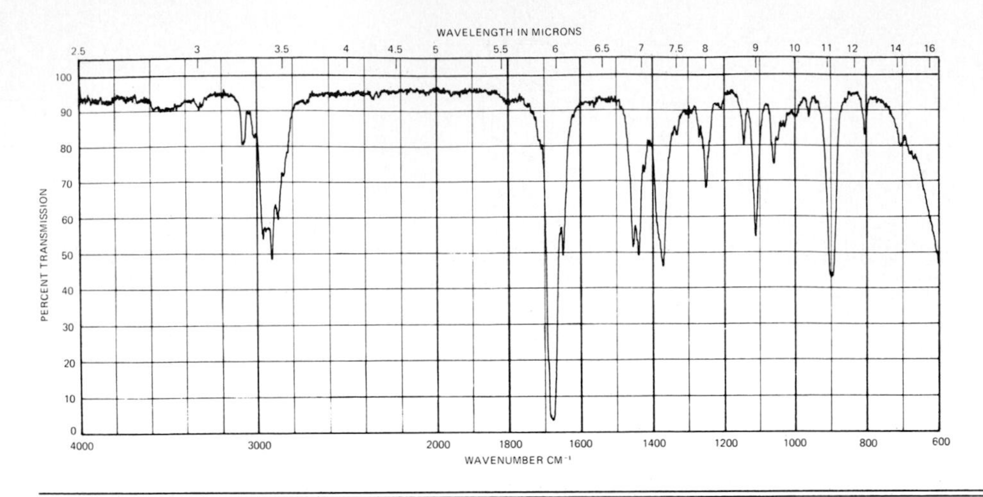

Figure 47.2. Infrared spectrum of (S)-(+)-carvone; thin film.

ous values between 224 and 231°C; limonene: 176°C; α-phellandrene: 176°C; β-phellandrene: 172°C.

Notes

1. By omitting the Claisen adapter and using a 10-mL flask, you can distill as little as 5 mL of oil. If the distillation is done carefully, the separation is quite good.
2. In order to drive the carvone over, it will probably be necessary to insulate the apparatus by wrapping the adapters with a single layer of aluminum foil. We did not experience the difficulty with foaming reported in Reference 3.

Time: less than 3 hours.

Distillation under Reduced Pressure

Reduced-pressure distillation: Section 10

Procedure (Reference 3). Place 10 mL of one of the oils in a 25-mL round-bottom flask. Assemble the apparatus as shown in Figure 10.2, but use a fraction cutter (Figure 10.5) or a cow (Figure 10.3) in place of the simple vacuum adapter. Connect the fraction cutter or cow to a source of vacuum (Section 10.3) and a manometer (Section 10.4). It is preferable to use a magnetically stirred oil bath for heating, as shown in Figure 33.1. If magnetic stirring is available, place a very small magnetic stirring bar in the flask instead of boiling stones. If the flask must be heated with a flame, add a few boiling stones and clamp the flask in position on a wire gauze supported by an iron ring. Reduce the pressure

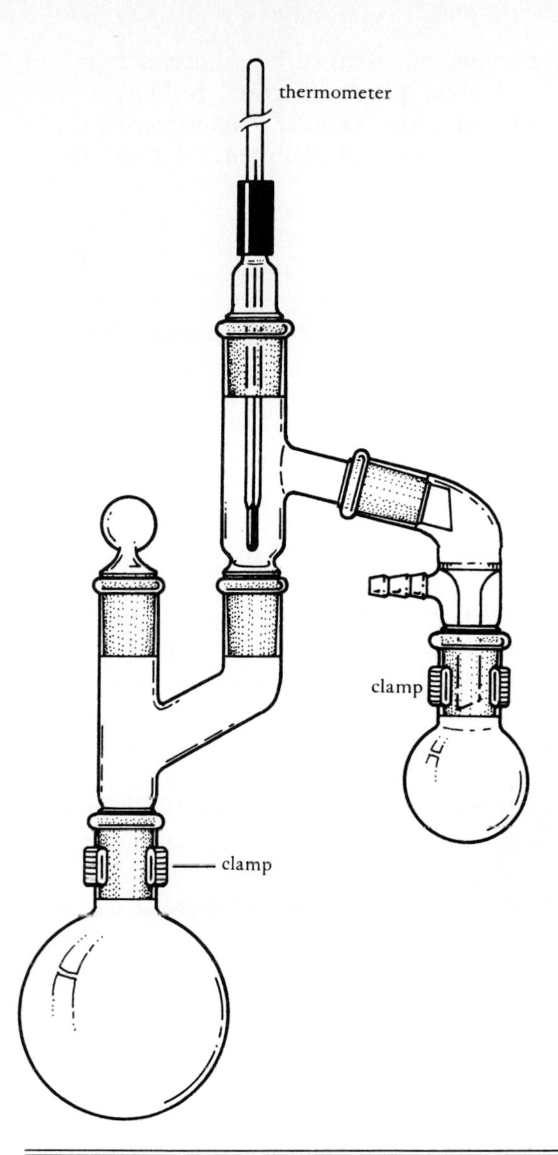

Figure 47.3. Apparatus for the fractional distillation of oil of spearmint or oil of caraway.

of the system by turning on the water or oil pump, and allow the pressure to stabilize at 15–30 Torr. It is vital to the success of the distillation that the pressure of the system does not fluctuate. Heat the flask until the oil starts to boil (Note 1); continue heating until distillation begins. At this time, adjust the rate of heating so that the distillate accumulates at the rate of 1 drop every 3–5 seconds (Sections 9.6 and 10.5). Collect the distillate in at least 3 fractions: $C_{10}H_{16}$ isomers, an intermediate fraction, and the carvone fraction.

The expected boiling points will have to be estimated from the reported reduced-pressure boiling points (Section 10.1): Carvone: 115°C (20 Torr), 100°C (10 Torr), 91°C (6 Torr); limonene: 64°C (15 Torr); α-phellandrene: 61°C (11 Torr); β-phellandrene: 57°C (11 Torr).

Note

1. The air-saturated oil will degas under reduced pressure long before it starts to boil.

Time: less than 3 hours.

Column Chromatography

Column chromatography: Sections 14.1, 14.2

Procedure (Reference 4). Prepare a chromatographic column as described in Section 14.2, using a 22-mm-diameter tube and 25 grams of chromatographic-grade silica gel. Fit the bottom of the tube with a short length of plastic tubing with an adjustable clamp to control the rate of flow. Add the silica gel as a slurry in petroleum ether. After draining the petroleum ether from the column so that the level of the liquid is even with the top of the sand above the silica gel, add 2 mL of one of the oils. Use a little petroleum ether to assist in transferring the oil and to rinse down the sides of the tube.

For elution of the column use four portions of solvent:

1. 25–30 mL of petroleum ether
2. 50–60 mL of 1 : 9 dichloromethane : petroleum ether
3. 25 mL of 1 : 4 dichloromethane : petroleum ether
4. 100–130 mL of 1 : 1 dichloromethane : petroleum ether

Collect the eluant in four fractions whose volumes correspond to the volumes of the four portions of solvent added. The second fraction should contain $C_{10}H_{16}$ hydrocarbons, the third should be an intermediate fraction, and the fourth should contain carvone. ■ Recover the material in each fraction by distilling off the solvent, using the steam bath for heating. The last traces of solvent can be removed by heating under vacuum (Section 36).

Concentration: Sections 9, 36

Time: 3 hours.

Question

1. The dextrorotatory isomers of limonene and the two phellandrenes are illustrated. However, the phellandrene isomers are assigned the S configuration, while the limonene isomer is assigned the R configuration. Explain.

References

1. G. F. Russell and J. I. Hills, *Science* **172,** 1043 (1971).
2. L. Friedman and J. G. Miller, *Science* **172,** 1044 (1971).
3. S. L. Murov and M. Pickering, *J. Chem. Educ.* **50,** 74 (1973).
4. R. H. Mitchell and P. R. West, *J. Chem. Educ.* **51,** 274 (1974).

48 Resolution of (R),(S)-α-Phenylethylamine with (R),(R)-(+)-Tartaric Acid

In this experiment, the enantiomeric forms of α-phenylethylamine react with the chiral reagent (R),(R)-(+)-tartaric acid to give diastereomeric salts that differ greatly in their solubility in methanol. The enantiomeric reactants have the same structure but different configurations, being mirror images of one another; they have identical physical and chemical properties, except for the direction of rotation of plane-polarized light and the interaction with chiral molecules. The diastereomeric products have the same structure but different configurations; they are not mirror images of one another. The (R)-(+)-amine (R),(R)-(+)-tartrate is quite soluble in methanol, whereas its diastereomeric companion (S)-(−)-amine (R),(R)-(+)-tartrate is relatively insoluble in methanol.

CH_3, H, C, NH_2, ϕ — (R)-(+)-amine + CH_3, C, H, H_2N, ϕ — (S)-(−)-amine + COOH, HO, H, HO, H, COOH — (R),(R)-(+)-tartaric acid $\xrightarrow{MeOH}$

CH_3, H, C, $\overset{+}{N}H_3$, ϕ; COO^-, HO, H, HO, H, COOH — (R)-(+)-amine (R),(R)-(+)-tartrate (more soluble) + CH_3, C, H, $H_3\overset{+}{N}$, ϕ; COO, HO, H, HO, H, COOH — (S)-(−)-amine (R),(R)-(+)-tartrate (less soluble)

When a hot methanolic solution of the racemic amine and an equimolar portion of (R),(R)-(+)-tartaric acid is allowed to cool, the less soluble of the diastereomeric salts crystallizes from solution. Treatment of this salt with excess sodium hydroxide converts it to the free (S)-(−)-amine and disodium tartrate.

When the methanolic solution that remains after crystallization of the (S)-(−)-amine (R),(R)-(+)-tartrate is evaporated, the residue consists of the (R)-(+)-amine (R),(R)-(+)-tartrate, which did not crystallize, and that part of the (S)-(−)-amine (R),(R)-(+)-tartrate that remained in solution after crystallization. The amine recovered from this residue by treatment with sodium hydroxide will therefore contain more (R)-(+)-amine than (S)-(−)-amine.

Part of the excess (R)-(+)-amine can be isolated by treating a hot ethanolic solution of this amine mixture with an amount of sulfuric acid in methanol just great enough to convert the excess R-(+)-amine to its neutral sulfate salt:

$$\underset{x+y \text{ mole}}{\text{(R)-(+)-amine}} + \underset{x \text{ mole}}{\text{(S)-(−)-amine}} + \underset{\frac{1}{2}y \text{ mole}}{H_2SO_4} \xrightarrow{\text{ethanol}}$$

$$\underset{\substack{\frac{1}{2}y \text{ mole} \\ \text{crystallizes out}}}{[\text{(R)-(+)-amine}^+]_2SO_4^=} + \underset{x \text{ mole}}{\text{(R)-(+)-amine}} + \underset{x \text{ mole}}{\text{(S)-(−)-amine}}$$

(The experimental procedure for this step is described in detail under (R)-(+)-α-phenylethylamine.) The (R)-(+)-amine is then recovered from its neutral sulfate salt by treatment with excess sodium hydroxide.

The infrared spectrum of (R),(S)-α-phenylethylamine is shown in Figure 48.1. In contrast to the enantiomeric forms of carvone (Section 47), the enantiomeric forms of α-phenylethylamine appear to smell the same.

(S)-(−)-α-Phenylethylamine

Procedure. Add 31.25 grams of *d*-tartaric acid (0.208 mole) to 450 mL of methanol in a 1-liter Erlenmeyer flask, and heat the mixture almost to boiling. To the hot solution, add cautiously 25 grams of *d,l*-α-phenylethylamine (26.6 mL; 0.206 mole); too rapid an addition will cause the mixture to boil over. Since crystallization occurs slowly, the solution must be allowed to stand at room temperature for about 24 hours. ■ The (−)-amine-(+)-hydrogen tartrate separates as prismatic crystals (Note 1). Collect the product by suction filtration, and wash it with a little methanol (Note 2). Yield: 18.1 grams (65%).

Filtration: Section 7

A second crop (3.8 grams) may be obtained by concentrating the combined mother liquor and washings to 225 mL, and allowing crystallization to proceed at room temperature for another 24 hours. ■

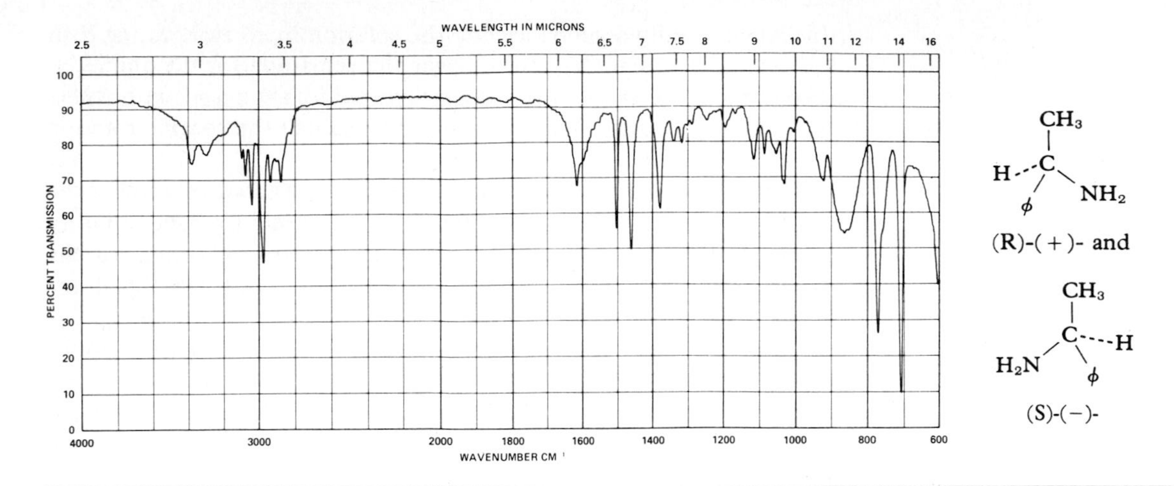

Figure 48.1. Infrared spectrum of (R),(S)-α-phenylethylamine; thin film.

Place the product (21.9 grams) in a 250-mL Erlenmeyer flask and add about 90 mL of water; the amine tartrate salt will partially dissolve. Add to this mixture 12.5 mL of 50% sodium hydroxide solution (0.24 mole) in order to convert the amine salt to the free base. Then add about 50 mL of ethyl ether and swirl the mixture in order to extract the free amine into the ether. Separate the ethereal extract using a separatory funnel (Section 13.3) and then dry the extract over anhydrous magnesium sulfate for about 10 minutes (Section 15.2). ■ Remove the drying agent from the ethereal extract by gravity filtration (Section 7.1); collect the filtrate in a 100-mL round-bottom flask. Add a boiling stone to the flask and, heating with the steam bath, remove most of the ether by distillation. Transfer the residue to a 25-mL round-bottom flask, using a little ether for rinsing. Add a boiling stone to the flask and set up an apparatus for distillation like that shown in Figure 9.3. Remove the last of the ether by heating the flask strongly with the steam bath. ■ Now, being sure you are in a part of the lab that is free of ether, distill the residue to obtain (S)-(−)-α-phenylethylamine by heating the flask with a flame (Note 3). Yield: 6.9 grams (55%); b.p., 184–186°C; d_4^{25}, 0.953; $[\alpha]_D^{27}$, −38.2° (lit.: $[\alpha]_D^{22}$, −40.3°).

Extraction: Section 13

Drying: Section 15

Concentration: Sections 9, 36

Distillation: Section 9

Optical rotation: Section 20.1

Time: 3 hours plus ½ hour of a previous laboratory period to prepare the initial solution.

(R)-(+)-α-Phenylethylamine

Procedure. Allow the methanolic solution remaining from the isolation of the (−)-amine-(+)-hydrogen tartrate to evaporate to dryness (this is most easily done by leaving the solution in an evaporating dish in the hood overnight ■), and recover the remainder of the amine by treating the residual salt with sodium hydroxide, extracting with ether, and distilling as previously described. Determine the specific rotation of the recovered amine.

The (+)-amine is isolated by treating a hot ethanolic solution of the recovered amine with an amount of sulfuric acid in ethanol slightly greater than that necessary to convert the excess (+)-amine to the neutral sulfate salt. The resulting crystals of (+)-α-phenylethylamine sulfate yield the (+)-amine (Note 4).

Dissolve 12.5 grams (0.103 mole) recovered amine ($[\alpha]_D^{27}$; +24.5°) in 88 mL of 95% ethanol. (This solution contains 0.062 mole excess (+)-amine, based on a $[\alpha]_D^{22}$ of +40.7° for the pure (+)-amine, and 0.041 mole racemic amine.) Heat the solution to boiling and add to it a solution of conc. sulfuric acid (3.2 grams of 98% H_2SO_4; 0.032 mole H_2SO_4) in 180 mL of 95% ethanol. ■ After allowing the mixture to cool slowly to room temperature, collect the crystalline (+)-amine sulfate by suction filtration and wash it thoroughly with ethanol. Yield: 7.8 grams (74%).

Isolate the free (+)-amine as described (treatment with sodium hydroxide, extraction, and distillation), using 40 mL of water and 5 mL of 50% sodium hydroxide for each 10 grams of the sulfate salt. Yield: 4.4 grams (59%); b.p., 184–186°C; $[\alpha]_D^{27}$, +38.3° (lit.: $[\alpha]_D^{22}$, +40.67°).

Notes

1. Sometimes a substance separates in the form of very fine needles. In this event, warm the mixture to dissolve the crystals, and allow it to cool again. The needle-like crystals dissolve much more rapidly than the prismatic crystals. If possible, seed the solution with prismatic crystals. Workup of the needle-like crystals gives α-phenylethylamine of a $[\alpha]_D^{25}$ of −19 to −21°.
2. As an alternative, the mother liquor can be decanted from the crystals and the crystals washed by adding a little methanol, swirling, and decanting again. If no second crop is to be taken, the experiment can be continued by adding the water and the base to these crystals in the original flask.
3. Sometimes there is a problem with foaming. The conditions that cause foaming are not known.
4. You will have to calculate the amount of conc. sulfuric acid that you will need as the amount required depends both upon the yield of the recovered amine and upon its optical purity. The directions that are given illustrate one possibility.

Time: 3 hours to recover the amine enriched in the (R)-(+)-isomer and to prepare the solution. Two hours to isolate the (+)-isomer.

Questions

1. What would be the result of the experiment if (S),(S)-(−)-tartaric acid were used in place of (R),(R)-(+)-tartaric acid?
2. If 0.100 mole "recovered amine" had $[\alpha]_D^{25} = +20.35°$:
   a. How many moles of (+)-amine and how many moles of (−)-amine would there be in the sample?
   b. How many moles of excess (+)-amine and how many moles of racemic amine would there be in the sample?
   c. How much sulfuric acid (both moles and grams) would be needed to convert the excess (+)-amine to its neutral sulfate salt?
3. a. Are the (+)-amine sulfate and (−)-amine sulfate salts of α-phenylethylamine enantiomeric or diastereomeric?
   b. Are their solubilities different in 95% ethanol?
   c. Why is it possible to isolate the (+)-amine in this experiment?
   d. What might happen if more sulfuric acid is used than the amount necessary to convert the excess (+)-amine to its neutral sulfate salt?
4. What might be the molecular composition of the needle-like crystals?
5. What experiments might be done in order to determine the conditions for foaming (see Note 3)?
6. How would you expect the infrared spectrum of either separate enantiomer of α-phenylethylamine to compare with that of the racemic mixture (shown in Figure 48.1)?

References

1. A. Ault, *J. Chem. Educ.* **42,** 269 (1965).
2. A. Ault, *Organic Syntheses,* Vol. 49, Wiley, New York, 1969, p. 93.

Transformations

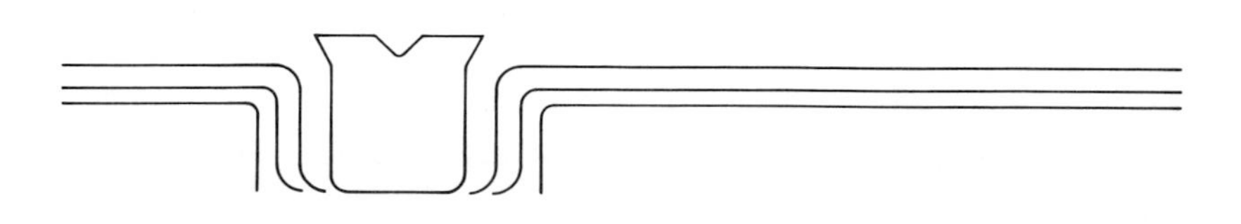

In most of these experiments, a substance is prepared by a single reaction. The experiments are grouped by type of reaction, and, in general, reactions typical of aliphatic compounds are presented before the reactions of aromatic compounds.

Often, several examples of each kind of reaction are set forth. By comparing the procedures, you can see which features are essential to that type of reaction and which are incidental to the particular compounds involved. Amounts of materials are given in moles, as well as in the units normally used to measure them, so that comparisons between procedures can be made more easily.

Occasionally, more than one procedure is given for the preparation of a particular compound. Sometimes your choice of procedure will be determined by the starting materials, time, or equipment available. At other times, however, experience and understanding will be used to decide which procedure, or combination of procedures, will provide the best results (yield, purity) with the least effort.

49 Isomerizations

These first two experiments are isomerizations, processes in which one substance is converted to a second substance that has the same molecular formula. In the first of these, one constitutional isomer of molecular formula $C_{10}H_{16}$ is converted to another, more stable, constitutional isomer of the same molecular formula (constitutional isomerism). In the second experiment, a more subtle change takes place: the starting materials and products differ only in molecular configuration (configurational isomerism).

49.1 ADAMANTANE FROM *endo*-TETRAHYDRODICYCLOPENTADIENE VIA THE THIOUREA CLATHRATE

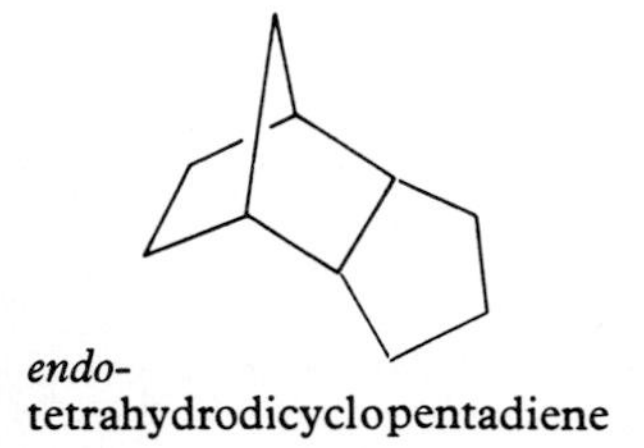

endo-tetrahydrodicyclopentadiene

$\downarrow$ $AlCl_2$

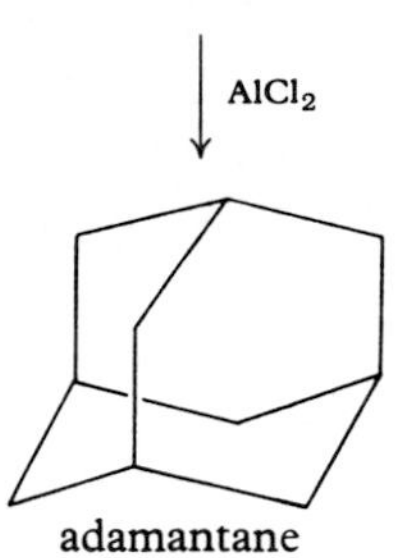

adamantane

In this experiment, adamantane is formed by the aluminum chloride–catalyzed rearrangement of *endo*-tetrahydrodicyclopentadiene. Because of its symmetry and freedom from angle strain, adamantane is the constitutional isomer of molecular formula $C_{10}H_{16}$ that would be expected to predominate under equilibrating conditions. The adamantane is recovered from solution in hexane by treatment with a methanolic solution of thiourea to form a crystalline inclusion complex in which the molecules of adamantane occupy channels in the crystal lattice of thiourea. When the inclusion complex is shaken with a mixture of water and ether, it breaks down. The thiourea dissolves in the water and the adamantane in the ether, from which it is recovered by evaporation.

The infrared spectrum of adamantane is shown in Figure 49.1.

Procedure (Reference 1). Weigh out 4 grams of aluminum chloride and store it until ready for use in a stoppered sample vial. Place 10.0 grams of *endo*-tetrahydrodicyclopentadiene in a 50-mL 19/22 Erlenmeyer flask, and fit the flask with a well-greased inner joint that will serve as an air condenser. Clamp the flask so that it can be heated by a small Bunsen burner flame. Heat the flask gently to melt the tetrahydrodicyclopentadiene. When it has melted, add about one-fourth of the aluminum chloride down the air condenser and suspend a thermometer in the mixture so that the bulb dips into the liquid as far as possible but does not touch the bottom of the flask. Adjust the flame of the burner so that the mixture is heated just barely to boiling (180–185°C) and is maintained in that condition. Add the remaining aluminum chloride down the condenser in 3 or 4 portions at about 5-minute intervals, and continue the heating for a total of 1 hour. During this time, aluminum chloride will sublime into the condenser tube. It should be pushed down into the reaction mixture from time to time with a spatula.

$$H_2N-\overset{\overset{\displaystyle S}{\|}}{C}-NH_2$$

thiourea

At the end of the heating period, remove the thermometer and allow the black mixture to cool for an hour. ■ During this time, prepare a solution of 10 grams of thiourea in 150 mL of methanol (Note 1).

At the end of the cooling period, extract the product from the flask

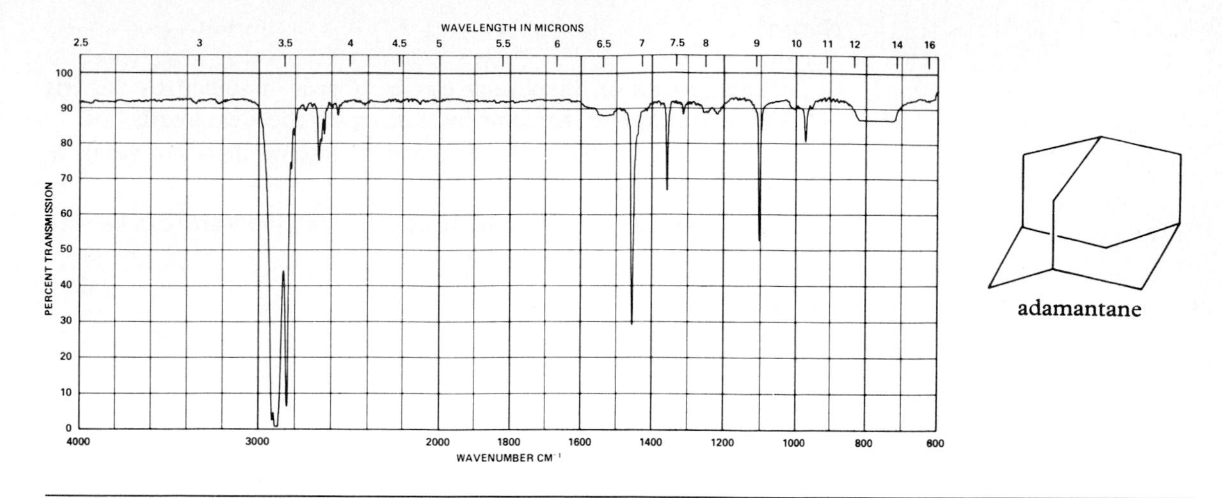

Figure 49.1. Infrared spectrum of adamantane; CCl_4 solution.

by adding hexane, swirling the mixture, and decanting, being careful not to pour out any of the black, tarry material; use a total of 70–80 mL of hexane in 4 or 5 portions. Add 1 gram of chromatography-grade alumina to the combined hexane extracts and swirl occasionally for about a minute. Filter the extract by gravity into the methanolic solution of thiourea, using 10–20 mL of hexane to rinse the flask and filter paper. Stir the two layers together for 2 or 3 minutes to permit the beautifully crystalline inclusion complex to complete its formation (Note 2). ■ Collect the complex by suction filtration and wash it with about 20 mL of hexane. The yield, after drying to constant weight, is 6–7 grams.

Transfer the crystals (which need not be dry) to a 125-mL separatory funnel and add 80 mL of water and 40 mL of ether. Shake the funnel vigorously for about 5 minutes, releasing the pressure occasionally, and then allow the layers to separate. If any solid remains, draw off as much of the lower layer as is convenient, add 50 mL more water, and shake vigorously until all the solid has disappeared (Note 3). Dry the ethereal extract over anhydrous magnesium sulfate and evaporate it to dryness in a tared (weighed) 50-mL Erlenmeyer flask. The last of the ether can be removed by keeping the flask on its side for about 20 minutes. The yield is 1.45–1.6 grams (14.5–16%) of material with a melting range of 258–265°C (sealed capillary).

The product can be recrystallized from isopropyl alcohol, using 13 mL per gram, with a recovery of about 60%, to give an overall yield of 0.93 gram (9.3%) of adamantane with a melting point of 268–270°C (sealed capillary).

Notes

1. The process of dissolution can be greatly hastened by stirring with a magnetic stirrer or by heating on the steam bath.
2. Saturation of a similar solution with hexane does not result in crystal formation.
3. If a larger separatory funnel is available, all the water can be used at once.

Time: 4 hours.

Reference

1. A. Ault and R. Kopet, *J. Chem. Educ.* **46,** 612 (1969).

49.2 *cis*-1,2-DIBENZOYLETHYLENE FROM *trans*-1,2-DIBENZOYLETHYLENE

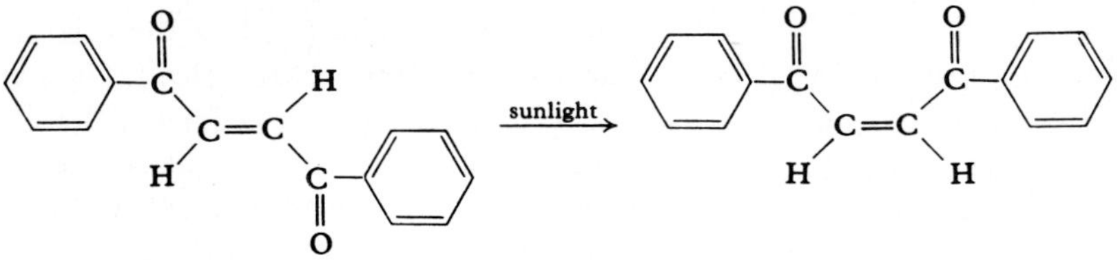

sunlight →

trans-1,2-dibenzoylethylene
m.p. 113–115°C; yellow

cis-1,2-dibenzoylethylene
m.p. 137–138°C; colorless

In this experiment, the colorless *cis* isomer can be prepared from the yellow *trans* isomer if the hot solution of the *trans* isomer is allowed to stand in the sun during recrystallization. If the solution is not placed in the sun, the *trans* isomer recrystallizes without change.

A detailed procedure for the preparation of *trans*-1,2-dibenzoylethylene is given in Reference 1.

Recrystallization of *trans*-1,2-Dibenzoylethylene

Procedure (Reference 2). Place 3 grams of *trans*-1,2-dibenzoylethylene in a 125-mL Erlenmeyer flask and add 100 mL of 95% ethanol. Heat the mixture almost to boiling on the steam bath and swirl the flask to dissolve the solid. Remove the flask from the heat and add as much decolorizing carbon as would cover a nickel. Filter the hot solution by gravity and collect the filtrate in another 125-mL Erlenmeyer flask.

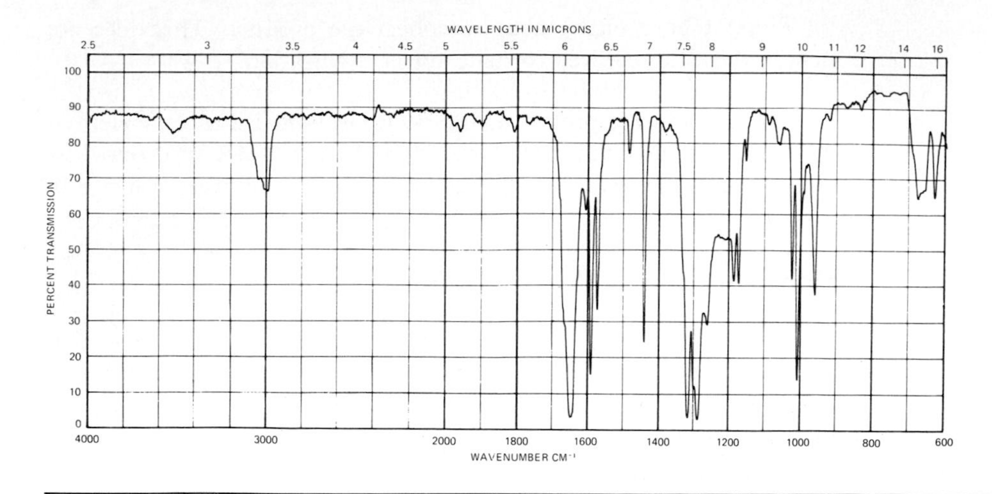

Figure 49.2. Infrared spectrum of *trans*-1,2-dibenzoylethylene; $CHCl_3$ solution.

Reheat the filtrate if necessary to dissolve any crystals that may have formed during the filtration. Set the flask aside to cool slowly. ■ After several hours the long, canary-yellow needles of the *trans* isomer may be collected by suction filtration, using a small amount of ethanol for rinsing and washing. Recovery: 2.2 grams (73%). Melting point: 113–115°C (Reference 2).

trans: yellow needles; m.p. = 113–115°

Figure 49.3. NMR spectrum of *trans*-1,2-dibenzoylethylene; $CDCl_3$ solution.

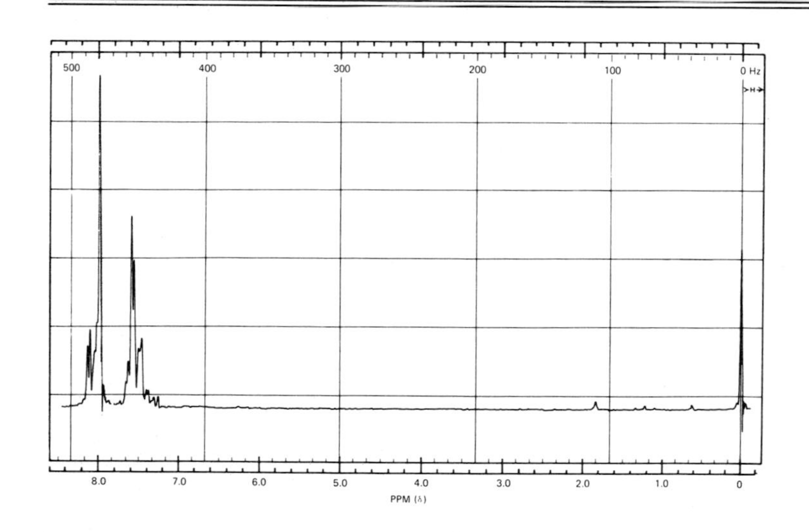

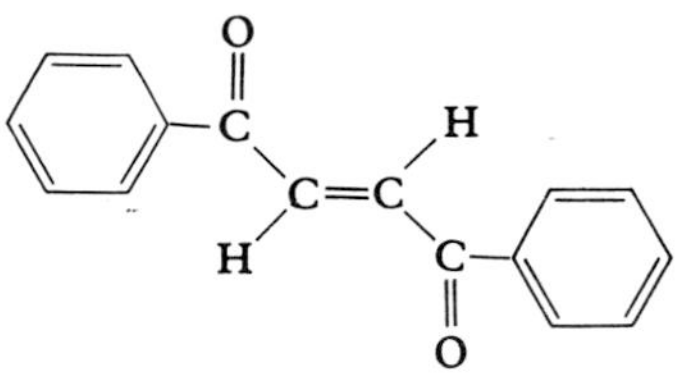

trans-1,2-dibenzoylethylene

Time: 1 hour plus $\frac{1}{2}$ hour to collect the product. This does not include the time required to stand for crystallization.

The infrared and proton NMR spectra of *trans*-1,2-dibenzoylethylene are shown in Figures 49.2 and 49.3. In the NMR spectrum, the singlet due to the resonance of the olefinic protons appears at $\delta = 8.0$.

Formation of *cis*-1,2-Dibenzoylethylene

Procedure (Reference 2). Prepare a solution of 3 grams of *trans*-1,2-dibenzoylethylene in 100 mL of 95% ethanol as described in the previous procedure. After removing the decolorizing carbon by gravity filtration and reheating to dissolve any crystals that may have formed during the filtration, place the loosely stoppered flask in bright sunlight (Note 1). ■ After several hours, very fine, colorless needles start to form (Note 2). When the flask appears to be about two-thirds full of crystals, they may be collected by suction filtration and washed with a little ethanol. Yield: 2.2 grams (73%). Melting point: 137–138°C (Reference 2).

cis: colorless needles; m.p. = 137–138°

Notes

1. A 275-watt sunlamp can be used instead of the sun (References 1 and 2).

Figure 49.4. Infrared spectrum of *cis*-1,2-dibenzoylethylene; $CHCl_3$ solution.

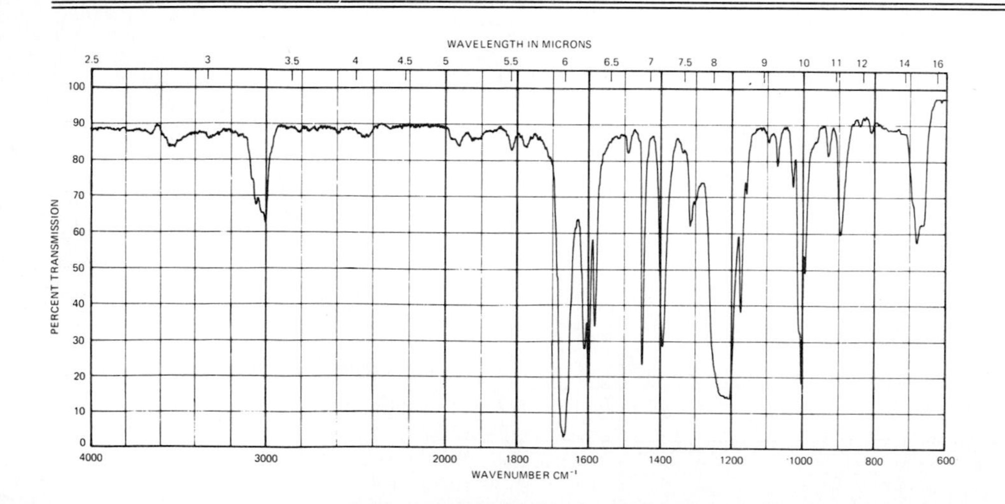

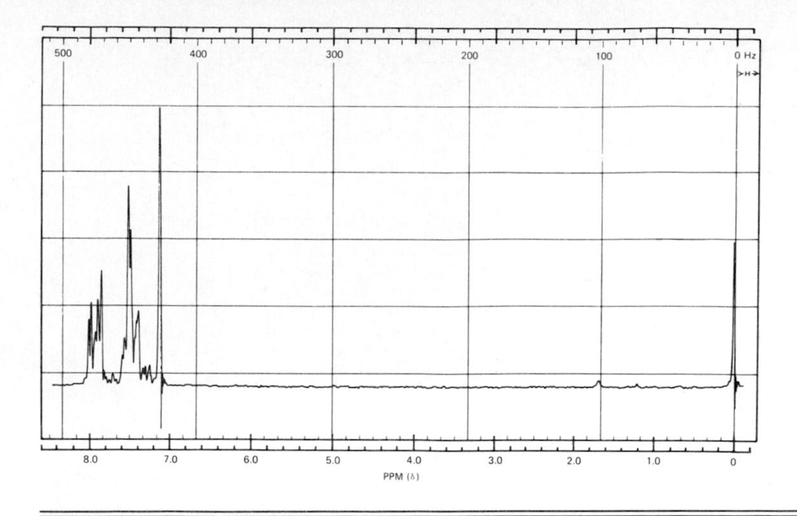

cis-1,2-dibenzoylethylene

Figure 49.5. NMR spectrum of *cis*-1,2-dibenzoylethylene; $CDCl_3$ solution.

2. If yellow needles of the *trans* isomer form, they will redissolve and give way to the colorless needles of the *cis* isomer on further irradiation.

Time: 1 hour plus $\frac{1}{2}$ hour to collect the product. This does not include the time required for irradiation and crystallization.

The infrared and proton NMR spectra of *cis*-1,2-dibenzoylethylene are shown in Figures 49.4 and 49.5. In the NMR spectrum, the singlet due to the resonance of the olefinic protons appears at $\delta = 7.1$.

References

1. D. J. Pasto, J. A. Duncan, and E. F. Silversmith, *J. Chem. Educ.* **51,** 227 (1974).
2. E. F. Silversmith and F. C. Dunson, *J. Chem. Educ.* **50,** 568 (1973).

50 Alkenes from Alcohols

Alkenes can often be prepared from alcohols when the alcohol is heated with a strong acid whose conjugate base is both a weak oxidizing agent and a weak base. The transformation begins in this way:

$$\mathrm{H{-}\overset{|}{\underset{|}{C}}{-}\overset{|}{\underset{|}{C}}{-}\ddot{\underset{..}{O}}{-}H + H{-}A \underset{}{\overset{\text{rapid}}{\rightleftarrows}} H{-}\overset{|}{\underset{|}{C}}{-}\overset{|}{\underset{|}{C}}{-}\overset{..+}{\underset{\underset{H}{|}}{O}}{-}H + A{:}^-}$$

Then Path 1 is taken:

$$\mathrm{H{-}\overset{|}{\underset{|}{C}}{-}\overset{|}{\underset{|}{C}}{-}\overset{..+}{\underset{\underset{H}{|}}{O}}{-}H \xrightarrow{\text{slow}} H{-}\overset{|}{\underset{|}{C}}{-}\overset{|}{\underset{|}{C}}{+} + H_2O \longrightarrow H^+ + {>}C{=}C{<}}$$

or Path 2:

$$\mathrm{H{-}\overset{|}{\underset{|}{C}}{-}\overset{|}{\underset{|}{C}}{-}\overset{..+}{\underset{\underset{H}{|}}{O}}{-}H \xrightarrow{\text{slow}} H^+ + {>}C{=}C{<} + H_2O}$$

Path 1 represents the reactions of tertiary alcohols and other alcohols that can form a relatively stable carbonium ion. Path 2 represents the reactions of primary alcohols.

Substitution versus elimination

If $A{:}^-$ is a good nucleophile (for example, Br^- or Cl^-), $A{:}^-$ can either compete successfully for the carbonium ion of Path 1 or displace H_2O from the conjugate acid of the alcohol of Path 2 to form the substitution product:

$$\mathrm{H{-}\overset{|}{\underset{|}{C}}{-}\overset{|}{\underset{|}{C}}{}^{+} + A^- \longrightarrow H{-}\overset{|}{\underset{|}{C}}{-}\overset{|}{\underset{|}{C}}{-}A}$$

$$\mathrm{H{-}\overset{|}{\underset{|}{C}}{-}\overset{|}{\underset{|}{C}}{-}\overset{+}{O}H_2 + A^- \longrightarrow H{-}\overset{|}{\underset{|}{C}}{-}\overset{|}{\underset{|}{C}}{-}A + H_2O}$$

Examples of reactions in which substitution predominates are given in Section 51 (cyclohexyl bromide from cyclohexanol), and Section 56 (alkyl halides from alcohols).

Oxidation

If $A{:}^-$ is a good oxidizing agent, the alcohol (or products derived from it) may undergo oxidation. In the experiment of Section 54, where sodium dichromate and sulfuric acid are present, oxidation of the secondary alcohol to the ketone is the predominant reaction. Sulfuric acid alone causes some oxidation when it is used to effect the dehydration of cyclohexanol.

50.1 CYCLOHEXENE FROM CYCLOHEXANOL

Apparatus: Fig. 50.1

Procedure (Reference 1). Introduce 20.0 g (21 mL; 0.20 mole) of cyclohexanol, 5 mL of 85% phosphoric acid (Note 1), and a boiling stone into a 50-mL boiling flask; swirl to mix the layers. Attach the flask to a fractionating column (Note 2) fitted with a distilling adapter and condenser (Figure 50.1), and distill until the volume of the residue

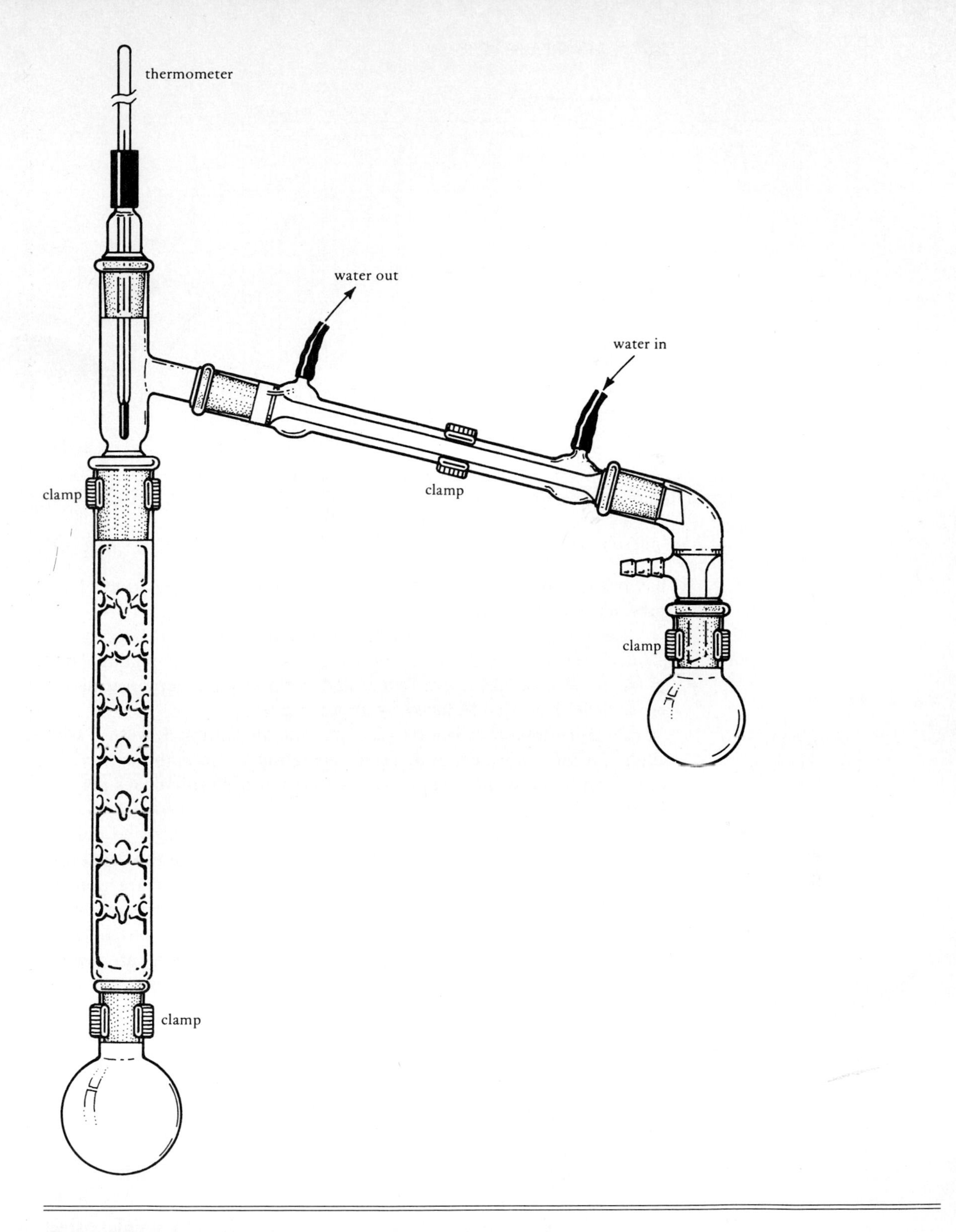

Figure 50.1. Apparatus for the dehydration of cyclohexanol to cyclohexene.

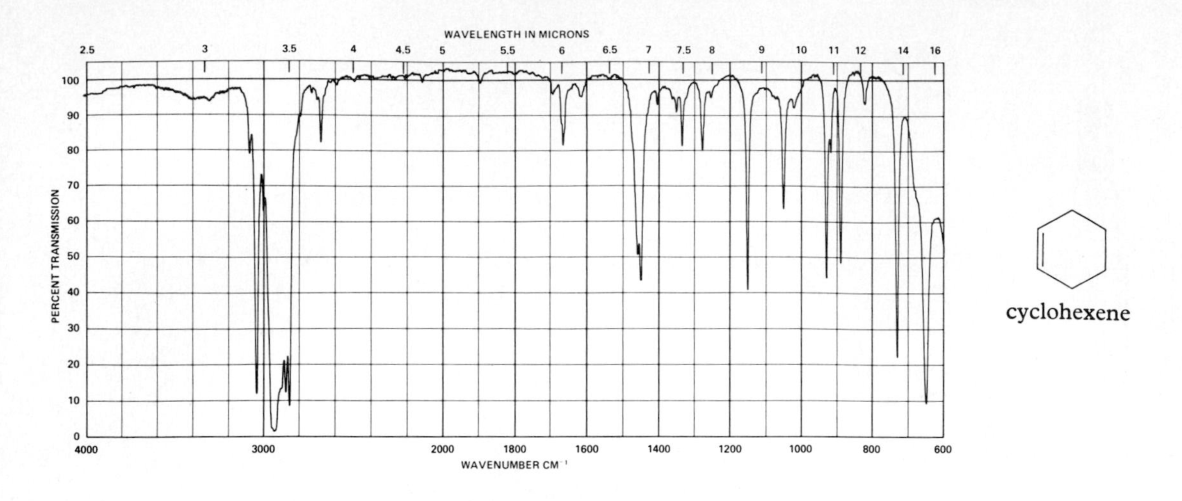

Figure 50.2. Infrared spectrum of cyclohexene; thin film.

OH

cyclohexanol

H_3PO_4 | distill

+ H_2O

cyclohexene

has been reduced to 5–10 mL. To reduce loss by evaporation of the product, cool the receiver in an ice bath.

Let the boiling flask cool a little and then pour 20 mL of xylene (Note 3) down through the fractionating column into the boiling flask. Note the size of the upper layer, and continue distillation until the upper layer has been reduced by about one-half.

Pour the total distillate into a small separatory funnel, wash it with about 20 mL of water, and, after separating it from the water and decanting into a small Erlenmeyer flask, dry it over anhydrous magnesium sulfate. ■

While waiting for the distillate to dry, clean and dry the boiling flask, fractionating column, and adapters. After filtering the dried distillate into the boiling flask, reassemble the apparatus and fractionally distill the product, taking precautions against losses by evaporation. A typical yield is about 13 grams.

Analyze your product

Gas chromatography (Note 4) and infrared spectrometry are recommended as methods of determining the purity of your product with respect to starting material and chaser solvent.

The infrared spectrum of the product of this transformation is shown in Figure 50.2.

Notes

1. Sulfuric acid can be used in place of phosphoric acid, but it causes some oxidation of the organic materials.
2. A Vigreux column or a tube packed loosely with stainless steel sponge is suitable.

3. Xylene functions as a "chaser solvent." That is, since it boils considerably higher than the compound of interest (in this case, cyclohexene), it serves to displace the product from the boiling flask and distilling column in the first distillation, and thus serves to reduce losses due to holdup. It is fairly easy to contaminate the product with chaser solvent in the final distillation unless it is done slowly, especially toward the end, and with even heating.
4. Suitable conditions for gas chromatographic analysis are column, Apiezon L; column temperature, 180°C; helium pressure, 25 lb; flow rate, 120 mL/min. Under these conditions, the analysis takes 2 minutes.

Time: 3 hours.

Questions

1. Why do you suppose that the dehydration of cyclohexanol to cyclohexene has been the traditional experiment to illustrate the acid-catalyzed dehydration of an alcohol to an alkene?
2. Suppose 2-methylcyclohexanol were subjected to the conditions of this experiment instead of cyclohexanol.
   a. What alkenes would you expect to be formed?
   b. In what relative amounts would you expect them to be formed?

Figure 50.3. NMR spectrum of a mixture of alkenes produced by boiling 3,3-dimethyl-2-butanol with 85% H_3PO_4. The peak at $\delta = 1.6$ was recorded at one-tenth the amplitude used to record the rest of the spectrum.

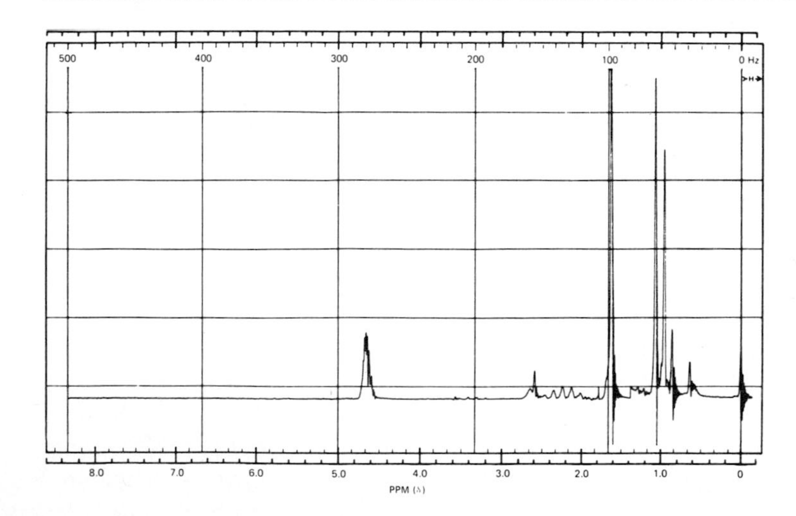

3. When 3,3-dimethyl-2-butanol is boiled with 85% H_3PO_4 according to the procedure of this section, an elimination reaction takes place. If a carbonium ion mechanism is involved and the initial carbonium ion undergoes elimination before rearrangement, the product should contain only 3,3-dimethyl-1-butene:

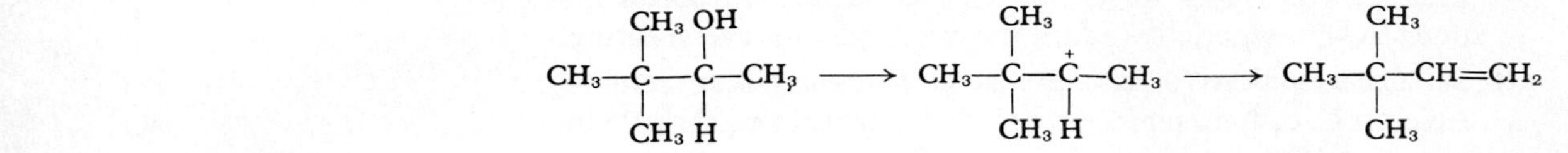

$$CH_3-C(CH_3)_2-CH(OH)-CH_3 \longrightarrow CH_3-C(CH_3)_2-\overset{+}{C}H-CH_3 \longrightarrow CH_3-C(CH_3)_2-CH{=}CH_2$$

3,3-dimethyl-1-butene

If the initial carbonium ion should rearrange by methyl migration, two other alkenes could be formed:

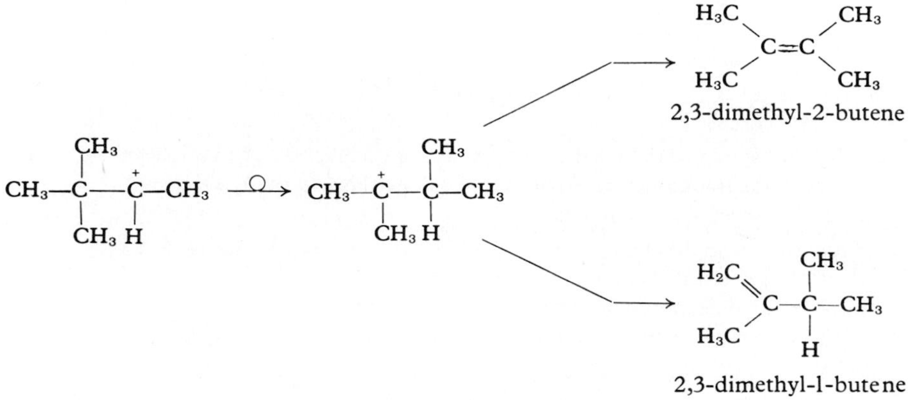

$$CH_3-C(CH_3)_2-\overset{+}{C}H-CH_3 \longrightarrow CH_3-\overset{+}{C}(CH_3)-CH(CH_3)-CH_3$$

$$\longrightarrow (H_3C)_2C{=}C(CH_3)_2$$

2,3-dimethyl-2-butene

$$\longrightarrow H_2C{=}C(H_3C)-CH(CH_3)-CH_3$$

2,3-dimethyl-1-butene

Figure 50.3 shows the proton NMR spectrum of the product mixture obtained in this elimination reaction. The integral of the multiplet at $\delta = 4.6$ is 4.2 units, and the integral of the remainder of the spectrum is 124 units. From these data, estimate the relative amounts of the three products formed and interpret the results in terms of the relative stability of the products and the mechanism of the reaction.

Reference

1. L. F. Fieser, *Experiments in Organic Chemistry*, 3rd edition, revised, Heath, Boston, 1955, p. 62.

50.2 A VARIATION: THE DEHYDRATION OF 2-METHYLCYCLOHEXANOL

Procedure. In the *Journal of Chemical Education* **44,** 620 (1967), Taber and Champion reported that 2-methylcyclohexanol can be dehydrated under conditions that are essentially the same as those that serve to convert cyclohexanol to cyclohexene in the preceding experiment (20 grams of compound boiled with 5 mL of 85% phosphoric acid). They did not suggest the use of a chaser solvent, and said that the rate of heating should be such that the temperature of the distillate does not rise much above 95°C, since higher temperatures will result in distillation of too much unreacted alcohol. After distillation has become very slow, they suggest that the distillate be worked up by washing it with water, 10% sodium bicarbonate, and finally with water again. It should then be dried over a little anhydrous calcium chloride and filtered.

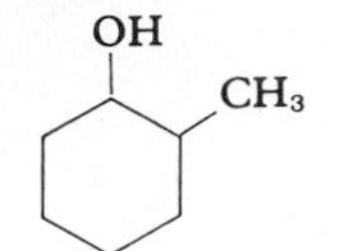

2-methylcyclohexanol

H_3PO_4 ↓ distill

alkenes

The composition of the product is then determined by gas chromatography (Sections 14.7 and 14.8). The authors suggest the use of a silicone grease column at 75°C.

As an alternative to gas chromatographic analysis, they say that one can determine the refractive index of the mixture and estimate the composition by assuming a linear relationship between molar concentration and refractive index.

Time: 3 hours.

Questions

1. What are the most likely products of this reaction?
2. a. How many stereoisomers can there be of 2-methylcyclohexanol?
   b. To what extent would you expect that these isomers could be separated by fractional distillation or gas chromatography? Explain.
   c. Which isomers would be expected to give exactly the same results in this experiment if each could be tested separately? Explain.
3. In the gas chromatographic analysis of the product of this reaction,
   a. How do you determine what peak corresponds to which compound?
   b. Does it matter whether you assume that the ratio of peak areas corresponds to the weight ratio or the mole ratio of the alkenes?
   c. How can one demonstrate that the ratio of peak areas is a valid measure of product composition?
4. In the analysis of the product by index of refraction,
   a. How can one determine the validity of the assumption of a linear relationship between molar concentration and index of refraction?
   b. Would you expect a linear relationship between weight concentration and index of refraction?
   c. What assumption has been made about the number of components in the product mixture?
5. How can you tell whether the relative amounts of alkenes found in the product of this experiment are determined by their relative rates of formation or by their relative stability?

51 Cyclohexyl Bromide from Cyclohexanol

OH

cyclohexanol

NaBr | H_2SO_4

Br

cyclohexyl bromide

A general procedure that serves to convert an alcohol to the corresponding alkyl chloride or bromide is treatment of the alcohol with a solution that is strongly acidic and contains a high concentration of halide ion. A number of reagents have been used, including a solution containing conc. sulfuric acid and sodium bromide (as in this experiment and the experiment of Section 56.1, *n*-butyl bromide from *n*-butyl alcohol), and conc. hydrochloric acid (Section 56.3, *tert*-butyl chloride from *tert*-butyl alcohol).

The course of the reaction is usually represented by these equations:

$$\mathrm{H{-}\overset{|}{\underset{|}{C}}{-}\overset{|}{\underset{|}{C}}{-}\ddot{\underset{..}{O}}{-}H + H{-}X \xrightleftharpoons{\text{rapid}} H{-}\overset{|}{\underset{|}{C}}{-}\overset{|}{\underset{|}{C}}{-}\overset{..+}{\underset{\underset{H}{|}}{O}}{-}H}$$

Then Path 1:

$$\mathrm{H{-}\overset{|}{\underset{|}{C}}{-}\overset{|}{\underset{|}{C}}{-}\overset{..+}{\underset{\underset{H}{|}}{O}}{-}H \xrightarrow{\text{slow}} H{-}\overset{|}{\underset{|}{C}}{-}\overset{|}{\underset{|}{C}}{+} + H_2O \xrightarrow{+X:^-} H{-}\overset{|}{\underset{|}{C}}{-}\overset{|}{\underset{|}{C}}{-}X}$$

or Path 2:

$$\mathrm{H{-}\overset{|}{\underset{|}{C}}{-}\overset{|}{\underset{|}{C}}{-}\overset{..+}{\underset{\underset{H}{|}}{O}}{-}H + X:^- \xrightarrow{\text{slow}} H{-}\overset{|}{\underset{|}{C}}{-}\overset{|}{\underset{|}{C}}{-}X + H_2O}$$

Path 1 represents the reaction of tertiary alcohols and other alcohols that can form a relatively stable carbonium ion, whereas Path 2 represents the bimolecular substitution reaction to be expected of primary alcohols. Presumably, the attack of the halide ion as a nucleophile on either the carbonium ion in Path 1 or the conjugate acid of the alcohol in Path 2 is sufficiently fast to compete successfully with elimination to form the alkene.

The infrared spectrum of cyclohexyl bromide is shown in Figure 51.1.

Procedure. Place a magnetic stirring bar in a 100-mL boiling flask, and add to the flask 20 mL of water, 20 g (0.19 mole) of sodium bromide, and 10 g (10.5 mL; 0.10 mole) of cyclohexanol. Fit the flask with a Claisen adapter and put the bottom joint of the separatory funnel in the center joint of the adapter. Add 20 mL (37 g; 0.37 mole) of conc. sulfuric acid to the separatory funnel.

Support the flask over the magnetic stirring motor, and adjust the speed of the motor so that the contents of the flask are stirred quite vigorously. While the contents of the flask are being stirred, add the sulfuric acid drop by drop but otherwise as rapidly as possible; the addition should take about 2 minutes.

After completing the addition of the acid, let the mixture stir for 5

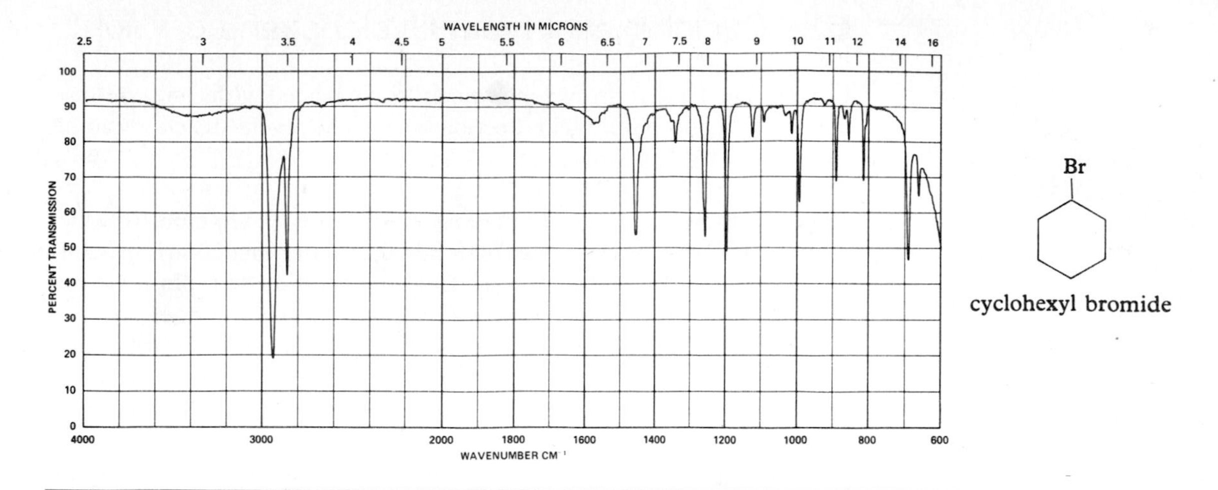

Figure 51.1. Infrared spectrum of cyclohexyl bromide; thin film.

more minutes. At this time, pour the hot (Note 1) mixture into the separatory funnel and, after allowing sufficient time for separation of the layers, draw off the aqueous phase. To the cyclohexyl bromide, which remains in the separatory funnel, add a solution of about 5 grams potassium carbonate in about 50 mL of water. Mix the layers thoroughly, and, after separating the layers, dry the crude cyclohexyl bromide over a mixture of anhydrous potassium carbonate and anhydrous magnesium sulfate for about 10 minutes.

Warning: see Note 1!

Filter the crude product from the drying agent into a 25-mL boiling flask, add a pinch of anhydrous potassium carbonate, and distill (it is not necessary to use a condenser, Figure 9.17). Collect as product the material that boils between 155 and 165°C. Yield: about 9 grams.

Note

1. Be very careful when working with the hot reaction mixture, as it contains sulfuric and hydrobromic acids: it will cause painful burns if it is spilled on your hands, and it will burn holes in your clothes.

Warning!

Time: 3 hours.

Question

1. It was assumed in the discussion of the mechanism of this reaction that cyclohexyl bromide is not formed via elimination to give cyclohexene followed by addition of hydrogen bromide to the cyclohexene.
   a. Is there any evidence to support this assumption?
   b. What kinds of experiments might be done to test this assumption?

52 Cyclohexanol from Cyclohexene

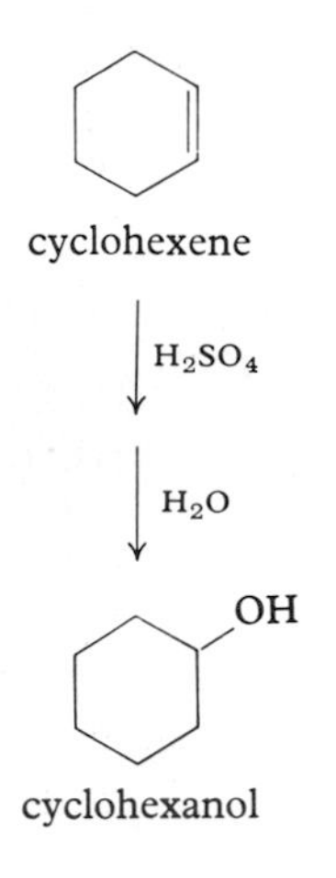

Cyclohexanol will form cyclohexene on being heated with conc. sulfuric acid or 85% phosphoric acid (Section 50.1). The reverse reaction can be carried out by this procedure.

Procedure (Reference 1). Cautiously add 7.0 mL (0.126 mole) of conc. sulfuric acid to 3.4 mL (3.4 g; 0.19 mole) of water in a 50-mL ground-glass-stoppered Erlenmeyer flask. Cool the solution to room temperature. Add 10.1 mL (8.2 g; 0.10 mole) of cyclohexene. Stopper the flask and shake to mix; the mixture should be shaken or stirred until a clear homogeneous solution is formed (Note 1 ■). At this point, pour the mixture into a 250-mL boiling flask and rinse the Erlenmeyer flask with a total of about 120 mL of water, adding the rinsings to the boiling flask. Distill the mixture until 50 or 60 mL has been collected. Saturate the distillate with sodium chloride and separate the cyclohexanol by extracting with ether. Dry the ethereal extract over anhydrous potassium carbonate, filter, and distill the ether on the steam bath. Distill the residue and collect the fraction boiling between 155–162°C as cyclohexanol.

Note

1. Some time can be saved by allowing the mixture to stand between laboratory periods.

Time: 3 hours.

The infrared spectrum of cyclohexanol is shown in Figure 52.1.

Questions

1. Write balanced equations for the reactions that take place in this experiment.
2. At which point in the experiment does each reaction take place?
3. Compare this procedure with that for the conversion of cyclohexanol to cyclohexene (Section 50.1). Explain why the reactions go in opposite directions under the different conditions.

Reference

1. G. H. Coleman, S. Wawzonek, and R. E. Buckles, *Laboratory Manual of Organic Chemistry*, 2nd edition, Prentice-Hall, Englewood Cliffs, N.J., 1962, p. 52.

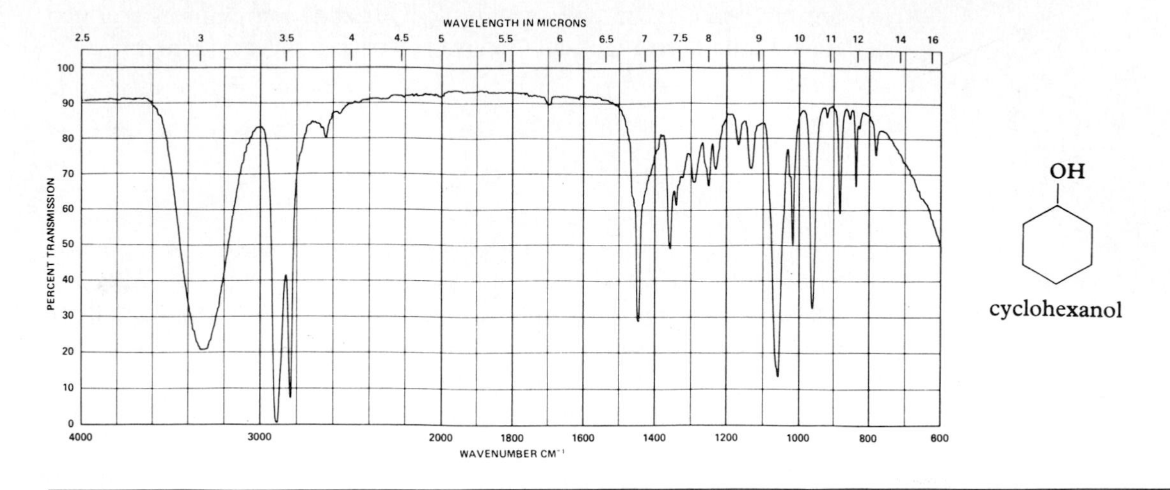

Figure 52.1. Infrared spectrum of cyclohexanol; thin film.

53 Addition of Dichlorocarbene to Alkenes by Phase Transfer Catalysis

Cl
:C
Cl
dichlorocarbene

One method for the formation of a cyclopropane ring is the addition of a carbene to a carbon–carbon double bond. If the carbene bears substituents, such as a pair of chlorine atoms, a substituted cyclopropane ring can be formed, and the addition of dichlorocarbene to an alkene can be represented in general as

alkene + dichlorocarbene (:CCl_2) → dichlorocyclopropane

Although carbenes are uncharged, the valence shell of the carbon atom is electron deficient since it contains only six electrons. For this reason, carbenes act as powerful electrophiles. In order to avoid reaction with water, a nucleophile, carbene reactions are usually run under anhydrous conditions. In this procedure, however, dichlorocarbene is generated in a water-free environment in an unusual way. The reaction mixture consists of two phases: an aqueous phase containing sodium hydroxide and a catalytic amount of tetra-*n*-butylammonium bromide, and an organic phase containing chloroform and cyclohexene. Apparently tetra-*n*-butylammonium ions and hydroxide ions migrate into the organic phase as ion pairs where the hydroxide ion removes a proton from a chloroform molecule to give the trichloromethyl anion:

$$HO^- + H{-}CCl_3 \longrightarrow H_2O + {:}C^-Cl_3$$

The trichloromethyl anion then undergoes an alpha elimination, in which a chloride ion is lost, to form dichlorocarbene:

$$^-{:}CCl_3 \longrightarrow {:}CCl_2 + Cl^-$$

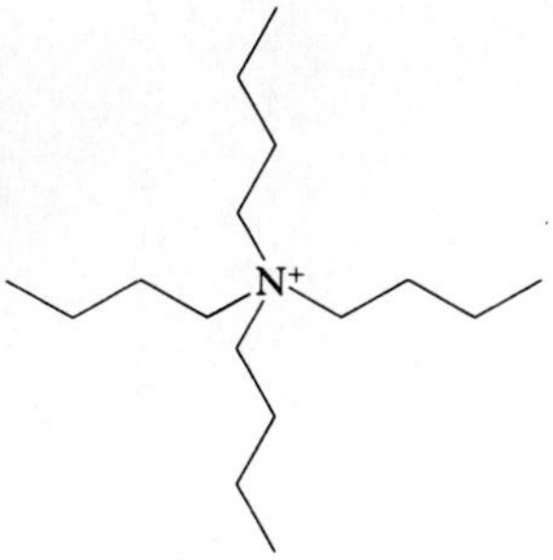

The dichlorocarbene, formed in the organic layer where the water concentration is very low and the cyclohexene concentration very high, will usually react to give dichloronorcarane. The chloride ion formed and the tetra-*n*-butylammonium ion then migrate as an ion pair into the water layer where the ammonium cation can again pick up a hydroxide ion to take into the organic layer. The catalytic role of the tetra-*n*-butylammonium cation can be summarized in this way:

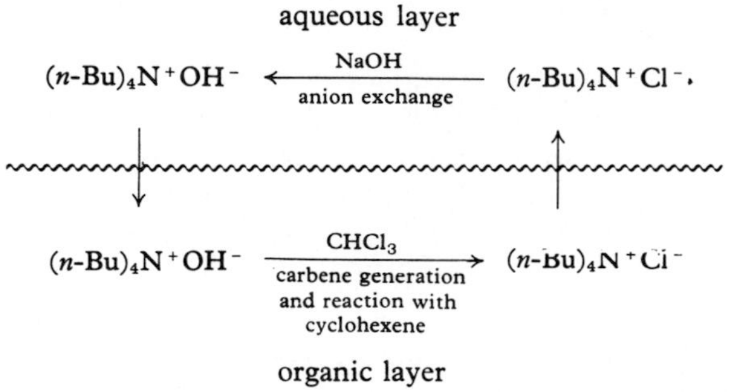

N+

tetra-*n*-butylammonium cation

aqueous layer

$(n\text{-Bu})_4N^+OH^- \xleftarrow[\text{anion exchange}]{\text{NaOH}} (n\text{-Bu})_4N^+Cl^-$

$(n\text{-Bu})_4N^+OH^- \xrightarrow[\text{carbene generation and reaction with cyclohexene}]{CHCl_3} (n\text{-Bu})_4N^+Cl^-$

organic layer

The catalytic function of the tetra-*n*-butylammonium salt in this reaction has been called *phase transfer catalysis*. Other quaternary ammonium salts have the same effect, and no reaction takes place in the absence of the salt.

The first experiment in this section gives a procedure for the addition of dichlorocarbene to cyclohexene, and the two variations that follow suggest that a similar procedure could be developed for the addition of dichlorocarbene to styrene and to 1,5-cyclooctadiene.

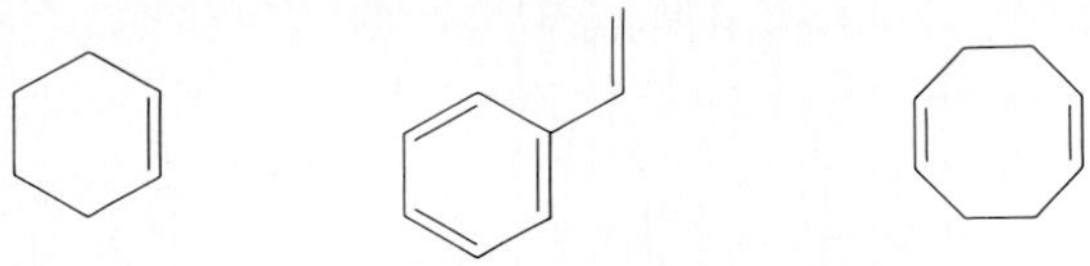

cyclohexene   styrene   1,5-cyclooctadiene

53.1 ADDITION OF DICHLOROCARBENE TO CYCLOHEXENE

The addition of dichlorocarbene to cyclohexene gives 2,2-dichlorobicyclo[4.1.0]heptane, or dichloronorcarane. The reaction can be represented in this way:

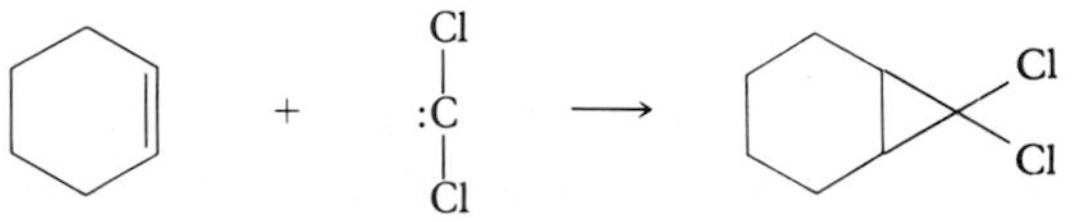

cyclohexene + dichlorocarbene ($:CCl_2$) ⟶ dichloronorcarane

The infrared spectrum of the product, dichloronorcarane, is shown in Figure 53.1.

Procedure. Put a magnetic stirring bar in a 250-mL round-bottom boiling flask. Place a steam bath on a magnetic stirrer and clamp the flask in position in the rings of the steam bath so that the contents of the flask can be heated by steam and magnetically stirred at the same time. Add to the flask 19 mL of chloroform (27.5 grams; 230 mmole), 7.7 mL of cyclohexene (6.2 grams; 75 mmole), and 250 mg (0.78 mmole) of tetra-*n*-butylammonium bromide. Fit the flask with a reflux condenser and then pour, in 1 portion, down the condenser into the flask, a mixture of 30 mL of 50% sodium hydroxide (570 mmole) and 22 mL of water. Gently heat and vigorously stir the resulting mixture for 30 minutes (Note 1). At the end of the heating period, add ice in order to cool the mixture to room temperature. Then add to the flask 50 mL of ether and vigorously stir the resulting mixture for three minutes (Note 2). Then transfer the mixture to a separatory funnel, allow the layers to separate, and draw off the lower, aqueous, layer, returning it to the original 250-mL boiling flask. Save the ether extract in a 125-mL Erlenmeyer flask. Add 25 mL of ether to the aqueous layer in the boiling flask and again vigorously stir the mixture for three minutes. Again, using the separatory funnel, separate the layers and then dry the combined ether extracts over anhydrous magnesium sulfate. Filter the dried ethereal solution by gravity into a round-bottom

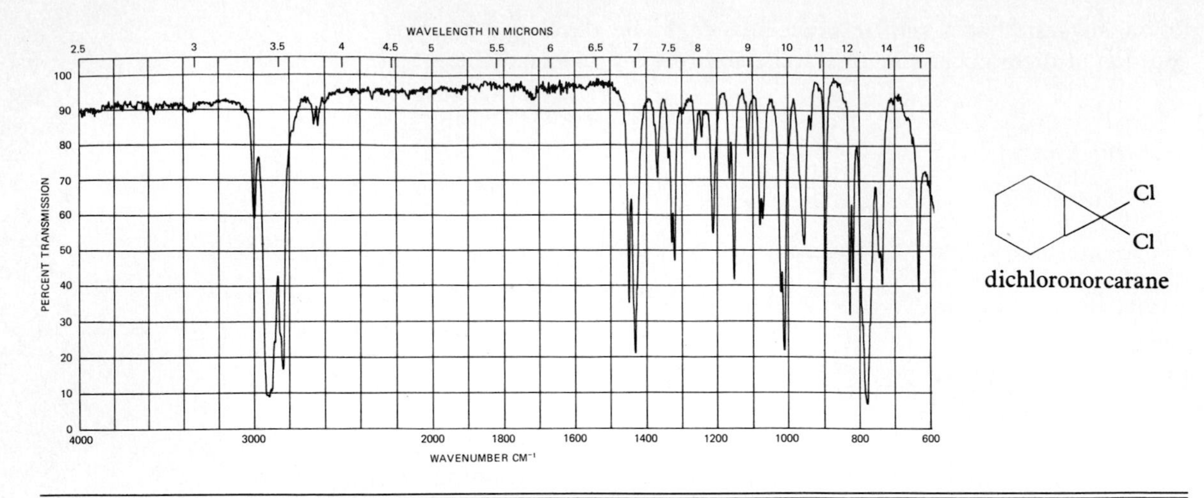

Figure 53.1. Infrared spectrum of dichloronorcarane; thin film.

boiling flask and remove the ether by distillation on the steam bath. Distill the residue at atmospheric pressure using a flame. Collect the fraction boiling between 192 and 197°C as dichloronorcarane. Yield: about 7 grams (about 55%).

Notes

1. Heat the mixture just to boiling so that an occasional drop falls from the reflux condenser.
2. The yield is critically dependent on the thoroughness of the extraction.

Time: 3 hours.

Question

1. What product(s) would you expect to get by a similar treatment of
   a. styrene?
   b. 1,5-cyclooctadiene?

References

1. A. Ault and B. Wright, *J. Chem. Educ.* **53,** 486 (1976).
2. M. Makosza and M. Wawrzyniewicz, *Tetrahedron Lett.* 4659 (1969).
3. G. C. Joshi, N. Singh, and L. M. Pande, *Tetrahedron Lett.* 1461 (1972).
4. K. Isagawa, Y. Kimura, and S. Kwon, *J. Org. Chem.* **39,** 3171 (1974).

53.2 A VARIATION: ADDITION OF DICHLOROCARBENE TO STYRENE

The addition of dichlorocarbene to styrene should give 1,1-dichloro-2-phenylcyclopropane:

styrene + :CCl_2 ⟶ 1,1-dichloro-2-phenylcyclopropane

styrene	1,1-dichloro-2-phenylcyclopropane
m.w. = 104	b.p. = 103°/10 Torr.

An obvious way to attempt to prepare 1,1-dichloro-2-phenylcyclopropane is to use the procedure of Section 53.1, substituting an equimolar amount of styrene for the cyclohexene.

As the equation implies, the "double bonds" of the aromatic ring of styrene are expected to be inert to dichlorocarbene.

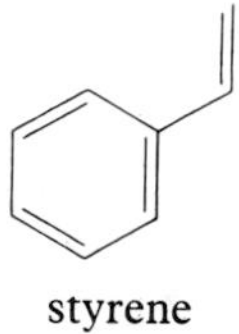
styrene

53.3 ANOTHER VARIATION: ADDITION OF DICHLOROCARBENE TO 1,5-CYCLOOCTADIENE

When dichlorocarbene is generated in the presence of 1,5-cyclooctadiene, *two* equivalents of the carbene should be consumed per mole of the diene, with formation of a *bis*-adduct:

1,5-cyclooctadiene + 2 :CCl_2 ⟶ *bis*-adduct

Thus, in adapting the procedure of Section 53.1 to the use of 1,5-cyclooctadiene, the cyclohexene should be replaced by only half as many moles of the diene.

The reaction of two equivalents of dichlorocarbene with 1,5-cyclooctadiene can, in principle, take place in two ways. Both reactions can take place at, say, the top face of the ring, to form a *cis-bis*-adduct, or one reaction can take place at the top face and the other at the bottom face, to form a *trans-bis*-adduct.

cis-bis-adduct	*trans-bis*-adduct
m.p. = 176°	m.p. = 230°

diastereomers

Another way to view these possibilities is to consider the addition of the second equivalent of dichlorocarbene to the product of the addition of the first. The second equivalent can attack either the same face as that attacked by the first, or the face opposite that attacked by the first.

Cl Cl H H H H

attack at diastereotopic faces to give diastereomeric products

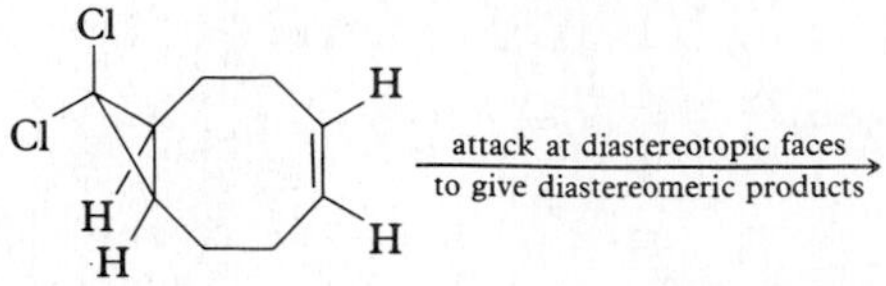

Since the diastereomeric products have quite different melting points, it should be easy to determine which isomer is formed in the reaction.

A suggestion

A suggestion: After the ether that was used for the extraction has been removed by heating on the steam bath, continue to heat the residue, using the steam bath and wrapping a towel around the flask so that it forms a kind of tent to contain the steam, until sufficient volatile material has been driven off that the residue starts to crystallize. Then cool the flask in cold water to complete the crystallization of the residue. Wash the crude product by adding 5 mL of methanol, crushing the lumps, and decanting the liquid. Repeat the washing with a second 5-mL portion of methanol. Recrystallize the crude product from ethyl acetate and determine its melting point.

54 Cyclohexanone from Cyclohexanol

One of the common methods for the preparation of ketones is the oxidation of a secondary alcohol by a sulfuric acid solution of a chromate or dichromate salt:

$$3\ R{-}\underset{H}{\overset{OH}{C}}{-}R + H_2Cr_2O_7 + 6\ H^+ \rightarrow 3\ R{-}\overset{O}{\overset{\|}{C}}{-}R + 2\ Cr^{+++} + 7\ H_2O$$

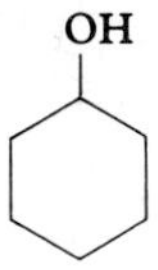

OH

cyclohexanol

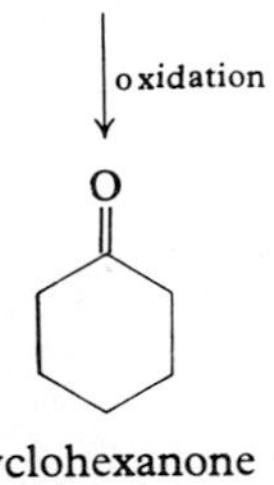

oxidation

O

cyclohexanone

Chromate salts are relatively expensive and may be somewhat hazardous. Therefore, an alternative procedure that uses sodium hypochlorite as the oxidizing agent has been presented by chemists at Loyola College, Baltimore, Maryland, in *The Journal of Chemical Education* (Reference 2). In this experiment, Procedure a is an example of a chromic acid oxidation, and Procedure b illustrates the hypochlorite method. The infrared spectrum of the product is shown in Figure 54.1.

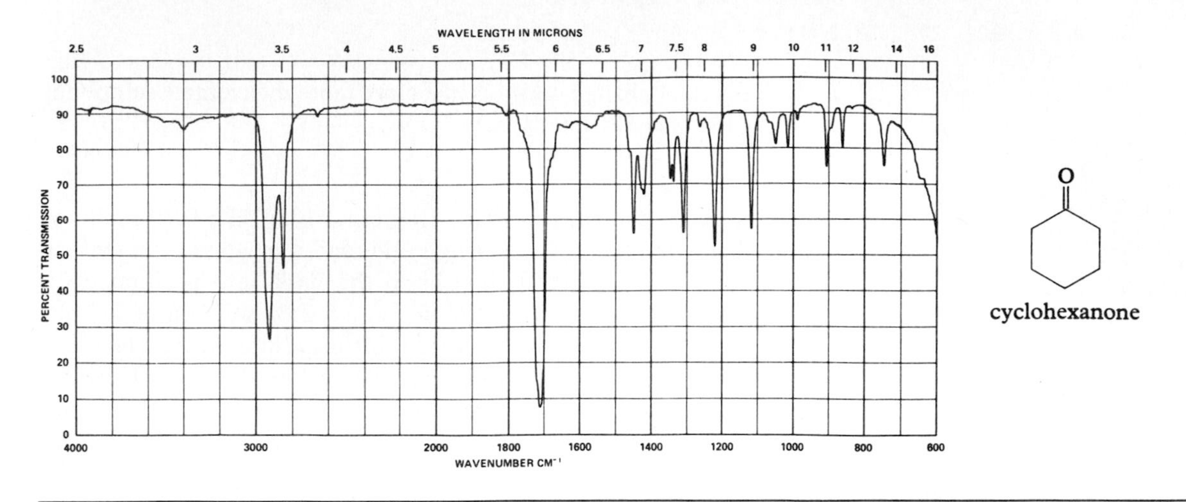

Figure 54.1. Infrared spectrum of cyclohexanone; thin film.

Procedure a (Reference 1). Cautiously add 20 mL (37 g; 0.37 mole) of conc. sulfuric acid to 60 grams of crushed ice. After mixing thoroughly, pour the solution into a 250-mL boiling flask and add a stirring bar. Fit the flask with a Claisen adapter, and put the bottom joint of a separatory funnel in the outer neck of the adapter. Add 20 g (21.0 mL; 0.20 mole) of cyclohexanol to the flask and suspend a thermometer in the center neck of the adapter so that the bulb is entirely in the liquid, but not in the way of the stirring bar. Don't make a closed system! Place a solution of 21 g (0.070 mole) of sodium dichromate dihydrate in 10 mL of water in the separatory funnel. While the mixture is being stirred, add about 1 mL of the dichromate solution from the separatory funnel. The temperature should increase, and the color should change from orange to green (Note 1). Add the remainder of the dichromate solution at such a rate that the temperature remains between 25 and 35°C. After the addition is complete (30–40 minutes), continue stirring until the temperature spontaneously falls 1 or 2 degrees. At this time, add about 1 gram of oxalic acid in order to reduce any excess dichromate.

Make sure the reaction starts

Now remove the thermometer, and, after rinsing the separatory funnel with water, add 100 mL of water to it and place it in the center neck of the Claisen adapter. Fit the other neck of the Claisen adapter with a distilling adapter and condenser. Distill the reaction mixture until about 70–90 mL has been collected, running in water from the separatory funnel to replace the material removed by distillation (see Section 11). ■

"Salting out"

Saturate the distillate with sodium chloride (about 15–18 grams), and extract it with ether. Dry the ethereal extract over anhydrous magnesium sulfate, filter and distill. Collect as cyclohexanone the portion boiling between 150 and 155°C.

Note

1. This color change must occur before more dichromate solution is added. See Section 31.

Procedure b (Reference 2). Place 10 grams (10.4 mL; 0.10 mole) of cyclohexanol, 25 mL of glacial acetic acid, and a magnetic stirring bar in a 250-mL Erlenmeyer flask. Clamp the flask into position on a magnetic stirring motor. Place 157 mL of a 5.25% solution of sodium hypochlorite (0.11 mole NaOCl; Note 1) in an addition funnel (Note 2), and clamp the funnel into position over the neck of the 250-mL Erlenmeyer flask so that the bleach solution can be gradually added while the mixture in the flask is stirred by the magnetic stirrer. Position a thermometer in the Erlenmeyer flask so that the bulb touches the bottom of the flask but does not interfere with the magnetic stirring bar. While the solution is magnetically stirred, add the bleach at such a rate that the temperature of the reaction mixture is maintained between 30 and 35°C.

After the addition is complete, test the solution with potassium iodide–starch test paper (Note 3). If the test is *not* positive (*not* blue), add an additional 5 mL of bleach and stir the solution at room temperature for another 15 minutes; an excess of bleach solution should then be present, as indicated by a greenish-yellow color, and the oxidation of all the alcohol is thereby assured. Reduce the excess bleach by adding concentrated sodium bisulfite until the greenish-yellow color is gone and the solution no longer gives a positive potassium iodide–starch test (Note 3; between 1 and 5 mL will be needed).

Bleach should be in excess

Transfer the contents of the flask to a 500-mL boiling flask, and distill the solution until about 50 mL of distillate has been collected in a 125-mL Erlenmeyer flask (Note 4). Add about 7 grams of anhydrous sodium carbonate to the distillate in order to neutralize the acetic acid that has distilled along with the cyclohexanone and water; stir the mixture until the sodium carbonate has all dissolved.

Steam distillation: Section 11

Transfer this solution from the flask to a separatory funnel, rinse the flask with 25 mL of ether, and add the ether to the separatory funnel. Extract the cyclohexanone into the ether, and separate the layers. Dry the ethereal extract over anhydrous magnesium sulfate, filter, and distill. Collect as cyclohexanone the portion of the distillate that boils between 150 and 155°C. Yield: about 8 grams (80%).

Extraction: Section 13

Notes

1. A commercial bleach, available at the supermarket, can be used. Alternatively, a 5% solution of sodium hypochlorite is available from the Aldrich Chemical Company.

2. You can use a 125-mL addition funnel, adding the bleach solution to the funnel in two portions.
3. When sodium hypochlorite is present, it will oxidize iodide ion to elemental iodine. In the presence of starch, iodine gives an intense blue color. Perform the test by transferring a drop of the solution to the test paper with a stirring rod.
4. This is a steam distillation; see Section 11.

References

1. G. K. Helmkamp and H. W. Johnson, Jr., *Selected Experiments in Organic Chemistry,* 2nd edition, W. H. Freeman, San Francisco, 1968, p. 108.
2. N. M. Zuczek and P. S. Furth, *J. Chem. Educ.* **58,** 824 (1981).

55 Reduction of Ketones to Secondary Alcohols

Sodium borohydride is an efficient and specific reducing agent for aldehydes, ketones, and acid chlorides, converting these substances to the corresponding alcohols. Other reducible groups such as carbon–carbon double or triple bonds, nitro or cyano groups, and other carbonyl-containing functional groups such as esters or lactones and amides are not reduced under conditions whereby aldehydes and ketones can be reduced. Sodium borohydride is fairly soluble in water, methanol, and ethanol. In these solvents, in the presence of base, it hydrolyzes to form hydrogen very slowly; acidification, however, leads to rapid evolution of hydrogen. The reaction proceeds by hydride transfer from the borohydride ion to the carbonyl group:

$$4\,R_2C{=}O + Na^+BH_4^- \longrightarrow [(H{-}CR_2{-}O)_4{-}B]^-Na^+$$

The borate ester can then be hydrolyzed by addition of water:

$$[(H{-}CR_2{-}O)_4{-}B]^-Na^+ + 2\,H_2O \longrightarrow 4\,H{-}CR_2{-}O{-}H + Na^+BO_2^-$$

55.1 CYCLOHEXANOL FROM CYCLOHEXANONE

Procedure. Dissolve 9.8 g (10.3 mL; 0.10 mole) cyclohexanone in 25 mL of methanol. To this solution, add a solution of about 0.5 g sodium methoxide and 1.0 g (0.026 mole) sodium borohydride in 25 mL of methanol. After about 5 minutes, pour the mixture into 100 mL of ice water and add 10 mL of dilute hydrochloric acid. Extract the aqueous mixture with 50 mL of ether, and wash the ethereal extract with three 15-mL portions of water. After drying the extract over anhydrous magnesium sulfate, remove the drying agent by gravity filtration and isolate the product by distillation. Collect as cyclohexanol the portion boiling between 155 and 165°C.

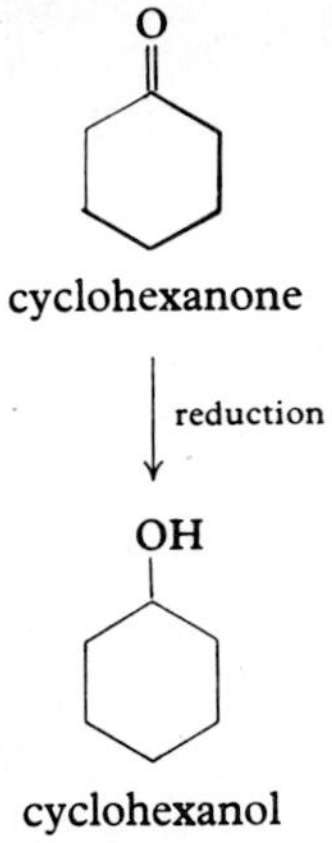

O
cyclohexanone
reduction
OH
cyclohexanol

Time: less than 3 hours.

The IR spectra of cyclohexanol and cyclohexanone are presented in Figures 52.1 and 54.1.

55.2 A VARIATION: *cis*- AND *trans*-4-*tert*-BUTYLCYCLOHEXANOL FROM 4-*tert*-BUTYLCYCLOHEXANONE

Reduction of 4-*tert*-butylcyclohexanone under the conditions of the preceding experiment leads to a mixture of the *cis*- and *trans*-4-*tert*-butylcyclohexanols:

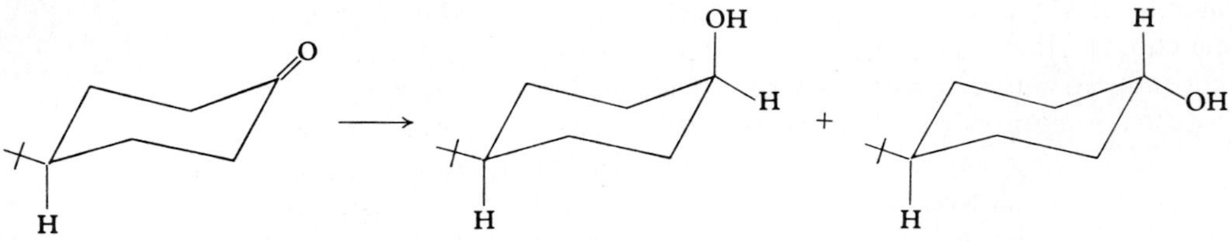

O H → OH H H + H OH H

The ratio of *cis* to *trans* isomers can be determined by gas chromatographic analysis of the dried ethereal extract on a 20% Carbowax(polyethylene glycol)-on-firebrick column at 150°C; the *cis* isomer has the shorter retention time (Reference 1).

Time: less than 3 hours.

Questions

1. How do you determine that the *cis* isomer has the shorter retention time?
2. What assumption is made when taking the peak area ratio as the isomer ratio?
3. How do you explain the observed isomer ratio?

Reference

1. *Organic Syntheses,* Wiley, New York, 1973, Collective Volume V, p. 175.

55.3 VANILLYL ALCOHOL FROM VANILLIN

The aldehyde vanillin, the major flavor component of vanilla extract, is easily reduced to the corresponding alcohol by treatment with sodium borohydride in aqueous base.

vanillin $\xrightarrow{NaBH_4}$ vanillyl alcohol

Procedure. Add to a 250-mL Erlenmeyer flask 7.6 grams (0.050 mole) of vanillin and 55 mL of 1 *M* aqueous NaOH solution (0.055 mole of NaOH). Swirl the flask to dissolve the vanillin, and then cool the contents of the flask to between 10 and 15°C by swirling the flask in an ice bath.

Why should the vanillin dissolve?

Now, in several portions over a period of 5 minutes and with good mixing, add 1.0 gram (0.026 mole) of sodium borohydride, and then let the flask stand at room temperature for about 30 minutes.

Again cool the contents of the flask by briefly swirling it in the ice bath. Following this, add in 2-mL to 3-mL portions most of a solution of 5 mL of conc. HCl (0.060 mole of HCl) in 15–20 mL of water, stopping when the solution in the flask is slightly acidic to litmus (the pH is on the acidic side of 7). If at this point the product has not separated from solution, it can be encouraged to do so by scratching the walls of the flask near the bottom with a stirring rod. Thoroughly cool the contents of the flask and then collect the vanillyl alcohol by suction filtration. After as much water as possible has been sucked from the product, scrape it out of the suction funnel and set it aside to dry. Recrystallize the dried vanillyl alcohol from ethyl acetate, using about 5 mL of ethyl acetate per gram of dried vanillyl alcohol. It may be necessary to filter the hot solution so as to remove some insoluble material. The melting point of vanillyl alcohol is reported to be 113–115°C.

Scratch to induce crystallization

Suck product dry!

Reference

1. D. Todd, *Experimental Organic Chemistry*, Prentice-Hall, Englewood Cliffs, N.J., 1979, p. 137.

56 Alkyl Halides from Alcohols

Sections 49 and 50 explained how treatment of an alcohol with a strong acid in the presence of chloride or bromide ion could serve to convert the alcohol to the corresponding alkyl halide. The five experiments in this section illustrate this type of reaction.

56.1 *n*-BUTYL BROMIDE FROM *n*-BUTYL ALCOHOL

n-Butyl bromide is prepared by boiling *n*-butyl alcohol with a mixture of sodium bromide, water, and conc. sulfuric acid. It is likely that the reaction proceeds by a nucleophilic displacement of water by bromide ion from the conjugate acid of the alcohol:

$$CH_3CH_2CH_2CH_2{-}OH + \text{"}H^+\text{"} \xrightleftharpoons{\text{rapid}} CH_3CH_2CH_2CH_2{-}\overset{+}{O}H_2$$

$$CH_3CH_2CH_2CH_2{-}OH_2 + Br^- \xrightarrow{\text{slow}} CH_3CH_2CH_2CH_2{-}Br + H_2O$$

The infrared spectrum of *n*-butyl bromide is shown in Figure 56.1.

Procedure. Add to a 250-mL boiling flask 27.0 g (0.26 mole) sodium bromide, 30 mL of water, and 20 mL (16.2 g; 0.22 mole) of *n*-butyl alcohol. Cool the mixture in an ice bath, and add, with swirling and cooling, 23 mL (42.3 g; 0.44 mole) of conc. sulfuric acid. Fit the flask with a condenser and heat the mixture under reflux for one-half hour

Figure 56.1. Infrared spectrum of *n*-butyl bromide; thin film.

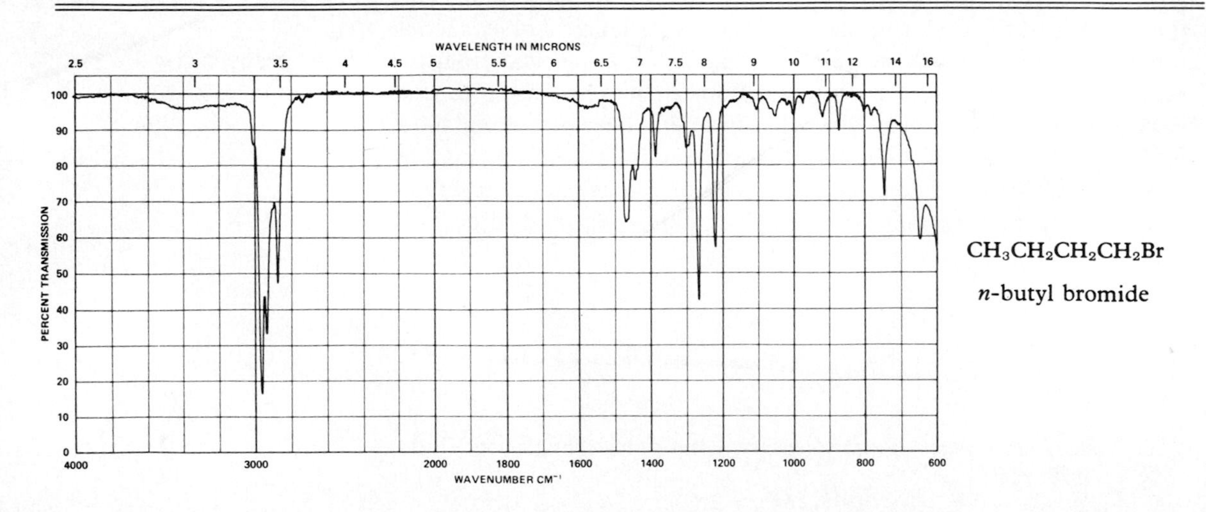

(Notes 1 and 2). At the end of the heating period, distill the mixture until no more water-insoluble material comes over (see Section 11). Pour the distillate into a separatory funnel and add 20 mL of water. Add a pinch of sodium bisulfite to remove any free bromine and shake the funnel to mix the contents. Remove the lower organic layer, and wash it successively with a 20-mL portion of ice-cold sulfuric acid, 20 mL of water, and 20 mL of 10% aqueous sodium carbonate solution (Note 3). After drying the *n*-butyl bromide over anhydrous calcium chloride, distill it and collect as product the fraction that boils between 99 and 103°C.

Apparatus: Fig. 32.1

Notes

1. A longer heating time increases the yield only slightly.
2. Some hydrogen bromide is evolved. Usually the amount is not objectionable, but it can be trapped by absorption in water if desired (Section 35).
3. Except in the wash with conc. sulfuric acid, the organic layer will be the bottom layer. Thus, the bottom layer will have to be removed, the separatory funnel rinsed with water, and the organic layer added again to the funnel. It is easy to lose material in these transfers.

Time: 3 hours.

56.2 A VARIATION: ISOAMYL BROMIDE FROM ISOAMYL ALCOHOL

It should be possible to convert isoamyl alcohol, 3-methyl-1-butanol, to isoamyl bromide by the procedure for the conversion of *n*-butyl alcohol to *n*-butyl bromide that is described in Section 56.1:

$$\underset{\substack{\text{isoamyl alcohol} \\ \text{m.w.} = 88}}{CH_3\underset{\substack{|\\CH_3}}{C}HCH_2CH_2—OH} \xrightarrow{\text{Section 56.1}} \underset{\substack{\text{isoamyl bromide} \\ \text{b.p.} = 121°C}}{CH_3\underset{\substack{|\\CH_3}}{C}HCH_2CH_2—Br}$$

Isoamyl bromide can be converted to isoamyl acetate, one of the components of the alarm pheromone of the honey bee, by the procedure of Section 58.2.

56.3 *tert*-BUTYL CHLORIDE FROM *tert*-BUTYL ALCOHOL

tert-Butyl chloride is prepared by treating *tert*-butyl alcohol with conc. hydrochloric acid. Apparently, the *tert*-butyl carbonium ion is formed at a convenient rate near room temperature and without additional Lewis acid catalysis:

$$\begin{array}{c}CH_3\\|\\CH_3\text{—}C\text{—}OH\\|\\CH_3\end{array} + HCl \underset{}{\overset{\text{rapid}}{\rightleftharpoons}} \begin{array}{c}CH_3\\|\\CH_3\text{—}C\text{—}\overset{+}{O}H_2\\|\\CH_3\end{array} + Cl^-$$

$$\begin{array}{c}CH_3\\|\\CH_3\text{—}C\text{—}\overset{+}{O}H_2\\|\\CH_3\end{array} \xrightarrow{\text{slow}} \begin{array}{c}CH_3\\|\\CH_3\text{—}C^+\\|\\CH_3\end{array} \xrightarrow[\text{fast}]{+\,Cl^-} \begin{array}{c}CH_3\\|\\CH_3\text{—}C\text{—}Cl\\|\\CH_3\end{array}$$

The infrared and NMR spectra of *tert*-butyl chloride are shown in Figures 56.2 and 56.3.

Procedure. Place 60 mL (71.5 g; 0.72 mole) conc. hydrochloric acid, which has been cooled in an ice bath, in a 125-mL separatory funnel. To this, add 20 mL (15.8 g; 0.21 mole) *tert*-butyl alcohol. Shake the mixture occasionally during twenty minutes, relieving the internal pressure by inverting the funnel and cautiously opening the stopcock. Allow the mixture to stand until the layers have separated cleanly, and then remove and discard the lower, hydrochloric acid, layer. Wash the product with 10 mL of water and then with 10 mL of aqueous sodium bicarbonate solution. Dry the product over anhydrous calcium chloride and distill, collecting the fraction boiling between 49 and 52°C as *tert*-butyl chloride.

Time: less than 3 hours.

Figure 56.2. Infrared spectrum of *tert*-butyl chloride; thin film.

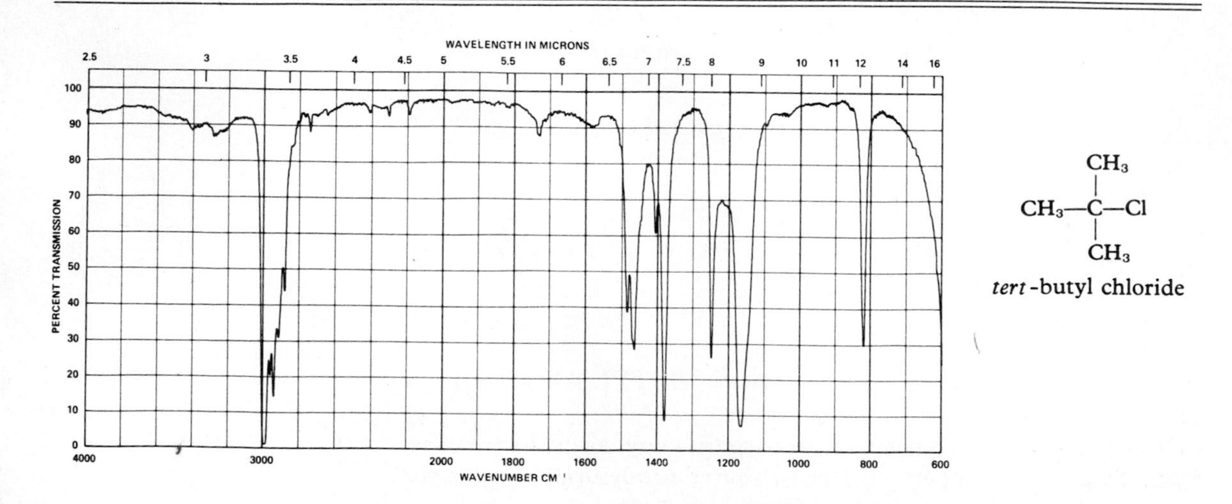

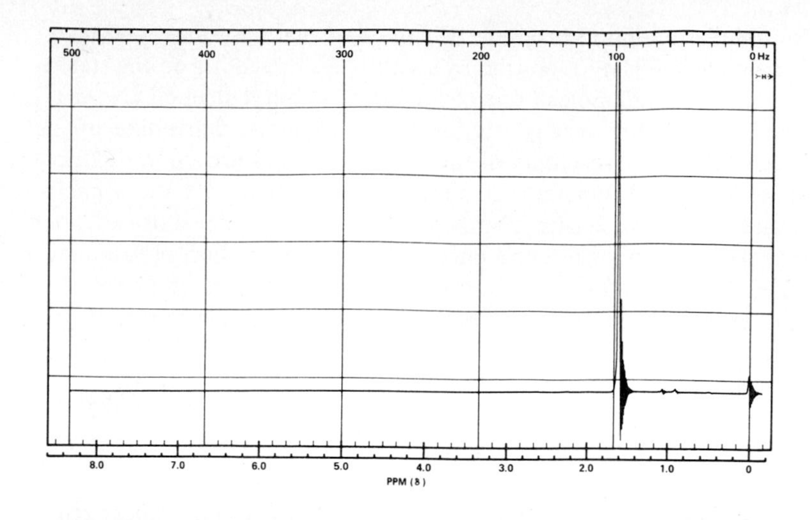

$$\begin{array}{c} CH_3 \\ | \\ CH_3-C-Cl \\ | \\ CH_3 \end{array}$$

tert-butyl chloride

Figure 56.3. NMR spectrum of *tert*-butyl chloride; neat.

56.4 A VARIATION: *tert*-AMYL CHLORIDE FROM *tert*-AMYL ALCOHOL

Treatment of *tert*-amyl alcohol by the same procedure used for the conversion of *tert*-butyl alcohol to *tert*-butyl chloride should serve to convert *tert*-amyl alcohol to *tert*-amyl chloride.

$$\begin{array}{c} OH \\ | \\ CH_3CCH_2CH_3 \\ | \\ CH_3 \end{array} \xrightarrow{\text{Section 56.3}} \begin{array}{c} Cl \\ | \\ CH_3CCH_2CH_3 \\ | \\ CH_3 \end{array}$$

tert-amyl alcohol, m.w. = 88 → *tert*-amyl chloride, b.p. = 86°C

56.5 COMPETITIVE NUCLEOPHILIC SUBSTITUTION OF BUTYL ALCOHOLS BY BROMIDE AND CHLORIDE ION

The relative reactivities of two species can sometimes be determined by allowing them to compete for a limited amount of a second reagent. In this experiment, bromide ion and chloride ion compete to form the corresponding alkyl halides from one of the isomeric butyl alcohols. The relative nucleophilicity of the two ions is inferred from the relative amounts of the two alkyl halides formed.

Procedure (Reference 1). Cautiously pour 50 mL of conc. sulfuric acid over 60 grams of ice. Pour the resulting solution into a 250-mL boiling flask. To this, add 13.5 g (0.25 mole) ammonium chloride and 24.5 g

(0.25 mole) ammonium bromide. Fit the flask with a reflux condenser and bring the solids into solution by swirling and warming on the steam bath. Now add 0.20 mole of one of the isomeric butyl alcohols (Note 1) and boil the mixture very gently for 15–30 minutes, depending upon your estimate of the reactivity of the alcohol. Cool the reaction mixture in an ice bath, and transfer it to a separatory funnel. Save the organic layer and wash it twice with 20-mL portions of cold conc. sulfuric acid, once with 50 mL of water, and once with aqueous sodium bicarbonate solution. Dry the organic phase over anhydrous calcium chloride.

Determine the ratio of alkyl bromide to alkyl chloride by one of the following methods:

1. Determination of the index of refraction (Section 19), assuming a linear relation between the refractive index and molar composition of the sample (Note 2).
2. Determination of the density (Section 18), assuming a linear relation between the density and molar composition of the sample (Note 2).
3. Vapor-phase chromatography on a Carbowax column (Sections 14.7 and 14.8), assuming that the ratio of the peak areas equals the molar ratio.

Notes

1. Normal, secondary, and tertiary butyl alcohol appear to work satisfactorily in this experiment. When isobutyl alcohol was used, the index of refraction of the product was greater than that of isobutyl bromide.
2. The following values have been reported for the density and the index of refraction of isomeric butyl chlorides and bromides. (The data are from A. I. Vogel, *Practical Organic Chemistry*, 3rd edition, Wiley, New York, 1957, pp. 293 and 294. The values in parentheses were estimated from the other values.)

		n-Butyl	*i*-Butyl	*s*-Butyl	*t*-Butyl
d_4^{20}	Cl	0.886	0.881	0.874	0.846
	Br	1.274	1.253	1.256	(1.226)
n_D^{20}	Cl	1.402	1.398	1.397	1.386
	Br	1.440	1.435	1.437	(1.424)

Time: 3 hours.

Questions

1. What effect would the following procedural variations have on the numerical value of the observed ratio, compared to what the value would be if the variation had not occurred?
   a. The mixture was boiled so vigorously that some of the product escaped from the condenser.
   b. Some of the product evaporated during work-up.
   c. The product contained substances other than the two alkyl halides. (Consider each method of analysis.)
   d. The alcohol was added before the ammonium salts had completely dissolved.
2. In interpreting the results, it will be assumed that the observed ratio of R—Br to R—Cl is the result of the relative rates of formation of the two compounds, and not of their relative stability. How could one show experimentally whether or not this is a valid assumption?
3. One possible result is that the ratios would be the same for each alcohol. How would this result be interpreted?
4. Another result might be that the ratio of R—Br to R—Cl would be different for each alcohol. What would be a reasonable variation in this ratio with the structure of the alcohol?
5. Assuming that class results are available from the reaction of normal, secondary, and tertiary butyl alcohols, interpret them by the presently accepted mechanisms for this type of reaction.
6. What difference would it make if the initial concentrations of ammonium chloride and ammonium bromide were not equal? How would you take this into account?
7. What effect would the following factors have on the numerical value of the observed product ratio, compared to what the value would be otherwise? Assume an observed ratio of 2.0.
   a. 0.20 mole of each ammonium salt was used.
   b. 0.10 mole of alcohol was used.
   c. The reaction was not complete (some alcohol remained unreacted).
   d. Some of the alcohol was converted to the alkene.
8. Answer Question 7 assuming an observed ratio of 1.0.
9. Answer Question 7 assuming an observed ratio of 0.5.
10. What would be the result in this experiment if 0.50 mole of alcohol were used?
11. It should now be apparent that taking the observed ratio of R—Br to R—Cl as a measure of the relative nucleophilicity of bromide and chloride ions in this reaction is an approximation. In fact, in an experiment using 0.25 mole each of bromide and chloride and 0.20 mole of alcohol, a relative nucleophilicity of 2.80 is required to account for an observed ratio of 2.0. If only 0.10 mole of alcohol had been used in an experiment that was otherwise the same, a relative nucleophilicity of 2.23 would account for an observed ratio of 2.0. Derive an exact expression, verify the examples given, and show under what conditions the exact expression reduces to the approximation.

Reference

1. G. K. Helmkamp and H. W. Johnson, *Selected Experiments in Organic Chemistry*, 2nd edition, Freeman, San Francisco, 1968, p. 59.

57 Kinetics of the Hydrolysis of *tert*-Butyl Chloride

When *tert*-butyl chloride is dissolved in aqueous acetone, it reacts to give *tert*-butyl alcohol and hydrogen chloride:

$$\underset{\textit{tert}\text{-butyl chloride}}{CH_3-\overset{\displaystyle CH_3}{\underset{\displaystyle CH_3}{C}}-Cl} + H-O-H \longrightarrow \underset{\textit{tert}\text{-butyl alcohol}}{CH_3-\overset{\displaystyle CH_3}{\underset{\displaystyle CH_3}{C}}-OH} + H^+\ Cl^-$$

The mechanism of this reaction is slow loss of chloride ion from the halide, to give the *tert*-butyl carbonium ion, followed by a more rapid reaction of the carbonium ion with water to form the conjugate acid of the alcohol:

$$CH_3-\overset{\displaystyle CH_3}{\underset{\displaystyle CH_3}{C}}-Cl \xrightarrow{\text{slow}} CH_3-\overset{\displaystyle CH_3}{\underset{\displaystyle CH_3}{C}}+ \; + \; Cl^-$$

$$CH_3-\overset{\displaystyle CH_3}{\underset{\displaystyle CH_3}{C}}+ \; + \; H-\ddot{\underset{\cdot\cdot}{O}}-H \xrightarrow{\text{fast}} CH_3-\overset{\displaystyle CH_3}{\underset{\displaystyle H_3C}{C}}-\underset{\displaystyle H}{\overset{\cdot\cdot +}{O}}-H$$

Although water and *tert*-butyl alcohol are similar in basic strength, the fact that there is a very large excess of water present means that most of the molecules of the conjugate acid of *tert*-butyl alcohol will transfer a proton to water:

$$CH_3-\overset{\displaystyle CH_3}{\underset{\displaystyle H_3C}{C}}-\underset{\displaystyle H}{\overset{\cdot\cdot +}{O}}-H + H-\ddot{\underset{\cdot\cdot}{O}}-H \overset{\text{fast}}{\rightleftarrows} CH_3-\overset{\displaystyle CH_3}{\underset{\displaystyle CH_3}{C}}-\ddot{\underset{\cdot\cdot}{O}}H + H-\underset{\displaystyle H}{\overset{\cdot\cdot +}{O}}-H$$

Since, according to this mechanism, the rate-limiting step of the reaction is the ionization of *tert*-butyl chloride to form the *tert*-butyl carbonium ion, the rate of the reaction should be proportional to the concentration of *tert*-butyl chloride:

$$\text{rate} = \frac{d[\text{R–Cl}]}{dt} \; \alpha \; [\text{R–Cl}] \tag{57-1}$$

or

$$\frac{d[\text{R–Cl}]}{dt} = -k[\text{R–Cl}] \tag{57-2}$$

where k is the proportionality constant, or rate constant, and the minus sign indicates that the concentration of *tert*-butyl chloride decreases with time. Separation of the variables

$$\frac{d[\text{R–Cl}]}{[\text{R–Cl}]} = -k\;dt \tag{57-3}$$

followed by integration gives

$$\log[\text{R–Cl}] = \frac{-kt}{2.303} + C \tag{57-4}$$

where C is the constant of integration of the indefinite integral. Since at the start of the experiment, when $t = 0$, the concentration of *tert*-butyl chloride will have its initial value, $[\text{R–Cl}]_0$, the value of the constant of integration can be determined to be $[\text{R–Cl}]_0$:

$$\log[\text{R–Cl}]_0 = -k(0) + C = C$$

Thus Equation (57-4) can be written as

$$\log[\text{R–Cl}] = \frac{-kt}{2.303} + \log[\text{R–Cl}]_0 \tag{57-5}$$

or as

$$\log\frac{[\text{R–Cl}]}{[\text{R–Cl}]_0} = \frac{-kt}{2.303} \tag{57-6}$$

Since $[\text{R–Cl}]/[\text{R–Cl}]_0$ is the fraction of *tert*-butyl chloride remaining at any time, t, Equation (57-6) states that a graph of the log of the fraction of *tert*-butyl chloride remaining at any time, t, versus t will give a straight line whose slope equals $-k/2.303$. The rate constant, k, which is a measure of the intrinsic rate of a particular reaction under specified conditions of temperature and solvent composition, can be determined from the slope of this line.

In this experiment, the fraction of *tert*-butyl chloride remaining at any particular time will be determined by following the production of acid during the course of the reaction. Measured amounts of 0.010 M sodium hydroxide solution will be added to the mixture in order to make it basic. Since hydrogen chloride is produced during the course of the reaction, the base will be consumed; and when the base is gone, the solution will become acidic. The change in color of the indicator, bromthymol blue, from blue (basic) to yellow (acidic) will signal the time at which the base has been just used up. The original concentration of *tert*-butyl chloride will be proportional to the total volume of base that is consumed over the entire course of the reaction, V_∞:

$$[\text{R–Cl}]_0 = (\text{constant})V_\infty \tag{57-7}$$

and the concentration of *tert*-butyl chloride remaining at the intermediate time, t, will be proportional to the *difference* between this total volume of base and the volume of base that has been consumed up to the intermediate time:

$$[\text{R–Cl}] = (\text{constant})(V_\infty - V_t) \tag{57-8}$$

Thus, the fraction of *tert*-butyl chloride remaining at any time will equal the following ratio:

$$\frac{[\text{R–Cl}]}{[\text{R–Cl}]_0} = \frac{(\text{constant})(V_\infty - V_t)}{(\text{constant})V_\infty} = \frac{V_\infty - V_t}{V_\infty} \tag{57-9}$$

Substituting in Equation (57-6) then gives

$$\log \frac{V_\infty - V_t}{V_\infty} = \frac{-kt}{2.303} \tag{57-10}$$

which states that a graph of the log of the volume of base remaining to be consumed divided by the total that is ultimately consumed versus the time elapsed to that point will be a straight line with a slope equal to $-k/2.303$.

The details of the ways by which the rate constant, k, can be calculated will be presented after the procedure.

Materials.

Approximately 0.095 *M* *tert*-butyl chloride in acetone, prepared by adding 0.52 mL of *tert*-butyl chloride to 50 mL of reagent grade acetone (1 mL per run)

Approximately 0.010 *M* sodium hydroxide solution (10 mL per run)*

Water : acetone, 9 : 1, by volume (50 mL per run)*

* The 0.010 *M* sodium hydroxide solution and the 9 : 1 water : acetone solution must be prepared ahead of time from boiled, distilled water; they must be carbonate-free.

Bromthymol blue indicator solution, 0.04%
125-mL Erlenmeyer flask
Shallow glass or plastic pan
Three 5-mL plastic syringes, with needle (21 gauge)
Thermometer
Magnetic stirring bar
Magnetic stirring motor
Timer

Procedure. Add 50 mL of the 9 : 1 water : acetone solution to the 125-mL Erlenmeyer flask; add a magnetic stirring bar. Adjust the temperature of the solvent to the desired value (Note 1). Clamp the flask in position over a magnetic stirring motor, and start the stirrer. Fill each of two 5-mL plastic syringes with exactly 5.0 mL of 0.010 *M* sodium hydroxide solution (Note 2). Add 3 drops of the bromthymol blue indicator solution to the flask and then introduce as accurately as possible 2.0 mL of 0.010 *M* sodium hydroxide from the first syringe; the solution should now have a pale blue color. Next, using a third plastic syringe, add exactly 1.0 mL of the solution of *tert*-butyl chloride in acetone (Note 2). Express the *tert*-butyl chloride solution from the syringe in less than 1 second, if possible, and simultaneously start the timer. Record the time (without stopping the timer) at which the solution becomes acidic, as indicated by a change in color of the indicator from blue to yellow, and then quickly add another 2.0-mL portion of base from the first syringe; the solution should then become light blue. Again, record the time at which the color changes from blue to yellow, and then add the final 1.0 mL of sodium hydroxide solution from the first syringe; again the color of the solution should revert to blue. After recording the time of the third change in color from blue to yellow, add portions of base from the second syringe until a total of 9.0 mL of base has been added. After each addition the solution should be blue, and in each case the time at which the color change from blue to yellow occurs should be recorded. Three 1-mL portions followed by two 0.5-mL portions would be appropriate. Three or four minutes after the last change in color from blue to yellow, add base drop by drop from the syringe until the solution just turns blue; if it fades to yellow, add another drop of base. When the blue color from the addition of 1 drop of base persists for 30 seconds (Note 3), the reaction can be assumed to be essentially complete. If the solutions have exactly the concentrations specified, the addition of a total of 9.5 mL of base will be required (Note 4).

Notes

1. It is easiest to run the reaction at room temperature. If the solvent has been prepared at least a day ahead of time, it will be at room temperature and it can be used at the temperature at which

it comes from the container. Determine this temperature with your thermometer.

If you wish to run the reaction at a temperature that is only a few degrees above or below that of the lab (at 25°C, for example), you can adjust the temperature of the solvent in the 125-mL flask by swirling the flask in a pan of water (warm or cool, depending on whether you wish to raise or lower the temperature) until your thermometer indicates that the solvent has attained the desired temperature.

If you wish to run the reaction at a temperature that is more than about 5° different from that of the room, you will have to clamp the flask in place within a shallow glass or plastic dish, which is then supported by the magnetic stirrer. The dish should contain water of the desired temperature. The water in the dish serves as a thermal buffer between the Erlenmeyer flask and the air of the room.

2. Fill the syringe by drawing in somewhat more than the desired volume. While holding the syringe with the needle pointing up, withdraw the plunger slightly to remove the liquid from the needle, tap the barrel to cause all the air bubbles to rise to the top, and then slowly push the plunger in until liquid just starts to come out the tip of the needle. Adjust the volume of the solution in the syringe to that desired by slowly sliding the plunger in, expressing the excess solution into a spare beaker.
3. A longer waiting period will be required at lower temperatures, since at such temperatures the reaction will proceed more slowly. The reaction will be approximately 99.9% complete when the total elapsed time is ten times that required to reach the 5-mL point. For example, if the color change from blue to yellow that corresponds to the final addition from the first syringe takes place after an elapsed time of 50 seconds, the reaction should be about 99.9% complete after a total elapsed time of 500 seconds, or about 8.5 minutes.
4. A volume of 1.0 mL of 0.095 *M* *tert*-butyl chloride contains 0.095 mmole of *tert*-butyl chloride, and the hydrogen chloride produced by its hydrolysis will consume the base contained in 9.5 mL of 0.010 *M* sodium hydroxide solution. However, the concentrations of the solutions probably will not be exactly equal to those specified, and therefore the volume of base required probably will not be exactly 9.5 mL. Since the results of the experiment are interpreted in terms of the fractional extent of reaction, the actual concentrations of *tert*-butyl chloride and of base need not be accurately known.

Analysis of results. One way to approach the analysis of the results is to prepare a table for each run in which each row corresponds to a time, and each column to a value measured at that time or calculated for that time. The headings for such a table could be these:

Time	Volume of Base Added	Volume of Base Remaining	Fraction of Base Remaining	Log Fraction of Base Remaining
t	V_t	$V_\infty - V_t$	$(V_\infty - V_t)/V_\infty$	$\log(V_\infty - V_t)/V_\infty$

Then a graph of the values in the last column (log of the fraction of base remaining) versus the value of the first column (time) should give a straight line of a slope equal to $-k/2.303$. The rate constant, k, can then be calculated by multiplying the slope (rise over run) by -2.303. The units of the rate constant are time^{-1}, and the magnitude of the number indicates that fraction of the amount present that will undergo reaction during the next unit of time, neglecting any decrease in rate due to consumption of starting material.

Alternatively, the time at which only $\frac{1}{2}$ of the initial amount is left can be estimated from the line (the time corresponding to log ($\frac{1}{2}$), or -0.301). The rate constant can be calculated from this time, the half-life or half-time $t_{1/2}$, by the following relationship:

$$k = \frac{\ln 2}{t_{1/2}} = \frac{0.693}{t_{1/2}} \qquad (57\text{-}11)$$

For example, if the half-time is 50 seconds, the first-order rate constant is 0.0139 sec^{-1}. This relationship between half-time and rate constant is valid only for first-order reactions.

Questions

1. Given the following data, calculate the half-time and the first-order rate constant for the hydrolysis of *tert*-butyl chloride in 90% water : acetone at 25°C:

Time[a] t	Volume of[b] Base Added V_t	Volume of[b] Base Remaining $V_\infty - V_t$	Fraction of Base Remaining $(V_\infty - V_t)/V_\infty$	Log Fraction of Base Remaining $\log(V_\infty - V_t)/V_\infty$
0	0			
14	2			
33	4			
46	5			
62	6			
82	7			
112	8			
137	8.5			
177	9.0			
∞	9.6			

[a] in seconds
[b] in mL

2. Derive Equation (57-11) from Equation (57-10).
3. Show that if a first-order reaction runs for ten half-times, only 0.00098 of the original amount of starting material will remain.
4. Sometimes first-order processes are described by a *characteristic time*, or *time constant*, τ. This number is the reciprocal of the first-order rate constant:

$$\tau = \frac{1}{k} \qquad (57\text{-}12)$$

We have seen that a half-time of 50 seconds corresponds to a first-order rate constant of 0.0139 sec^{-1}. According to Equation (57-12), this rate constant corresponds to a characteristic time, or time constant, τ, of 72 seconds. What fraction remains after the elapse of the characteristic time?

58 Isoamyl Acetate: A Component of the Alarm Pheromone of the Honey Bee

Esters of carboxylic acids can be prepared either by forming the acyl-to-oxygen bond (acyl transfer) or by forming the alkyl-to-oxygen bond (alkyl transfer).

$$\mathrm{R{-}\overset{\overset{\displaystyle O}{\|}}{C}{-}O{-}R}$$

acyl-to-oxygen bond ↑ alkyl-to-oxygen bond ↑

a carboxylic acid ester

Acylation of an alcohol by an acyl chloride (cholesteryl benzoate from cholesterol and benzoyl chloride, Section 59.1), an anhydride (acetylsalicylic acid, Section 59.4; glucose pentaacetate, Section 59.5 and 59.6) or by the carboxylic acid itself (methyl benzoate, Section 59.2; methyl salicylate, Section 59.3) is the approach that is most often followed for the preparation of an ester. The first synthesis of isoamyl acetate, Section 58.1, is an example of ester formation by acyl transfer.

Acyl transfer

It is also possible, however, to prepare isoamyl acetate by an alkyl transfer reaction, in which the alkyl-to-oxygen bond is formed. The second synthesis of isoamyl acetate, Section 58.2, in which potassium acetate is alkylated by isoamyl bromide, is an example of this alternative approach.

Alkyl transfer

Isoamyl acetate has been shown to be one of the active compounds in the alarm pheromone of the honey bee (Reference 1). Cotton balls containing freshly excised honey bee stings and placed near the hive entrance were seen to be more frequently stung than control balls. Gas

Alarm pheromone

chromatographic and infrared analysis of 700 stings indicated that approximately one microgram of isoamyl acetate can be isolated from each sting. Other substances in addition to isoamyl acetate must be present in the stings, however. While cotton balls treated with isoamyl acetate do alert and agitate the guard bees, other balls containing an equivalent number of stings incite the bees to sting as well.

58.1 ISOAMYL ACETATE FROM ISOAMYL ALCOHOL AND ACETIC ACID; FISCHER ESTERIFICATION

Fischer esterification

Isoamyl acetate, sometimes called pear oil or banana oil, can be prepared as previously described by heating together acetic acid and isoamyl alcohol in the presence of concentrated sulfuric acid. An excess of acetic acid is used to shift the equilibrium toward product formation, since the acid is a little less expensive and is easier to remove from the product.

$$CH_3{-}\overset{\overset{\large O}{\|}}{C}{-}OH + HO{-}CH_2CH_2CH(CH_3)_2 \underset{}{\overset{H^+}{\rightleftharpoons}} CH_3{-}\overset{\overset{\large O}{\|}}{C}{-}O{-}CH_2CH_2CH(CH_3)_2 + H_2O$$

The infrared and proton NMR spectra of isoamyl acetate are shown in Figures 58.1 and 58.2.

Procedure. Place in a 100-mL round-bottom boiling flask 10.9 mL of isoamyl alcohol (3-methyl-1-butanol; 8.8 grams; 0.1 mole), 23 mL of

Figure 58.1. Infrared spectrum of isoamyl acetate; thin film.

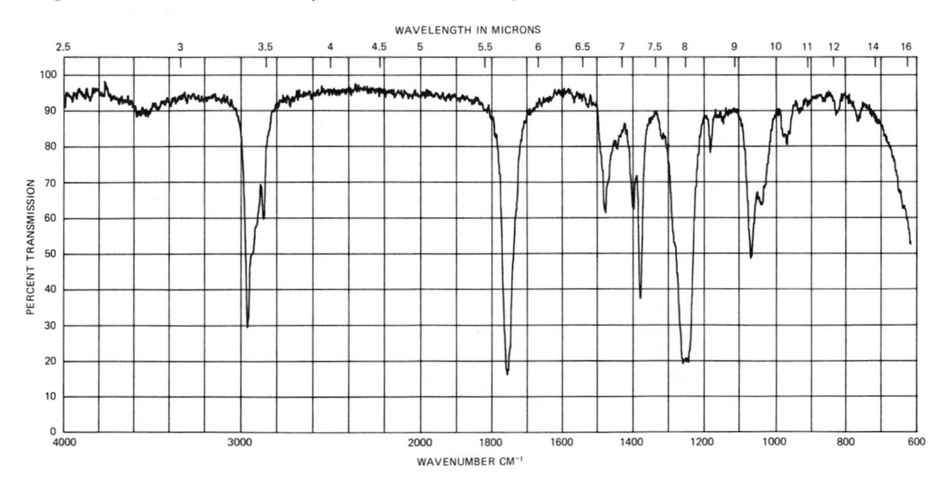

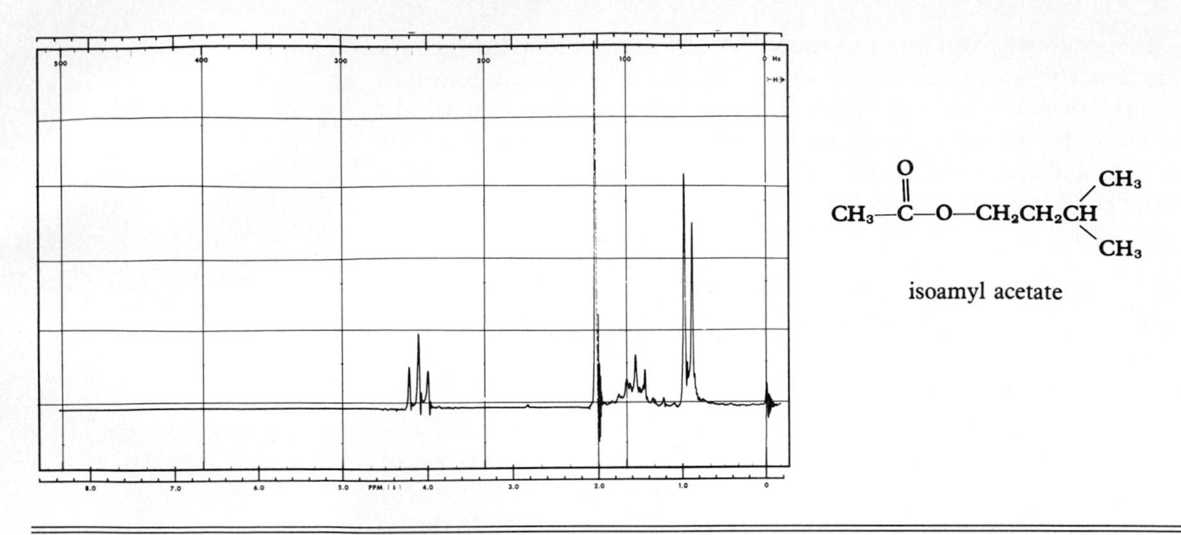

Figure 58.2. NMR spectrum of isoamyl acetate; CCl_4 solution.

Apparatus: Fig. 32.1

glacial acetic acid (24 grams; 0.4 mole), and 3 mL of conc. sulfuric acid (5.5 grams; 0.054 mole). Add a few boiling chips, fit the flask with a condenser, and boil the mixture gently, using a flame, for 1 to 2 hours (Note 1). At the end of the heating period, cool the contents of the flask by swirling the flask in a beaker of cold water for 2 or 3 minutes and then adding 50 grams of ice (Notes 2 and 3). Transfer the reaction mixture to a separatory funnel. Add to the separatory funnel 40 mL of diethyl ether, using some of the ether to rinse the boiling flask. Thoroughly mix the contents of the separatory funnel, allow the layers to separate, and draw off the lower, aqueous, layer. Wash the ether layer with a 50-mL portion of cold water and then extract the ether layer with a solution of 3 grams of sodium carbonate dissolved in 50 mL of water, using this solution in two 25-mL portions (Note 4). Dry the ether extract over anhydrous magnesium sulfate, remove the drying agent by gravity filtration, and strip off the ether by distillation on the steam bath. Distill the residue, using a flame, to isolate the isoamyl acetate. The boiling point of isoamyl acetate is reported to be 142°C at 756 Torr. Yield (Note 1): 8 to 9 grams (60 to 70%).

Notes

1. A slightly higher yield is obtained with the longer reaction time.
2. Fifty grams of ice is approximately the amount of crushed ice that can be contained in a 100-mL beaker.
3. The point is to cool the mixture well below the boiling point of the solvent that will be used for extraction. Since the solvent will

be diethyl ether, b.p. 35°C, the mixture should be cooled to below 25°C. See Section 13.3.

4. Carbon dioxide will be produced. Take care not to build up excessive pressure in the separatory funnel.

Warning!

Time: 3 to 4 hours.

Question

1. How is the excess acetic acid removed from the product mixture?

Reference

1. R. Boch, D. A. Shearer, and B. C. Stone, *Nature* **195**, 1018 (1962).

58.2 ISOAMYL ACETATE FROM ISOAMYL BROMIDE AND POTASSIUM ACETATE

Isoamyl acetate can also be prepared by heating a mixture of potassium acetate and isoamyl bromide in dimethylformamide. The mechanism of the reaction is an S_N2-like displacement of bromide by acetate. Apparently, the inability of the polar aprotic solvent, dimethylformamide (DMF), to form hydrogen bonds with the oxygen atoms of acetate ion permits acetate to manifest its nucleophilicity.

Alkylation of acetate

DMF: a dipolar, aprotic solvent

$$CH_3-\overset{\overset{\displaystyle O}{\|}}{C}-\ddot{\underset{\cdot\cdot}{O}}:^-\ K^+ + Br-CH_2CH_2CH(CH_3)_2 \xrightarrow{DMF} CH_3-\overset{\overset{\displaystyle O}{\|}}{C}-\ddot{\underset{\cdot\cdot}{O}}-CH_2CH_2CH(CH_3)_2 + K^+\ Br^-$$

potassium acetate   isoamyl bromide   isoamyl acetate

Procedure. Place 12.0 mL (15.1 grams; 0.10 mole) of isoamyl bromide (Note 1), 12 grams (0.12 mole) of potassium acetate, 20 mL of dimethylformamide, and a magnetic stirring bar in a 100-mL boiling flask. Fit the flask with a condenser, and support the flask in a steam bath so that the flask can be heated by steam and magnetically stirred at the same time. Heat and stir the mixture for 2 hours. At the end of the heating period, cool the contents of the flask by swirling the flask in cold water, and then add 25 mL of cold water down the condenser. Swirl the mixture so as to dissolve all of the solid material. Transfer the mixture to a separatory funnel, add 25 mL of ethyl ether, and mix the contents of the separatory funnel. After allowing the layers to separate, draw off the lower, aqueous, layer, and wash the remaining, ether, layer with one 10-mL portion of water. After drying the ether layer over anhydrous magnesium sulfate, transfer the ether solution to a 50-mL boiling flask, strip off the ether, using the steam bath as a source of heat, and then distill the residue, using a flame. Collect the fraction that boils between 137° and 145°C as isoamyl acetate. Yield: about 10 grams (about 75%). According to the proton NMR spectrum, the product contains a trace of dimethylformamide.

Note

1. Isoamyl bromide can be prepared from isoamyl alcohol (Section 56.2).

Time: 3 to 4 hours.

59 Preparation of Esters

Esters can be prepared either by acyl transfer to the oxygen atom of an alcohol:

$$R-\overset{\overset{O}{\|}}{C}-X + H-O-R' \xrightarrow{\text{acyl transfer}} \underset{\text{ester}}{R-\overset{\overset{O}{\|}}{C}-O-R'} + H-X$$

or by alkyl transfer to an oxygen atom of the carboxylate anion:

$$R-\overset{\overset{O}{\|}}{C}-O^- + X-R' \xrightarrow{\text{alkyl transfer}} \underset{\text{ester}}{R-\overset{\overset{O}{\|}}{C}-O-R'} + X^-$$

as explained in Section 58.

The most common acylating agents used for the preparation of esters are the carboxylic acids themselves (benzoic acid, Section 59.2; salicylic acid, Section 59.3), the acid halides (benzoyl chloride, Section 59.1), and the carboxylic acid anhydrides (acetic anhydride, Sections 59.4, 59.5, and 59.6).

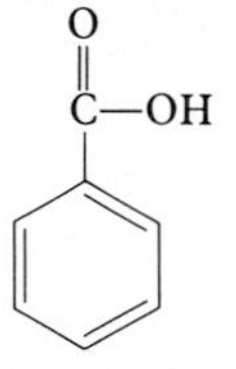

$C_6H_5-\overset{\overset{O}{\|}}{C}-OH$

benzoic acid

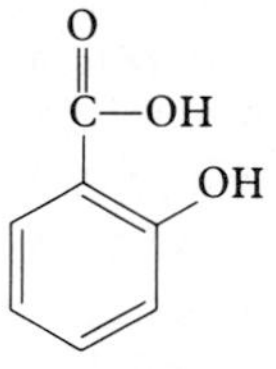

$o\text{-}HOC_6H_4-\overset{\overset{O}{\|}}{C}-OH$

salicylic acid

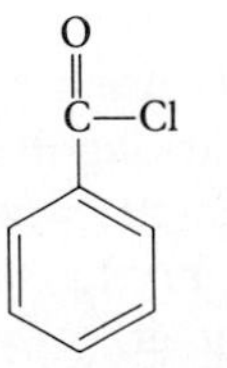

$C_6H_5-\overset{\overset{O}{\|}}{C}-Cl$

benzoyl chloride

$CH_3-\overset{\overset{O}{\|}}{C}-O-\overset{\overset{O}{\|}}{C}-CH_3$

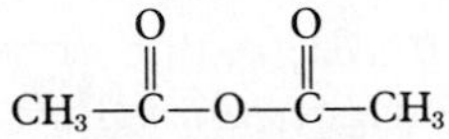

acetic anhydride

59.1 CHOLESTERYL BENZOATE FROM CHOLESTEROL; LIQUID CRYSTALS

In 1888, an Austrian botanist prepared cholesteryl benzoate and noticed that at about 150°C it "melted" to give a cloudy liquid that became clear only on further heating to above 180°C. Since this first observation of the phenomenon, which has come to be called the formation of a liquid crystal, many other compounds have been found that show similar behavior. The liquid crystal phase differs from the ordinary liquid phase in this way. Although the liquid crystal phase takes the shape of its container, its molecules are still ordered to a certain extent, and so the liquid still exhibits some of the properties of the crystalline state. In cholesteryl benzoate, the molecules are associated in layers, with their long axes parallel to one another, and the layers are stacked in a spiral arrangement. This is the so-called *twisted nematic* form, which was earlier called the *cholesteric* form. All substances that give the twisted nematic form of liquid crystal are made up of chiral molecules; it is the chirality of the molecules that leads to the large-scale spiral organization. This spiral organization makes it possible for certain cholesterol derivatives, though colorless in themselves, to selectively scatter light into different colors. As cholesteryl benzoate enters and leaves the liquid crystal phase, reddish purple colors can be seen if the sample is strongly illuminated from the side. An easy way to see this is described below. Reference 1 discusses the phenomenon of liquid crystals in more detail.

It is often desirable to obtain only one member of a pair of enantiomers. One way to do this is by separating a mixture of enantiomers, as in the resolution of α-phenylethylamine (Section 48). Another is to start the synthesis with only one member of a pair of enantiomers rather than starting with the racemic mixture. However it is done, *something* chiral must be involved. A particularly attractive possibility is the use of a chiral catalyst, and much research is being done to find ways to form or consume exclusively only one member of a pair of enantiomers in a process that makes use of a chiral catalyst. One recent experiment of this general nature was to heat phenyl ethyl malonic acid in the liquid crystal phase of cholesteryl benzoate in order to form 2-phenylbutyric acid by thermal decarboxylation (Reference 2):

ϕ—C(COOH)(COOH)—CH_3CH_2 (phenyl ethyl malonic acid) $\xrightarrow{\text{liquid crystal phase of cholesteryl benzoate}}$ ϕ—C(COOH)(H)—CH_3CH_2 ((R)-(−)-2-phenylbutyric acid, 59%) + ϕ—C(H)(COOH)—CH_3CH_2 ((S)-(+)-2-phenylbutyric acid, 41%)

Apparently the two carboxyl groups, which reside in enantiomeric environments, react at slightly different rates in the chiral reaction medium, although there is disagreement about the relative importance in this experiment of the chirality of the individual molecules and the chirality of the large-scale spiral structure (Reference 3).

Although this approach has been applied to synthetic organic chemistry only recently, enzymes have been doing similar things for millions of years. For example, the alcohol dehydrogenases react exclusively with either the R or the S proton in the alpha-methylene group of ethanol:

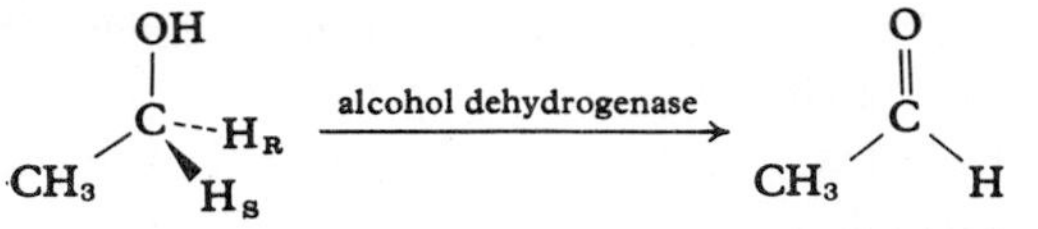

OH
CH3 C---HR HS —alcohol dehydrogenase→ CH3 C(=O) H

acetaldehyde that contains H_S or H_R exclusively

One class of alcohol dehydrogenases removes exclusively H_R, and the other removes exclusively H_S (Reference 4).

In this experiment, the ester is prepared by treating the alcohol with an acid chloride in pyridine solution:

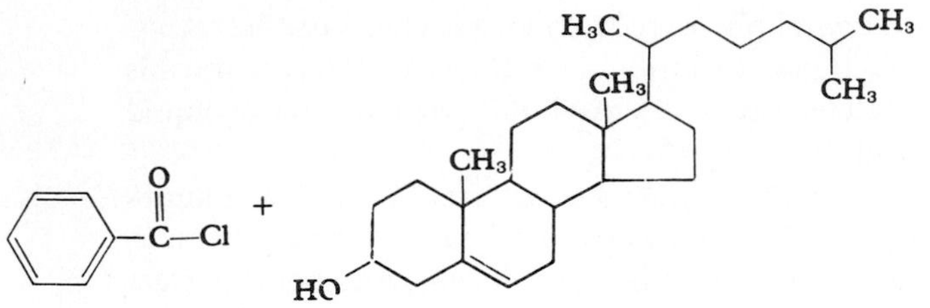

C6H5–C(=O)–Cl + HO–(cholesterol)

cholesterol

—pyridine→ C6H5–C(=O)–O–(cholesteryl)

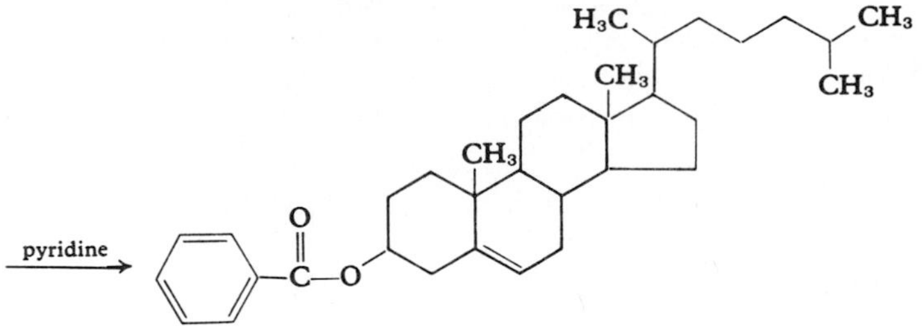

cholesteryl benzoate

The infrared spectrum of cholesteryl benzoate is shown in Figure 59.1.

Procedure. Place 1.0 gram (2.6 mmole) of cholesterol in a 50-mL Erlenmeyer flask. To this add 3 mL of pyridine and swirl the mixture to dissolve the cholesterol. Then add 0.4 mL (0.48 gram; 3.5 mmole) of benzoyl chloride (Note 1) and heat the resulting mixture on the steam bath for about 10 minutes. At the end of the heating period, cool the mixture somewhat by swirling the flask in a beaker of cold water. Then

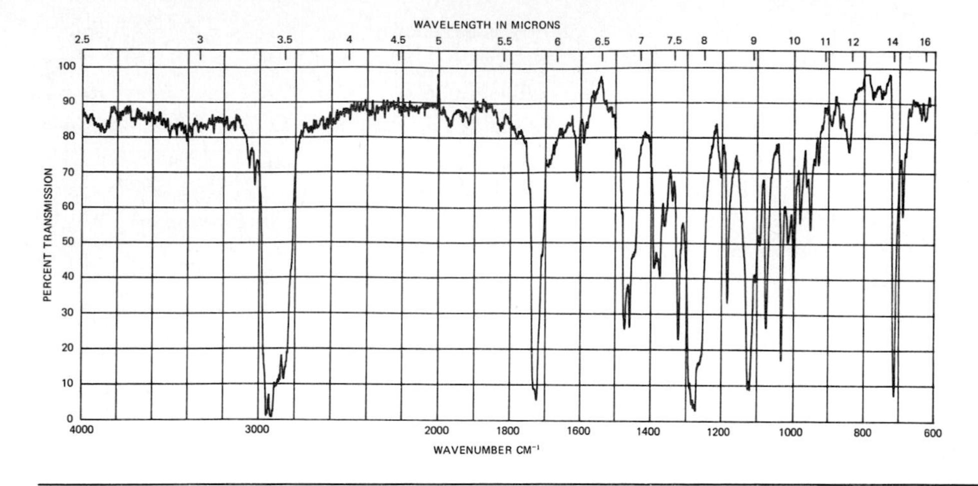

Figure 59.1. Infrared spectrum of cholesteryl benzoate; CCl_4 solution.

dilute the mixture with 15 mL of methanol and collect the solid cholesteryl benzoate by suction filtration, using a little methanol to rinse the flask and to wash the crystals. Recrystallize the cholesteryl benzoate by heating it in an Erlenmeyer flask with 20 mL of ethyl acetate until it has dissolved, filtering the hot solution by gravity, allowing the filtrate to cool to room temperature, and collecting the crystals by suction filtration (Note 2). Yield: from 0.6 to 0.8 grams (45 to 65%).

Notes

1. Benzoyl chloride is a lachrymator. It should be stored in the hood and you should take your flask to the hood and work there while making the addition. A graduated pipet should be supplied to use to transfer the benzoyl chloride.
2. The recovery can be improved somewhat by cooling the mixture in an ice bath before collecting the crystals.

Formation of liquid crystals

The phenomenon of the formation of the liquid crystal phase of cholesteryl benzoate can be easily seen by placing 100–200 milligrams of the compound on the end of a microscope slide and heating the sample by holding the slide with a pair of tongs above a small burner flame. The solid will turn first to a cloudy liquid and then with further heating to a clear melt. On cooling, the cloudy liquid will first appear, and then it will change to a hard, crystalline solid. With strong lighting from the side, purple and red colors will be seen as the sample changes phases on both heating and cooling. The more cautious the heating, the better you can see the changes. You can repeat the heating and cooling many times with the same sample.

Time: 3 hours.

References

1. L. Verbit, *J. Chem. Educ.* **49,** 37 (1972).
2. L. Verbit, T. R. Halbert, and R. B. Patterson, *J. Org. Chem.* **40,** 1649 (1975).
3. W. H. Pirkle and P. L. Rinaldi, *J. Am. Chem. Soc.* **99,** 3510 (1977).
4. W. L. Alworth, *Stereochemistry and Its Application in Biochemistry,* Wiley-Interscience, 1972, p. 77.

59.2 METHYL BENZOATE

$$C_6H_5\text{—}C(=O)\text{—}OH$$

benzoic acid

↓

$$C_6H_5\text{—}C(=O)\text{—}O\text{—}CH_3$$

methyl benzoate

In this section, two related methods are presented for the preparation of methyl benzoate from benzoic acid and methanol. The first makes use of conc. sulfuric acid as catalyst, while the second uses the Lewis acid boron trifluoride:

$$\phi\text{—}\overset{O}{\overset{\|}{C}}\text{—}O\text{—}H + CH_3\text{—}O\text{—}H \underset{}{\overset{\text{acid}}{\rightleftharpoons}} \phi\text{—}\overset{O}{\overset{\|}{C}}\text{—}O\text{—}CH_3 + H_2O$$

The second method, that of Procedure b, seems to be suitable for the preparation of the methyl esters of a large number of aromatic carboxylic acids.

The infrared and NMR spectra of the product methyl benzoate are shown in Figures 59.2 and 59.3.

Catalysis by Conc. H_2SO_4 (Procedure a)

Apparatus: Fig. 32.1

Procedure. Add to a 100-mL boiling flask 12.2 g (0.1 mole) of benzoic acid, 25 mL (19.7 g; 0.62 mole) of methanol, and 3 mL conc. sulfuric acid (Note 1). Fit the flask with a reflux condenser and boil the mixture for about 45 minutes. Cool the mixture to room temperature and pour it into a separatory funnel that contains 50 mL of cold water. Rinse the flask with 25 mL of ether and pour this into the separatory funnel. Mix the contents of the separatory funnel, allow them to settle, and then draw off the aqueous layer. Wash the ethereal layer with 25 mL of water and then 25 mL of 5% aqueous sodium carbonate solution (Note 2). Dry the ethereal extract over anhydrous magnesium sulfate, filter, and remove the ether by distillation on the steam bath. Complete the distillation with a flame, collecting as methyl benzoate the material boiling above 190°C. Yield: about 70%.

Notes

1. Add the sulfuric acid cautiously, allowing it to run down the wall of the flask. Swirl to mix.
2. Add the sodium carbonate solution in portions; swirl to mix; watch for foaming.

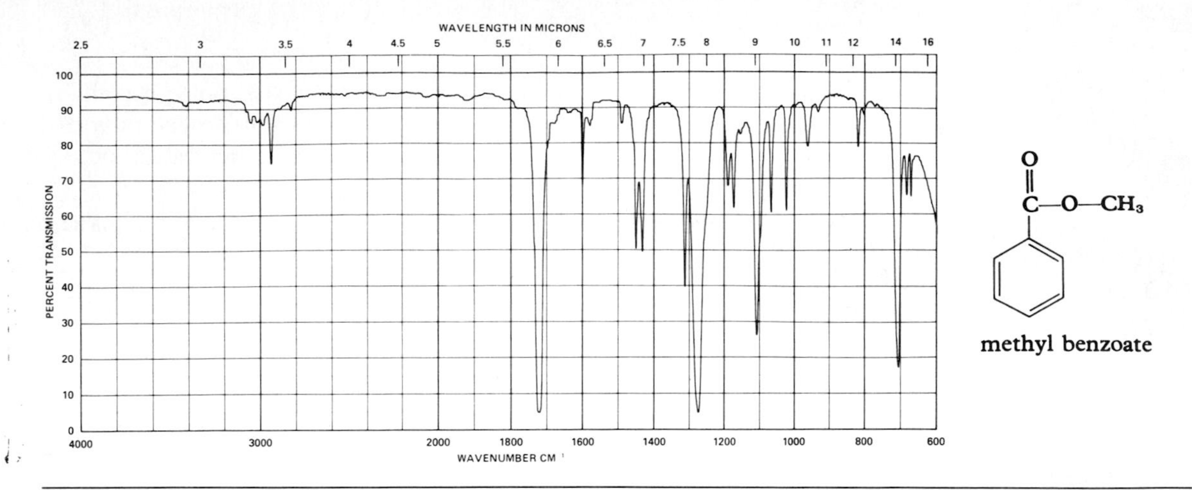

Figure 59.2. Infrared spectrum of methyl benzoate; thin film.

Time: 3 hours.

Catalysis by BF_3 (Procedure b)

Procedure (Reference 1). Add to a 100-mL boiling flask 2.44 g (0.020 mole) of benzoic acid, 22 mL of anhydrous methanol (0.54 mole), and 4.4 mL of the boron trifluoride/methanol complex (0.040 mole of $BF_3 \cdot 2MeOH$). Fit the flask with a condenser and boil the mixture for one hour. Cool the mixture to room temperature and pour it into a saturated solution of sodium bicarbonate (Note 1). Isolate the ester by extraction with ether and recover it as described in Procedure a. Yield: about 96%.

Figure 59.3. NMR spectrum of methyl benzoate; CCl_4 solution.

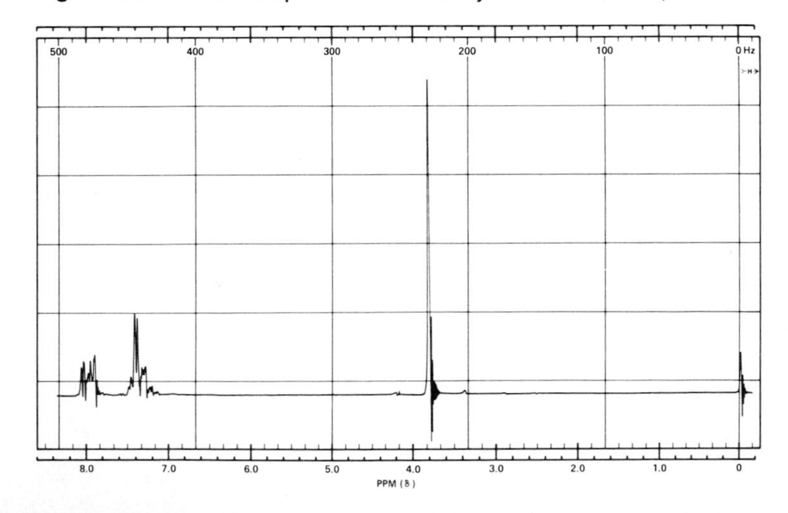

methyl benzoate

Note

1. The amount of sodium bicarbonate required is not specified. If all the boron trifluoride is hydrolyzed to give hydrogen fluoride, at least 0.12 mole of base will be required for neutralization.

Time: 3 hours.

Questions

1. Write a mechanism for the BF_3-catalyzed reaction of Procedure b.
2. Is it possible that the BF_3 is doing more than just catalyzing the reaction that takes place in Procedure b? If so, what else might it be doing?

Reference

1. G. Hallas, *J. Chem. Soc.* **1965,** 5770.

59.3 METHYL SALICYLATE; OIL OF WINTERGREEN

O=C—OH, OH (benzene ring) ⟶ O=C—O—CH_3, OH (benzene ring)

salicylic acid ⟶ methyl salicylate

Figure 59.4. Infrared spectrum of methyl salicylate; thin film.

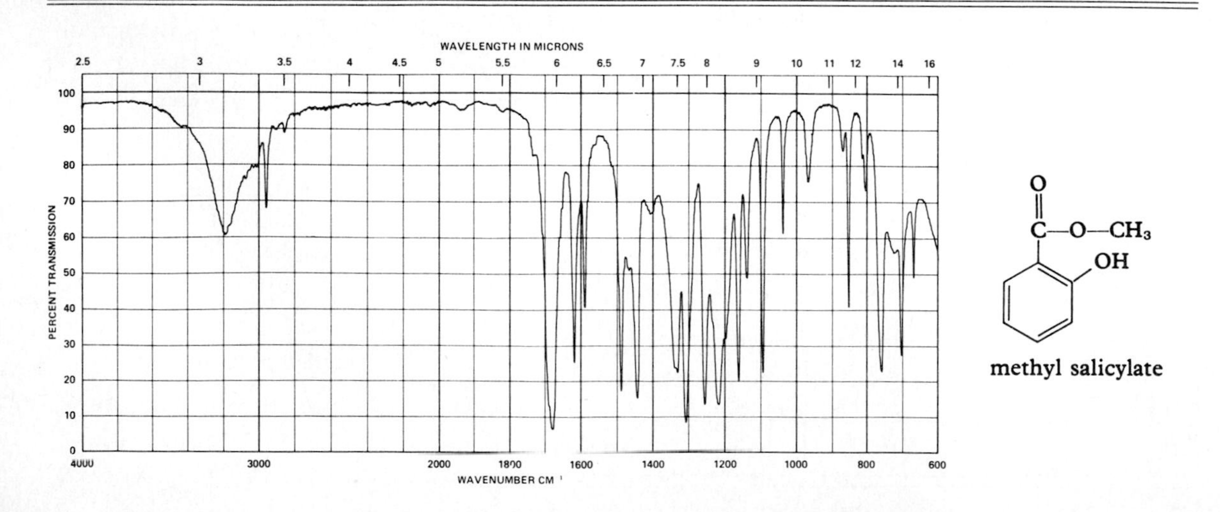

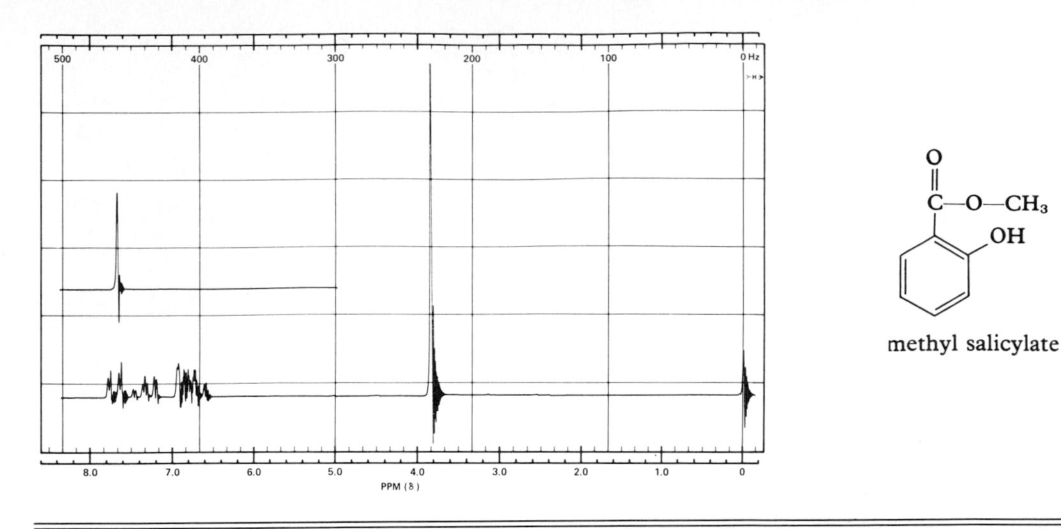

Figure 59.5. NMR spectrum of methyl salicylate; CCl_4 solution.

Using Procedure b of Section 59.2 with salicylic acid and a reflux time of three hours would give a 91% yield of methyl salicylate.

Methyl salicylate is widely used as a perfume and flavor ingredient; it smells like wintergreen. Its infrared and NMR spectra are shown in Figures 59.4 and 59.5.

59.4 ACETYLSALICYLIC ACID; ASPIRIN

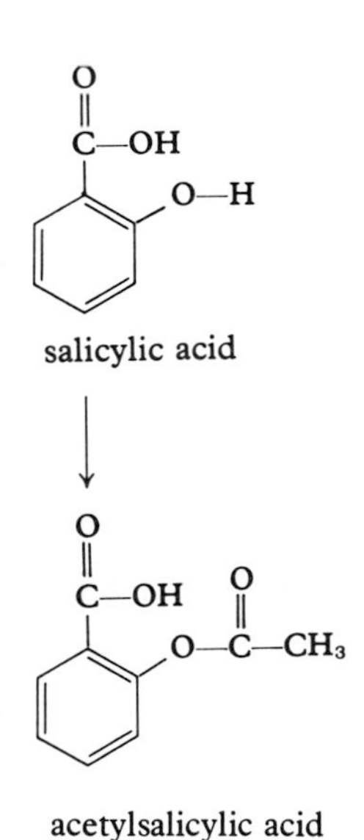

In the preceding experiment, salicylic acid participates in an esterification reaction through its carboxyl group. In this experiment, it takes part through the phenolic hydroxyl group.

The industrial method of production of acetylsalicylic acid is similar to the method given in this section. In 1983, industrial production in the United States was in the amount of about 31 million pounds, or about 185 5-grain tablets for each man, woman, and child in the United States.

Procedure. Add to a 125-mL Erlenmeyer flask 2.0 grams (14.5 mmole) of salicylic acid and 4 mL (4.3 grams; 42 mole) of acetic anhydride. To this mixture add 5 drops of 85% phosphoric acid and swirl to mix. Fit the flask with a reflux condenser and heat the mixture on the steam bath for about 5 minutes. Without cooling the mixture add 2 mL of water in one portion down the condenser. The excess acetic anhydride will hydrolyze and the contents of the flask will come to a boil. When the vigorous reaction has ended, add 40 mL of cold water, cool the mixture to room temperature, stir and rub the mixture with a stirring rod if necessary to induce crystallization, and finally allow the

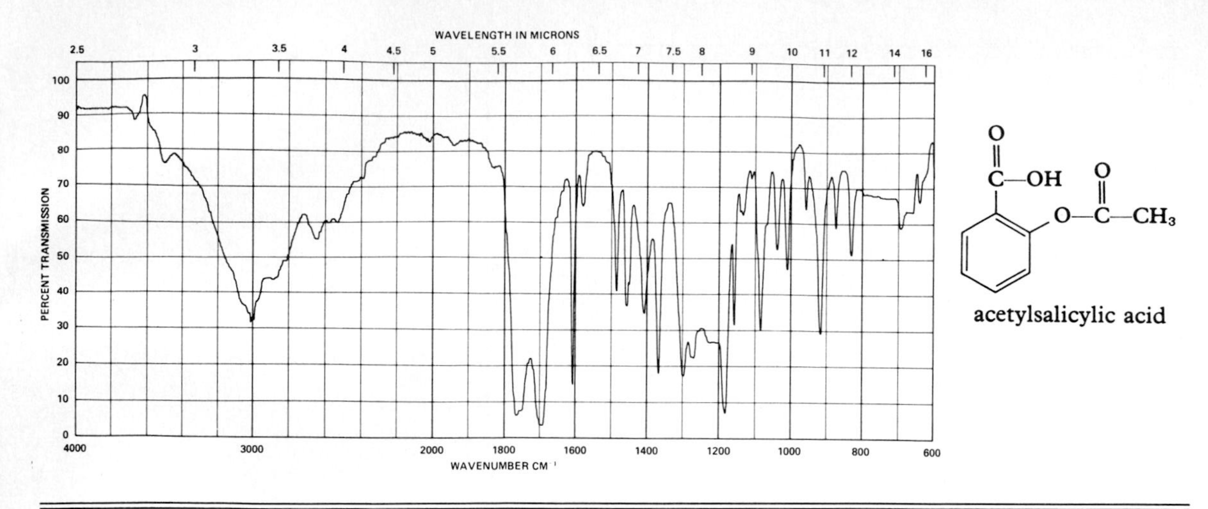

Figure 59.6. Infrared spectrum of acetylsalicylic acid; $CHCl_3$ solution.

mixture to stand in the ice bath to complete crystallization. Collect the product by suction filtration and wash it with a little water. The product can be recrystallized from water.

Time: 2 hours.

Figures 59.6 and 59.7 show the infrared and NMR spectra of acetylsalicylic acid.

Figure 59.7. NMR spectrum of acetylsalicylic acid; $CDCl_3$ solution.

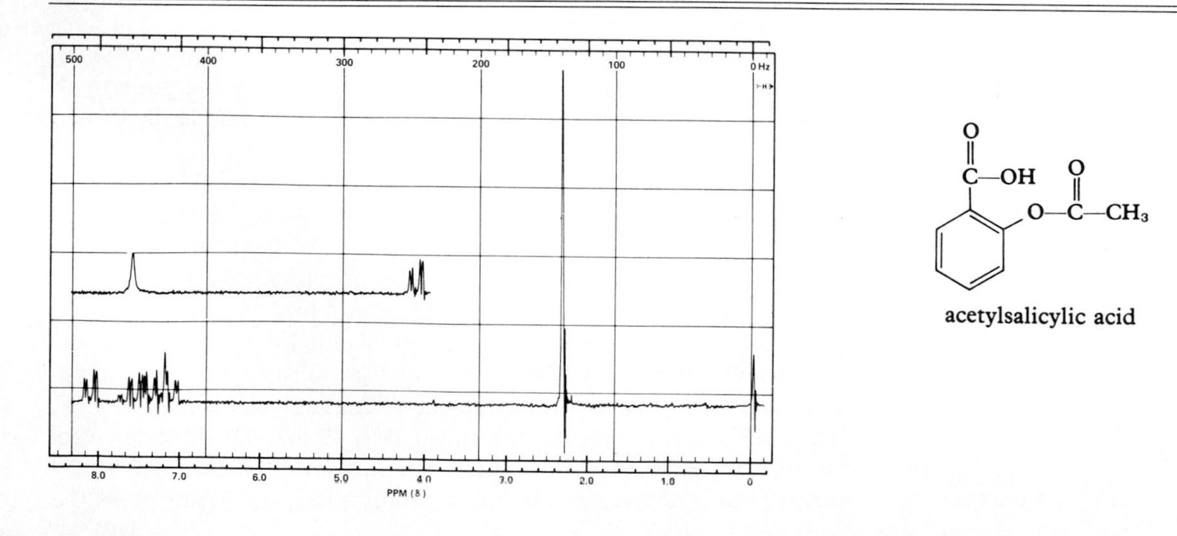

59.5 α- or β-D-GLUCOSE PENTAACETATE FROM D-GLUCOSE

When D-glucose is heated with acetic anhydride and zinc chloride, α-D-glucose pentaacetate is formed.

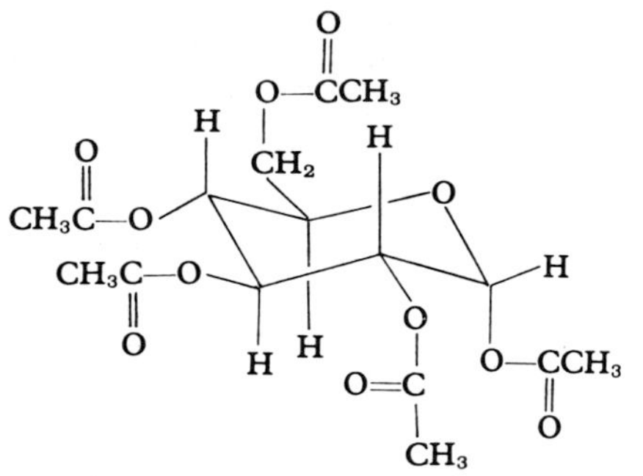

α-D-glucose pentaacetate

If sodium acetate is used in place of zinc chloride, the product is β-D-glucose pentaacetate.

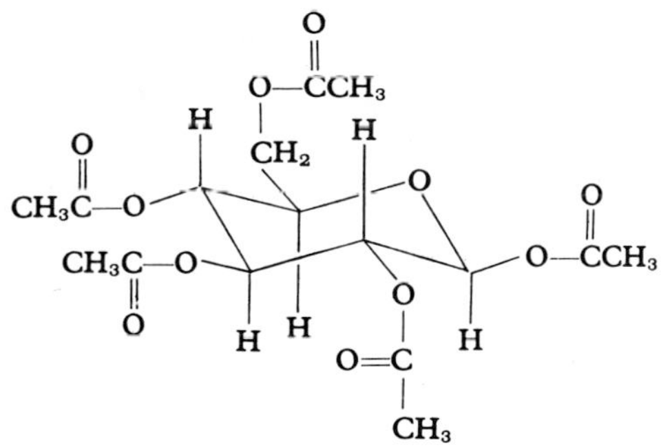

β-D-glucose pentaacetate

The two products differ only in the configuration at C-1, the anomeric carbon.* The scheme of Figure 59.8 is proposed to account for the different results of these two procedures.

In both experiments, β-D-glucose pentaacetate is the initial product, being formed more rapidly than α-D-glucose pentaacetate (kinetic control). However, when zinc chloride is the catalyst, the initially

* Since these diastereoisomers differ in configuration at only a single chiral center, they can be called *epimers*. Epimers of this type, created by formation of a new chiral center upon cyclization of an aldose, can also be called *anomers*.

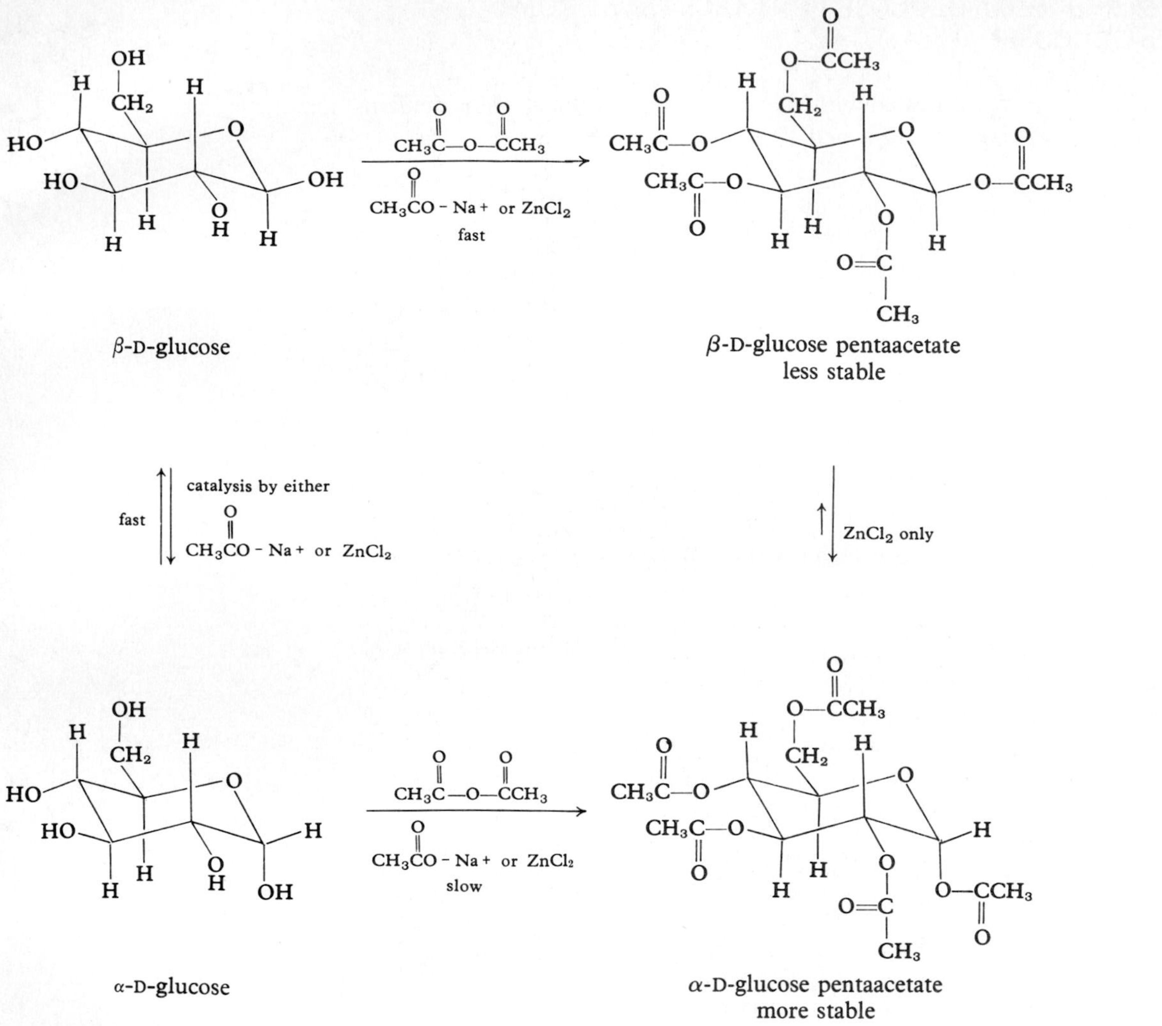

β-D-glucose

β-D-glucose pentaacetate
less stable

α-D-glucose

α-D-glucose pentaacetate
more stable

Figure 59.8. Scheme for the formation of α- or β-D-glucose pentaacetate from D-glucose.

formed β-D-glucose pentaacetate is rapidly isomerized to the more stable α isomer (thermodynamic control). Apparently, sodium acetate is not an effective catalyst for this isomerization.

It may be surprising that the α anomer is the more stable since in this isomer the anomeric acetoxyl group has the axial configuration. However, as required by this scheme, when β-D-glucose pentaacetate is treated with acetic anhydride in the presence of zinc chloride, the α isomer can be isolated from the reaction mixture in good yield.

α-D-Glucose Pentaacetate

Procedure. Place 0.5 g anhydrous zinc chloride (3.7 mmole), 12 mL acetic anhydride (13.0 g; 127 mmole), and 2.0 g anhydrous D-glucose (11.1 mmole) in a 50-mL round-bottom boiling flask. Add a boiling stone, fit the flask with a condenser and heat the flask cautiously with a burner flame until the contents start to boil. Remove the flask from the heat until the brief exothermic reaction is over, and then, by further heating with the flame, gently boil the mixture for about two more minutes. While it is still hot, pour the solution in a thin stream with good stirring into about 200 mL of a mixture of water and ice. Stir the resulting suspension until the oil has solidified. Break up any lumps and collect the solid by suction filtration. Recrystallize the crude α-D-glucose pentaacetate from 10 mL of methanol. The melting point of purified α-D-glucose pentaacetate is reported to be 112–113°C. Yield: about 1.9 grams (about 50%).

Time: 2 hours.

β-D-Glucose Pentaacetate

Procedure. The procedure is the same as that just given for the preparation of α-D-glucose pentaacetate except that 1.2 g anhydrous sodium acetate (14.7 mmole) is used in place of the anhydrous zinc chloride. The melting point of purified β-D-glucose pentaacetate is reported to be 132–134°C. Yield: about 2.4 grams (about 62%).

Time: 2 hours.

Questions

1. The scheme proposed to account for these results specifies explicitly that the equilibrium between the α and the β isomers of D-glucose pentaacetate favors the α isomer. Why (or under what conditions) is the position of the equilibrium between the α and the β forms of D-glucose irrelevant?
2. An alternative explanation for the results of these two reactions is that when zinc chloride is used as catalyst the α isomer is formed more rapidly, but when sodium acetate is the catalyst the β isomer is formed more rapidly. Explain how this interpretation is or is not consistent with the experimental facts.
3. Why is it reasonable that the —OH of the anomeric carbon atom of β-D-glucose is acetylated more rapidly than that of α-D-glucose?
4. Why is it reasonable that β-D-glucose pentaacetate can be converted to the α isomer by zinc chloride catalysis but not by sodium acetate catalysis?

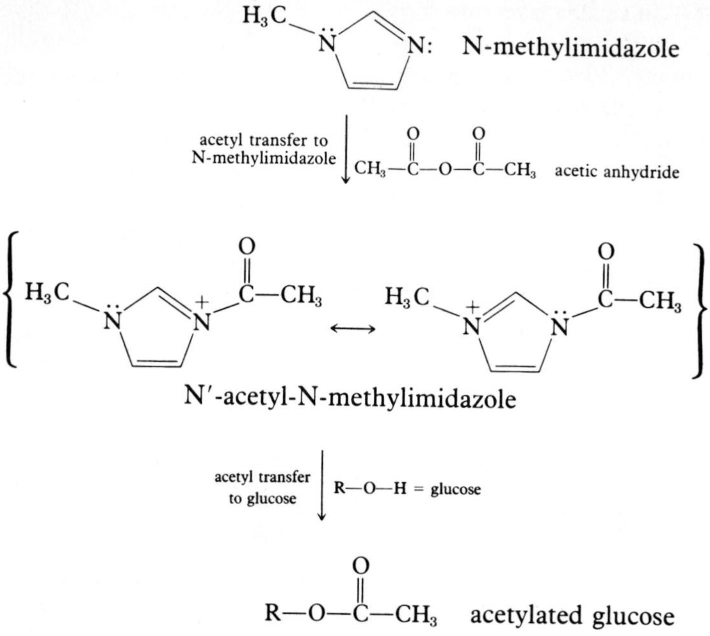

N-methylimidazole

acetyl transfer to N-methylimidazole

$CH_3-\overset{O}{\overset{\|}{C}}-O-\overset{O}{\overset{\|}{C}}-CH_3$ acetic anhydride

N′-acetyl-N-methylimidazole

acetyl transfer to glucose

R—O—H = glucose

$R-O-\overset{O}{\overset{\|}{C}}-CH_3$ acetylated glucose

Figure 59.9. Catalysis of the acetylation of glucose by N-methylimidazole.

59.6 A VARIATION: ACETYLATION OF GLUCOSE IN N-METHYLIMIDAZOLE

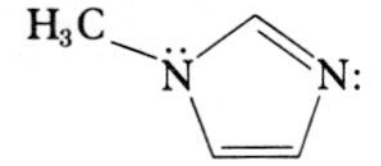

N-methylimidazole

Wachowiak and Connors, Reference 1, report that when 0.5 gram of glucose is dissolved with heating in 5 mL of N-methylimidazole (Note 1), 1.5 mL of acetic anhydride is added, and, after 15 minutes, 5 mL of water is added, glucose pentaacetate crystallizes from solution during the next 5 to 10 minutes in a yield of more than 90%. They report that they recrystallized their product from ethanol.

Catalysis by N-methylimidazole occurs by acyl transfer to N-methylimidazole, to give N′-acetyl-N-methylimidazole, followed by acyl transfer from N′-acetyl-N-methylimidazole to the —OH groups of glucose, as summarized in Figure 59.9.

The fact that N-methylimidazole catalyzes the acetylation of glucose by acetic anhydride implies that (1) N-methylimidazole is acetylated by acetic anhydride more rapidly than glucose; (2) the product of acetylation of N-methylimidazole, N′-acetyl-N-methylimidazole, acetylates glucose more rapidly than acetic anhydride; (3) N′-acetyl-N-methylimidazole is less stable than the product of the acetylation of glucose. The origins of catalytic effects such as these are not yet understood.

Note

1. N-Methylimidazole (1-methylimidazole) is available from the Aldrich Chemical Company.

Questions

1. Which isomer, α-D-glucose pentaacetate or β-D-glucose pentaacetate, is formed by this procedure?
2. During the recrystallization of the product from ethanol, the crystals that form upon cooling sometimes *disappear* if the flask is allowed to stand overnight! What might be the reason for this?

Reference

1. R. Wachowiak and K. A. Connors, *Anal. Chem.* **51,** 27 (1979).

60 The Grignard Reaction: Preparation of Aliphatic Alcohols

One important method for the formation of a carbon–carbon bond is the Grignard reaction. The first step of this three-step reaction is to prepare the Grignard reagent by allowing an ethereal solution of an alkyl halide to react with magnesium:

$$\text{R—X} + \text{Mg} \xrightarrow{\text{dry ether}} \text{R—Mg—X}$$

Dry equipment and reagents

The reaction is strongly inhibited by traces of water and therefore dry reagents and equipment are essential (Note 1). Although there is disagreement as to the molecular structure of the active Grignard reagent, it behaves as though it were a source of $R{:}^-$. Thus, it is a very good nucleophile and a very strong base.

The second step is to treat the ethereal solution of the Grignard reagent with an aldehyde, ketone, or ester. The three equations illustrate the mode of reaction of the reagent with each of these three types of compound:

$$CH_3-\overset{\overset{O}{\|}}{C}-H + R-Mg-X \xrightarrow{\text{ether}} CH_3-\underset{\underset{R}{|}}{\overset{\overset{O-Mg-X}{|}}{C}}-H$$

$$CH_3-\overset{\overset{O}{\|}}{C}-CH_3 + R-Mg-X \xrightarrow{\text{ether}} CH_3-\underset{\underset{R}{|}}{\overset{\overset{O-Mg-X}{|}}{C}}-CH_3$$

$$CH_3-\overset{\overset{O}{\|}}{C}-O-Et + 2\,R-Mg-X \xrightarrow{\text{ether}} CH_3-\underset{\underset{R}{|}}{\overset{\overset{O-Mg-X}{|}}{C}}-R + Et-O-Mg-X$$

The third step is the hydrolysis of the halomagnesium complex with aqueous acid and isolation of the alcohol that is produced.

$$\underset{\text{halomagnesium complex}}{-\underset{|}{\overset{\overset{:\ddot{O}-Mg-X}{|}}{C}}-} \quad H-O-H \longrightarrow \underset{\text{alcohol}}{-\underset{|}{\overset{\overset{:\ddot{O}-H}{|}}{C}}-} + H-O-Mg-X$$

60.1 ALIPHATIC ALCOHOLS

In this experiment, you are given a general procedure for the preparation of an aliphatic alcohol by the Grignard synthesis.

For your synthesis, choose a combination of alkyl halide and aldehyde, ketone, or ester from the following lists. Your product should have more than five carbon atoms so that it will not be too soluble in water. Higher-boiling alcohols, especially those that can form a tertiary carbonium ion, may have to be distilled under reduced pressure in order to prevent elimination to give an alkene.

- *Alkyl halides:* methyl iodide, ethyl iodide, ethyl bromide, *n*-propyl bromide, *n*-butyl bromide.
- *Aldehydes:* acetaldehyde, propionaldehyde, *n*-butyraldehyde, isobutyraldehyde (Note 2).
- *Ketones:* acetone, 2-butanone, 2-pentanone, 3-pentanone, cyclopentanone, cyclohexanone.
- *Esters:* ethyl acetate, ethyl propionate.

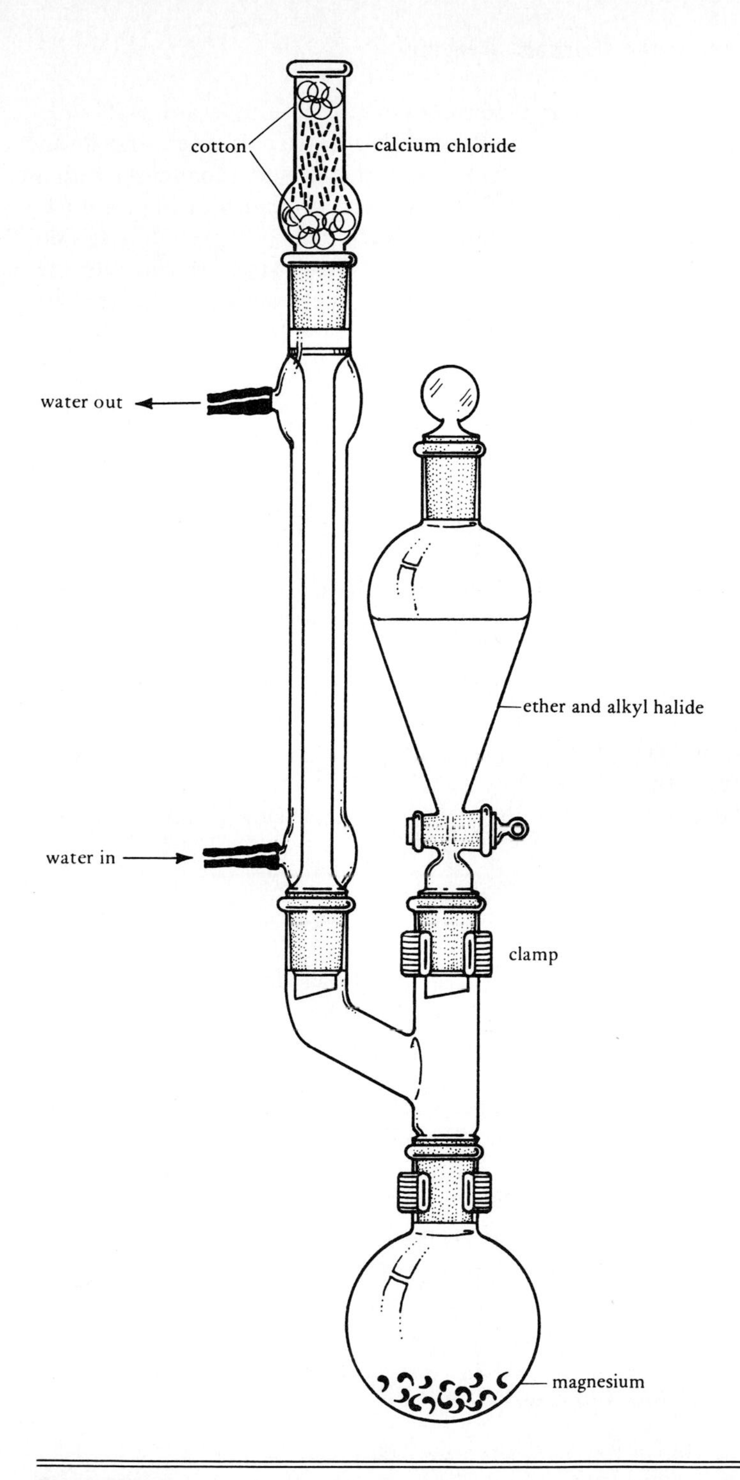

Figure 60.1. Apparatus for the preparation of the Grignard reagent.

Preparation of the Grignard Reagent

Apparatus: Fig. 60.1

Procedure a. Place 4.0 g (0.16 mole) of magnesium in a dry 250-mL boiling flask (Note 3). Fit the flask with a Claisen adapter, and fit an addition funnel in the center neck of the adapter and a condenser with a drying tube in the side neck of the adapter, as shown in Figure 60.1. Add to the addition funnel a mixture of 0.16 mole of alkyl halide and 60 mL of absolute ether. Allow 8–10 mL of the solution to run into the flask with the magnesium and wait until the reaction has started as indicated by the boiling of the reaction mixture (Note 4). Then add the remainder of the mixture of alkyl halide and ether at such a rate that the reaction mixture boils gently from the heat of the reaction (Note 5). After all of the alkyl halide/ether mixture has been added, allow the reaction mixture to stand for about 20 minutes with occasional swirling.

Flask 39.15

Procedure b. Place 4.0 g (0.16 mole) of magnesium in a dry 250-mL boiling flask (Note 3). Fit the flask with a Claisen adapter, and fit an addition funnel in the center neck of the adapter and a condenser with a drying tube in the side neck of the adapter, as shown in Figure 60.1. Add to the flask 50 mL of anhydrous ether and 0.04 mole of alkyl halide. The reaction should start immediately (Note 4). When the reaction has moderated, add via the addition funnel another 0.04-mole portion of the alkyl halide, and repeat with two more 0.04-mole portions. When the reaction has moderated from the last addition of alkyl halide, gently reflux the mixture on the steam bath for about 15 minutes.

Treatment of the Grignard Reagent with Aldehyde, Ketone, or Ester

Procedure. Cool the flask containing the Grignard reagent with an ice bath. With cooling and swirling or stirring, add drop by drop from the addition funnel a solution of 0.15 mole of the aldehyde or ketone (or 0.075 mole of the ester) in 20 mL of dry ether. The reaction is often quite exothermic and the addition should therefore be made with caution. After all the solution has been added, remove the ice bath and allow the mixture to stand at room temperature for about 20 minutes. ■

Hydrolysis of the Addition Product

Procedure. Pour the reaction mixture slowly, with stirring, onto a mixture of chipped ice and dilute sulfuric acid (prepared by adding

6 mL of conc. sulfuric acid to about 50 grams of ice and then adding about 50 mL of water). Any addition complex remaining in the reaction flask can be hydrolyzed by pouring the aqueous acid/ether mixture back in and swirling. After separating the ether layer, extract the water layer with two 25-mL portions of ordinary solvent ether. Combine all the ether solutions, wash with aqueous sodium bisulfite (if necessary; Note 4) and aqueous sodium bicarbonate, and dry over anhydrous magnesium sulfate. Isolate the product alcohol by evaporation of the ether on the steam bath and distillation of the residue.

Notes

1. Ether and alkyl halide can be dried satisfactorily over anhydrous magnesium sulfate.
2. Unless fresh aldehydes are used, they must be purified by distillation.
3. If the humidity is not high, the flask need not be dried by any special method. If desired, the flask can be dried by heating it with a flame, fitting it with a calcium chloride drying tube, and then allowing it to cool to room temperature.
4. If the reaction does not start within a few minutes, the following may be tried: (a) With a clean, dry stirring rod (not fire-polished) crush two or three pieces of the magnesium under the surface of the solution in order to break the magnesium and expose fresh surfaces. If this is successful, little bubbles will appear where the magnesium has been crushed, and the reaction mixture will become slightly cloudy. (b) Add a tiny crystal of iodine. If this is done, the ethereal solution of the *final* reaction product should be treated with a solution of sodium bisulfite in order to remove the iodine. (c) Warm the mixture on the steam bath and see if boiling will continue when the steam bath is removed. (d) Add about 1 mL of someone else's successfully formed Grignard reagent. (e) Start over, taking more care to see that the apparatus and reagents are dry.
5. Have a beaker or pan of cold water available in case the reaction mixture boils so vigorously that ether starts to escape from the condenser. If this happens, the reaction can be moderated by immersing the flask in the cold water.

Time: 3 hours to the end of the second step; 3 hours to work up the reaction mixture and isolate the product.

Questions

1. Which combinations of alkyl halide and aldehyde or ketone could be used to prepare 2-butanol by a Grignard synthesis?

2. Which combinations of alkyl halide and aldehyde or ketone could be used to prepare 2-methyl-2-butanol by a Grignard synthesis?
3. **a.** What would be the final product of the Grignard synthesis between methyl propionate and two equivalents of ethyl magnesium iodide?
   b. What would ethyl propionate give when treated as in **a**?

Reference

A good general reference on the Grignard reagent and its uses is

1. M. S. Kharasch and O. Reinmuth, *Grignard Reactions of Nonmetallic Substances,* Prentice-Hall, Englewood Cliffs, N.J., 1964.

60.2 A VARIATION: SYNTHESIS OF PHEROMONES: 4-METHYL-3-HEPTANOL AND 4-METHYL-3-HEPTANONE

It is now very clear that many organisms communicate with members of their own species and with other species by means of chemical signals (Reference 1, 2). For example, we all understand the message conveyed by the smell of the skunk, and bees are known to make use of quite an elaborate system of chemical signals. Substances that are used to carry messages between members of a particular species are often called *pheromones,* and several classes of pheromones have been recognized. These include sex attractants, alarm pheromones, trail marking pheromones, and aggregation pheromones. Some pheromones are quite ordinary compounds. For example, *n*-undecanal is a sex attractant for the female greater wax moth, valeric acid attracts the male sugar beet wireworm, and isoamyl acetate is one component of the alarm pheromone of the honeybee (see Section 58). Other pheromones are more complex, and the molecules may include one or more rings and several sites of unsaturation. Examples include (R)-(+)-limonene (Section 46), a substance that triggers a frenzy of biting and snapping among Australian harvester termites, and phenylacetic acid, one component of the stink of the stink pot turtle. Long-chain unsaturated, aliphatic alcohols, aldehydes, ketones, and esters seem to be the most common structures for most insect pheromones.

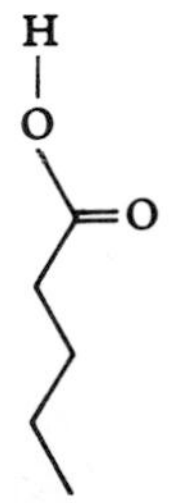

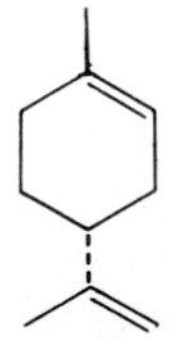

undecanal

valeric acid

(R)-(+)-limonene

$CH_2-C(=O)-OH$

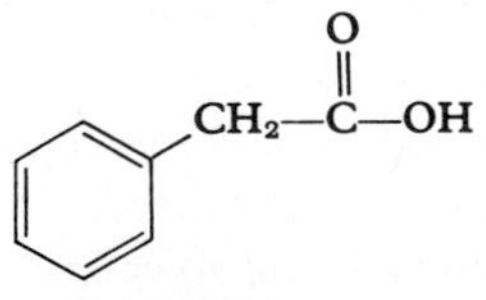

phenylacetic acid

One of the most interesting aspects of pheromone communication is the structural and stereochemical specificity of the messenger molecules. For example, the sex attractant of the silkworm moth is 10,12-hexadecadien-1-ol. Of the four possible diasteromeric forms of this compound, the 10E,12Z isomer is at least 10^9 times more active than any of the other three isomers. Another example of specificity is that the New York version of the European corn borer is attracted to (E)-11-tetradecenyl acetate, while the Iowa version is attracted to the (Z) isomer.

O

O

isoamyl acetate

OH

10E,12Z-hexadecadien-1-ol

O

O

(Z)-11-tetradecenyl acetate

O

O

(E)-11-tetradecenyl acetate

Specificity also extends to enantiomeric forms of pheromones. Only the S-(+) isomer of 4-methyl-3-heptanone functions as an alarm pheromone for Texas leaf-cutting ants, and it appears that the (7R,8S)-(+) isomer of *cis*-7,8-epoxy-2-methyloctadecane (disparlure) is far more active as a sex attractant to the female gypsy moth than the enantiomeric (7S,8R)-(−) isomer. Enantiomeric specificity such as this should not be surprising, however, as there are known many similar examples of this kind of biological specificity. Quite frequently it has been shown that only one stereoisomeric form of a drug is physiologically active, or that only one enantiomer of a substance, such as penicillamine, is toxic. Although S penicillamine is toxic, the R isomer is used as a chelating agent to accelerate the elimination from the body of normal heavy metals present in too high, nonphysiological concentrations (Wilson's disease) and of nonphysiological heavy metals (heavy metal poisoning). The R and S forms of amino acids generally have different tastes; only the S form of monosodium glutamate (MSG) has a meaty taste, the enantiomer of glucose is bitter, only the R enantiomer of luciferin (shown in Section 75) reacts enzymatically with the production of light, and, of course, the R and S isomers of carvone smell like spearmint and caraway (Section 47).

O

4-methyl-3-heptanone

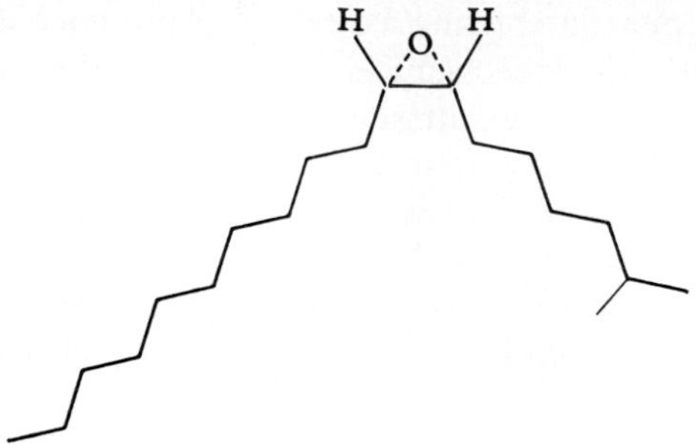

7R,8S-(+)-disparlure

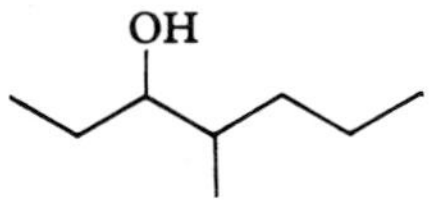

4-methyl-3-heptanol

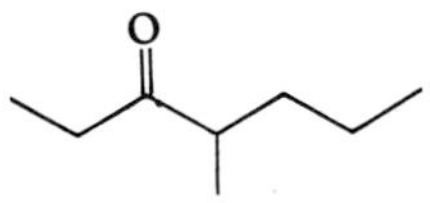

4-methyl-3-heptanone

Einterz, Ponder, and Lenox of Wabash College described the preparation of 4-methyl-3-heptanol and 4-methyl-3-heptanone in an article in *The Journal of Chemical Education*, Reference 3. They said that the alcohol could be prepared by a Grignard reaction between propionaldehyde and the Grignard reagent that can be prepared from 2-bromopentane. The ketone can be made by chromic acid oxidation of the alcohol. Perhaps these two compounds can be made by the procedures of Section 60.1 (the Grignard synthesis of aliphatic alcohols) and Section 54 (the oxidation of cyclohexanol to cyclohexanone).

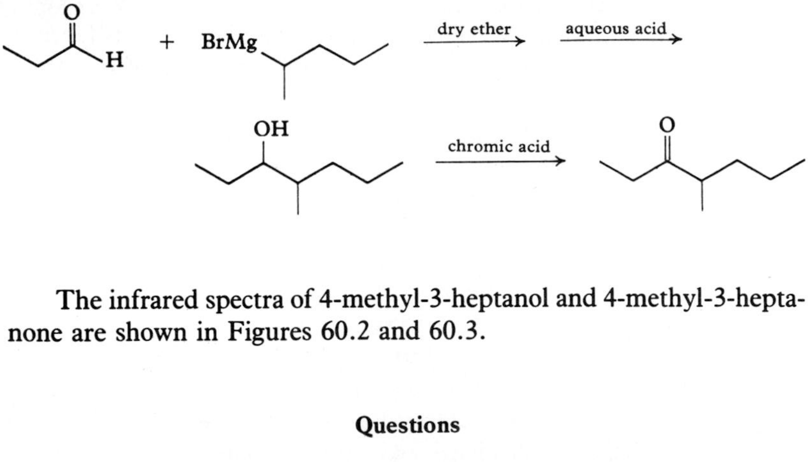

O, H, + BrMg, dry ether, aqueous acid

OH, chromic acid, O

The infrared spectra of 4-methyl-3-heptanol and 4-methyl-3-heptanone are shown in Figures 60.2 and 60.3.

Questions

1. a. In how many stereoisomeric forms can 4-methyl-3-heptanone exist?
   b. Which of these isomers will be produced in this experiment?
   c. In what relative amounts will these isomers be formed?

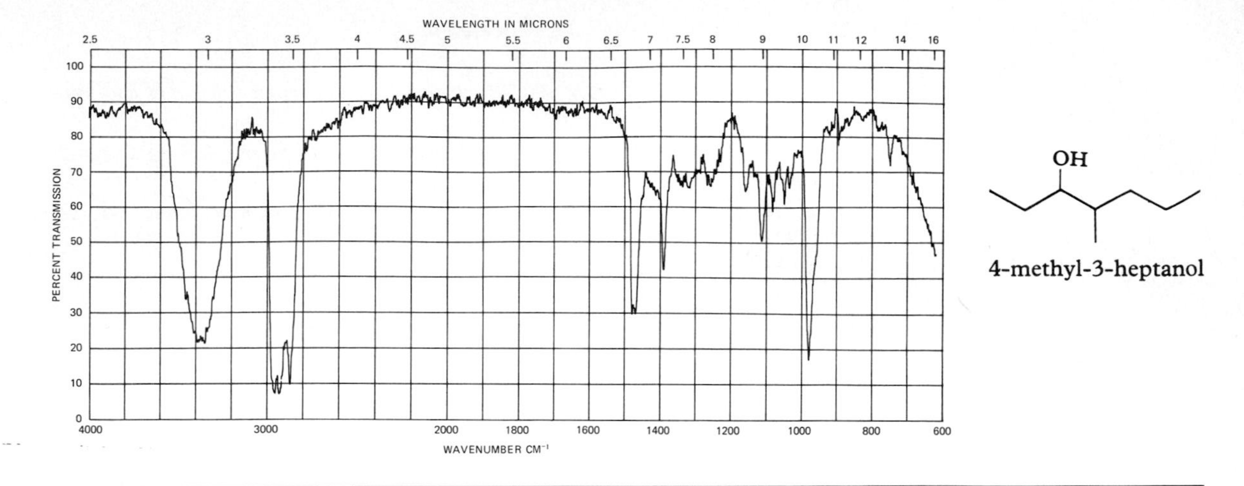

Figure 60.2. Infrared spectrum of 4-methyl-3-heptanol; thin film.

- **d.** To what extent can these isomers be separated by distillation or chromatography?
- **e.** Only the S-(+)-isomer is active toward the Texas leaf-cutting ant. Draw a representation of the configuration of this isomer.

2.
- **a.** In how many stereoisomeric forms can 4-methyl-3-heptanol exist?
- **b.** Which of these isomers will be produced in this experiment?
- **c.** In what relative amounts will these isomers be formed?
- **d.** To what extent can these isomers be separated by distillation or chromatography?
- **e.** Only the S,S-(+)-isomer is active toward the European elm beetle. Draw a representation of this isomer.

Figure 60.3. Infrared spectrum of 4-methyl-3-heptanone; thin film.

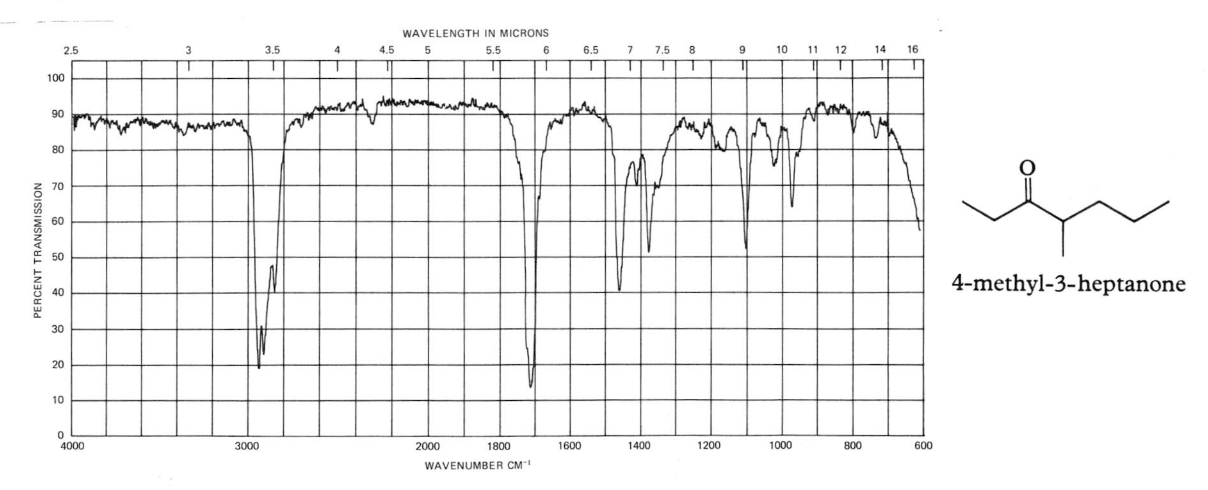

References

1. J. H. Law and F. E. Regnier, "Pheromones," in *Annual Review of Biochemistry,* **40,** 533 (1971).
2. D. A. Evans and C. L. Green, "Insect Attractants of Natural Origin," *Chemical Society Reviews* **2,** 75 (1973).
3. R. M. Einterz, J. W. Ponder, and R. S. Lenox, *J. Chem. Educ.* **54,** 382 (1977).
4. R. V. Hoffman et al., *J. Chem. Educ.* **60,** 78 (1983).

61 The Grignard Reaction: Preparation of Triphenylmethanol

The tertiary alcohol triphenylmethanol is traditionally prepared from methyl benzoate (Section 61.1), using a Grignard reagent: ester ratio of 2 : 1 (2 phenyl groups from the Grignard reagent and 1 from the ester), or from benzophenone (Section 61.2), using a Grignard reagent: ketone ratio of 1 : 1 (1 phenyl group from the Grignard reagent and 2 from the ketone). It should also be possible to prepare triphenylmethanol from diethyl carbonate, using a Grignard reagent: carbonate ester ratio of 3 : 1 (all 3 phenyl groups from the Grignard reagent). This possibility is presented in Section 61.3.

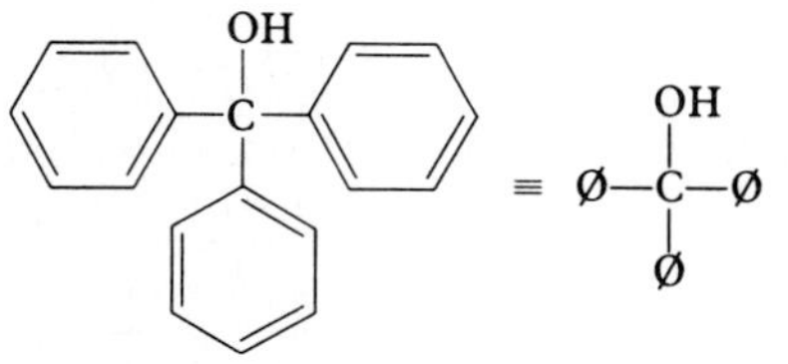

OH ≡ Ø—C(OH)(Ø)—Ø

triphenylmethanol
m.p. = 160–163°C

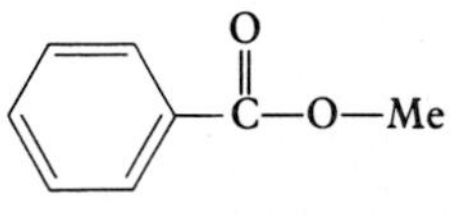

C(=O)—O—Me

methyl benzoate

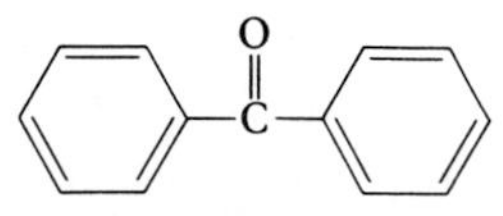

C(=O)

benzophenone

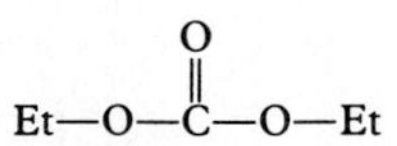

Et—O—C(=O)—O—Et
diethyl carbonate

61.1 TRIPHENYLMETHANOL FROM PHENYLMAGNESIUM BROMIDE AND METHYL BENZOATE

Triphenylmethanol can be prepared by treating the Grignard reagent formed from bromobenzene and magnesium with methyl benzoate:

$$\phi\text{—Br} + \text{Mg} \xrightarrow{\text{dry ether}} \phi\text{—Mg—Br}$$

$$2\,\phi\text{—Mg—Br} + \underset{\text{methyl benzoate}}{\phi\text{—}\overset{\text{O}}{\overset{\|}{\text{C}}}\text{—O—CH}_3} \xrightarrow{\text{dry ether}} \phi\text{—}\underset{\phi}{\underset{|}{\overset{\text{O—Mg—Br}}{\overset{|}{\text{C}}}}}\text{—}\phi + \text{CH}_3\text{—O—Mg—Br}$$

$$\phi\text{—}\underset{\phi}{\underset{|}{\overset{\text{O—Mg—Br}}{\overset{|}{\text{C}}}}}\text{—}\phi + \text{H}_2\text{O} \xrightarrow{\text{aqueous acid}} \underset{\text{triphenylmethanol}}{\phi\text{—}\underset{\phi}{\underset{|}{\overset{\text{OH}}{\overset{|}{\text{C}}}}}\text{—}\phi} + \text{HO—Mg—Br}$$

Presumably the reaction takes place by formation of benzophenone as an intermediate, which then adds a second molecule of the Grignard reagent:

$$\phi\text{—Mg—Br} + \phi\text{—}\overset{\text{O}}{\overset{\|}{\text{C}}}\text{—O—CH}_3 \longrightarrow \phi\text{—}\underset{\phi}{\underset{|}{\overset{\text{O—Mg—Br}}{\overset{|}{\text{C}}}}}\text{—O—CH}_3$$

$$\phi\text{—}\underset{\phi}{\underset{|}{\overset{\text{O—Mg—Br}}{\overset{|}{\text{C}}}}}\text{—O—CH}_3 \longrightarrow \underset{\text{benzophenone}}{\phi\text{—}\overset{\text{O}}{\overset{\|}{\text{C}}}\text{—}\phi} + \text{CH}_3\text{—O—Mg—Br}$$

$$\phi\text{—}\overset{\text{O}}{\overset{\|}{\text{C}}}\text{—}\phi + \phi\text{—Mg—Br} \longrightarrow \phi\text{—}\underset{\phi}{\underset{|}{\overset{\text{O—Mg—Br}}{\overset{|}{\text{C}}}}}\text{—}\phi$$

Preparation of the Grignard Reagent

Procedure. Prepare an ethereal solution of phenylmagnesium bromide according to Procedure b of Section 60.1 using 2.0 grams (0.08 mole) of magnesium, 8.4 mL (12.5 grams; 0.08 mole) of bromobenzene, and 25 mL of anhydrous ether. At first, add about one quarter of the bromobenzene and then, after the initial reaction has moderated, add the remaining three quarters of the bromobenzene drop by drop from the addition funnel at such a rate that vigorous boiling is maintained but not so fast that ether escapes from the top of the condenser.

Reaction with Methyl Benzoate

Procedure. Cool the ethereal solution of phenylmagnesium bromide by swirling or stirring it in a pan of cold water. Then, while continuing to swirl or stir the mixture, add from the addition funnel a solution of 4.9 mL (5.4 grams; 0.04 mole) of methyl benzoate in 15 mL of anhydrous ether. Add the solution of methyl benzoate at such a rate that the reaction mixture boils gently. When the addition is complete, boil the mixture under reflux on the steam bath for about 15 minutes.

Hydrolysis of the Addition Product

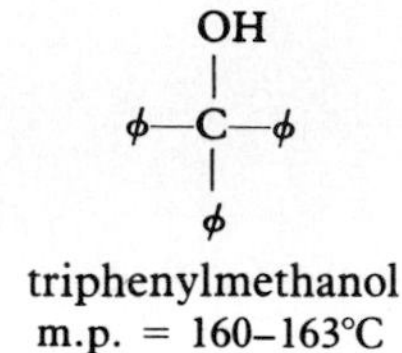

triphenylmethanol
m.p. = 160–163°C

Procedure. Carry out the hydrolysis as described in the procedure in Section 58.1, but extract the water layer with only one 25-mL portion of solvent ether. Dry the combined ether solutions of the product over anhydrous magnesium sulfate and filter the solution into a 100-mL boiling flask. Add to the flask 25 mL of hexane and concentrate the solution by distilling off the majority of the ether on the steam bath. Condense and collect the ether, and put the recovered ether in the container reserved for it. Continue the distillation until crystals start to appear. At this time, remove the flask from the distillation apparatus and set the flask aside to cool to room temperature. Complete the crystallization by cooling the flask in an ice bath and collect the produce by suction filtration, using a small amount of cold hexane to rinse the flask and to wash the crystals. The melting point of the product is 160–163°C.

Questions

1. Explain how triphenylmethanol could be prepared from bromobenzene and benzophenone. How would the procedure given in this section have to be modified if benzophenone were used?
2. Explain how triphenylmethanol could be prepared from bromobenzene and ethyl benzoate. How would the procedure given in this section have to be modified if ethyl benzoate were used?
3. a. What would be the product of reaction of methyl benzoate with two equivalents of ethyl magnesium bromide?
   b. Show how this substance could be prepared by two other Grignard syntheses.

Reference

A good general reference on the Grignard reagent and its uses is

1. M. S. Kharasch and O. Reinmuth, *Grignard Reactions of Nonmetallic Substances*, Prentice-Hall, Englewood Cliffs, N.J., 1964.

61.2 A VARIATION: TRIPHENYLMETHANOL FROM PHENYLMAGNESIUM BROMIDE AND BENZOPHENONE

If benzophenone is an intermediate in the formation of triphenylmethanol from methyl benzoate (Section 61.1), it should be possible to prepare triphenylmethanol from phenylmagnesium bromide and benzophenone. One way to do this would be to prepare 0.08 mole of phenylmagnesium bromide, as described in Section 61.1, and add to this a solution of 0.08 mole (14.6 g) of benzophenone in about 50 mL of anhydrous ether. Other than substituting an ethereal solution of benzophenone for an ethereal solution of methyl benzoate, the procedure of Section 61.1 could be followed.

$$\text{Ø—Mg—Br} + \underset{\text{benzophenone}}{\text{Ø—}\overset{\text{O}}{\overset{\|}{\text{C}}}\text{—Ø}} \xrightarrow{\text{dry ether}} \text{Ø—}\underset{\underset{\text{Ø}}{|}}{\overset{\overset{\text{O—Mg—Br}}{|}}{\text{C}}}\text{—Ø} \xrightarrow{\text{aqueous acid}} \underset{\text{triphenylmethanol}}{\text{Ø—}\underset{\underset{\text{Ø}}{|}}{\overset{\overset{\text{OH}}{|}}{\text{C}}}\text{—Ø}}$$

Question

1. If 0.08 mole of phenylmagnesium bromide is used in the reaction with benzophenone, what will be the theoretical yield of triphenylmethanol?

61.3 ANOTHER VARIATION: TRIPHENYLMETHANOL FROM PHENYLMAGNESIUM BROMIDE AND DIMETHYL CARBONATE

If triphenylmethanol can be prepared from benzophenone and one equivalent of phenylmagnesium bromide (Section 61.2), and from a benzoate ester and two equivalents of phenylmagnesium bromide (Section 61.1), it should be possible to prepare triphenylmethanol from a carbonate ester and *three* equivalents of phenylmagnesium bromide.

$$\underset{\text{benzophenone}}{\text{Ø—}\overset{\text{O}}{\overset{\|}{\text{C}}}\text{—Ø}} \qquad \underset{\text{a benzoate ester}}{\text{Ø—}\overset{\text{O}}{\overset{\|}{\text{C}}}\text{—O—R}} \qquad \underset{\text{a carbonate ester}}{\text{R—O—}\overset{\text{O}}{\overset{\|}{\text{C}}}\text{—O—R}}$$

The first equivalent of phenylmagnesium bromide would react with the carbonate ester to give a benzoate ester, which would react with a second equivalent of phenylmagnesium bromide to give benzophenone, which would react with the third equivalent of the Grignard reagent to give, after hydrolysis, triphenylmethanol.

$$\text{Ø—Mg—Br} + \text{R—O—}\overset{\text{O}}{\overset{\|}{\text{C}}}\text{—O—R} \longrightarrow \text{Ø—}\overset{\text{O}}{\overset{\|}{\text{C}}}\text{—O—R} + \text{R—O—Mg—Br}$$

Write out a detailed procedure for the preparation of triphenylmethanol, using 0.08 mole of phenylmagnesium bromide (Section 61.1) and either dimethyl carbonate (Aldrich; 500 grams for \$13.90) or diethyl carbonate (Aldrich; 1000 grams for \$9.20).

$$\underset{\text{dimethyl carbonate}}{\text{Me—O—}\overset{\text{O}}{\overset{\|}{\text{C}}}\text{—O—Me}} \qquad \underset{\text{diethyl carbonate}}{\text{Et—O—}\overset{\text{O}}{\overset{\|}{\text{C}}}\text{—O—Et}}$$

62 Aniline from Nitrobenzene

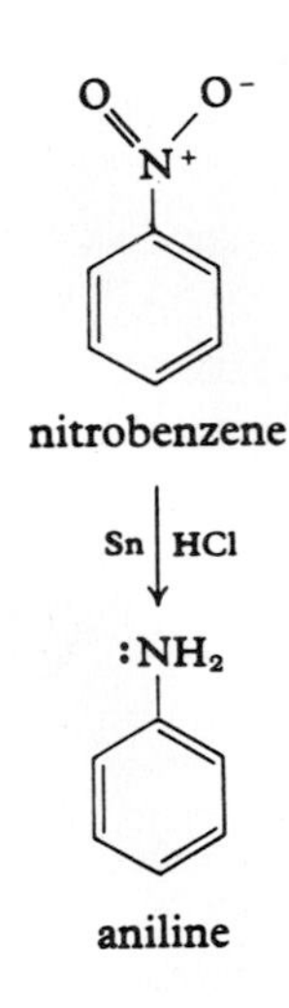

Aromatic nitro compounds can be reduced to the corresponding aromatic amine by many reagents. In this experiment, aniline is produced from nitrobenzene by tin and concentrated hydrochloric acid. The distillate obtained in the course of the procedure, a mixture of aniline and water, can be used directly for the preparation of acetanilide (Section 61.1); assume that the mixture contains 0.09 mole of aniline.

The infrared spectrum of aniline is shown in Figure 62.1.

Procedure. Place 25 g (0.212 mole) of granulated tin and 10.3 mL (12.3 g; 0.10 mole) of nitrobenzene in a 250-mL boiling flask. Make ready an ice bath, and add to the tin and nitrobenzene mixture 55 mL (0.66 mole) of conc. hydrochloric acid. Insert a thermometer and swirl the mixture. Keep the temperature of the reaction between 55 and 60°C by swirling and occasional immersion of the flask in the ice bath. After 15 minutes, fit the flask with a condenser and heat the reaction mixture on the steam bath with frequent swirling until the condensate in the condenser shows the absence of oily drops of unreacted nitrobenzene (about 15 minutes). Cool the flask in an ice bath and add slowly with swirling and cooling 50 mL of 50% aqueous sodium hydroxide (0.95 mole), followed by 75 mL of water. Fit the flask with a Claisen adapter (Note 1), and fit the center neck with a separatory funnel and the side neck with an adapter and condenser set for downward distillation (Note 2). Steam distill the aniline from the flask by heating it with a flame, adding water by means of the separatory funnel to keep the volume

Steam distillation

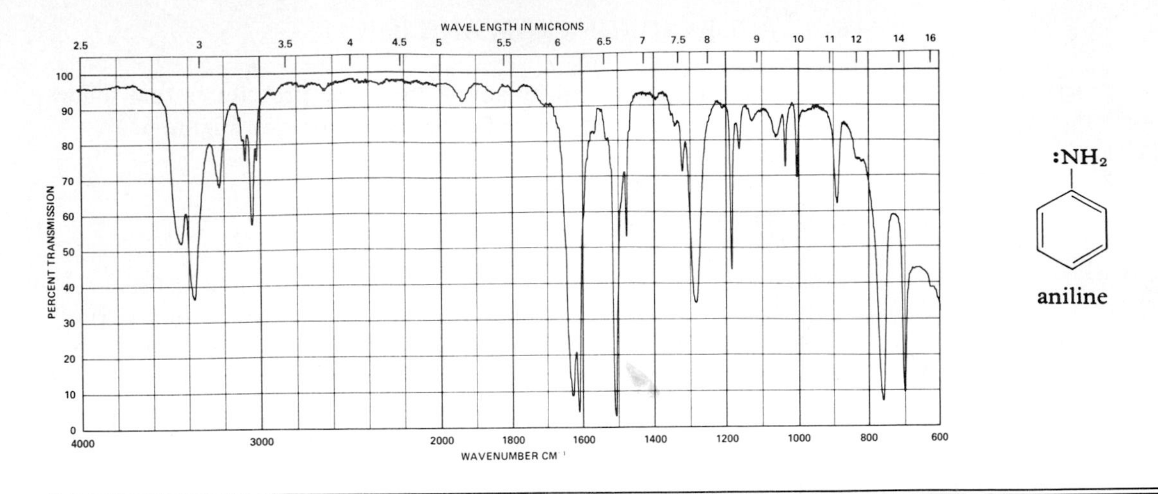

Figure 62.1. Infrared spectrum of aniline; thin film.

constant. After the distillate begins to come over clear, continue to distill; distill about 50 mL more. ■

Aniline can be isolated from the distillate by salting it out (using 20 g sodium chloride per 100 mL distillate), extraction with ether or dichloromethane, drying over sodium sulfate, and distilling, collecting the fraction boiling between 180 and 185°C as aniline.

Notes

1. The joints involving the Claisen adapter must be well greased, and the apparatus must be disassembled immediately after the distillation is stopped. Otherwise, the joints will freeze tight (Section 30).
2. The apparatus should look like that shown in Figure 11.3 except that the separatory funnel takes the place of the steam addition tube in the center neck of the Claisen adapter.

Time: a long 3 hours if the aniline is isolated and distilled.

63 Preparation of Amides

Amines can be converted to the corresponding amides of acetic acid in several ways. In the first experiment of this section, three procedures are given for the acetylation of aniline to form acetanilide. The second experiment presents one method for the preparation of *p*-ethoxyacetanilide, or phenacetin. The third experiment illustrates the preparation of an amide via the acid chloride by the synthesis of the insect repellent "Off."

63.1 ACETANILIDE FROM ANILINE

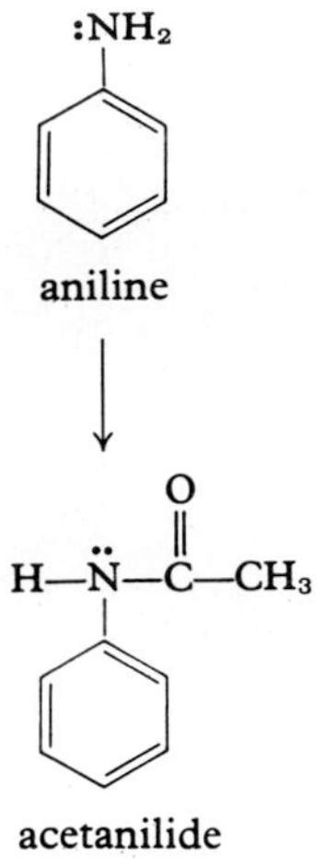

:NH_2

aniline

acetanilide

The acetanilide produced by any of the three procedures that follow should have an infrared spectrum like that shown in Figure 63.1.

Acetylation with Acetic Acid (Procedure a)

Procedure. Place 5.0 g (4.9 mL; 0.054 mole) aniline, 21 g (20.0 mL; 0.35 mole) acetic acid, and a pinch of zinc dust (Note 1) in a 125-mL boiling flask. Fit the flask with an air condenser, and heat the mixture under reflux for about four hours. Pour the hot mixture (Note 2) in a thin stream into 200 mL of cold water. ▪ After cooling the mixture in an ice bath for about 10 minutes, collect the product by suction filtration, and wash it with cold water (Note 3).

Acetylation with Acetic Anhydride (Procedure b)

Choice of procedures

Procedure. Place 5.0 g (4.9 mL; 0.054 mole) aniline, 5.4 g (5.0 mL; 0.053 mole) acetic anhydride, and a pinch of zinc dust (Note 1) in a 125-mL boiling flask. Fit the flask with a condenser and boil the mixture gently for 30 minutes. Pour the hot mixture in a thin stream into 200 mL of cold water. ▪ After cooling the mixture in an ice bath for about 10 minutes, collect the product by suction filtration and wash it with cold water (Note 3).

Acetylation with Acetic Anhydride and Sodium Acetate in Water (Procedure c)

Procedure. Dissolve 5.0 g (4.9 mL; 0.054 mole) aniline in 135 mL of water and 4.5 mL (0.054 mole) of conc. hydrochloric acid. If the solution is highly colored, treat it with decolorizing carbon and filter. Prepare a solution of 5.3 g (0.065 mole) anhydrous sodium acetate in 30 mL of water. To the solution of aniline hydrochloride, add 6.6 g (6.2 mL; 0.065 mole) acetic anhydride and, as soon as this has been brought into solution by swirling or stirring, add the solution of sodium acetate. Cool the mixture in an ice bath for about 10 minutes, collect the product by suction filtration, and wash it with cold water (Note 3).

Notes

1. The zinc reduces the colored impurities in the aniline and helps to prevent its oxidation during the reaction.
2. If the mixture is allowed to cool, it will set to a solid cake.
3. Acetanilide can be recrystallized from water.

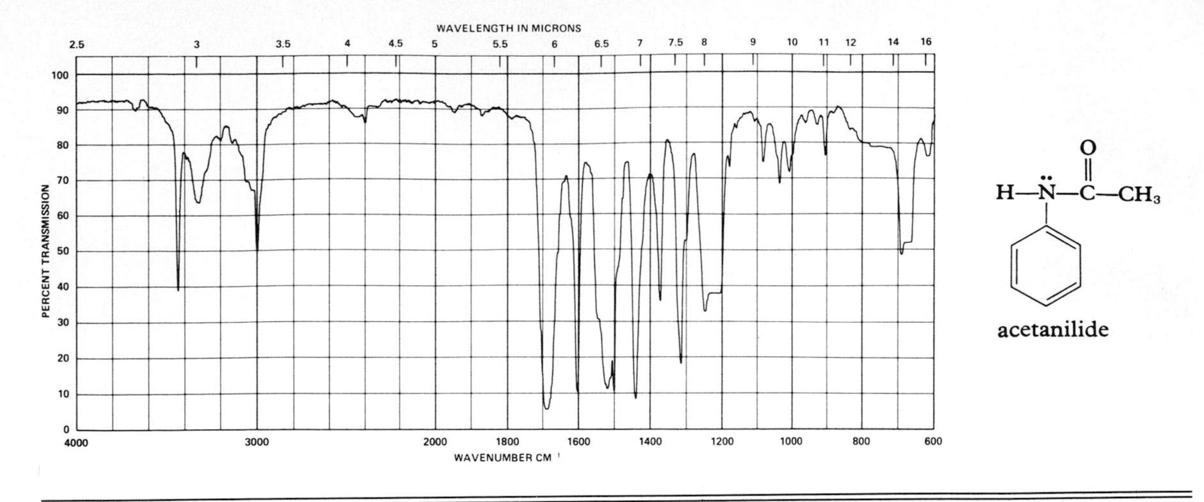

Figure 63.1. Infrared spectrum of acetanilide; $CHCl_3$ solution.

Questions

1. Justify your choice of procedure.
2. Why is a slight excess of acetic anhydride called for in Procedure c?
3. Why is a large excess of acetylating agent called for in Procedure a as compared to b and c?
4. What is the function of sodium acetate in Procedure c?
5. Could sodium acetate trihydrate be used in Procedure c? If so, what changes, if any, would be necessary in the experiment?

63.2 *p*-ETHOXYACETANILIDE FROM *p*-PHENETIDINE

$^+NH_3\ Cl^-$ … $O—CH_2—CH_3$

p-phenetidine hydrochloride

acetic anhydride | sodium acetate ↓

H—N̈—C(=O)—CH_3 … $O—CH_2—CH_3$

p-ethoxyacetanilide

p-Ethoxyacetanilide, or phenacetin, is described by the *Merck Index* as an analgesic and antipyretic (pain reliever and fever reducer). It is a component of "APC" tablets, along with aspirin (Section 41) and caffeine (Section 42). Phenacetin can be prepared by treatment of *p*-phenetidine hydrochloride with sodium acetate and acetic anhydride.

Figures 63.2 and 63.3 show the infrared and NMR spectra of *p*-ethoxyacetanilide.

Procedure. Add 150 mL of water and 7.4 grams (7.0 mL; 0.054 mole) of *p*-phenetidine to a 250-mL Erlenmeyer flask. Dissolve the amine by adding 4.5 mL (0.054 mole) of conc. hydrochloric acid. If the solution is very dark, treat it with decolorizing carbon, and filter. Prepare a solution of 5.3 grams (0.065 mole) of anhydrous sodium acetate in 30 mL of water. To the solution of *p*-phenetidine hydrochloride, add 6.6 grams (6.2 mL; 0.065 mole) of acetic anhydride. Swirl the mixture

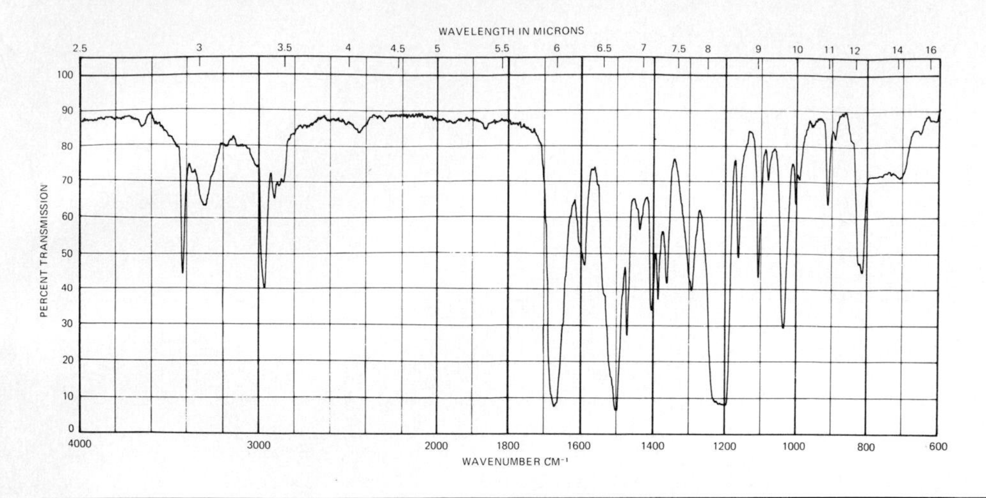

Figure 63.2. Infrared spectrum of *p*-ethoxyacetanilide; $CHCl_3$ solution.

for a few seconds and then add the solution of sodium acetate. Phenacetin forms and precipitates immediately. Collect the product by suction filtration and wash it with cold water. Recrystallize the damp material from about 35 mL of 95% ethanol. On slow cooling, phenacetin will crystallize as long spars. Yield after recrystallization: 7.5 grams (78%); m.p.: 137–138°C.

Time: 2 hours.

Figure 63.3. NMR spectrum of *p*-ethoxyacetanilide; $CDCl_3$ solution.

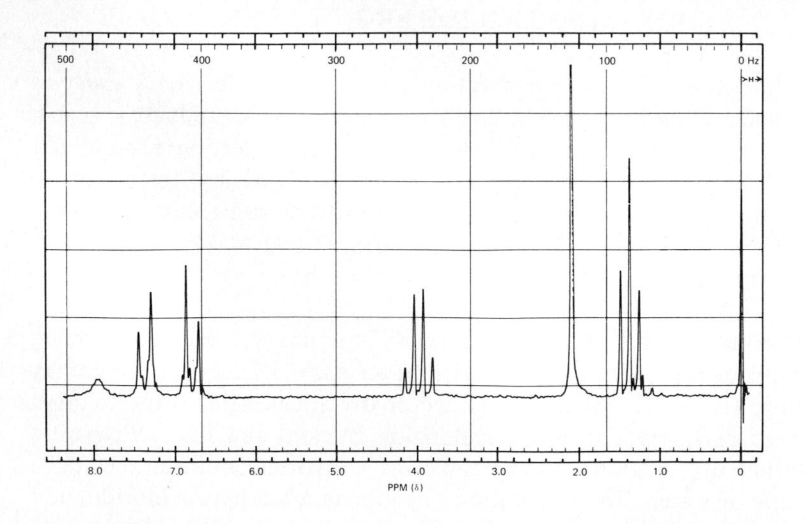

p-ethoxyacetanilide

63.3 N,N-DIETHYL-*m*-TOLUAMIDE FROM *m*-TOLUIC ACID; "OFF"

One of the more effective insect repellents is the N,N-diethylamide of *m*-toluic acid. It can be prepared by converting *m*-toluic acid to the acid chloride with thionyl chloride and then treatment of the product with diethylamine:

$$m\text{-}CH_3C_6H_4\text{-}C(=O)\text{-}OH \xrightarrow{SOCl_2} m\text{-}CH_3C_6H_4\text{-}C(=O)\text{-}Cl \xrightarrow{Et_2\ddot{N}H} m\text{-}CH_3C_6H_4\text{-}C(=O)\text{-}\ddot{N}(CH_2CH_3)_2$$

The amide is an ingredient of a number of commercial insect repellents, but it is probably most familiar as the active ingredient of "Off." Its infrared spectrum is shown in Figure 63.4.

Procedure. Place 8.4 grams (60 mmole) of *m*-toluic acid in a dry 250-mL boiling flask and add to this 9.0 mL (14.5 grams; 122 mmole) of thionyl chloride (Note 1). Fit the flask with a reflux condenser and heat the mixture on the steam bath for about 20 minutes. Since SO_2 and HCl will be evolved during the heating period you must either work in the hood or make provision for the absorption of these gases (Section 35).

At the end of the heating period, remove the condenser and add to the flask 40 mL of toluene. Add a magnetic stirring bar to the flask and

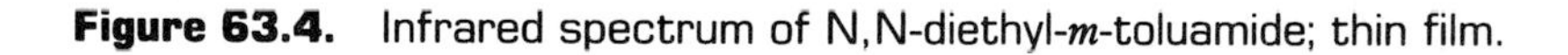

Figure 63.4. Infrared spectrum of N,N-diethyl-*m*-toluamide; thin film.

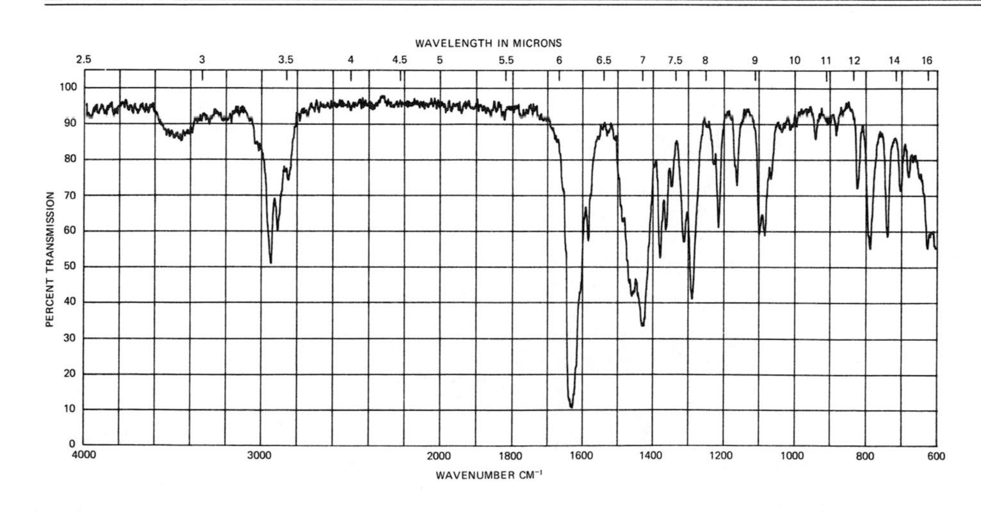

clamp the flask in position over a magnetic stirring motor. Fit the flask with a Claisen adapter and place in the center neck a dropping funnel that contains a solution of 20 mL (14.2 grams; 196 mmole) of diethylamine in 25 mL of toluene. Add the diethylamine solution dropwise while stirring the contents of the flask; the addition should take from 30 to 40 minutes.

When the addition of the diethylamine solution is complete, add to the flask 25 mL of cold water. Continue stirring in order to thoroughly mix the contents of the flask. Transfer the dark reaction mixture to a separatory funnel and draw off the lower, aqueous, layer (Note 2). Wash the toluene solution again with a second 25-mL portion of water. Now wash the toluene solution first with a solution prepared by diluting 5 mL of 50% sodium hydroxide with water to 50 mL (use this solution in two 25-mL portions) and then with a solution prepared by diluting 10 mL of conc. hydrochloric acid with water to 50 mL (use this solution also in two 25-mL portions). Finally wash the toluene solution with two more 25-mL portions of water.

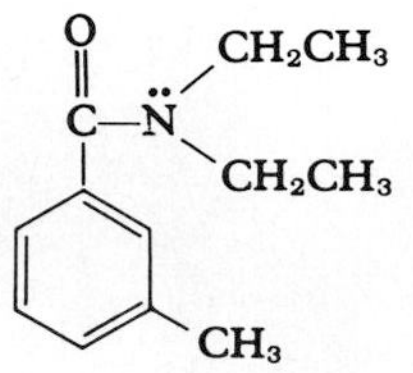

N,N-diethyl-*m*-toluamide

Transfer the brown toluene solution of N,N-diethyl-*m*-toluamide to a 125-mL Erlenmeyer flask and dry it for 5 to 10 minutes over anhydrous magnesium sulfate. Filter the dried solution into a 100-mL boiling flask and remove the toluene by distillation. Since the boiling point of the product is quite high, it should be distilled under reduced pressure (Section 10) (Note 3). Although it is colorless when pure, the product is usually obtained as a light brown oil. The boiling point has been reported to be 160°C at 20 Torr and 173°C at 24 Torr. Yield: about 7 grams (60%).

Notes

1. Thionyl chloride is a lachrymator. You should work in the hood while you measure it out and add it to your flask.
2. It may be impossible to distinguish the layers. If so, draw off 25 mL of the liquid.
3. The product can also be purified by chromatography on alumina (B. J.-S. Wang, *J. Chem. Educ.* **51,** 631 (1974)).

Time: 4 hours.

Reference

1. E. T. McCabe, W. F. Barthels, S. I. Gertler, and S. A. Hall, *J. Org. Chem.* **19,** 493 (1954).

64 Electrophilic Substitution Reactions of Benzene Derivatives

Most electrophilic substitution reactions of benzene and of substituted benzenes are thought to take place according to the following general mechanism:

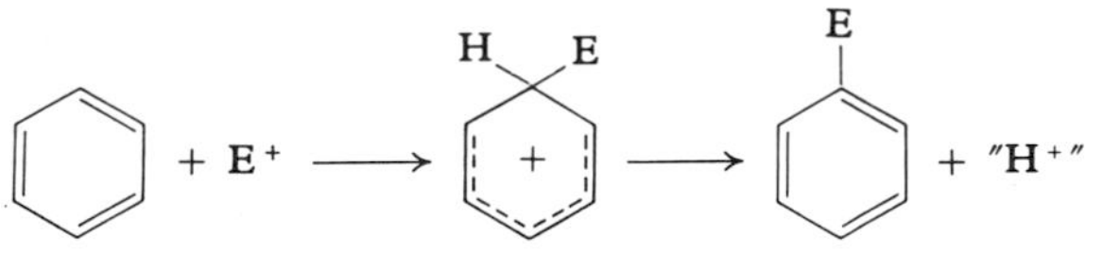

H E E

+ E^+ ⟶ + ⟶ + "H^+" (64.1)

sigma complex

In some cases, the experimental evidence is interpreted best by proposing that E^+, the electrophile, is completely formed as a cation before it reacts with the aromatic compound, as implied by Equation 64.1. In other cases, the data are best interpreted in terms of a simultaneous attack of the aromatic system on E—X and loss of the leaving group X^-:

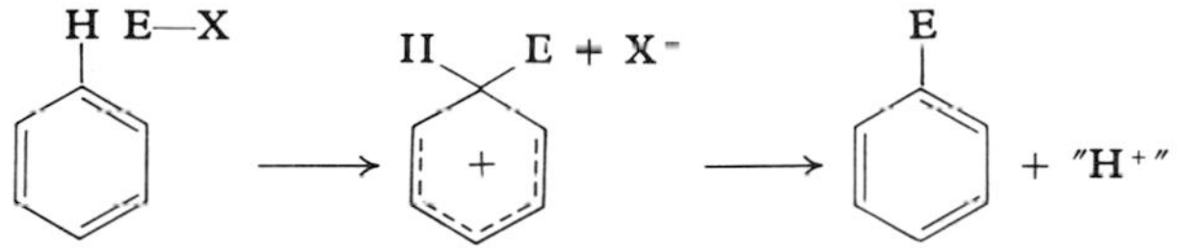

H E—X H E + X^- E

(64.2)

⟶ + ⟶ + "H^+"

For convenience, however, electrophilic aromatic substitution reactions are often interpreted in terms of attack by previously formed E^+, with the mental reservation that this may not always be the best representation of the reaction.

A complete discussion of the variety of mechanistic possibilities for electrophilic aromatic substitution reactions would have to include the possibility of formation of pi complexes before and after the sigma complex. In some cases, the formation of the first pi complex appears to be the rate-determining step.

The reactivity toward electrophilic substitution of a monosubstituted benzene may be either greater or less than that of benzene itself; the substituent is said to be either activating or deactivating. The new substituent may enter either the *ortho* and *para* positions, or the *meta* position; the original substituent is said to be either *ortho-para* directing, or *meta* directing. Substituents of the general type —A=O have always been found to be deactivating and *meta* directing; examples are the nitro group, $—NO_2$, and the carbonyl group, —C=O. Substituents of the general type —A: have, with the exception of bromine and chlorine, been found to be activating and *ortho-para* directing. Bromine and chlorine have been found to be *ortho-para* directing but slightly deactivating, presumably because of greater electron withdrawal through an inductive mechanism rather than electron release by a resonance mechanism.

64.1 METHYL *m*-NITROBENZOATE FROM METHYL BENZOATE

$:\ddot{O}=\overset{+}{N}=\ddot{O}:$
nitronium ion

The actual nitrating agent in the nitration of most aromatic compounds is the nitronium ion, NO_2^+. The source of the nitronium ion can be a nitronium salt, such as nitronium fluoroborate, $NO_2^+BF_4^-$, or, as in this experiment, the ion can be derived from nitric acid by the action of sulfuric acid:

$$H-O-S(=O)_2-O-H + H-O-\overset{+}{N}(-O^-)(=O) \rightleftharpoons H-O-S(=O)_2-O^- + H-\overset{+}{O}(H)-\overset{+}{N}(-O^-)(=O)$$

$$H-\overset{+}{O}(H)-\overset{+}{N}(-O^-)(=O) \longrightarrow O=\overset{+}{N}=O + H_2O$$

$$H_2O + H-O-S(=O)_2-O-H \rightleftharpoons H_3O^+ + H-O-S(=O)_2-O^-$$

The infrared spectrum of the product methyl *m*-nitrobenzoate is shown in Figure 64.1.

$C_6H_5-C(=O)-O-CH_3$
methyl benzoate

↓

$m-O_2N-C_6H_4-C(=O)-O-CH_3$
methyl *m*-nitrobenzoate

Procedure. Place 14.5 mL (25 g; 0.250 mole) of conc. sulfuric acid in a 125-mL Erlenmeyer flask, cool it to 0°C in an ice bath, and then add 6.3 mL (6.8 g; 0.050 mole) of methyl benzoate with swirling. Prepare a mixture of 5 mL (9 g; 0.090 mole) of conc. sulfuric acid and 5 mL (7 g; 0.078 mole) of conc. nitric acid, and cool it in an ice bath. Add the cold acid mixture drop by drop over a period of about 5 minutes to the methyl benzoate solution, which is constantly swirled in the ice bath. Allow the resulting mixture to stand at room temperature for an additional 10 minutes with occasional swirling and then pour it with stirring over about 50 grams of crushed ice. Collect the resulting solid by suction filtration and wash it thoroughly with water to remove the acids. Finally, wash the product with two 5-mL portions of ice-cold methanol. The product can be recrystallized from a small amount of methanol.

Time: 2 hours.

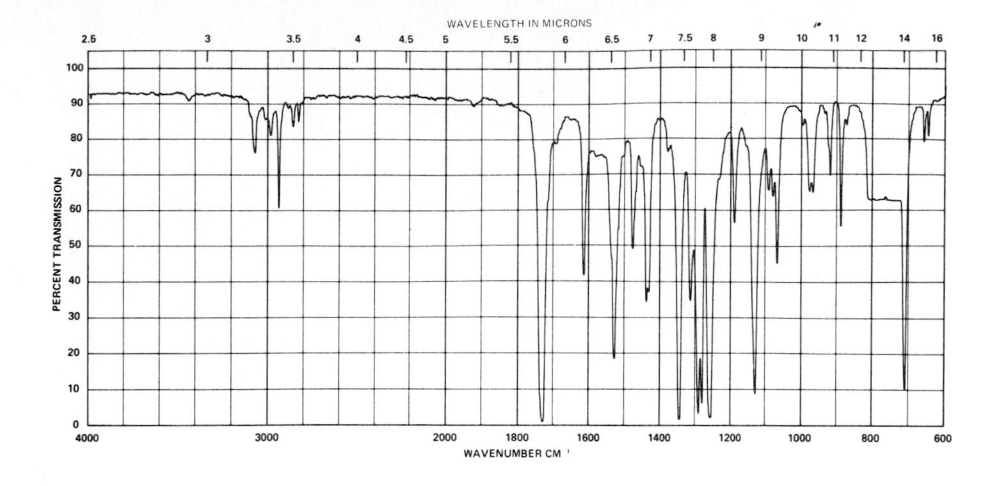

Figure 64.1. Infrared spectrum of methyl *m*-nitrobenzoate; CCl_4 solution.

64.2 *p*-BROMOACETANILIDE FROM ACETANILIDE

The electrophilic agent in many electrophilic aromatic bromination reactions is elemental bromine; bromide ion is displaced from the bromine molecule as the electrophilic attack is made upon the aromatic ring (see Equation 64.2). When the reaction is catalyzed by ferric bromide, the effective electrophile can be considered to be formed in the following way:

$$Br{-}Br + FeBr_3 \longrightarrow Br{-}\overset{+}{Br}{-}\overset{-}{Fe}Br_3$$

$$Br{-}\overset{+}{Br}{-}\overset{-}{Fe}Br_3 \longrightarrow Br^+ + Br{-}\overset{-}{Fe}Br_3$$

$H{-}\ddot{N}{-}C(=O){-}CH_3$ (on benzene ring)

acetanilide

$\downarrow$ Br_2 | HOAc

$H{-}\ddot{N}{-}C(=O){-}CH_3$ (on benzene ring, para-Br)

p-bromoacetanilide

A catalytic amount of ferric bromide apparently may be generated in the reaction mixture from bromine and elemental iron in the form of an iron tack or iron filings. Figure 64.2 shows the infrared spectrum of *p*-bromoacetanilide.

Procedure. Dissolve 6.7 g (0.050 mole) acetanilide in 25 mL of glacial acetic acid in a 250-mL flask, and add slowly, with stirring, a solution of 8.1 g (2.60 mL; 0.051 mole) of bromine in 5 mL of glacial acetic acid

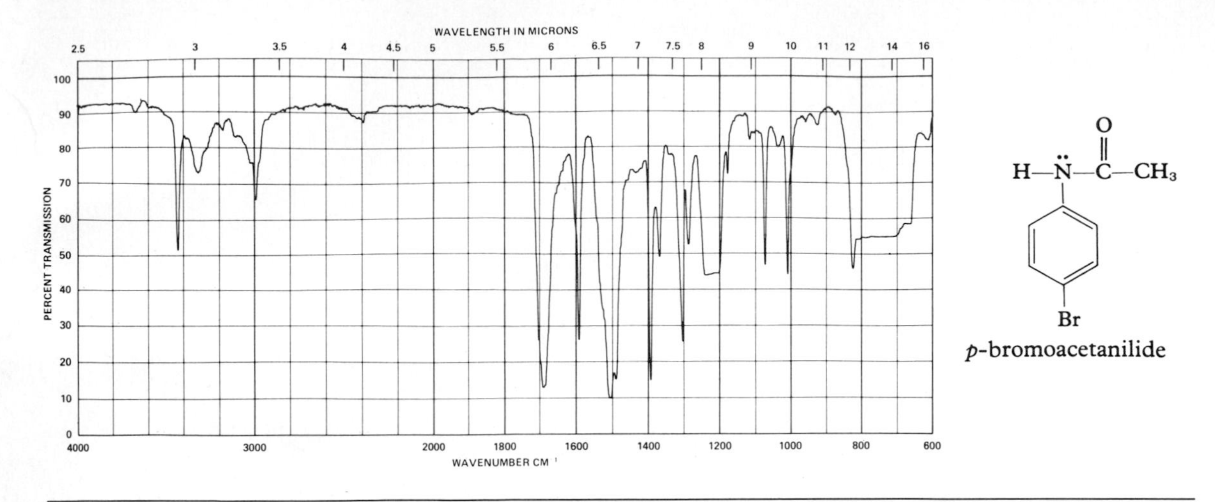

Figure 64.2. Infrared spectrum of *p*-bromoacetanilide; $CHCl_3$ solution.

(*caution:* see Note 1). Stir the mixture for 2 or 3 minutes more, and then add slowly with stirring 200 mL of water. Add enough concentrated sodium bisulfite solution to discharge the yellow color, and then collect the product by suction filtration, washing it well with water and sucking it as dry as possible. Yield: 10.2 g (96%). The product can be recrystallized from ethanol, using 3.3 mL per gram, with a recovery of 90%; m.p., 168–169°C.

Note

1. See Section 1.7 for precautions to be observed when working with bromine.

Time: 2 hours.

64.3 2,4-DINITROBROMOBENZENE FROM BROMOBENZENE

Br
NO_2
NO_2
2,4-dinitrobromobenzene

The 2,4-dinitrobromobenzene produced should have infrared and NMR spectra like those of Figures 64.3 and 64.4.

Procedure. Heat a mixture of 15 mL (27 g; 0.27 mole) of conc. sulfuric acid and 5 mL (7 g; 0.08 mole) of conc. nitric acid in a 50-mL Erlenmeyer flask to 85–90°C. Add 2.0 mL (3.0 g; 0.019 mole) of bromobenzene in 3 or 4 portions during a minute, and swirl the mixture

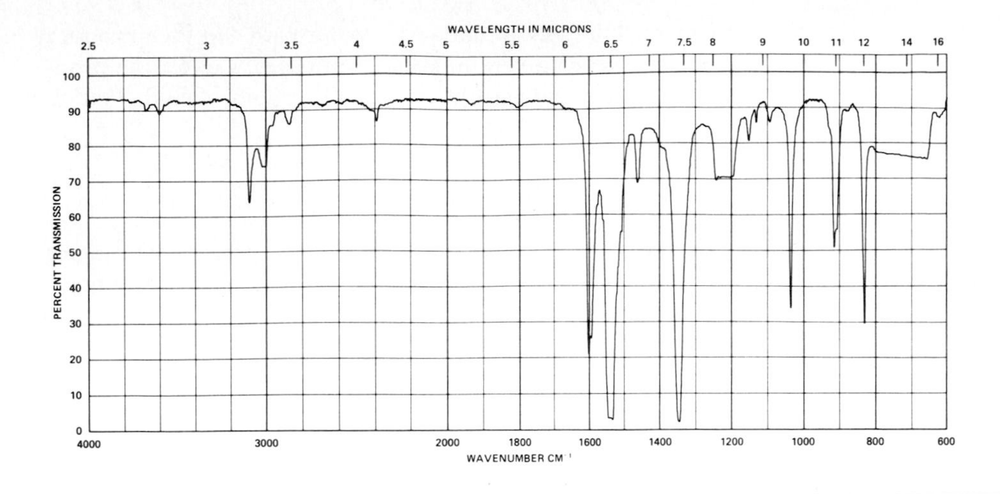

Figure 64.3. Infrared spectrum of 2,4-dinitrobromobenzene; $CHCl_3$ solution.

well after each addition. The temperature will rise to 130–135°C. Allow the mixture to stand with occasional swirling for 5 minutes, then cool it nearly to room temperature and pour it over about 100 g of ice. Stir the resulting mixture until the product has solidified, crush the lumps, and collect the crude product by suction filtration.

Recrystallize the product by dissolving it in 15 mL of hot 95% ethanol and allowing the solution to cool (Note 1). After crystals have

Figure 64.4. NMR spectrum of 2,4-dinitrobromobenzene; $CDCl_3$ solution.

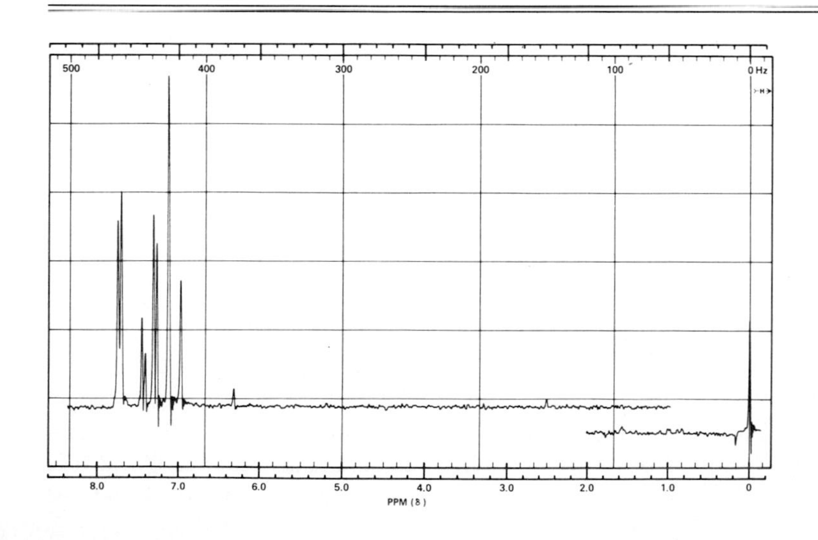

Br

NO_2

NO_2

2,4-dinitrobromobenzene

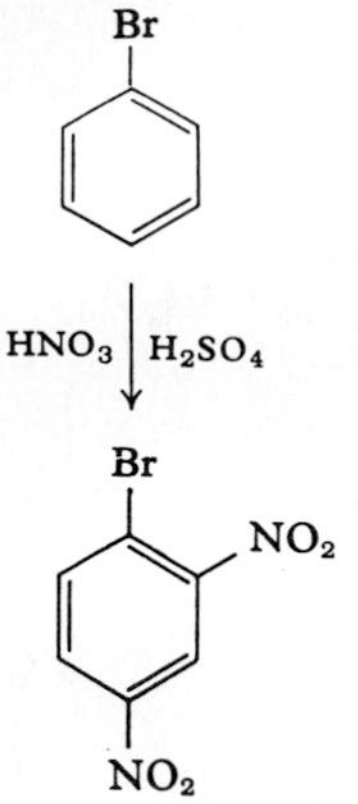

Br

HNO_3 | H_2SO_4

Br
NO_2
NO_2

started to form, cool the solution in a cold-water bath and then in an ice bath. Crystallization is complete in about 5 minutes. Collect the product by suction filtration and wash with a small amount of cold ethanol.

Note

1. The product usually separates as an oil, but vigorous swirling of the mixture when the oil appears will promote crystallization.

Time: 2 hours.

65 Nucleophilic Aromatic Substitution Reactions of 2,4-Dinitrobromobenzene

Certain benzene derivatives undergo nucleophilic substitution reactions. The larger the number of electron-withdrawing substituents on the ring, especially in the positions *ortho* and *para* to the point of substitution, the greater the rate of the reaction. This effect is explained by a two-step addition-elimination mechanism in which the role of the electron-withdrawing substituent is to stabilize the negative charge on the benzene ring in the intermediate:

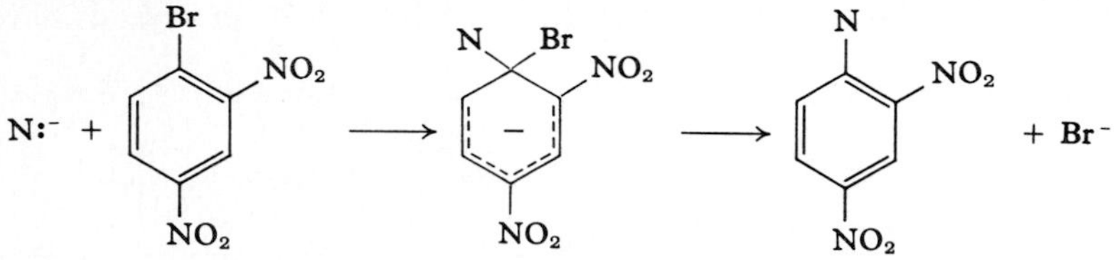

N:⁻ + (Br, NO_2, NO_2) ⟶ (N, Br, −, NO_2, NO_2) ⟶ (N, NO_2, NO_2) + Br^-

The same groups, for example the —NO_2 group, are activating for nucleophilic aromatic substitution and deactivating for electrophilic aromatic substitution; both effects are explained by the electron-withdrawing ability of the groups.

65.1 2,4-DINITROANILINE

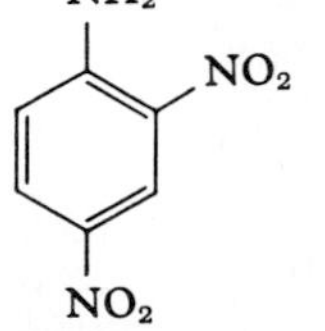

NH_2
NO_2
NO_2
2,4-dinitroaniline
(yellow)

Procedure. Place 25 mL of ethanol and 2.47 g (0.010 mole) of 2,4-dinitrobromobenzene in a 250-mL boiling flask. Heat the mixture gently to dissolve the solid. Add 20 mL (18 g; 0.32 mole) of conc. ammonium hydroxide. Relieve the resulting cloudiness of the solution by adding another 20 mL of ethanol and warming slightly. Allow the solution to stand for at least 96 hours (Note 1). ■ Then collect the yellow solid by suction filtration and recrystallize it from a mixture of 7.5 mL of water and 20 mL of ethanol. Yield: about 60%.

Note

1. A shorter period of standing will give a lower yield. Heating the mixture seems to speed loss of ammonia more than the desired reaction.

Time: 1 hour plus 4-day reaction time.

65.2 2,4-DINITROPHENYLHYDRAZINE

Procedure. Prepare a solution of 0.5 g (2 mmole) of 2,4-dinitrobromobenzene in 7.5 mL of 95% ethanol. Heat the solution almost to boiling and add to it a solution of 0.5 mL (10 mmole) of 64% hydrazine in 2.5 mL of 95% ethanol. Allow the light orange solution, which rapidly turns a deep red-purple, to cool undisturbed for 15–20 minutes. Collect the resulting crystals by suction filtration and wash them with a little ethanol. The product sometimes crystallizes as red-purple prisms alone, and sometimes as red-purple prisms and orange blades. Recrystallization of either form from boiling ethyl acetate (50 mL per gram) affords a product in the form of orange plates.

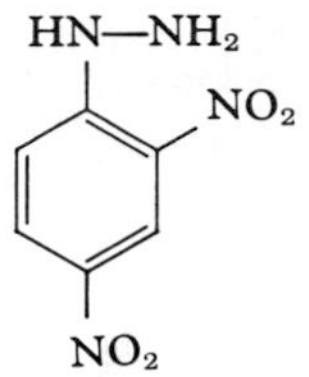

HN—NH_2
NO_2
NO_2
2,4-dinitrophenylhydrazine
(orange plates)

Time: 2 hours.

65.3 2,4-DINITRODIPHENYLAMINE

Procedure. Place 20 mL of ethanol, 1.84 g (7.5 mmole) of 2,4-dinitrobromobenzene, and 1.5 g (1.5 mL; 16 mmole) of aniline in a 250-mL boiling flask. Gently boil the mixture under reflux until all the solid has been dissolved. After allowing the mixture to cool slowly to room temperature (Note 1), collect the red needles by suction filtration. The product may be recrystallized from boiling ethanol (Note 2).

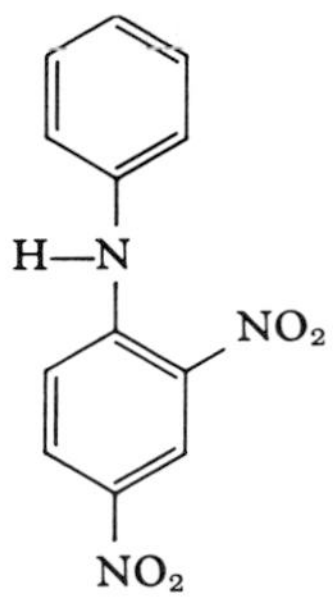

H—N
NO_2
NO_2
2,4-dinitrodiphenylamine
(red needles)

Notes

1. The solution should be allowed to stand for at least an hour before filtering.
2. The recrystallization will require 100–135 mL. The product dissolves slowly in boiling ethanol.

Time: 2 hours.

65.4 2,4-DINITROPHENYLPIPERIDINE

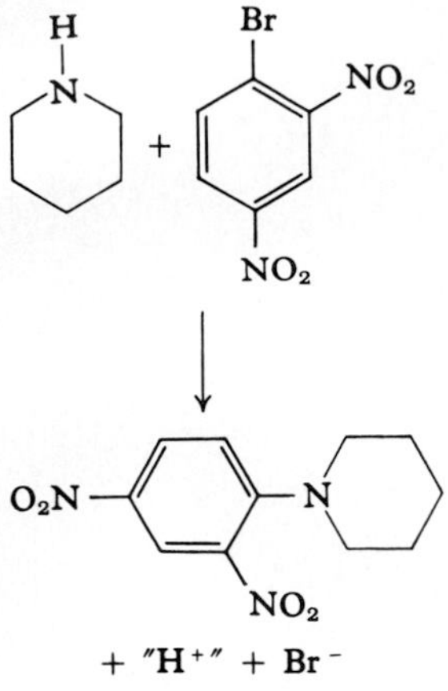

Procedure (Reference 1). Place 25 mL of ethanol and 1.84 g (7.5 mmole) of 2,4-dinitrobromobenzene in a 250-mL boiling flask. Heat the mixture gently to dissolve the solid. Then add 1.72 g (2.0 mL; 20 mmole) of piperidine and boil the mixture under reflux for 10 minutes. After allowing the solution to cool slowly to room temperature (Note 1) and finally cooling it in an ice bath, collect the orange needles by suction filtration. The product may be recrystallized from 20 mL of ethanol.

H N + Br NO_2 NO_2 → O_2N N NO_2 + "H^+" + Br^-

(orange needles)

Note

1. The solution tends to become supersaturated. Crystallization may be induced by adding a seed crystal or by tapping the flask with a spatula.

Time: 2 hours.

Reference

1. J. W. McFarland, *Organic Laboratory Chemistry*, C. V. Mosby, St. Louis, 1969, p. 209.

65.5 A VARIATION: 4′-SUBSTITUTED 2,4-DINITROPHENYLANILINES

If a *p*-substituted aniline is used instead of aniline in the procedure of Section 65.3, the reaction with 2,4-dinitrobromobenzene should give a 4′-substituted 2,4-dinitrophenylaniline.

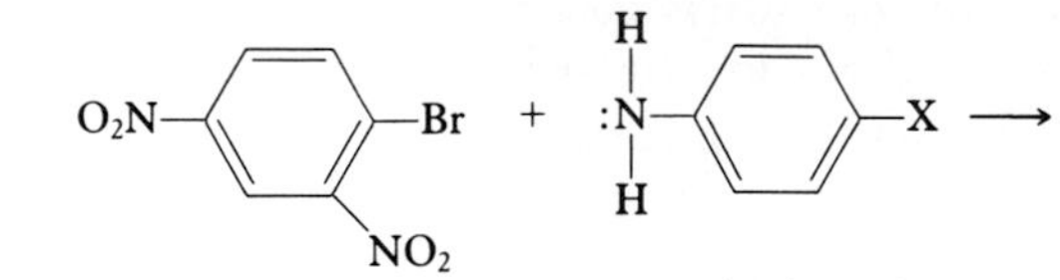

O_2N—Br + H–:N(H)—X ⟶ (NO_2)

2,4-dinitrobromobenzene *p*-substituted aniline

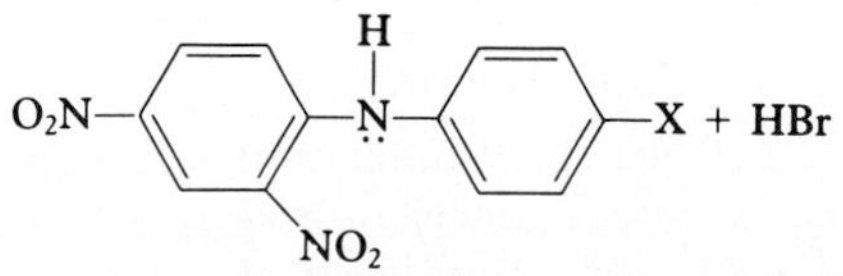

O_2N—N(H)—X + HBr (NO_2)

4′-substituted 2,4-dinitrophenylaniline

Some *p*-substituted anilines that could be used include: *p*-methylaniline (*p*-toluidine); *p*-bromoaniline; *p*-ethoxyaniline (*p*-phenetidine); and *p*-methoxyaniline (*p*-anisidine), which gives a beautifully crystalline red product.

66 Diazonium Salts of Aromatic Amines

Diazonium salts are prepared from the corresponding aromatic amine by treatment of an acidic solution of the amine with one equivalent of aqueous sodium nitrite solution:

$$\underset{X^-}{Ar{-}\overset{+}{N}H_3} + \underset{\text{nitrous acid}}{H{-}O{-}\ddot{N}{=}O} \longrightarrow \underset{\text{diazonium salt}}{\{Ar{-}\overset{+}{N}{\equiv}N{:} \longleftrightarrow Ar{-}\ddot{N}{=}\overset{+}{N}{:}\}} + 2H_2O$$

The first step of the mechanism is electrophilic attack of the conjugate acid of nitrous acid on the nitrogen atom of the free amine, which is in equilibrium with its conjugate acid:

$$Ar{-}\overset{+}{N}H_3 \rightleftharpoons H^+ + \underset{\text{amine}}{Ar{-}\ddot{N}H_2} \xrightarrow[\text{slow}]{H{-}\overset{+}{O}H{-}\ddot{N}{=}O} Ar{-}\overset{+}{N}H_2{-}\ddot{N}{=}O + H_2O$$

The final steps are fast; they are enolization of the N-nitrosoamine and ionization of the resulting diazohydroxide:

$$Ar{-}\overset{+}{N}H_2{-}\ddot{N}{=}O \rightleftharpoons H^+ + \underset{\text{N-nitrosoamine}}{Ar{-}\ddot{N}H{-}\ddot{N}{=}O} \rightleftharpoons \underset{\text{diazohydroxide}}{Ar{-}\ddot{N}{=}\ddot{N}{-}O{-}H}$$

$$\underset{\text{diazohydroxide}}{Ar{-}\ddot{N}{=}\ddot{N}{-}O{-}H} + H^+ \rightleftharpoons \underset{\text{diazonium ion}}{\{Ar{-}\ddot{N}{=}\overset{+}{N}{:} \longleftrightarrow Ar{-}\overset{+}{N}{\equiv}N{:}\}} + H_2O$$

The diazotization is carried out as near 0°C as possible in order to minimize the rate of the reaction of the diazonium salt with water to form the corresponding phenol:

$$\underset{\text{diazonium ion}}{\{Ar{-}\ddot{N}{=}\overset{+}{N}{:} \longleftrightarrow Ar{-}\overset{+}{N}{\equiv}N{:}\}} + H_2O \longrightarrow \underset{\text{phenol}}{Ar{-}OH} + {:}N{\equiv}N{:} + H^+$$

Any phenol that forms will react with another molecule of the diazonium salt to form a highly colored azo compound (compare Section 68.2). If the aqueous solution of the diazonium salt is allowed to warm to room temperature, the formation of the phenol can become a significant reaction.

Since diazotization is an exothermic reaction, one might attempt to minimize the temperature rise and the subsequent formation of phenol by very slow addition of the sodium nitrite solution. However, the diazonium salt can react with undiazotized amine to form an azo compound (Section 68.4). Because of this, it is desirable to diazotize as quickly as possible in order to minimize the time during which the diazonium salt and undiazotized amine are present together. As a compromise, diazotization is usually carried out as fast as possible with good stirring and cooling, but slowly enough that the temperature of the solution does not rise above about 5°C. The lower the temperature is, the slower, also, will be the reaction between the diazonium salt and undiazotized amine.

A compromise

An excess of sodium nitrite is to be avoided since the product of a subsequent reaction can react with excess nitrous acid.

66.1 BENZENEDIAZONIUM CHLORIDE FROM ANILINE

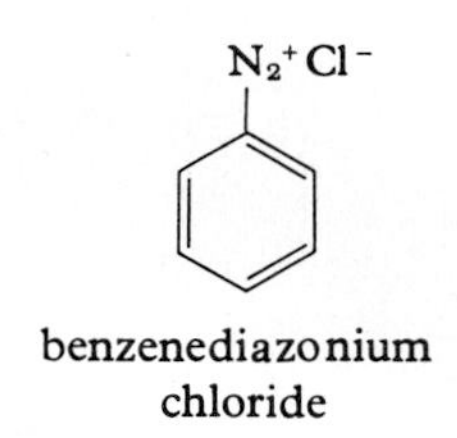

benzenediazonium chloride

Procedure. Dissolve 9.1 mL (9.3 g; 0.10 mole) of aniline in a mixture of 25 mL (0.30 mole) of conc. hydrochloric acid and 25 mL of water. Cool the solution to 0°C in an ice bath, and add with good stirring and cooling (Note 1) a solution of 7.5 g (0.108 mole) of sodium nitrite in 25 mL of water at such a rate that the temperature of the solution does not exceed 5°C. Add only enough of this solution to give a slight excess of nitrous acid (Note 2). The resulting solution of benzenediazonium chloride must be stored in the ice bath and used fairly quickly.

Time: 1 hour.

66.2 *p*-TOLUENEDIAZONIUM CHLORIDE FROM *p*-TOLUIDINE

$N_2^+ Cl^-$

CH_3

p-toluenediazonium chloride

Procedure. Dissolve 10.7 g (0.10 mole) of *p*-toluidine in 15 mL of water by adding 10 mL (0.12 mole) of conc. hydrochloric acid and gently heating the mixture. Now add another 15 mL (0.18 mole) of conc. hydrochloric acid and cool the mixture to 0°C in an ice bath (Note 3). With good stirring and cooling (Note 1), add a solution of 7.5 g (0.108 mole) of sodium nitrite in 25 mL of water at such a rate that the temperature of the solution does not exceed 5°C. Add only enough of this solution to give a slight excess of nitrous acid (Note 2). The resulting solution of *p*-toluenediazonium chloride must be stored in an ice bath and used fairly soon.

Time: 1 hour.

66.3 *p*-NITROBENZENEDIAZONIUM SULFATE FROM *p*-NITROANILINE

Procedure. Make a solution of 20 mL (0.36 mole) of conc. sulfuric acid in 100 mL of water. Add to this 13.8 g (0.10 mole) of *p*-nitroaniline and heat the mixture gently to bring the amine into solution. Cool the mixture to about 10°C in an ice bath and add to the suspension of *p*-nitroaniline sulfate, with stirring and cooling, a solution of 6.9 g (0.10 mole) of sodium nitrite in 20 mL of water at such a rate that the temperature does not rise above about 10°C. Store the resulting solution of *p*-nitrobenzenediazonium sulfate in the ice bath.

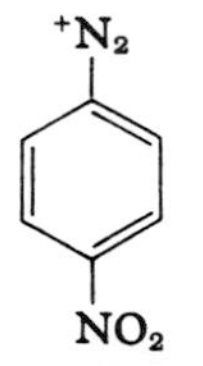

$^{+}N_2$

NO_2

p-nitrobenzene-diazonium ion

Notes

1. The efficiency of cooling can be increased by adding pieces of ice to the reaction mixture.
2. Since nitrous acid oxidizes iodide to iodine, the presence of nitrous acid in the mixture can be detected by placing a drop of the solution on a piece of starch potassium iodide test paper. An immediate blue color will be produced if nitrous acid is present. If the quantities of reagents have been carefully measured, one need not start testing for excess nitrous acid until most of the solution of sodium nitrite has been added. The test for nitrous acid should be positive for at least 1–2 minutes after the last addition has been made in order to safely assume that slightly more than an equivalent of sodium nitrite has been added.
3. *p*-Toluidine hydrochloride will crystallize out of solution. As the diazotization of the dissolved salt proceeds, the precipitated salt goes into solution.

Time: 1 hour.

67 Replacement Reactions of Diazonium Salts

The two most useful reactions of aryl diazonium salts are the replacement reactions and the coupling reactions. The procedure of this section illustrates one of the replacement reactions, and Section 68 presents examples of coupling reactions.

67.1 CHLOROBENZENE FROM BENZENEDIAZONIUM CHLORIDE

$$\{\phi\text{—}\overset{+}{N}\equiv N\text{:} \longleftrightarrow \phi\text{—}\ddot{N}=\overset{+}{N}\text{:}\} + CuCl_2^- \xrightarrow{\text{cold}} \phi\text{—}\overset{+}{N}_2\ CuCl_2^-$$

benzenediazonium ion — cuprous diazonium chloride

$$\phi\text{—}\overset{+}{N}_2\ CuCl_2^- \xrightarrow{\text{warm}} \phi\text{—}Cl + CuCl + N_2$$

chlorobenzene

The mechanism of the decomposition of cuprous diazonium chloride is not completely agreed upon, but the reaction does appear to involve free radical intermediates.

Figure 67.1 shows the infrared spectrum of chlorobenzene.

Preparation of cuprous chloride

Procedure. Dissolve 30 g (0.12 mole) of copper sulfate pentahydrate in 100 mL of hot water in a 500-mL boiling flask. Then add and dissolve 10 g (0.17 mole) of sodium chloride. Prepare a solution of sodium sulfite by dissolving 7 g (0.067 mole) of sodium bisulfite and 4.5 g (0.11 mole) of sodium hydroxide in 50 mL of water. Reduce the copper by adding the solution of sodium sulfite in several portions over 2–3 minutes with good mixing. Allow the mixture containing the pre-

Figure 67.1. Infrared spectrum of chlorobenzene; thin film.

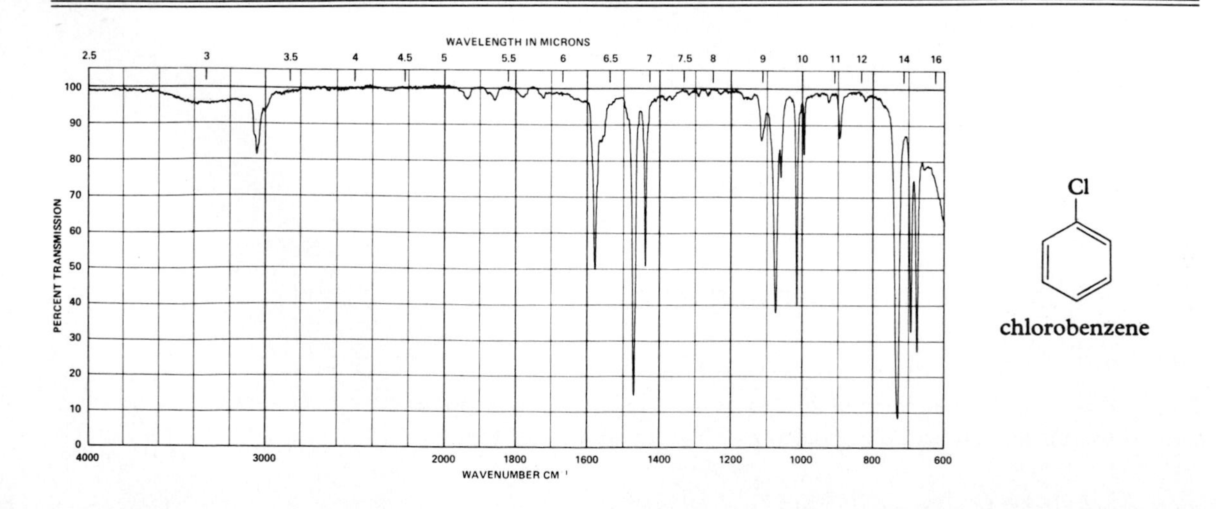

cipitated cuprous chloride to stand and settle in a pan of cold water while you prepare a solution of benzenediazonium chloride as described in Section 66.1.

When the diazotization is complete, decant the supernatant liquid from the precipitated cuprous chloride, and wash the precipitate by adding water, swirling, allowing to settle, and decanting. Then dissolve the cuprous chloride by adding 45 mL (0.54 mole) of conc. hydrochloric acid and swirling the mixture. Cool this solution in the ice bath and slowly add the solution of benzenediazonium chloride. Allow the mixture to stand at room temperature with occasional swirling for about 10 minutes. During this time, the initially formed cuprous diazonium chloride begins to decompose to give nitrogen and chlorobenzene. Cautiously heat the mixture to 50°C on the steam bath to complete the decomposition. Steam distill the mixture to isolate the chlorobenzene (Section 11). Separate the product from the distillate by extraction with ether, washing with dilute sodium hydroxide solution, drying, and distilling.

N_2^+

CuCl; heat

Cl

Time: 3 hours.

67.2 A VARIATION: *p*-CHLOROTOLUENE FROM *p*-TOLUENEDIAZONIUM CHLORIDE

p-Chlorotoluene can be prepared by the procedure described in Section 67.1, substituting *p*-toluenediazonium chloride (Section 66.2) for benzenediazonium chloride.

The infrared spectrum of *p*-chlorotoluene appears in Figure 67.2.

Figure 67.2. Infrared spectrum of *p*-chlorotoluene; thin film.

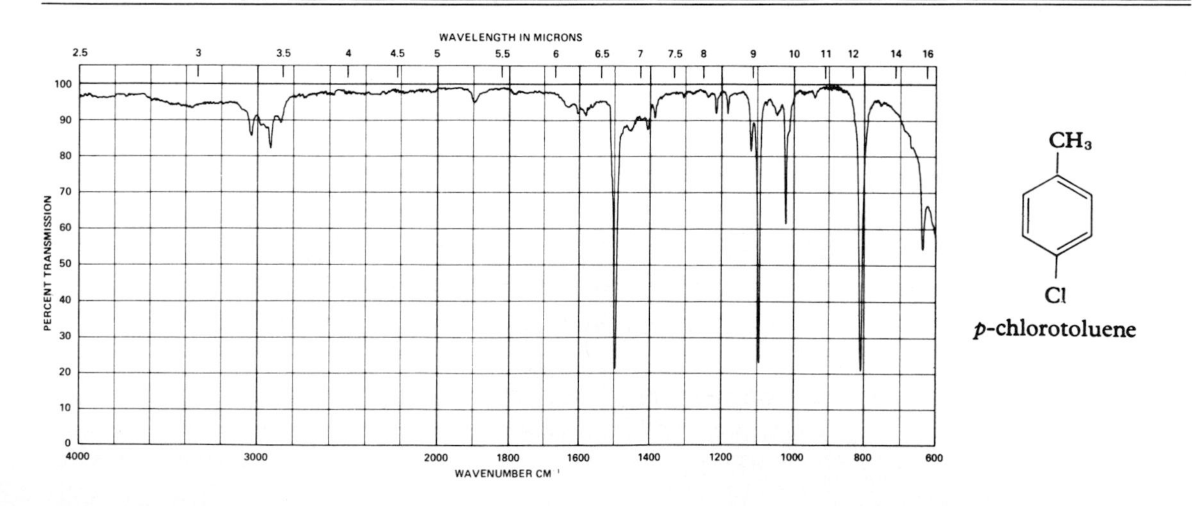

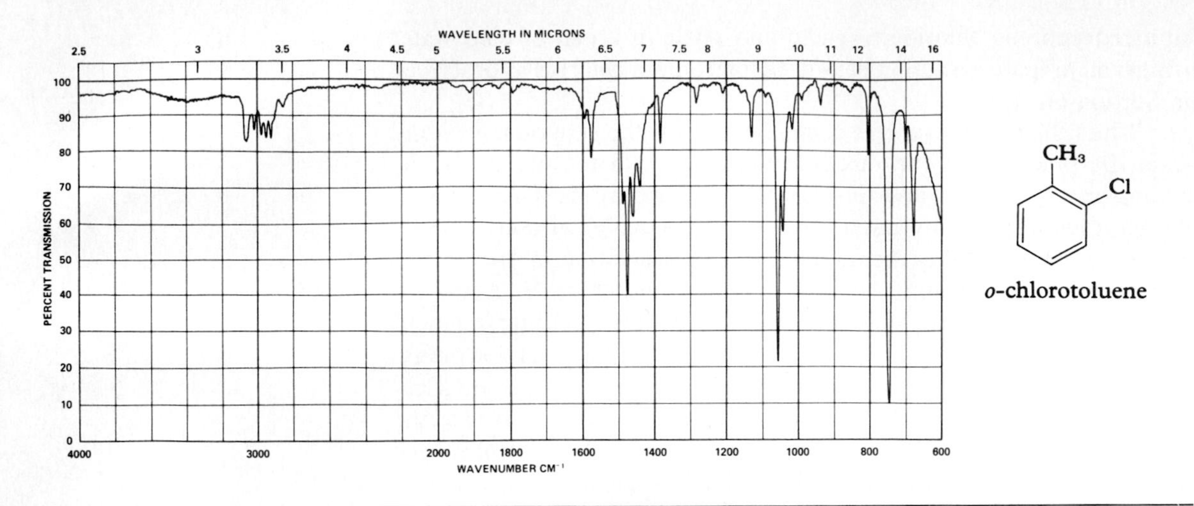

Figure 67.3. Infrared spectrum of *o*-chlorotoluene; thin film.

67.3 ANOTHER VARIATION: *o*-CHLOROTOLUENE FROM *o*-TOLUENEDIAZONIUM CHLORIDE

o-Chlorotoluene can be prepared from *o*-toluidine by diazotizing *o*-toluidine according to the procedure of Section 66.1 (diazotization of aniline) or of Section 66.2 (diazotization of *p*-toluidine), and then converting the resulting *o*-toluenediazonium chloride to *o*-chlorotoluene by the procedure of Section 67.1 (chlorobenzene from benzenediazonium chloride).

NH_2, CH_3 (*o*-toluidine) $\xrightarrow[\text{Section 66.2}]{\text{Section 66.1 or}}$ $N_2^+ Cl^-$, CH_3 $\xrightarrow{\text{Section 67.1}}$ Cl, CH_3 (*o*-chlorotoluene)

o-toluidine

o-chlorotoluene

Figure 67.3 shows the infrared spectrum of *o*-chlorotoluene.

68 Electrophilic Aromatic Substitution by Diazonium Ions; Coupling Reactions of Diazonium Salts

The positively charged diazonium ion can effect electrophilic substitution reactions on highly activated aromatic compounds such as amines and phenols. The molecules produced by such reactions have the

—N=N— (azo) system linking two aromatic rings. This system of extended conjugation causes light to be absorbed in the visible region of the spectrum (Section 24.3). The resulting partial reflection or transmission of white light is the reason that these compounds and their solutions appear colored.

68.1 BENZENEDIAZONIUM CHLORIDE AND β-NAPHTHOL: 1-PHENYLAZO-2-NAPHTHOL (SUDAN I)

$\{\phi-\overset{+}{N}\equiv N: \longleftrightarrow \phi-\ddot{N}=\overset{+}{N}:\}$ + β-naphthol (OH) → 1-phenylazo-2-naphthol

(orange-red)

Procedure. Prepare a solution of benzenediazonium chloride according to the procedure of Section 66.1, using one-tenth of the quantities (0.010-mole scale).

Prepare a solution of 1.44 g (0.010 mole) of β-naphthol in 5 mL of 3 *M* sodium hydroxide (0.015 mole). After cooling this solution to 5°C, add slowly and with good stirring the solution of benzenediazonium chloride. After allowing the reaction mixture to stand in the ice bath for 15 minutes, collect the product by suction filtration and wash it thoroughly with cold water. The product can be recrystallized from ethanol or from glacial acetic acid.

Time: 3 hours.

68.2 *p*-NITROBENZENEDIAZONIUM SULFATE AND PHENOL: *p*-(4-NITROBENZENEAZO)-PHENOL

$$O_2N-C_6H_4-N_2^+ + C_6H_5-OH \longrightarrow O_2N-C_6H_4-N=N-C_6H_4-OH$$

(orange-red)

Procedure. Prepare a solution of *p*-nitrobenzenediazonium sulfate according to the procedure of Section 66.3, using one-tenth of the quantities (0.010-mole scale).

Prepare a solution of 0.94 g (0.010 mole) of phenol in 5 mL of 1 *M* sodium hydroxide. Cool this mixture to 5°C and add to it the diazonium salt solution. Stir the mixture for 5 minutes and collect the product by suction filtration. The product can be recrystallized from toluene.

Time: 3 hours.

68.3 *p*-NITROBENZENEDIAZONIUM SULFATE AND β-NAPHTHOL: 1-(*p*-NITROPHENYLAZO)-2-NAPHTHOL (PARA RED; AMERICAN FLAG RED)

O_2N—C_6H_4—N_2^+ + β-naphthol (HO) ⟶ O_2N—C_6H_4—N=N—naphthyl (HO)

(red)

Procedure. Prepare a solution of *p*-nitrobenzenediazonium sulfate according to the procedure of Section 66.3, using one-tenth of the quantities (0.010-mole scale).

Prepare a solution of 1.44 g (0.010 mole) of β-naphthol in 25 mL of 10% sodium hydroxide solution (0.042 mole). When this solution has been cooled to 10°C, pour it into the diazonium salt solution. Acidify the mixture to litmus, and collect the red product by suction filtration. The product can be recrystallized from toluene or acetic acid.

Time: 3 hours.

68.4 *p*-NITROBENZENEDIAZONIUM SULFATE AND DIMETHYLANILINE: *p*-(4-NITROBENZENEAZO)-DIMETHYLANILINE

O_2N—C_6H_4—N_2^+ + C_6H_5—$N(CH_3)_2$ ⟶ O_2N—C_6H_4—N=N—C_6H_4—$N(CH_3)_2$

(reddish-brown)

Procedure. Prepare a solution of *p*-nitrobenzenediazonium sulfate according to the procedure of Section 66.3, using one-tenth of the quantities (0.010-mole scale).

Prepare a solution of 1.21 g (0.010 mole) of dimethylaniline in 1.5 mL of 1 *M* hydrochloric acid. Cool this solution in the ice bath and add the diazonium salt solution; mix well. Add 10 mL cold 1 *M* sodium hydroxide solution. Collect the product by suction filtration.

Time: 3 hours.

69 The Diels-Alder Reaction

The Diels-Alder reaction is one of the most interesting reactions in organic chemistry. It is a reaction between an alkene and a conjugated diene in which a six-membered ring is formed:

diene + dienophile ⇄ adduct

The reaction appears to involve only a single step (one transition state).

Many different combinations of diene and dienophile will undergo this reaction, and the experiments in this section illustrate a few of the great variety of structures that can result.

69.1 BUTADIENE (FROM 3-SULFOLENE) AND MALEIC ANHYDRIDE

Diels-Alder reactions with butadiene must generally be carried out in a pressure vessel, since the temperature at which reaction takes place readily is usually far above the boiling point of butadiene, −3°C. However, Sample and Hatch (1) have described a convenient way to avoid the use of a pressure vessel in the preparation of the Diels-Alder adduct between butadiene and maleic anhydride. In their procedure, butadiene is generated by the thermal decomposition of 3-sulfolene, a sort of reverse Diels-Alder reaction itself:

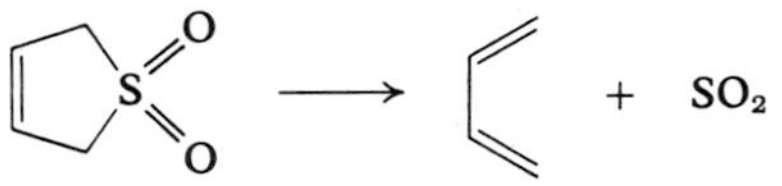

3-sulfolene → butadiene + SO_2

Apparently, at the temperature of the reaction, butadiene reacts with maleic anhydride more rapidly than it can be lost from the reaction mixture:

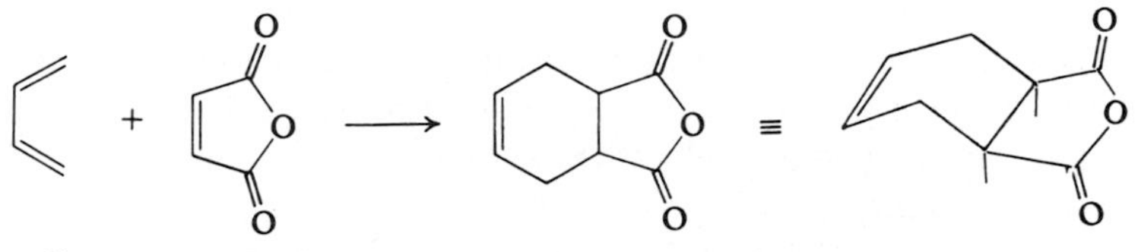

butadiene + maleic anhydride → Diels-Alder adduct

m.p. = 104°C

Procedure (Reference 1). Place 8.5 g (0.072 mole) of 3-sulfolene and 4.5 g (0.050 mole) of powdered maleic anhydride in a 125-mL Erlenmeyer flask. Add 4 mL of xylene, fit the flask with a condenser, and stopper the condenser with a plug of absorbent cotton (Note 1). Swirl the flask gently to effect partial solution. Remove the cotton plug and cautiously heat the mixture to a slow boil (Notes 2 and 3). After boiling the mixture for about one-half hour, allow it to cool for about 5 minutes (Note 4) and then add 50 mL of toluene (Note 5) and 0.5 g of decolorizing carbon. Heat the suspension on the steam bath and filter it by gravity. Reheat the filtrate until the product redissolves, and then add with swirling 20–25 mL of petroleum ether (Note 5) until a slight cloudiness persists. Rewarm the mixture until it is substantially clear, and finally allow the solution to cool in an ice bath for crystallization.

Use the hood

Notes

1. The mixture will become quite cool. The absorbent cotton will prevent water vapor from entering the flask and causing hydrolysis of the maleic anhydride.
2. Since sulfur dioxide will be evolved, the reaction must be run in the hood or some provision must be made to absorb the fumes.
3. Because the reaction is exothermic, some care must be taken to avoid overheating at the beginning of the heating period.
4. If the cooling period is too long, the product will separate as a hard cake that will be difficult to redissolve.
5. The solvents should be dry in order to avoid hydrolyzing the product to the corresponding diacid. Since the diacid is much less soluble, it cannot be removed by crystallization.

Time: 3 hours.

Reference

1. T. E. Sample, Jr., and L. F. Hatch, *J. Chem. Educ.* **45,** 55 (1968).

69.2 CYCLOPENTADIENE AND MALEIC ANHYDRIDE

In all Diels-Alder reactions, the structure of the product indicates that the reaction takes place with *cis* addition to the double bond of the dienophile. With cyclopentadiene, *cis* addition can take place in two different ways to give products with the same structure but with different configurations.

With butadiene (Section 69.1), the two different modes of reaction would give two equivalent (but presumably rapidly interconverting)

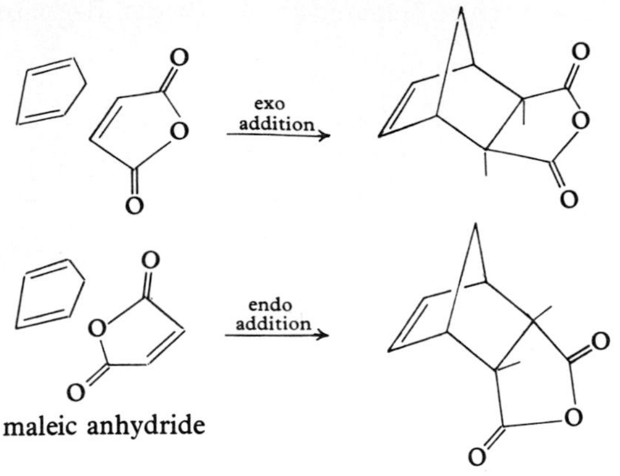

exo addition

endo addition

maleic anhydride

m.p. 165°C

conformations of the same molecule, and thus there is no way of knowing whether one path was followed in preference to the other. With cyclopentadiene, under the conditions of this experiment, the isomer that corresponds to endo addition can be isolated in high yield.

The predominance of the endo isomer in Diels-Alder reactions is generally observed, and, since the exo isomer is usually the more stable, the predominance of the endo isomer must be the result of kinetic control rather than thermodynamic control. The predominance of endo addition (Alder Rule) has been interpreted as the result of inductive effects, electrostatic effects of an initial charge transfer between diene and dienophile, "maximum accumulation of unsaturation," and, most recently, orbital symmetry effects.

When the exo isomer is more stable, high temperatures and long reaction times can give the exo isomer. Sometimes, the endo isomer can be converted to the exo isomer by heating. While it would be reasonable to propose that the exo isomer is formed via dissociation of the endo isomer to diene and dienophile, followed by recombination, the isomerization of the endo adduct of cyclopentadiene and maleic anhydride takes place at 190°C in an open flask without observable loss in weight. One would expect free cyclopentadiene to be lost under these conditions, although the success of the procedure of Section 69.1 indicates that this loss may not necessarily take place.

Cyclopentadiene dimerizes on standing via a Diels-Alder reaction. It must be prepared by thermal decomposition of the dimer:

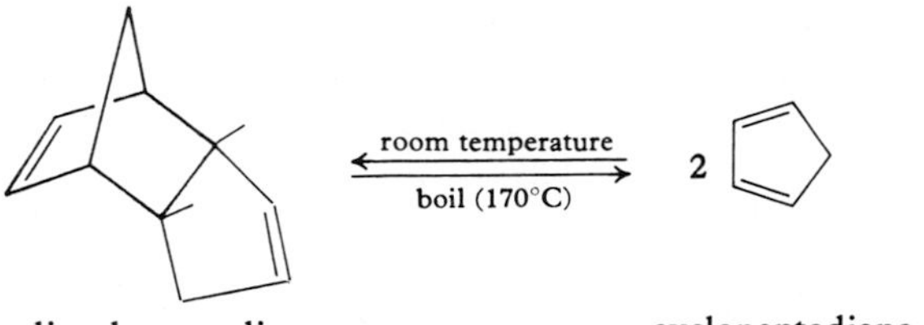

room temperature

boil (170°C)

2

endo-dicyclopentadiene

cyclopentadiene

The *endo-cis* diacid can be prepared as described in Exercise 1 at the end of Section 8.

Procedure. Place 15 mL of dicyclopentadiene in a 50-mL boiling flask. Fit the flask with a fractionating column and connect an adapter with a condenser set for downward distillation on the top of the column. Heat the dicyclopentadiene with a flame until it boils, and then continue to heat at such a rate that the cyclopentadiene distills at about its boiling point of 43°C (Note 1). Cool the receiver in an ice bath.

During the distillation, dissolve 6 g (0.061 mole) of maleic anhydride in 20 mL of ethyl acetate by heating on the steam bath. Add 20 mL of ligroin, and cool the mixture thoroughly in an ice bath (Note 2). Add 6 mL (4.8 g; 0.072 mole) of dry cyclopentadiene (Note 3) and swirl the mixture in the ice bath until the product separates as a white solid. Heat the mixture on the steam bath until it is all dissolved (Note 4), and allow it to stand for crystallization; slow crystallization will result in a beautiful display of crystal formation.

Notes

1. A fair fractionating column and some patience are needed. It may take 30 minutes to collect a little more than the 6 mL needed.
2. Some maleic anhydride may crystallize.
3. Water in the distilled cyclopentadiene will cause it to be cloudy. The cyclopentadiene may be dried with a little anhydrous calcium chloride.
4. Any diacid formed by water that got into the reaction mixture will remain undissolved at this point; it should be removed by gravity filtration.

Time: 3 hours.

Question

1. Propose an alternative procedure for this reaction in which the cyclopentadiene is produced in situ, in analogy with the in situ production of butadiene in Section 69.1.

References

1. W. J. Sheppard, *J. Chem. Educ.* **40,** 40 (1963).
2. L. F. Fieser, *Organic Experiments*, Heath, Boston, 1964, p. 83.

69.3 FURAN AND MALEIC ANHYDRIDE

In the Diels-Alder reaction between furan and maleic anhydride, the only adduct that has been isolated is one whose m.p. has been variously

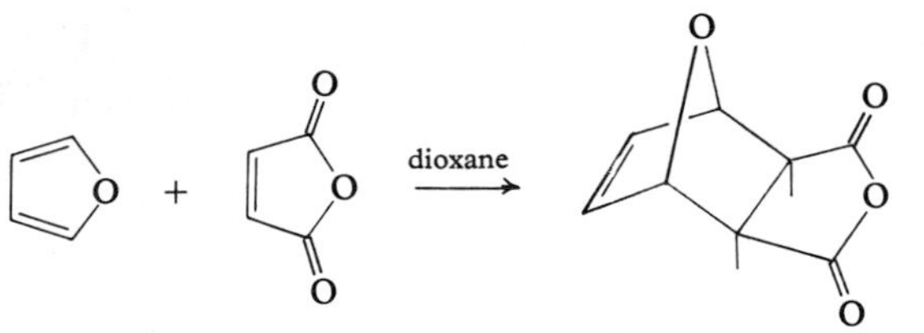

furan

dioxane

product of exo addition: m.p. 117°C

reported as 116–117°C; 122°C; and 125°C with foaming and decomposition. The stereochemistry of this product was shown to correspond to the exo mode of addition. It is suspected that this is the thermodynamically more stable isomer.

Procedure. Place 10.0 grams (0.1 mole) of maleic anhydride in a 125-mL ground-glass-stoppered Erlenmeyer flask. Add 25 mL of dioxane and swirl the mixture until the anhydride has all dissolved. Add to the flask 7.5 mL (7 grams; 0.1 mole) of furan, swirl to mix, and allow the solution to stand for at least 24 hours. Isolate the crystalline adduct by suction filtration, washing it with a little ether. Yield: about 11 grams (65%; Note 1).

Note

1. If 12.5 mL of dioxane is used the yield will be greater than 80% but the product will separate from solution in hard lumps.

Time: 1 hour plus a reaction time of at least 24 hours.

70 Aldol Condensations and Related Reactions

A generally useful type of reaction in which a new carbon–carbon double bond is formed occurs when a molecule of water is eliminated between an acidic methylene and a carbonyl group.

$$2\ -\overset{\text{H}}{\underset{\text{H}}{\text{C}}}-\overset{\text{O}}{\overset{\|}{\text{C}}}- \longrightarrow -\overset{\text{H}}{\underset{\text{H}}{\text{C}}}-\overset{\text{OH}}{\text{C}}(-\overset{\text{H}}{\text{C}}-\overset{\text{O}}{\overset{\|}{\text{C}}}-)- \longrightarrow -\overset{\text{H}}{\underset{\text{H}}{\text{C}}}-\text{C}=\text{C}-\overset{\text{O}}{\overset{\|}{\text{C}}}- + H_2O$$

Although acidic catalysis is possible, the reaction is usually carried out under basic catalysis:

Formation of the Carbanion

$$-\overset{\text{H}}{\underset{\text{H}}{\text{C}}}-\overset{\text{O}}{\overset{\|}{\text{C}}}- + \text{B:}^- \rightleftharpoons \left\{ {}^-\!:\overset{\text{H}}{\text{C}}-\overset{\text{O}}{\overset{\|}{\text{C}}}- \longleftrightarrow \overset{\text{H}}{\text{C}}=\overset{\text{O}^-}{\text{C}}- \right\} + \text{B—H}$$

resonance-stabilized carbanion

Addition of the Carbanion to the Carbonyl Group

$$\left\{ {}^-\!:\overset{\text{H}}{\text{C}}-\overset{\text{O}}{\overset{\|}{\text{C}}}- \longleftrightarrow \overset{\text{H}}{\text{C}}=\overset{\text{O}^-}{\text{C}}- \right\} + -\overset{\text{H}}{\underset{\text{H}}{\text{C}}}-\overset{\text{O}}{\overset{\|}{\text{C}}}- \longrightarrow -\overset{\text{H}}{\underset{\text{H}}{\text{C}}}-\overset{\text{O}^-}{\text{C}}(-\overset{\text{H}}{\text{C}}-\overset{\text{O}}{\overset{\|}{\text{C}}}-)- \xrightarrow{+\text{BH}^-} -\overset{\text{H}}{\underset{\text{H}}{\text{C}}}-\overset{\text{OH}}{\text{C}}(-\overset{\text{H}}{\text{C}}-\overset{\text{O}}{\overset{\|}{\text{C}}}-)- + \text{B:}^-$$

resonance-stabilized carbanion

Elimination of Water to Form the Carbon–Carbon Double Bond

$$-\overset{\text{H}}{\underset{\text{H}}{\text{C}}}-\overset{\text{OH}}{\text{C}}(-\overset{\text{H}}{\text{C}}-\overset{\text{O}}{\overset{\|}{\text{C}}}-)- + \text{B:}^- \longrightarrow \left\{ -\overset{\text{H}}{\underset{\text{H}}{\text{C}}}-\overset{\text{OH}}{\text{C}}(-\,{}^-\!:\text{C}-\overset{\text{O}}{\overset{\|}{\text{C}}}-)- \longleftrightarrow -\overset{\text{H}}{\underset{\text{H}}{\text{C}}}-\overset{\text{OH}}{\text{C}}(-\text{C}=\overset{\text{O}^-}{\text{C}}-)- \right\} \longrightarrow -\overset{\text{H}}{\underset{\text{H}}{\text{C}}}-\text{C}=\text{C}-\overset{\text{O}}{\overset{\|}{\text{C}}}- + \text{HO}^-$$

It is occasionally possible to isolate the initially formed β-hydroxy aldehyde or ketone (Aldol condensation).

If two different carbonyl compounds are used, as many as four different products of this type of reaction can be formed. In order for the reaction between two different carbonyl compounds to be useful in synthesis, usually one must have no alpha hydrogen atoms. In such cases, there remain only two possibilities: the desired mixed condensation, and the self-condensation of the other compound.

70.1 SELF-CONDENSATION OF PROPIONALDEHYDE: 2-METHYL-2-PENTENAL

$$2\ CH_3CH_2\overset{O}{\overset{\|}{C}}{-}H \xrightarrow{NaOH} \underset{CH_3CH_2}{\overset{H}{\ }}\!\!\!\!>C{=}C<\!\!\!\!\underset{CH_3}{\overset{\overset{O}{\|}}{C-H}} + H_2O$$

The infrared and NMR spectra of 2-methyl-2-pentenal are shown in Figures 70.1 and 70.2; Figures 70.3 and 70.4 show ultraviolet spectra.

Procedure (Reference 1). Place 5 mL of 10% sodium hydroxide in a 100-mL boiling flask, add a magnetic stirring bar, and fit the flask with a Claisen adapter. Fit a separatory funnel in the center neck of the Claisen adapter and a condenser in the other neck. Place 25 mL (20 g; 0.35 mole) of propionaldehyde (Note 1) in the separatory funnel, and add it slowly and with good stirring to the flask. After the addition is complete, stir the mixture until it has cooled nearly to room temperature. Remove as much of the aqueous layer as possible with a pipet or long medicine dropper and distill the residue, using a fractionating

Figure 70.1. IR spectrum of 2-methyl-2-pentenal; thin film.

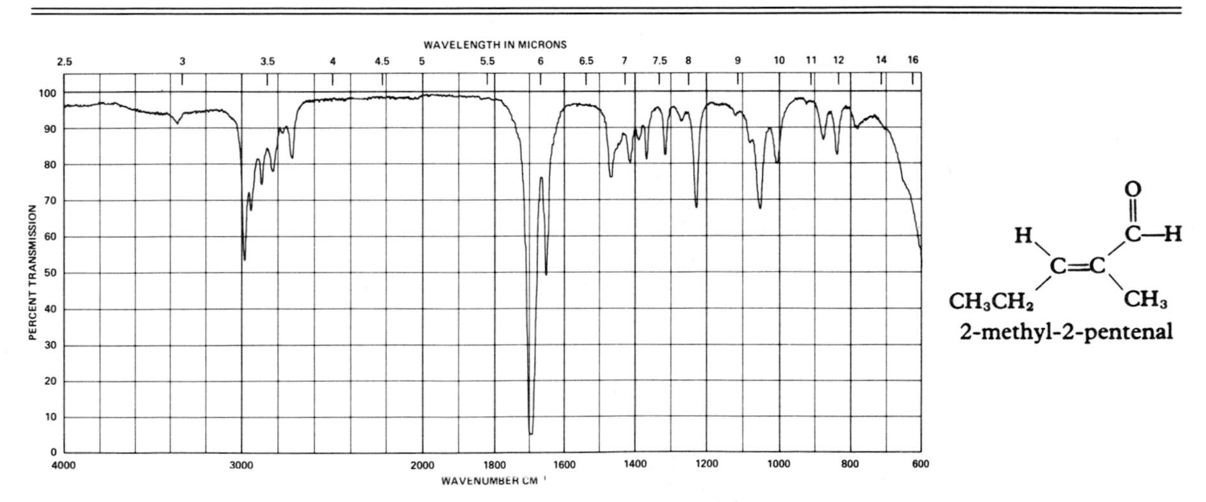

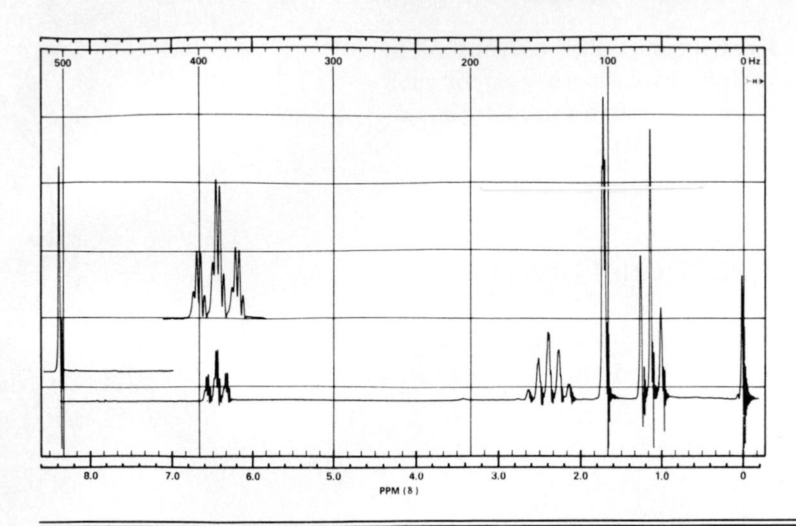

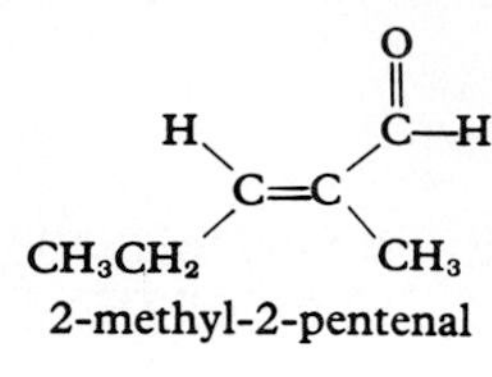

Figure 70.2. NMR spectrum of 2-methyl-2-pentenal; CCl_4 solution.

column, keeping the temperature at the still-head from rising above about 140°C. When the temperature at the still-head begins to fall, discontinue the distillation and remove the water from the distillate with a pipet or medicine dropper. Add a little anhydrous calcium chloride to the distillate, and over a period of about 15 minutes, remove the aqueous phase as it forms; add more solid calcium chloride if necessary. ▪ Redistill the product, collecting as 2-methyl-2-pentenal the material boiling between 133 and 137°C.

Figure 70.3. Ultraviolet spectrum of 2-methyl-2-pentenal; 5.3×10^{-5} molar in 95% ethanol.

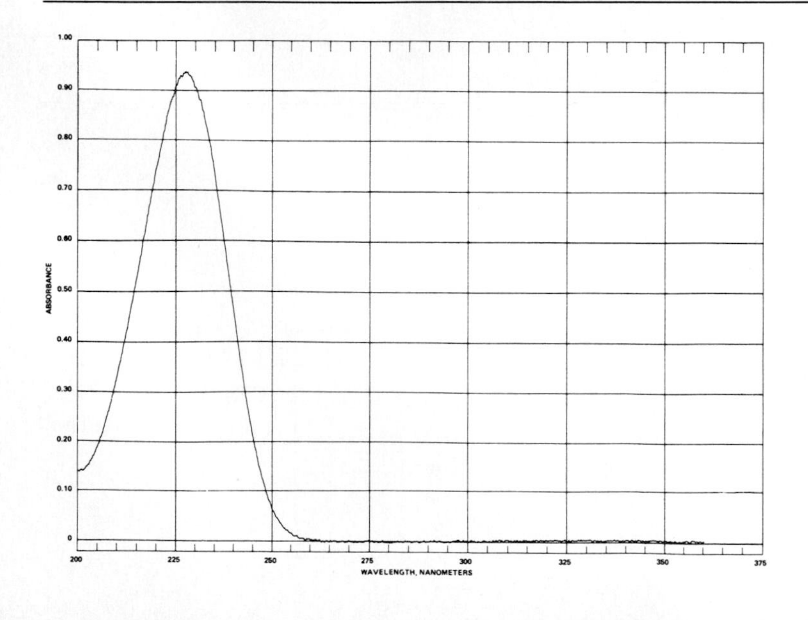

O
H
C—H
C=C
CH_3CH_2
CH_3
2-methyl-2-pentenal

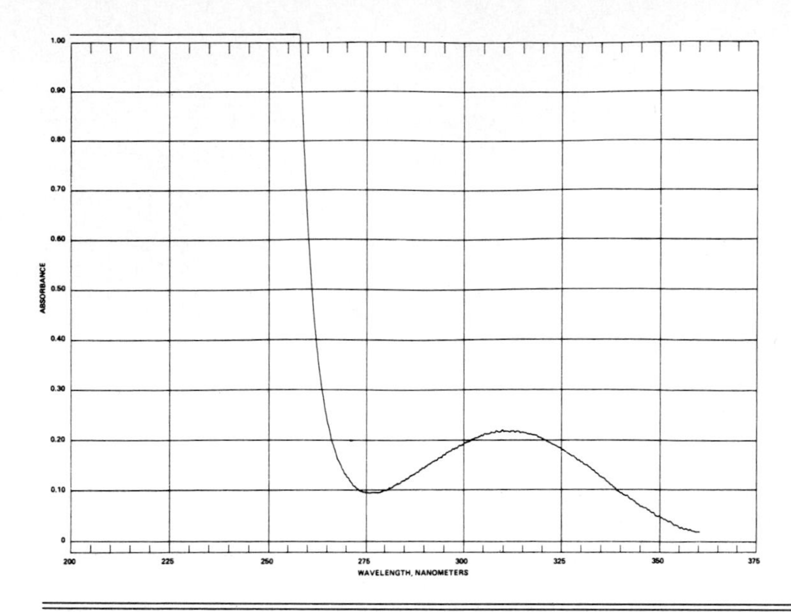

H, C=C, C(=O)—H, CH_3CH_2, CH_3

2-methyl-2-pentenal

Figure 70.4. Ultraviolet spectrum of 2-methyl-2-pentenal; 5.3×10^{-3} molar in 95% ethanol.

Note

1. The propionaldehyde must be free of propionic acid. It can be purified by distillation.

Time: 3 hours.

Reference

1. C. A. MacKenzie, *Experimental Organic Chemistry*, 3rd edition, Prentice-Hall, Englewood Cliffs, N.J., 1967, p. 161.

70.2 MESITYL OXIDE AND DIETHYL MALONATE: 5,5-DIMETHYL-1,3-CYCLOHEXANEDIONE; DIMEDON; METHONE

The initial base-catalyzed addition of diethyl malonate to the double bond of the α,β-unsaturated ketone is an example of the Michael condensation. This is followed by a Dieckmann, or cyclic Claisen, condensation. The final steps are the hydrolysis and acid-catalyzed decarboxylation of a β-ketoacid.

The product Dimedon is useful for characterizing aldehydes (Section 29.4).

$CH_3C(=O)CH=C(CH_3)_2$ + $H_2C(CO\text{—}O\text{—}Et)_2$ $\xrightarrow{EtO^-}$ $CH_3\text{—}C(=O)\text{—}CH_2\text{—}C(CH_3)_2\text{—}CH(CO\text{—}O\text{—}Et)_2$ $\xrightarrow{EtO^-}$

mesityl oxide diethyl malonate

$\xrightarrow{HO^-}$ $\xrightarrow{H^+}$ + CO_2

Procedure (Reference 1). Place 2.5 mL of methanol in a 25-mL boiling flask, and add, in portions and with cooling, 0.15 g (6.5 mmole) of sodium (*caution:* Note 1). When all the sodium has dissolved, add 1.0 mL (1.05 g; 6.5 mmole) of diethyl malonate. Fit the flask with a condenser, and heat the mixture to boiling on the steam bath. By means of a pipet, cautiously add down the condenser 0.7 mL (0.60 g; 6.1 mmole) of mesityl oxide (Note 2). After a vigorous reaction has occurred, heat the mixture under reflux for one-half hour. At this time, add 5 mL of 2 *M* sodium hydroxide and boil the mixture for another hour and a half. At the end of the heating period, set the condenser for downward distillation and remove most of the alcohol by distillation on the steam bath. Prepare a mixture of 2.5 mL conc. hydrochloric acid and 5 mL of water, and when the removal of the alcohol is complete, heat the mixture to boiling with a very small flame and cautiously add the hydrochloric acid solution until the mixture is acidic to methyl orange (Note 3). Cool the mixture in an ice bath, collect the product by suction filtration, and wash it with a little cold water. Dimedon can be recrystallized from acetone, using about 8 mL per gram; m.p., 147–148°C.

Notes

1. Sodium reacts violently with water, and the hydrogen formed usually ignites in air. See Section 1.7 for advice on handling sodium. An equivalent amount of sodium methoxide may be substituted for the sodium.
2. It appears that a recently redistilled sample of a good grade of mesityl oxide is required in this reaction.
3. The evolution of carbon dioxide will cause considerable foaming.

Time: 4 hours.

Question

1. Give the details of the mechanism of the reaction, taking care to specify the state of ionization of the intermediates.

Reference

1. J. H. Wilkinson, *Semi-micro Organic Preparations,* 2nd edition, Oliver and Boyd, Edinburgh, Scotland, 1958, p. 161.

71 The Wittig Reaction

$$\phi_3\overset{+}{P}\text{—}CH_2R \xrightarrow{\text{base}} \phi_3\overset{+}{P}\text{—}\overset{\cdot\cdot\,-}{C}HR$$

ylide

$$\phi_3\overset{+}{P}\text{—}\overset{\cdot\cdot\,-}{C}HR + R\text{—}\overset{O}{\overset{\|}{C}}\text{—}H \longrightarrow \phi_3\overset{+}{P}\text{—}\underset{H}{\overset{R}{C}}\text{—}\underset{R}{\overset{H}{C}}\text{—}O^-$$

ylide — intermediate (diastereomeric forms)

$$\text{diastereomeric intermediates} \longrightarrow RHC{=}CHR + \phi_3\overset{+}{P}\text{—}O^-$$

olefin (Z and E forms) — triphenylphosphine oxide

An important method for the synthesis of alkenes is the Wittig reaction. In one version of this reaction, a triphenyl alkyl phosphonium salt is treated with a strong base so as to form an inner salt, an *ylide.* Upon addition of a carbonyl compound the carbanion carbon of the ylide attacks the carbonyl carbon to form an intermediate that then eliminates triphenylphosphine oxide to give the olefin. The first intermediate then eliminates triphenylphosphine oxide via a second, cyclic, intermediate. The relative amounts of the Z and E forms of the alkene depend upon the relative amounts of the diastereomeric intermediates and upon their relative rates of elimination of triphenylphosphine oxide.

$$\underset{\text{first intermediate}}{\phi_3\overset{+}{P}\;:\overset{\cdot\cdot}{\underset{\cdot\cdot}{O}}:^-\ \text{(on C(a)(b)—C(c)(d))}} \longrightarrow \underset{\text{second intermediate}}{\phi_3P\text{—}\overset{\cdot\cdot}{O}:\ \text{(ring with C(a)(b)—C(c)(d))}} \longrightarrow \underset{a,b}{C}{=}\underset{c,d}{C} + \underset{\text{triphenylphosphine oxide}}{\phi_3\overset{+}{P}\text{—}O^-}$$

A modification of the Wittig reaction has been described by Seus and Wilson (Reference 1) in which the triphenyl phosphonium compound is replaced by diethyl benzylphosphonate, which is converted to the corresponding ylide by treatment with sodium methoxide.

$$\text{Et—O—}\overset{\overset{:\ddot{O}:^{-}}{|}}{\underset{\underset{\text{O—Et}}{|}}{\text{P}^{+}}}\text{—CH}_2\phi + \text{CH}_3\text{—O}^-\text{Na}^+ \longrightarrow \text{Et—O—}\overset{\overset{:\ddot{O}:^{-}}{|}}{\underset{\underset{\text{O—Et}}{|}}{\text{P}^{+}}}\text{—}\ddot{\text{C}}\text{H}^{-}\phi$$

diethyl benzylphosphonate — ylide

Addition of benzaldehyde, for example, then leads to the formation of *trans*-stilbene.

$$\text{ylide} + \phi(\text{H})\text{C=O} \longrightarrow \phi(\text{H})\text{C=C}(\text{H})\phi + \text{Et—O—}\overset{\overset{:\ddot{O}:^{-}}{|}}{\underset{\underset{\text{O—Et}}{|}}{\text{P}^{+}}}\text{—}\ddot{\underset{..}{\text{O}}}\text{:}^{-}$$

benzaldehyde — *trans*-stilbene

Section 71.1 gives a procedure for the preparation of diethyl benzylphosphonate, and describes the preparation of *trans*-stilbene from diethyl benzylphosphonate and benzaldehyde by a Wittig reaction.

Section 71.2 suggests that *trans,trans*-1,4-butadiene can be formed by a modification of the Wittig reaction in which cinnamaldehyde is used instead of benzaldehyde.

$$\phi(\text{H})\text{C=C}(\text{H})\text{—C}(\text{H})\text{=O} \longrightarrow \phi(\text{H})\text{C=C}(\text{H})\text{—C}(\text{H})\text{=C}(\text{H})\phi$$

cinnamaldehyde — *trans, trans*-1,4-diphenylbutadiene

71.1 THE PREPARATION OF *trans*-STILBENE

When triethyl phosphite and benzyl chloride are heated together, the initially formed phosphonium salt loses ethyl chloride in an elimination reaction to produce diethyl benzylphosphonate. The ethyl chloride that is formed during the reaction distills from the reaction mixture, and the boiling point of the mixture gradually rises from about 165° to more than 200°C. The crude diethyl benzylphosphonate can be used as soon as it has been cooled.

$$\underset{\text{triethyl phosphite}}{\text{Et—O—}\overset{\text{O—Et}}{\underset{\text{O—Et}}{\text{P:}}}} + \phi\text{—CH}_2\text{—Cl} \longrightarrow \text{Et—O—}\overset{\text{O—Et}}{\underset{\text{O—Et}}{\overset{+}{\text{P}}}}\text{—CH}_2\text{—}\phi \ \text{Cl}^- \longrightarrow \underset{\text{diethyl benzylphosphonate}}{\text{Et—O—}\overset{:\ddot{\text{O}}:^-}{\underset{\text{O—Et}}{\overset{+}{\text{P}}}}\text{—CH}_2\text{—}\phi} + \text{Et—Cl}$$

Addition of sodium methoxide to a solution of the phosphonate ester in dimethylformamide converts the phosphonate ester to the ylide, which then gives *trans*-stilbene upon addition of benzaldehyde.

$$\underset{\text{diethyl benzylphosphonate}}{\text{Et—O—}\overset{:\ddot{\text{O}}:^-}{\underset{\text{O—Et}}{\overset{+}{\text{P}}}}\text{—CH}_2\phi} + \text{CH}_3\text{—O}^-\text{Na}^+ \longrightarrow \underset{\text{ylide}}{\text{Et—O—}\overset{:\ddot{\text{O}}:^-}{\underset{\text{O—Et}}{\overset{+}{\text{P}}}}\text{—}\overset{..\,-}{\text{C}}\text{H}\phi}$$

$$\text{ylide} + \underset{\text{benzaldehyde}}{\overset{\phi}{\underset{\text{H}}{\text{C}}}\text{=O}} \longrightarrow \underset{\textit{trans}\text{-stilbene}}{\overset{\phi}{\underset{\text{H}}{\text{C}}}\text{=}\overset{\text{H}}{\underset{\phi}{\text{C}}}} + \text{Et—O—}\overset{:\ddot{\text{O}}:^-}{\underset{\text{O—Et}}{\overset{+}{\text{P}}}}\text{—}\ddot{\underset{..}{\text{O}}}:^-$$

The infrared spectrum of *trans*-stilbene is shown in Figure 71.1.

Figure 71.1. Infrared spectrum of *trans*-stilbene; $CHCl_3$ solution.

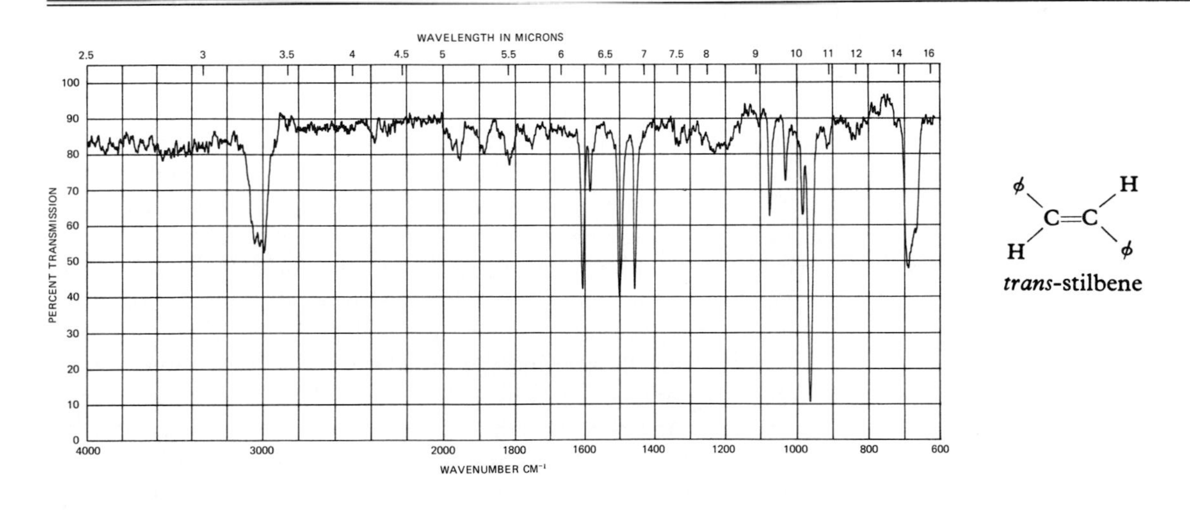

Preparation of Diethyl Benzylphosphonate from Benzyl Chloride and Triethyl Phosphite

Use the hood

Procedure (Note 1). Add to a 25-mL round-bottom flask 6.0 mL (6.6 g; 52 mmole) of benzyl chloride (Note 1) and 9.0 mL (8.7 g; 53 mmole) of triethyl phosphite. Add a boiling stone to the flask and fit the flask with a reflux condenser. Clamp the flask in position over a wire gauze supported by an iron ring so that the contents can be heated under reflux by a burner flame. Lower a thermometer down the condenser until the bulb rests on the bottom of the boiling flask so that the temperature of the boiling liquid can be determined. By means of a burner flame, heat the mixture to boiling (Note 1) and continue to boil it for 1 hour (Note 2). At the end of the heating period, remove the burner, the wire gauze, and the iron ring, and allow the flask to cool. After the temperature of the contents of the flask has fallen below 100°C, cooling can be hastened by immersing the flask in a beaker of cold water. The cool diethyl benzylphosphonate is suitable for use in the Wittig reaction.

Notes

1. All transfers of reagents and the reaction itself must be carried out in a good hood. Benzyl chloride is a potent lachrymator; it should be dispensed by means of a buret. Triethyl phosphite is not known to be particularly toxic, but it does have an obnoxious smell. Ethyl chloride is produced as a gas during the reaction.
2. A shorter heating period results in a lower yield. The temperature of the boiling liquid is initially about 165°; at the end of the hour it will be above 220°.

Preparation of *trans*-Stilbene from Diethyl Benzylphosphonate by a Wittig Reaction

Procedure. Add to a 125-mL Erlenmeyer flask 1.5 g (28 mmole) of sodium methoxide (Note 1) and 13 mL of dimethylformamide. Add to this suspension 25 mmole of diethyl benzylphosphonate and an additional 12 mL of dimethylformamide (Note 2). After the temperature of the resulting mixture has been adjusted to 20°C by brief swirling in an

Use an ice bath

ice bath, add 2.6 mL (2.7 g; 26 mmole) of benzaldehyde (Note 3). Swirl the flask to mix the contents. The temperature of the reaction mixture will start to rise; keep the temperature of the mixture between 30° and 40°C by occasionally swirling the flask in the ice bath. After about 5 minutes, the temperature will no longer tend to rise. At this point, allow the flask to stand at room temperature for another 5 minutes and then add 15 mL of water to precipitate the *trans*-stilbene. Collect the product by suction filtration, using 30 mL of a 1 : 1 mixture (by volume) of water and methanol for rinsing and washing. The product should be recrystallized from isopropyl alcohol to give clusters of diamond-shaped prisms (Note 4). The yield, after recrystallization, is about 2.5 grams (about 55%). The melting point of *trans*-stilbene is reported to be 124–126°C.

Notes

1. Sodium methoxide is caustic. Avoid spilling or breathing the finely divided powder. Sodium methoxide from a recently opened bottle should be used.
2. This second portion of dimethylformamide can be used to assist in the transfer of the diethyl benzylphosphonate.
3. Benzaldehyde from a recently opened bottle should be used.
4. The amount of isopropyl alcohol required will be between 30 and 50 mL, depending on how thoroughly the crude product has been sucked dry during suction filtration. Methanol can be used for rinsing and washing when the recrystallized material is collected by suction filtration.

71.2 A VARIATION: PREPARATION OF *trans,trans*-1,4-DIPHENYLBUTADIENE

ϕ(H)C=C(H)–C(H)=O → ϕ(H)C=C(H)–C(H)=C(H)ϕ

cinnamaldehyde → *trans, trans*-1,4-diphenylbutadiene

trans, trans-1,4-Diphenylbutadiene can be prepared from diethyl benzylphosphonate and cinnamaldehyde by the procedure described in Section 71.1 for the synthesis of *trans*-stilbene. In this variation, however, the final reaction mixture will gradually change in color to deep red, and the product will start to crystallize. When the temperature of the mixture shows no further tendency to rise, allow the mixture to stand at room temperature for 10 minutes and then add a mixture of 20 mL of water and 10 mL of methanol. Swirl the flask to break up the mass of crystals and to mix the contents, and collect the product by suction filtration. Pour the red filtrate back into the reaction flask to help transfer the remainder of the product to the funnel. Thoroughly wash the crystals with water until the red color is entirely removed, and then wash them with methanol until the washings are almost colorless so as to remove a yellow contaminant. Continue to apply suction to the crude product for about 5 minutes so as to remove as much methanol as possible. Then remove the product from the funnel, break up the

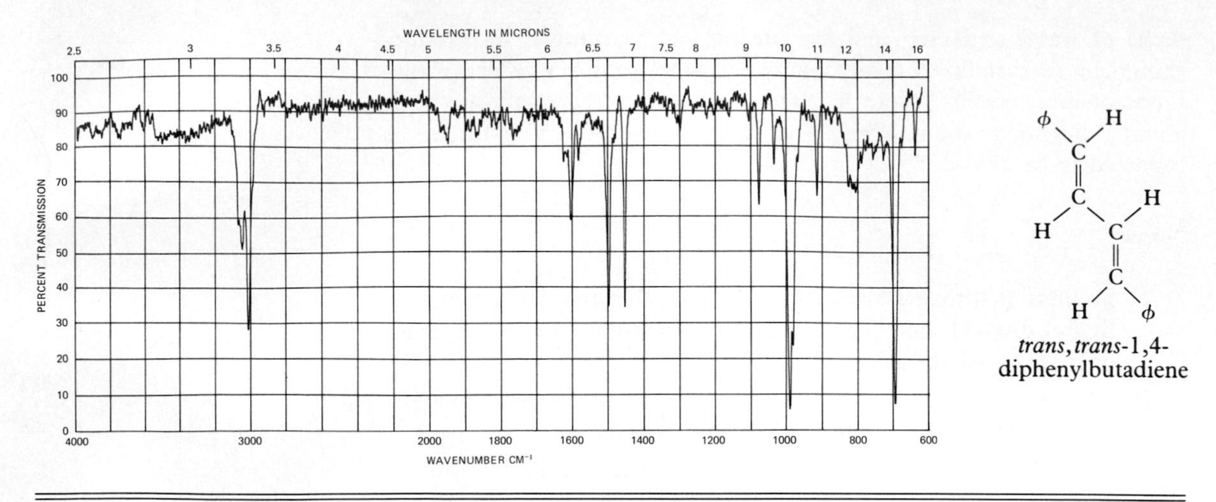

Figure 71.2. Infrared spectrum of *trans, trans*-1,4-diphenylbutadiene; CCl_4 solution.

lumps, and spread it out to dry some more. The dry material can be recrystallized from cyclohexane, using 10 mL per gram, to give light, faintly yellow plates of the purified hydrocarbon. *trans, trans*-1,4-Diphenylbutadiene is reported to melt at 151–153°C, and its infrared spectrum is shown in Figure 71.2.

Reference

1. E. J. Seus and C. V. Wilson, *J. Org. Chem.* **26,** 5243 (1961).

72 Two Thermochromic Compounds: Dixanthylene and Dianthraquinone

Thermochromism

Compounds whose color depends upon temperature are said to be *thermochromic*. Dixanthylene is a very pale yellow-green solid at room temperature, but it becomes dark blue when melted or heated in solution; at liquid nitrogen temperature it is completely colorless. Dianthraquinone is a bright canary-yellow at room temperature, but it becomes a brilliant parrot-green when heated in solution. The thermochromism of each compound can be observed very nicely during recrystallization from mesitylene.

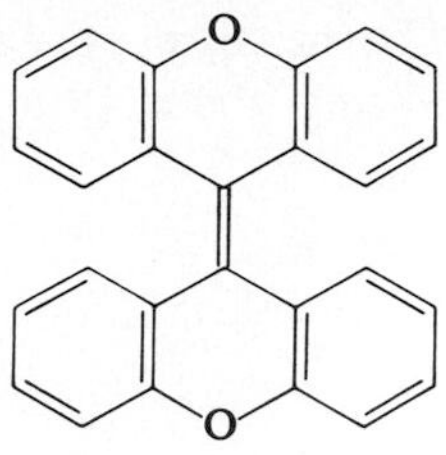

dixanthylene

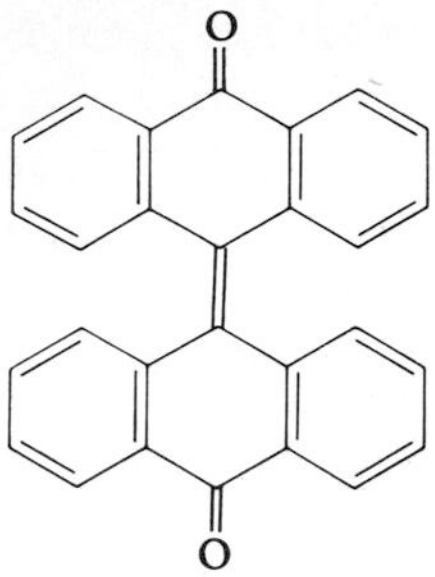

dianthraquinone

72.1 DIXANTHYLENE FROM XANTHONE

Dixanthylene can be prepared by the reductive dimerization of xanthone by zinc and acetic acid:

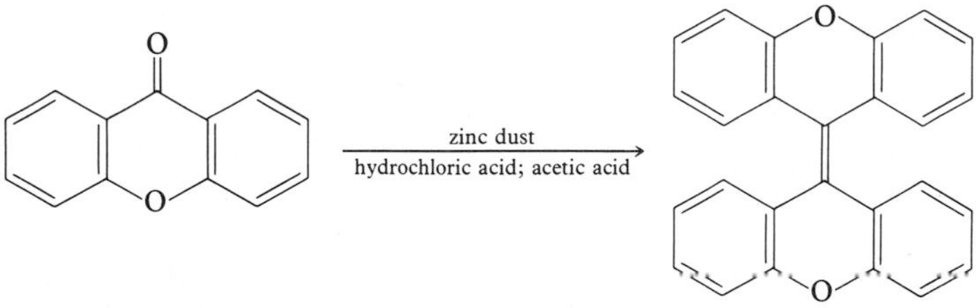

zinc dust
hydrochloric acid; acetic acid

xanthone

dixanthylene

Dixanthylene from Xanthone

Procedure. Place 2.5 g (12.7 mmole) of xanthone in a 250-mL boiling flask, add 100 mL of glacial acetic acid, and swirl until most of the solid has dissolved. Add 0.8 g (12 mmole) of zinc dust and 2 drops of conc. hydrochloric acid, and boil the mixture under reflux for about 90 minutes. Every 5 minutes or so, add 2 drops of conc. hydrochloric acid down the condenser, taking care not to add the acid so often that the solution takes on a red color. Add three more 0.8-g portions of zinc dust, one at a time at intervals of about 20 minutes, adding the first about 20 minutes after the mixture has been brought to a boil. At the end of the period of boiling under reflux, allow the solution to cool completely to room temperature, collect the solid by suction filtration, and wash it thoroughly with water. Separate any zinc from the product by boiling it with mesitylene to dissolve the dixanthylene (about 25 mL will be required), filtering the hot, blue solution by gravity, reheating the filtrate if necessary to redissolve any material that may have crystallized, and allowing the filtrate to cool to room temperature. Collect the very pale yellow-green crystals by suction filtration and wash them with a little benzene. Yield: about 0.8 g (35%).

Thermochromism: dark blue → pale green

The thermochromism of dixanthylene can be observed by heating a sample in a melting point capillary over a small burner flame, or, better, by recrystallization from boiling mesitylene, using about 19 mL per gram (recovery: about 90%).

Time: 3 hours.

References

1. A. Schonberg and A. Mustafa, *J. Chem. Soc.* **1944,** 67.
2. G. Gurgenjanz and S. von Kostanecki, *Chem. Ber.* **28,** 2310 (1895).
3. A. Ault, R. Kopet, and A. Serianz, *J. Chem. Educ.* **48,** 410 (1971).

72.2 DIANTHRAQUINONE FROM ANTHRONE VIA 9-BROMOANTHRONE

Dianthraquinone can be prepared by bromination of anthrone to give 9-bromoanthrone, followed by treatment of this product with diethylamine in chloroform:

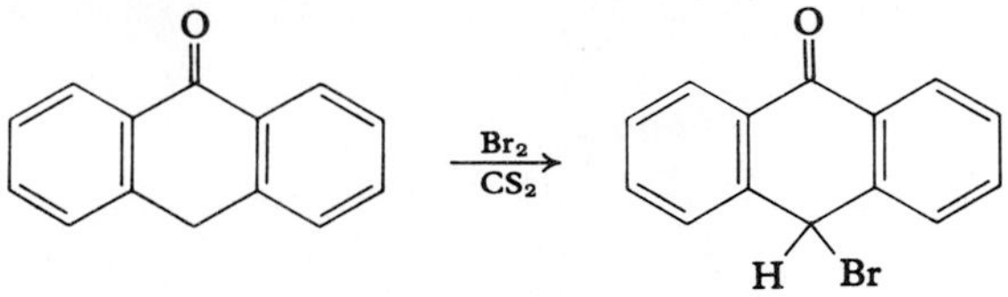

$$\text{anthrone} \xrightarrow[CS_2]{Br_2} \text{9-bromoanthrone}$$

anthrone 9-bromoanthrone

$$2\ \text{9-bromoanthrone} \xrightarrow[CHCl_3]{Et_2NH} \text{dianthraquinone}$$

9-Bromoanthrone from Anthrone

Procedure (Reference 1). Suspend 5.0 grams (26 mmole) of anthrone in 15 mL of carbon disulfide. Add bromine drop by drop with stirring over a period of about 15 minutes until the bromine color persists; about 4.1 grams (1.4 mL; 1 equivalent) will be required (Note 1). Collect the tan crystalline product by suction filtration and wash it with a little toluene. Yield: 6.0 grams (85%).

Time: 1 hour.

Dianthraquinone from 9-Bromoanthrone

Procedure (Reference 2). Dissolve 2.0 grams (7.3 mmole) of 9-bromoanthrone (Note 2) in 30 mL of chloroform (Note 3). Add 2.0 mL (1.4 g; 19 mmole) of diethylamine, swirl to mix, and allow the resulting warm solution to stand for 2 hours. During this time, the color of the solution will change from yellow to a dark red-brown. At this time, add 75 mL of ether slowly with swirling. After 5 minutes, collect the yellow precipitate by suction filtration and wash it thoroughly with ether. Suspend the precipitate in about 10 mL of 95% ethanol and then collect the canary-yellow dianthraquinone by suction filtration, washing it with a little ethanol. Yield: about 0.8 g (55%).

The thermochromism of dianthraquinone can be observed very nicely upon recrystallization from boiling mesitylene using about 50 mL per gram.

Thermochromism: parrot green → canary yellow

Notes

1. Since hydrogen bromide gas is evolved in this reaction, it must be carried out in the hood or some provision must be made to absorb the gas produced (see Section 35).
2. The 9-bromoanthrone must be prepared just before use; otherwise the yield of dianthraquinone is greatly reduced.
3. Any insoluble material should be removed at this point by gravity filtration.

Time: 3 hours.

References

1. E. Barnett, J. W. Cook, and M. A. Matthews, *J. Chem. Soc.* **123,** 1994 (1923); and K. H. Meyer, *Annalen* **379,** 62 (1911).
2. E. Barnett, J. W. Cook, and H. H. Grainger, *J. Chem. Soc.* **121,** 2059 (1922).
3. A. Ault, R. Kopet, and A. Serianz, *J. Chem. Educ.* **48,** 410 (1971).

73 An Analgesic: *p*-Ethoxyacetanilide from *p*-Aminophenol

Section 63.2 describes the preparation of *p*-ethoxyacetanilide (phenacetin), an analgesic and antipyretic, by the acetylation of *p*-ethoxyaniline (also called *p*-phenetidine). In that synthesis, *p*-aminophenol is, in effect, first ethylated and then acetylated. Here, we present an alternative synthesis in which *p*-aminophenol (*p*-hydroxyaniline) is first acetylated (Section 73.1) and then ethylated (Section 73.2).

73.1 *p*-ACETAMIDOPHENOL FROM *p*-AMINOPHENOL

$$\text{H—}\ddot{\text{N}}\text{—H (on benzene ring, para-OH)} \xrightarrow[\text{sodium acetate}]{\text{acetic anhydride}} \text{H—}\ddot{\text{N}}\text{—C(=O)—CH}_3 \text{ (on benzene ring, para-OH)}$$

p-aminophenol *p*-acetamidophenol

Why does the aminophenol dissolve?

Procedure. Place 5.9 grams (0.054 mole) of *p*-aminophenol (Note 1) in a 125-mL Erlenmeyer flask. To this flask add 50 mL of water and 4.5 mL (0.054 mole) of conc. HCl. Swirl the flask so as to dissolve the aminophenol (Note 2). Prepare a solution of 8.9 grams of sodium acetate trihydrate (Note 3) in 25 mL of water. To the solution of *p*-aminophenol in HCl now add 6.6 grams (6.2 mL; 0.065 mole) of acetic anhydride. Swirl the mixture to dissolve almost all of the acetic anhydride, add the solution of sodium acetate, swirl the mixture to mix it thoroughly, and set it aside for crystallization (Note 4). Collect the product by suction filtration, and wash it with a little cold water. Yield: about 6 grams; about 75%.

Notes

1. Commercial *p*-aminophenol is often quite dark. It can be somewhat cleaned up by suspending it in about twice its volume of methanol and then recovering it by suction filtration.
2. At this point, the solution can be treated with decolorizing carbon, which can then be removed by gravity filtration.
3. An equivalent amount of anhydrous sodium acetate can be used instead of the trihydrate.
4. The solution can be allowed to stand overnight, or longer, or it can be cooled more quickly after crystals have started to form by immersion first in cold water and then in an ice bath.

73.2 PHENACETIN FROM *p*-ACETAMIDOPHENOL

$$\text{H–}\ddot{\text{N}}\text{–C(=O)–CH}_3 \text{ (on benzene ring, para O–H)} \xrightarrow[\text{sodium hydroxide}]{\text{ethyl bromide}} \text{H–}\ddot{\text{N}}\text{–C(=O)–CH}_3 \text{ (on benzene ring, para O–CH}_2\text{CH}_3\text{)}$$

p-acetamidophenol → phenacetin

Procedure. Place 3.0 grams (0.02 mole) of *p*-acetamidophenol and 20 mL of methanol in a 100-mL boiling flask. To this mixture add 1.25 mL (0.024 mole) of 50% sodium hydroxide solution; swirl the mixture so as to dissolve the *p*-acetamidophenol. Fit the flask with a reflux condenser, and clamp the apparatus in position on a steam bath. Now add 3.0 mL (4.4 g; 0.04 mole) (Note 1) of ethyl bromide down the condenser, and boil the mixture gently for about 2 hours (Note 2). At the end of the heating period, add 40 mL of hot water down the condenser; crystals should begin to appear as the last of the water is added. Remove the condenser, and set the flask aside for crystallization (Note 3). Collect the glistening, colorless crystals by suction filtration, washing them with portions of cold water. Yield: about 2.9 grams; about 80%.

Why does the *p*-acetamidophenol dissolve?

Why is the product not soluble in base?

Notes

1. Only 1 equivalent of ethyl bromide is required. We use 2 equivalents in order to decrease the time needed for the reaction.
2. A reaction time of 1 hour gives a yield of about 60%.
3. The mixture can be allowed to stand overnight, or longer, or crystallization can be hastened by immersing the flask first in cold water and then in an ice bath.

Questions

1. Why does the acetylation of *p*-aminophenol with 1 equivalent of acetic anhydride in a solution with a pH of 4 give the product due to acetyl transfer to nitrogen rather than to oxygen?
2. What product would you expect if *p*-aminophenol were acetylated with 2 equivalents of acetic anhydride at a pH of 11?
3. What product would you expect if *p*-aminophenol were acetylated with 1 equivalent of acetic anhydride at a pH of 11?

74 A Photochromic Compound: 2-(2,4-Dinitrobenzyl)pyridine

Photochromism: colorless → dark blue

Crystals of the compound 2-(2,4-dinitrobenzyl)pyridine have the most unusual property of turning a very deep blue color when exposed to sunlight. During storage in the dark, the crystals revert to their original sandy color. Color formation takes only a few minutes in bright sunlight, but the loss of color takes about a day. The interconversion appears to be completely reversible any number of times. One explanation of the phenomenon proposes the formation of a tautomeric form by action of the light (Reference 1).

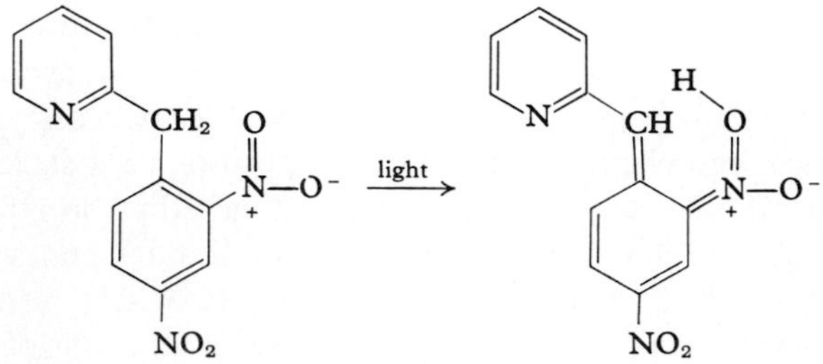

N CH$_2$ O N—O$^-$ + NO$_2$ —light→ N CH H O N—O$^-$ + NO$_2$

2-(2,4-dinitrobenzyl)pyridine

Procedure (Reference 2). Place 25 mL of conc. sulfuric acid (0.45 mole) in a 250-mL boiling flask. Cool the acid to 5°C or below by means of an ice bath. Arrange the flask so that it can be stirred or swirled in an ice bath and add gradually with good mixing 5.0 mL (5.3 g; 0.031 mole) of 2-benzylpyridine. To this cooled and well-stirred mixture add drop by drop, over a period of about 3 minutes, 3.0 mL of red fuming nitric acid (density, 1.5 g/mL; 0.070 mole). The addition of the first few drops of the nitric acid will cause a color change to deep brown, but the mixture will become lighter in color as the remainder of the acid is added. After all the nitric acid has been added, heat the mixture for about 20 minutes on the steam bath.

At the end of the heating period, pour the mixture onto about 400 g of ice in a 2-1 flask. Basify the solution to a pH of about 11 by adding almost all of a solution of 40 g (1 mole) of sodium hydroxide in about 500 mL of water. Toward the end of the addition of base, the product separates to give a milky yellow suspension. Add about 400 mL of ether and stir the mixture for 10–15 minutes to extract the product into the ether layer. Separate the ethereal solution, dry it over anhydrous magnesium sulfate for a few minutes, filter, and distill to reduce its volume to about 50 mL (Note 1). The product will sometimes crystallize during the concentration of the solution. Complete the crystallization by cooling the mixture in an ice bath. Collect the large sandy prisms by suction filtration, and wash them with a small amount of cold 95% ethanol. Yield: about 4 g (50%). The product can be recrystallized from 95% ethanol with 90% recovery using about 10 mL per gram; use of decolorizing carbon helps a little.

Note

1. Collect the ether and return it to the container provided so that it can be used by others.

Time: 3 hours.

References

1. J. A. Sousa and J. Weinstein, *J. Org. Chem.* **27,** 3155 (1962); and A. L. Bluhm, J. Weinstein, and J. A. Sousa, *J. Org. Chem.* **28,** 1989 (1963).
2. A. Ault and C. Kouba, *J. Chem. Educ.* **51,** 395 (1974).
3. K. Schofield, *J. Chem. Soc.* **1949,** 2411; and A. J. Nunn and K. Schofield, *J. Chem. Soc.* **1952,** 586.
4. A. E. Tschitschibabin, B. M. Kuindshi, and S. W. Benewolenskaja, *Chem. Ber.* **58,** 1580 (1925).

75 A Chemiluminescent Compound: Luminol

The production of "cold light" by the American firefly is a familiar example of light emission that is nonthermal in origin; it is luminescence rather than incandescence. The light of the firefly is produced during oxidation of luciferin to oxyluciferin by oxygen in the presence of adenosine triphosphate (ATP), magnesium ion, and the enzyme luciferase. It appears that oxyluciferin is produced in an electronically excited state, and that the energy released when the corresponding ground state is formed is given off as light.

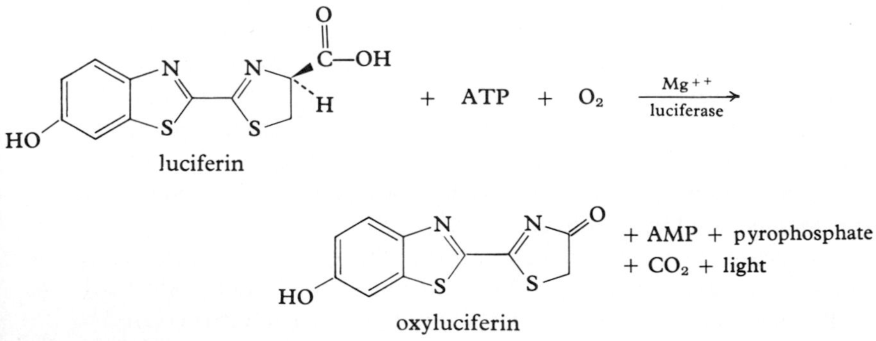

luciferin $+$ ATP $+$ O_2 $\xrightarrow[\text{luciferase}]{Mg^{++}}$

oxyluciferin $+$ AMP $+$ pyrophosphate $+$ CO_2 $+$ light

A number of other substances have been shown to undergo chemical reaction with simultaneous production of light. A good example that is easy to observe is that of luminol reacting with oxygen and base in dimethylsulfoxide solution.

$$\text{luminol} + 2OH^- + O_2 \xrightarrow{\text{dimethylsulfoxide}} \text{3-aminophthalate} + 2H_2O + N_2 + \text{light}$$

luminol 3-aminophthalate

Luminol can be prepared from 3-nitrophthalic anhydride by treatment with hydrazine followed by reduction of the first product by sodium dithionite.

$$\text{3-nitrophthalic acid} \xrightarrow[215°C]{H_2N{-}NH_2} \text{3-nitrophthalhydrazide} \xrightarrow{Na_2S_2O_4} \text{luminol}$$

3-nitrophthalic acid 3-nitrophthalhydrazide luminol

Preparation of Luminol from 3-Nitrophthalic Acid via 3-Nitrophthalhydrazide

Procedure. First put a flask containing 15 mL of water on the steam bath to get hot. Then add to a 20 mm × 150 mm test tube 1.0 gram (4.7 mmole) of 3-nitrophthalic acid and 2 mL of an 8% aqueous solution of hydrazine (5 mmole; Note 1). Heat the mixture over a small burner flame until all the solid has gone into solution. Now add 3 mL of triethylene glycol, clamp the tube in a vertical position over the burner, and insert both a thermometer and a tube connected to the water aspirator. Boil the solution vigorously to distill the excess water, removing the vapors by means of the aspirator. Let the temperature rise rapidly until (in 3–4 minutes) it reaches 215°C. Remove the burner, notice the time, and by intermittent gentle heating, maintain the temperature at 215–220°C for 2 minutes. At this time, remove the burner, allow the temperature of the mixture to fall to 100°C (crystals of product may appear), add the 15 mL of hot water (Note 2), and collect the precipitated 3-nitrophthalhydrazide by suction filtration.

Transfer the damp nitro compound to the test tube in which it was prepared; do not clean the tube before making the addition. Add 5 mL of 10% sodium hydroxide solution (14 mmole), stir to dissolve, and to the resulting deep brown-red solution add 3.3 grams of "90% practical" sodium dithionite (17 mmole). Wash the solid down from the walls of the test tube with a little water. Heat the mixture to the boiling point, stir it, and keep it hot for 5 minutes; during this time some of the reduction product may separate from the yellow-brown solution. Now add 2 mL of acetic acid (35 mmole), cool the mixture thoroughly by swirling the test tube in a beaker of cold water, and collect the resulting precipitate of light yellow luminol by suction filtration. The damp crude product is suitable for demonstration of chemiluminescence.

Notes

1. The 8% hydrazine solution can be prepared either by diluting 30.0 mL (31.2 grams) of 64% hydrazine to 250 mL or by diluting 22.3 mL (23.5 grams) of 85% hydrazine to 250 mL.
2. The reason for using hot water rather than cold is that the solid is then produced in a form that is more easily filtered.

Time: 2 hours.

Demonstration of the Chemiluminescence of Luminol

Procedure. Put 100 mL of dimethylsulfoxide in a 250-mL bottle or flask, add 1 mL of 50% aqueous sodium hydroxide, and swirl to mix. Then add 10–20 mg of luminol, stopper the container, and shake the contents vigorously. Within 60 seconds, the contents of the bottle or flask will emit a blue-green light, which will be sufficiently bright to make it possible to easily read this book in the dark. When the light fades it can be restored by shaking the bottle or flask again. When all the luminol has reacted, further portions can be added, even after several weeks.

Blue-green light!

References

1. L. F. Fieser, *Organic Experiments*, 2nd edition, Heath, Boston, 1968, p. 239.
2. E. H. White and D. F. Roswell, *Accounts Chem. Res.* **3,** 54 (1970); and E. H. White, *J. Chem. Educ.* **34,** 275 (1957).
3. H. W. Schneider, *J. Chem. Educ.* **47,** 519 (1970); and M. T. Beck and F. Joo, *J. Chem. Educ.* **48,** A559 (1971).
4. W. D. McElroy and M. DeLuca, in *Chemiluminescence and Bioluminescence,* Cormier, Hercules, and Lee, editors, Plenum Press, New York, 1973, p. 285.

76 Thiamine-Catalyzed Formation of Benzoin from Benzaldehyde

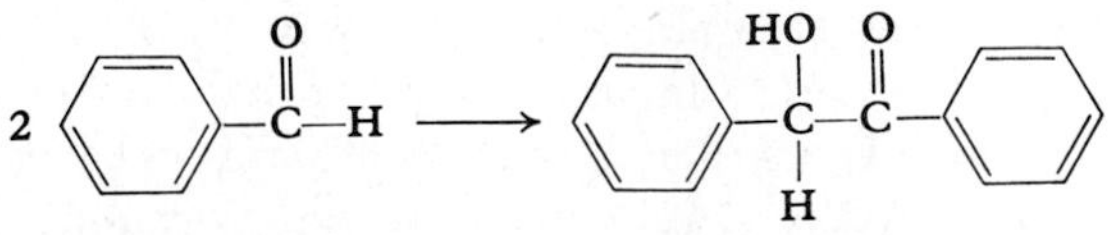

$$2\ C_6H_5\text{—}\overset{\overset{\displaystyle O}{\|}}{C}\text{—}H \longrightarrow C_6H_5\text{—}\underset{\underset{\displaystyle H}{|}}{\overset{\overset{\displaystyle HO}{|}}{C}}\text{—}\overset{\overset{\displaystyle O}{\|}}{C}\text{—}C_6H_5$$

benzaldehyde benzoin

$R\text{—}\ddot{C}\text{=}\ddot{\underset{\cdot\cdot}{O}}$

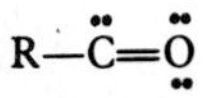

acyl carbanion

Thiamine pyrophosphate (TPP) is a necessary coenzyme for a number of enzyme-catalyzed reactions, all of which involve, in effect, the transfer of an acyl group, R—C=O, as a carbanion. Organic chemists try to avoid proposing acyl carbanions as free intermediates since

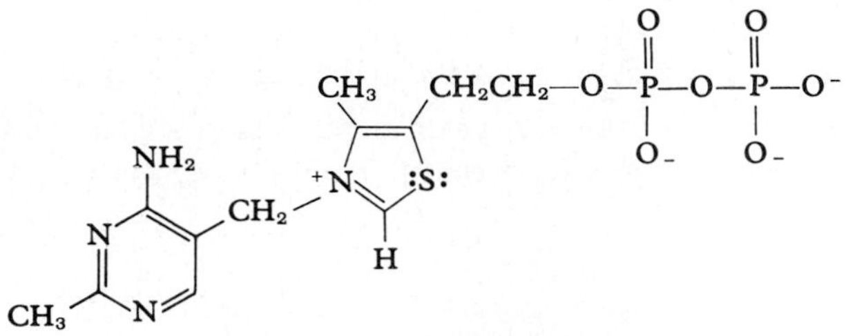

thiamine pyrophosphate (TPP)

they lack the structural features (electron accepting groups) that are thought to stabilize such ions. The way in which acyl groups are transferred in biochemical reactions at a pH of 7 and at 37°C, then, remained a mystery until Professor Breslow at Columbia University proposed a role for thiamine (Reference 1). Breslow remembered that a famous old reaction of organic chemistry, the cyanide-catalyzed condensation of benzaldehyde to form benzoin (Section 80.1) appeared to involve the equivalent of an acyl carbanion. Cyanide ion had long been known as a specific catalyst for this reaction and cyanide was thought to act by adding to the carbonyl group of benzaldehyde to form the cyanohydrin, which could then lose the aldehyde hydrogen to form a resonance-stabilized anion (Reference 2).

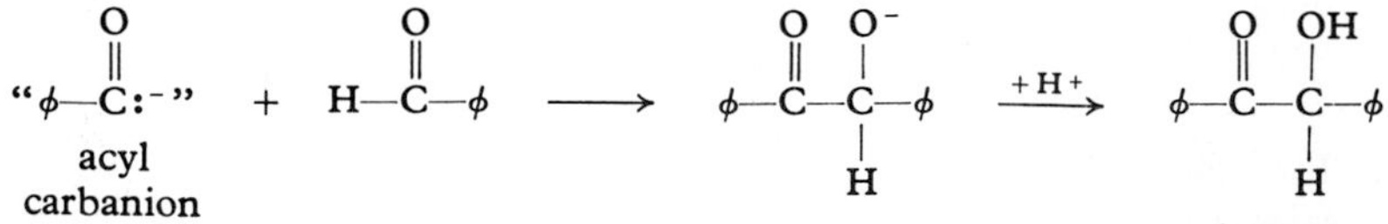

$$\text{“}\phi\text{—}\overset{\overset{\displaystyle O}{\|}}{C}\text{:}^-\text{”} + H\text{—}\overset{\overset{\displaystyle O}{\|}}{C}\text{—}\phi \longrightarrow \phi\text{—}\overset{\overset{\displaystyle O}{\|}}{C}\text{—}\underset{\underset{\displaystyle H}{|}}{\overset{\overset{\displaystyle O^-}{|}}{C}}\text{—}\phi \xrightarrow{+H^+} \phi\text{—}\overset{\overset{\displaystyle O}{\|}}{C}\text{—}\underset{\underset{\displaystyle H}{|}}{\overset{\overset{\displaystyle OH}{|}}{C}}\text{—}\phi$$

acyl carbanion benzoin

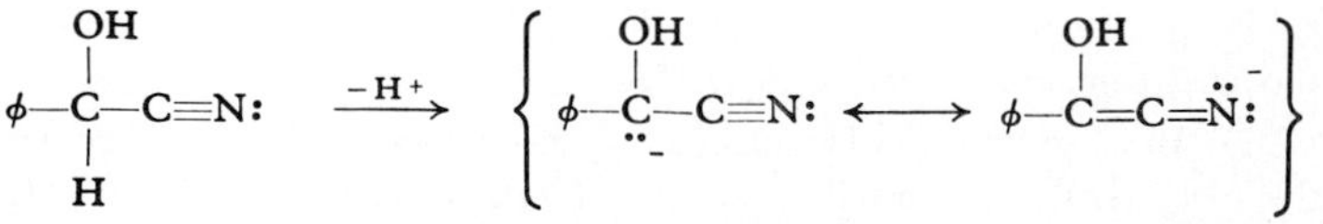

$$\phi\text{—}\underset{\underset{\displaystyle H}{|}}{\overset{\overset{\displaystyle OH}{|}}{C}}\text{—}C\equiv N\text{:} \xrightarrow{-H^+} \left\{ \phi\text{—}\underset{\cdot\cdot\,-}{\overset{\overset{\displaystyle OH}{|}}{C}}\text{—}C\equiv N\text{:} \longleftrightarrow \phi\text{—}\overset{\overset{\displaystyle OH}{|}}{C}\text{=}C\text{=}\ddot{N}\text{:}^- \right\}$$

anion

This anion could add to the carbonyl group of a second molecule of benzaldehyde to form the cyanohydrin of benzoin, which would then lose HCN to give benzoin and regenerate the catalyst.

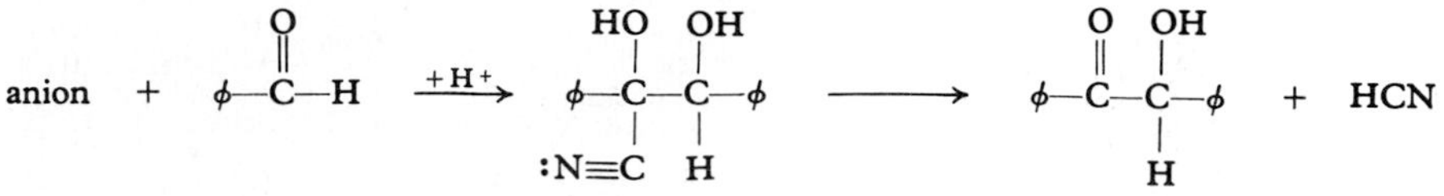

$$\text{anion} + \phi\text{—}\overset{\overset{\displaystyle O}{\|}}{C}\text{—H} \xrightarrow{+H^+} \phi\text{—}\underset{\underset{\displaystyle :N\equiv C}{|}}{\overset{\overset{\displaystyle HO}{|}}{C}}\text{—}\underset{\underset{\displaystyle H}{|}}{\overset{\overset{\displaystyle OH}{|}}{C}}\text{—}\phi \longrightarrow \underset{\text{benzoin}}{\phi\text{—}\overset{\overset{\displaystyle O}{\|}}{C}\text{—}\underset{\underset{\displaystyle H}{|}}{\overset{\overset{\displaystyle OH}{|}}{C}}\text{—}\phi} + \text{HCN}$$

Breslow also knew that a number of thiazolium salts, including thiamine, had been reported to catalyze the conversion of benzaldehyde to benzoin, and not to give the product that was originally expected. In addition, he had determined that the C-2 proton of the thiazolium ring of thiamine was rapidly exchanged for deuterium in D_2O at room temperature. For these and other reasons, he suggested that thiamine could catalyze the benzoin condensation in the following way.

Thiamine loses a proton to the solvent (or to the enzyme):

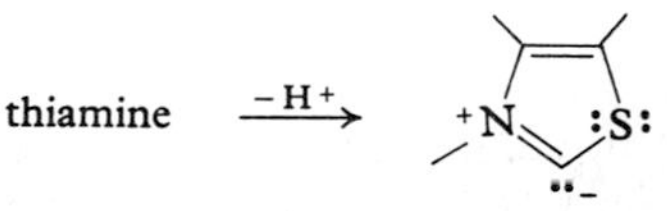

thiamine $\xrightarrow{-H^+}$ [thiazolium ring: $^+$N, :S:, C-2 carbanion $\ddot{}^-$]

conjugate base of thiamine

The conjugate base of thiamine adds to benzaldehyde:

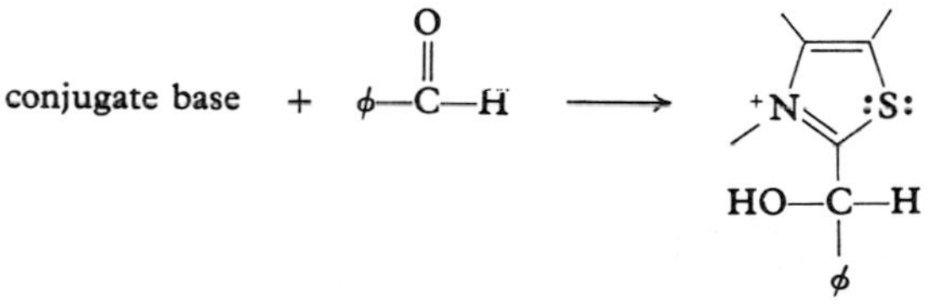

conjugate base + $\phi\text{—}\overset{\overset{\displaystyle O}{\|}}{C}\text{—H}$ $\longrightarrow$ [thiazolium ring: $^+$N, :S:, C-2 bonded to HO—C—H, with ϕ on C]

thiamine-benzaldehyde

Thiamine-benzaldehyde loses a proton to give a resonance-stabilized equivalent of an acyl carbanion:

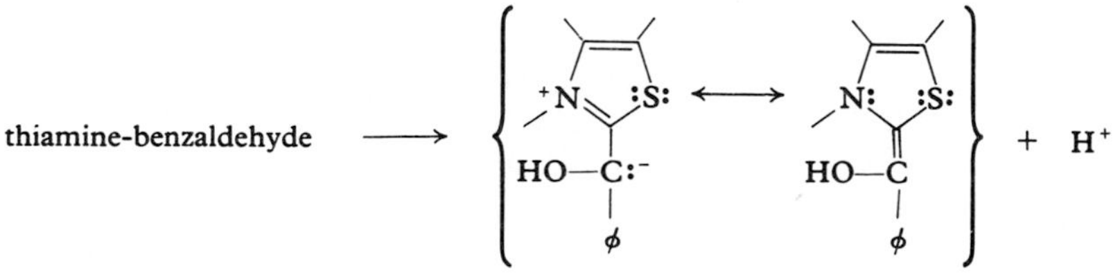

thiamine-benzaldehyde $\longrightarrow$ { [$^+$N, :S:, HO—C:$^-$, ϕ] $\longleftrightarrow$ [N:, :S:, =C(OH)ϕ] } + H^+

benzaldehyde anion equivalent

The equivalent of the anion of benzaldehyde then adds to a second molecule of benzaldehyde to give a product that eliminates thiamine to give benzoin and to regenerate the catalyst:

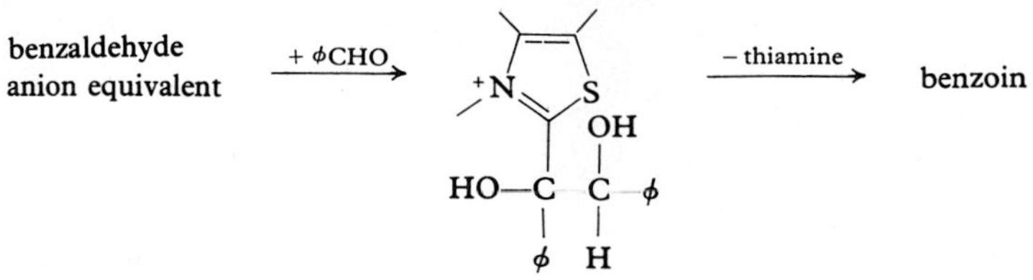

benzaldehyde anion equivalent $\xrightarrow{+\ \phi CHO}$ [thiazolium adduct: $^{+}$N, S ring; HO—C(ϕ)—C(OH)(H)—ϕ] $\xrightarrow{-\ \text{thiamine}}$ benzoin

Reasoning from this model, Breslow proposed a similar role for thiamine pyrophosphate in the enzyme-catalyzed systems: resonance stabilization of the carbanion equivalent of an aldehyde.

That is, "$R{-}\overset{O}{\overset{\|}{C}}{:}^{-}$" is

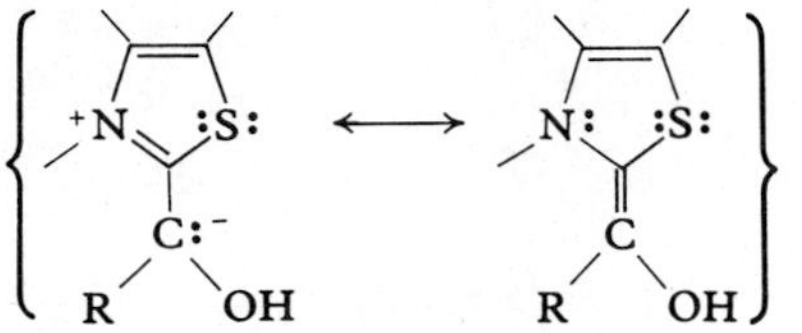

[resonance structures: $^{+}$N, :S: ring with C:$^{-}$(R)(OH) ⟷ N:, :S: ring with =C(R)(OH)]

Breslow was very careful not to draw conclusions that were not supported by the evidence. However, his proposal has now come to be accepted as representing the mechanism of thiamine action in enzymatic reactions in which thiamine pyrophosphate participates as a coenzyme. The role of the enzyme itself still remains to be determined.

The following procedure illustrates the use of thiamine as a catalyst in the benzoin condensation.

Thiamine-Catalyzed Benzoin Condensation

Procedure. Place 3.5 grams (0.01 mole) of thiamine hydrochloride in a 125-mL Erlenmeyer flask. Add 10 mL of water and swirl the mixture until the thiamine hydrochloride has all dissolved. Then add 25 mL of 95% ethanol, 10 mL of 2 *M* NaOH (0.02 mole), and 10 mL (10.4 grams; 0.1 mole) of benzaldehyde (Note 1) swirling the flask during and after each addition in order to thoroughly mix the contents of the flask. Allow the resulting mixture to stand at room temperature for at least 6 hours (Note 2). Collect the resulting crystals of benzoin by suction filtration, washing them with about 20 mL of a cold mixture of ethanol and water (5 : 1). Yield: about 7 grams (about 70%).

Notes

1. We have always used a fresh bottle of benzaldehyde.
2. The mixture may be allowed to stand for 24 hours or for a week. It is helpful to add a seed crystal. By avoiding supersaturation, crystallization starts sooner, and the resulting crystals are larger and are more easily washed on the filter paper after collection by suction filtration.

Time: 2 hours plus time required for reaction to occur.

Questions

1. Draw the structure of "thiamine hydrochloride" (thiamine chloride hydrochloride; Vitamin B_1).
2. The free acyl carbonium ion, $R{-}\overset{+}{C}{=}\ddot{O}{:}$, is sometimes invoked as an intermediate in the aluminum chloride catalyzed acylation of aromatic compounds. Does it have any resonance stabilization? Explain.
3. Thiamine pyrophosphate is a required coenzyme in the enzyme-catalyzed decarboxylation of pyruvic acid to form acetaldehyde. Show in detail how thiamine could be involved in this reaction.

$$\underset{\text{pyruvic acid}}{CH_3{-}\overset{O}{\overset{\|}{C}}{-}\overset{O}{\overset{\|}{C}}{-}OH} \longrightarrow \underset{\text{acetaldehyde}}{CH_3{-}\overset{O}{\overset{\|}{C}}{-}H} + CO_2$$

4. Thiamine catalyzes both the self-condensation of acetaldehyde to acetoin and the reaction of pyruvic acid with acetaldehyde to give acetoin. Show in detail how this might occur.

$$CH_3\ \overset{O}{\overset{\|}{C}}\ \underset{H}{\underset{|}{\overset{OH}{\overset{|}{C}}}}\ CH_3$$

acetoin

References

1. R. Breslow, *J. Am. Chem. Soc.* **80,** 3719 (1958).
2. A. Lapworth, *J. Chem. Soc.* **83,** 995 (1903).
3. T. S. Bruice and S. Benkovic, *Bioorganic Mechanisms,* Volume 2, W. A. Benjamin, Inc., New York, 1966, Chapter 8.

77 "Coconut Aldehyde"; γ-Nonanolactone

An early and continuing motivation for organic synthesis has been the search for compounds useful as flavors and fragrances. Sometimes the aim is to duplicate at lower cost a substance that occurs in nature, sometimes it is to prepare a similar or related compound with the hope that it will have an interesting property, and sometimes it is to produce a less costly substitute of unrelated structure. One such substance is the

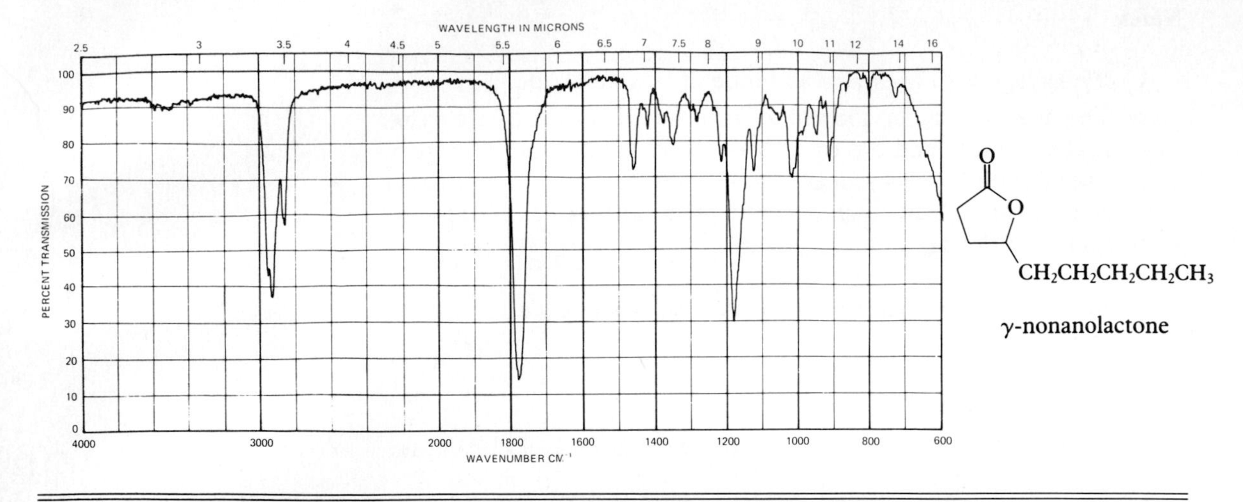

Figure 77.1. The infrared spectrum of γ-nonanolactone; thin film.

A coconut odor!

object of the synthesis presented in Sections 77.1 and 77.2. γ-Nonanolactone has been described as having "a strong coconut odor and flavor with a very faint anisaldehyde-like bynote" and "an odor indistinguishable from that of coconut." Up to this time, however, it has not been found to occur naturally.

Physical properties reported for this compound include: boiling point, 137–138°C/17 mm; $d_4^{19.5}$, 0.9672; $n_D^{19.5}$, 1.4462. The infrared spectrum is shown in Figure 77.1.

An interesting book about the sources, properties, and chemistry of flavors and fragrances is P. Z. Bedoukian, *Perfumery and Flavoring Synthetics* (New York: American Elsevier, 1967).

77.1 3-NONENOIC ACID FROM HEPTALDEHYDE AND MALONIC ACID

$$\underset{\text{heptaldehyde}}{CH_3CH_2CH_2CH_2CH_2CH_2\overset{O}{\overset{\|}{C}}H} + \underset{\text{malonic acid}}{HO\overset{O}{\overset{\|}{C}}-CH_2-\overset{O}{\overset{\|}{C}}OH} \xrightarrow{\text{triethylamine}}$$

$$CO_2\uparrow + \underset{\substack{\text{2-nonenoic acid}\\ \text{(minor product)}}}{CH_3CH_2CH_2CH_2CH_2CH_2-CH{=}CH-\overset{O}{\overset{\|}{C}}OH}$$

$$+ \underset{\substack{\text{3-nonenoic acid}\\ \text{(major product)}}}{CH_3CH_2CH_2CH_2CH_2-CH{=}CH-CH_2-\overset{O}{\overset{\|}{C}}OH}$$

Procedure. Place 5.2 grams (0.05 mole) of malonic acid and 6.7 mL (5.70 grams; 0.05 mole) of heptanal in a 125-mL Erlenmeyer flask. Add to the flask 10 mL (7.2 grams; 0.072 mole) of triethylamine. Heat this mixture on the steam bath, and swirl the flask until the solid has dissolved. Heat the solution on the steam bath for about 1 hour, or until the evolution of carbon dioxide has ceased.

At the end of the heating period, cool the contents of the flask by swirling it in cold water, and then pour the reaction mixture into a 125-mL separatory funnel. Next, rinse the reaction flask with 30 mL of ethyl ether, and pour the rinsings into the separatory funnel. Now pour 8 mL (0.096 mole) of conc. HCl over about 25 g of ice, stir the mixture until no more ice appears to dissolve, and then pour the cold acid into the separatory funnel (Note 1). Thoroughly mix the contents of the separatory funnel, and then draw off and discard the lower, aqueous, layer. Wash the ethereal layer, which has been retained in the separatory funnel, with one 25-mL portion of water. Transfer the ethereal solution to a 50-mL Erlenmeyer flask, and remove most of the ether by heating the flask on the steam bath (Note 2). The residue of crude 3-nonenoic acid can be converted to coconut aldehyde by treatment with 85% sulfuric acid (see Section 77.2).

Why is the mixture acidified?

Notes

1. The reaction of HCl with the material in the separatory funnel is only slightly exothermic.
2. Evaporate the ether in the hood, or fit the flask with a distillation adapter connected to a condenser so that the ether is condensed and ether fumes are not released into the room.

77.2 COCONUT ALDEHYDE FROM 3-NONENOIC ACID

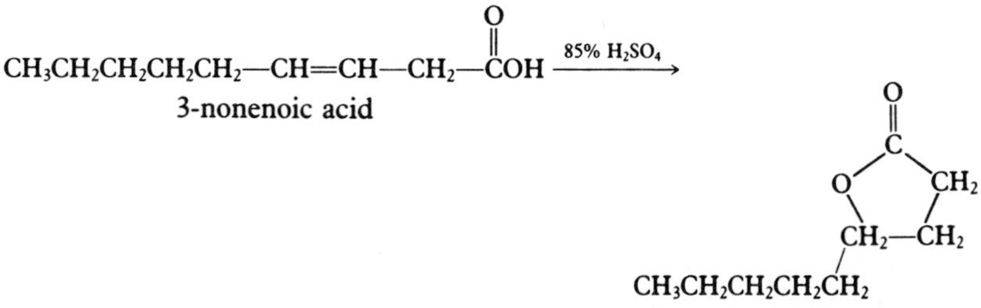

$$CH_3CH_2CH_2CH_2CH_2{-}CH{=}CH{-}CH_2{-}\overset{O}{\overset{\|}{C}}OH \xrightarrow{85\%\ H_2SO_4}$$

3-nonenoic acid

O
‖
C
O CH₂
CH₂—CH₂
$CH_3CH_2CH_2CH_2CH_2$

coconut aldehyde
(γ-nonanolactone)

Procedure. To the 50-mL flask that contains the crude 3-nonenoic acid from the reaction of 0.05 mole of malonic acid with 0.05 mole of heptanal (Note 1), add 10 mL of 85% sulfuric acid (0.16 mole) (Note 2).

Swirl the flask to mix the contents (Note 3), and heat the resulting solution on the steam bath for about 1 hour.

Caution: foaming!

At the end of the heating period, use a medicine dropper to transfer the reaction mixture in small portions to a 400-mL beaker that contains a mixture of 20 grams (0.19 mole) of sodium carbonate and about 90 mL of water (Note 4). After the transfer is complete, stir the mixture well to ensure that all the acid has had a chance to react with the base. Then pour about half of the oily suspension of coconut aldehyde in water into a 125-mL separatory funnel and extract it with a 30-mL portion of ether. After the layers have separated, draw off the lower, aqueous phase (Note 5) and add the remainder of the suspension of product and water. After thoroughly mixing the contents of the separatory funnel, again allow the layers to separate, and drain off the lower, aqueous layer (Note 5). Then wash the ethereal layer, which remains in the separatory funnel, with about 25 mL of water. Next, transfer the ethereal phase to a 50-mL Erlenmeyer flask, dry the solution with anhydrous magnesium sulfate (Note 6), and remove the ether by distillation on the steam bath. The residue of coconut aldehyde can be purified by distillation under reduced pressure (Note 7). Yield: about 3 grams (about 40% overall from malonic acid) of a very pale yellow oil that, at low concentrations, smells like coconut.

Notes

1. The preparation of 3-nonenoic acid is described in Section 77.1.
2. 85% sulfuric acid is prepared by adding 85 grams (46 mL) of conc. sulfuric acid to 15 grams of ice.
3. The reaction mixture quickly turns dark brown.
4. As the acidic reaction mixture is added to the mixture of sodium carbonate and water, carbon dioxide gas is produced. The gas causes the mixture to foam, and the large beaker serves to contain the foam.
5. If salts start to crystallize from the aqueous phase and interfere with removal of the water layer, add more water to the separatory funnel and swirl the contents gently so as to redissolve the salts.
6. Add anhydrous magnesium sulfate to the ethereal solution, allow the suspension to stand for about 10 minutes, and then remove the magnesium sulfate by gravity filtration. If you plan to purify the product by distillation, filter the solution into the boiling flask that you plan to use for the distillation. If the boiling flask is too small to hold all the filtrate at once, you can add part of it, concentrate it by distillation, and then add more.
7. Coconut aldehyde boils at 151°C under 23 Torr of pressure. The product can be distilled at atmospheric pressure, but it is then darker in color and has an acrid note in its smell.

References

1. *Chemical Abstracts* **56,** 8549 (1962).
2. *Chemical Abstracts* **59,** 11265 (1963).

78 A Model for the Biochemical Reducing Agent NADH

Many chemical oxidation-reduction reactions are rather brutal (see Sections 28.27, 54, 79.3, and 80.2). In contrast, biochemical redox reactions take place rapidly at about room temperature and at a pH of about 7. Chemists continue to be interested in trying to understand and reconstruct these gentle biochemical reactions. One approach is to study the reactions of related but less complex compounds, to study model reactions. In this experiment* we will work with a model that has been used to try to understand the mechanism of the biochemical reducing agent NADH (Nicotinamide Adenine Dinucleotide, reduced form). The reducing agent, NADH, is represented by formula (1)

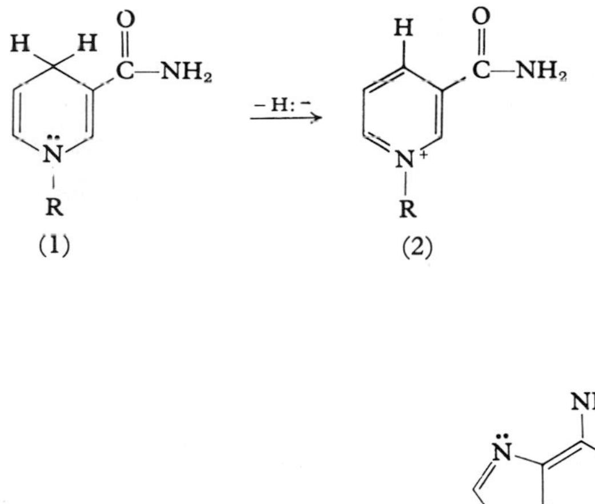

H H O C—NH_2 N̈ R (1) $\xrightarrow{-H:^-}$ H O C—NH_2 N^+ R (2)

where R is

NH_2 N̈ N: N N̈ O⁻ O⁻ CH_2O—P—O—P—O—CH_2 O OH HO O O O OH OH

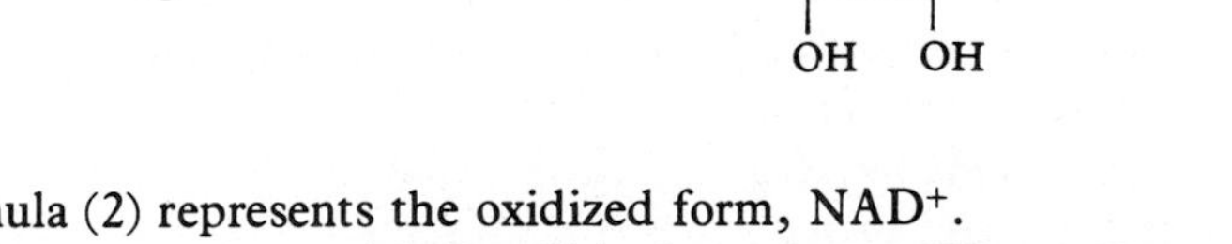

Formula (2) represents the oxidized form, NAD^+.

In our simple model, R will be benzyl: $\phi—CH_2—$. This model is

* This experiment was suggested by Dr. Bernard Golding, University of Warwick, Coventry, England.

one of several that were studied some years ago by Professor Westheimer at Harvard (Reference 1). In that research, deuterium labeling experiments showed that in the reduction of malachite green by 1-benzyldihydronicotinamide, the hydrogen lost by the reducing agent is transferred directly to the hydrogen acceptor:

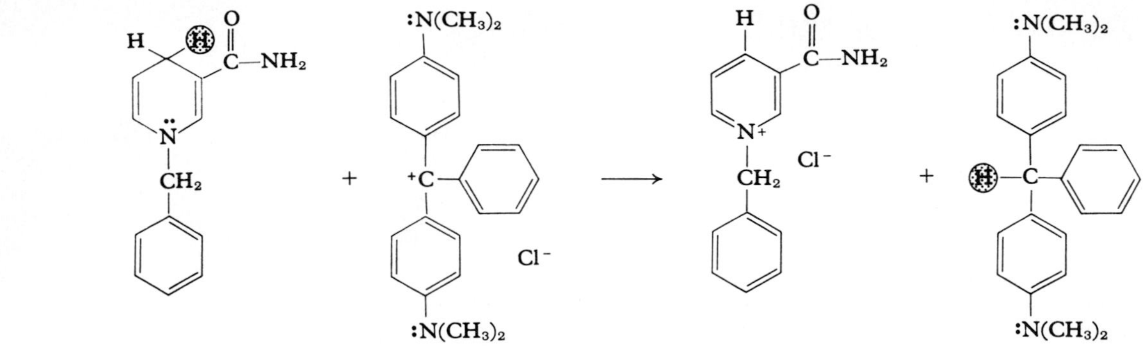

yellow
1-benzyldihydronicotinamide

green
malachite green

colorless
benzylnicotinamide
chloride

colorless
leuco base

This experiment includes the two-step preparation of 1-benzyldihydronicotinamide and a demonstration of its ability to reduce malachite green under mild conditions. The infrared spectrum of 1-benzyldihydronicotinamide is shown in Figure 78.1.

1-Benzylnicotinamide Chloride from Nicotinamide

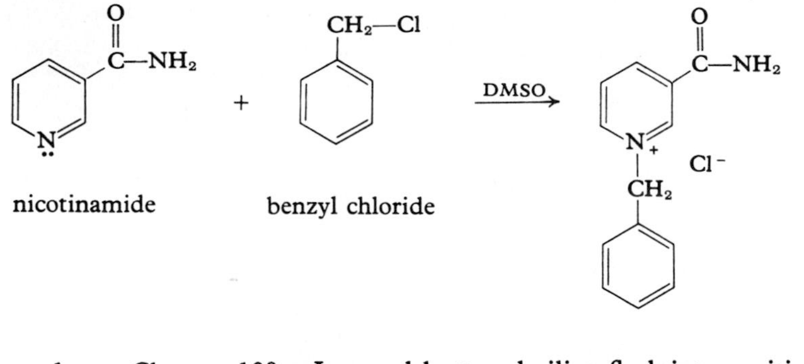

nicotinamide

benzyl chloride

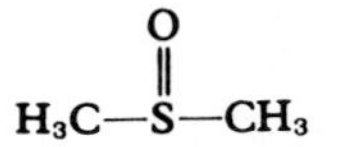

DMSO

$H_3C-\overset{O}{\overset{\|}{S}}-CH_3$
dimethylsulfoxide
(DMSO)

Procedure. Clamp a 100-mL round-bottom boiling flask into position over a wire gauze on an iron ring so that the flask can be heated with a burner flame. Add to the flask 5 mL of dimethylsulfoxide and 2 g (16.4 mmole) of nicotinamide. Heat the flask briefly with the burner and swirl the contents so as to dissolve the nicotinamide. When the solid has all dissolved, turn off the burner and add to the flask 10 mL (11 g; 88 mmole) of benzyl chloride (Note 1). Fit the flask with a Claisen adapter, and with a thermometer adapter in the center neck of the Claisen adapter position a thermometer so that the bulb dips into the solution in

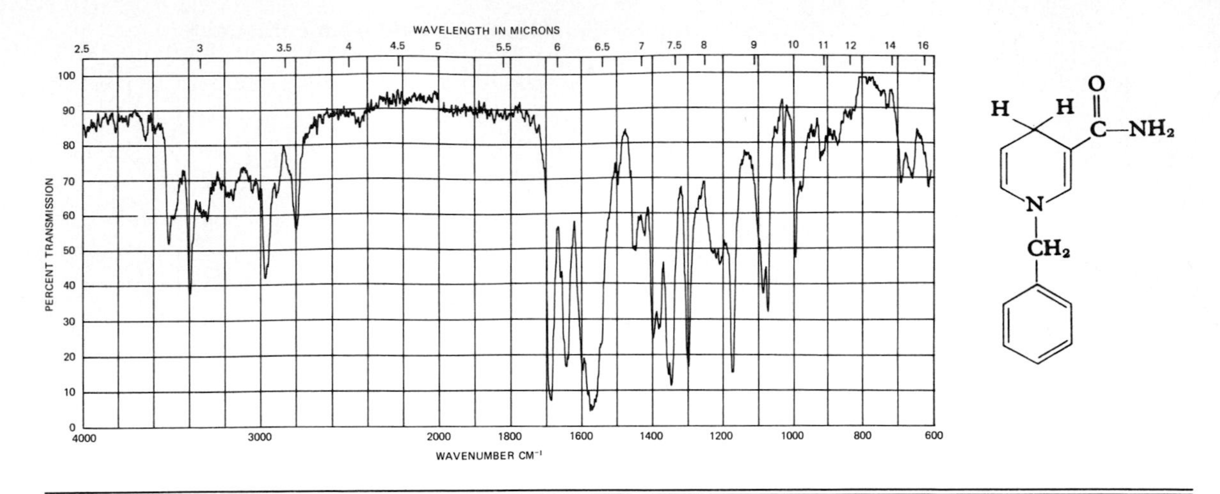

Figure 78.1. The infrared spectrum of 1-benzyldihydronicotinamide; $CHCl_3$ solution.

the flask. Fit the side arm of the Claisen adapter with a reflux condenser. Using the burner again, heat the flask until crystals begin to separate from the solution (Note 2). At this time, discontinue heating and allow the exothermic reaction to occur spontaneously (Note 3). After the reaction has taken place and the mixture has cooled to below 60°C (Note 4), remove the condenser, adapters, and thermometer and add 10 mL of isopropyl alcohol. Stopper the flask and mix the contents of the flask by shaking it vigorously. Collect the product by suction filtration and wash it first with a little isopropyl alcohol and then with some hexane.

Crystals form upon heating!

The crude product can be used to prepare 1-benzyldihydronicotinamide. It can also be recrystallized by heating it with 10 mL of 95% ethanol, adding just enough water (about 1 mL) to cause all the solid to dissolve, and allowing the solution to cool to room temperature.

Notes

1. Benzyl chloride is a lachrymator. It would be best to work in the hood while adding the benzyl chloride.
2. Crystals of the product start to appear when the temperature reaches about 135°C.
3. The temperature will rise to about 150–155°C before it starts to fall again.
4. When the temperature has fallen somewhat below 100°C, the flask may be placed in a beaker of cold water in order to cool it faster.

Time: about 1 hour.

1-Benzyldihydronicotinamide from 1-Benzylnicotinamide Chloride

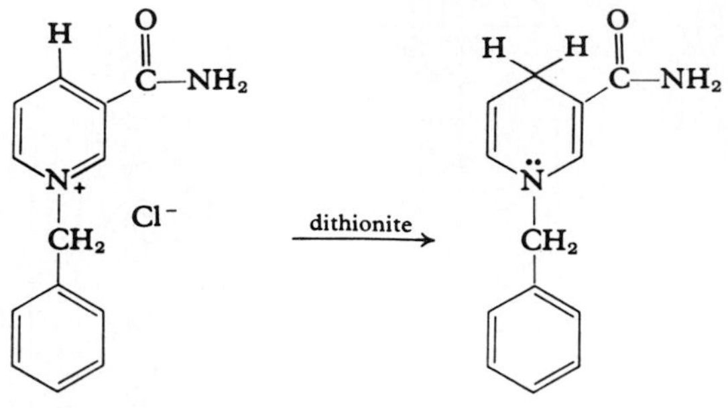

Procedure. Prepare a solution of 2.76 g of anhydrous sodium carbonate and 5.14 g (about 0.03 mole) of sodium dithionite in 40 mL of water in a 125-mL Erlenmeyer flask. Place a magnetic stirring bar in the flask and arrange for the solution to be stirred by a magnetic stirrer. Dissolve 2.0 g (8.0 mmoles) of 1-benzylnicotinamide chloride in 10 mL of water and add this solution all at once to the well-stirred solution of sodium carbonate and sodium dithionite. The resulting yellow-orange solution will suddenly produce a yellow precipitate in about 60 seconds. Continue to stir the mixture for another ten minutes and then collect the yellow solid by suction filtration using several small portions of water for rinsing and washing.

A sudden yellow precipitate

This product can be used directly for the reduction of malachite green, or it can be recrystallized to give beautiful yellow spars. To recrystallize, dissolve the crude product in 5 mL of hot 95% ethanol and filter the solution by gravity to remove a small amount of an insoluble impurity. Dilute the filtrate with 5 mL of warm water and, after swirling to mix, allow the solution to stand undisturbed for several hours (Note 1). Collect the crystals by suction filtration and wash them with a small amount of ice-cold 50% aqueous ethanol.

Note

1. It is helpful to seed the solution.

Time: less than 1 hour.

Reduction of Malachite Green by 1-Benzyldihydronicotinamide

Procedure. Dissolve about 10 milligrams (0.03 mmole) of malachite green in 1 mL of 95% ethanol in a small test tube. To this add 20–40 milligrams of 1-benzyldihydronicotinamide (0.1 to 0.2 mmoles). Swirl the mixture gently to dissolve the crystals. The intense green-blue color of the malachite green will give way to the yellow color of the excess reducing agent in 2 to 4 minutes.

Green → yellow

Time: about 30 minutes.

Questions

1. Malachite green is a carbonium ion salt. What accounts for its great stability?
2. Ethyl benzoylformate is not reduced by 1-benzyldihydronicotinamide alone at room temperature in the dark in acetonitrile solution. In the presence of an equimolar amount of magnesium perchlorate, however, racemic ethyl mandelate is formed in quantitative yield (2). Explain.

$$\phi\text{—}\overset{\displaystyle O}{\overset{\|}{C}}\text{—}\overset{\displaystyle O}{\overset{\|}{C}}\text{—OEt} \xrightarrow[\text{magnesium perchlorate}]{\text{1-benzyldihydronicotinamide}} \phi\text{—}\underset{\displaystyle H}{\underset{|}{\overset{\displaystyle HO}{\overset{|}{C}}}}\text{—}\overset{\displaystyle O}{\overset{\|}{C}}\text{—OEt}$$

ethyl benzoylformate | ethyl mandelate

3. When ethyl benzoylformate is treated with (R)-(−)-N-α-methylbenzyl-1-benzyldihydronicotinamide in the presence of an equimolar amount of magnesium perchlorate in acetonitrile solution, the ethyl mandelate produced is 60% R and 40% S.
   a. Draw formulas indicating the structure and configuration of (R)-(−)-N-α-methylbenzyl-1-benzyldihydronicotinamide and of the two enantiomers of ethyl mandelate.
   b. Explain the results of the experiment.

References

1. D. Mauzerall and F. H. Westheimer, *J. Am. Chem. Soc.* **77**, 2261 (1954).
2. Y. Ohnishi, M. Kagami, and A. Ohno, *J. Am. Chem. Soc.* **97**, 4766 (1975).

Synthetic Sequences: Synthesis Experiments That Use a Sequence of Reactions

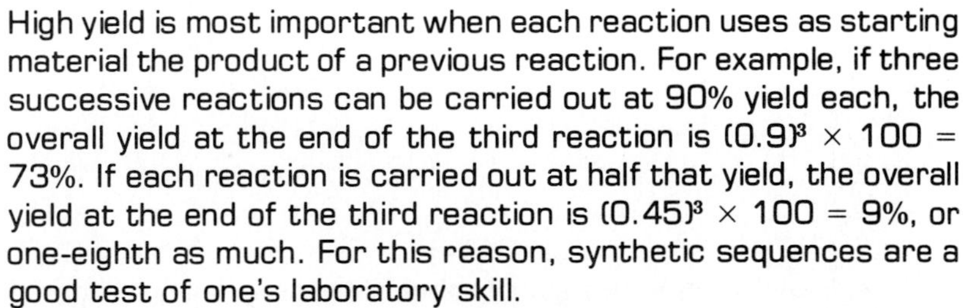

High yield is most important when each reaction uses as starting material the product of a previous reaction. For example, if three successive reactions can be carried out at 90% yield each, the overall yield at the end of the third reaction is $(0.9)^3 \times 100 =$ 73%. If each reaction is carried out at half that yield, the overall yield at the end of the third reaction is $(0.45)^3 \times 100 = 9\%$, or one-eighth as much. For this reason, synthetic sequences are a good test of one's laboratory skill.

Six synthetic sequences are outlined in this section. Many more can be made up from reactions in the preceding sections.

79 Steroid Transformations: Δ^4-Cholestene-3-one from Cholesterol via Cholesterol Dibromide, 5α,6β-Dibromocholestane-3-one, and Δ^5-Cholestene-3-one

As an initial step in exploring the chemistry of a rare and expensive substance, preliminary experiments can be done in which a less expensive substance of similar structure is used as a *model compound*. Cholesterol, which can be isolated in quantity from beef spinal cord and brains (and from human gallstones; see Section 39), has often served as a model compound for the exploration of the chemistry of a class of polycyclic aliphatic compounds known as *steroids*, the most interesting members of which are the sex hormones.

The Sex Hormones

The sex hormones are the substances that are directly responsible for the development of sex characteristics and for the sexual functioning of the human organism. These substances are produced in the ovaries and the testes in response to stimulation by the gonadotropic hormones secreted in the anterior lobe of the pituitary gland. In contrast to the proteinaceous gonadotropic hormones, the sex hormones are polycyclic aliphatic substances. The three primary human sex hormones are estradiol (the primary female sex hormone, or estrogen), testosterone (the primary male sex hormone, or androgen), and progesterone, a substance secreted by the corpus luteum of the female, which prepares the bed of the uterus for implantation of the fertilized ovum.

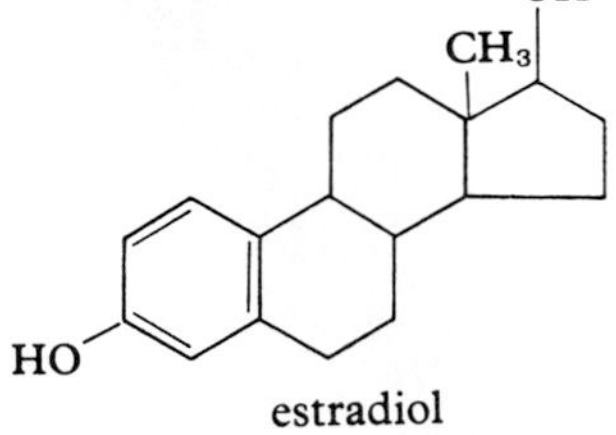

estradiol 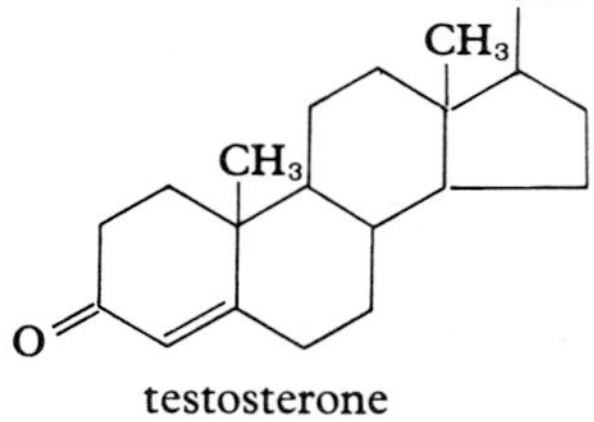 testosterone 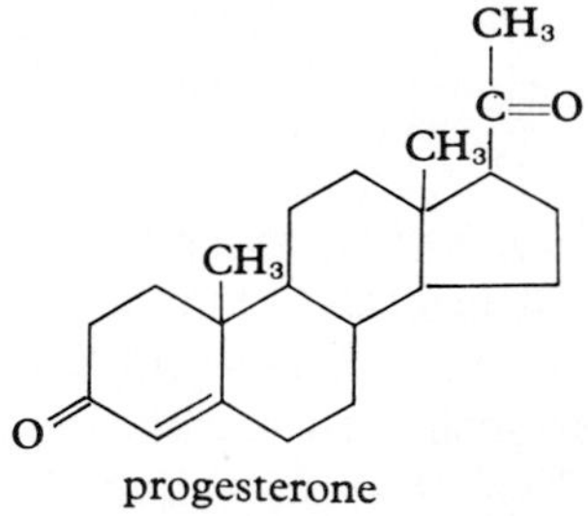 progesterone

These substances were originally obtained by isolation from natural sources. Estradiol was first isolated in 1934 from sow ovaries at a yield of about 12 mg from some 3600 kg (four tons) of ovaries, testosterone in 1935 from steer testes at a yield of 10 mg from 100 kg of steer testis tissue, and progesterone in 1934 from sow ovaries at a yield of 20 mg from 625 kg of ovaries, from 50,000 sows. Further development of isolation methods later gave somewhat larger yields, but there was an obvious need for vastly more efficient procedures for the production of these hormones.

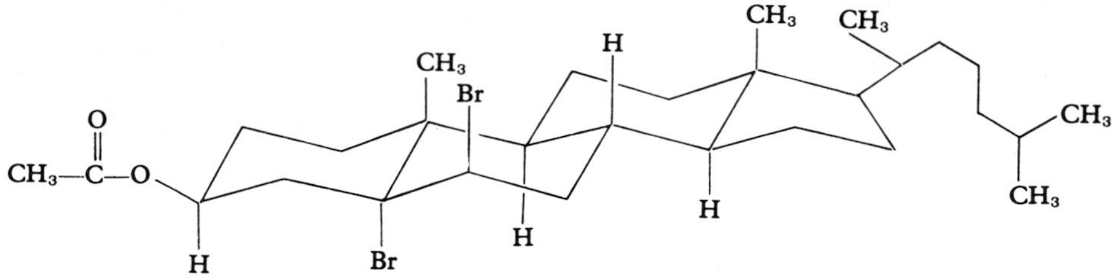

cholesterol acetate dibromide

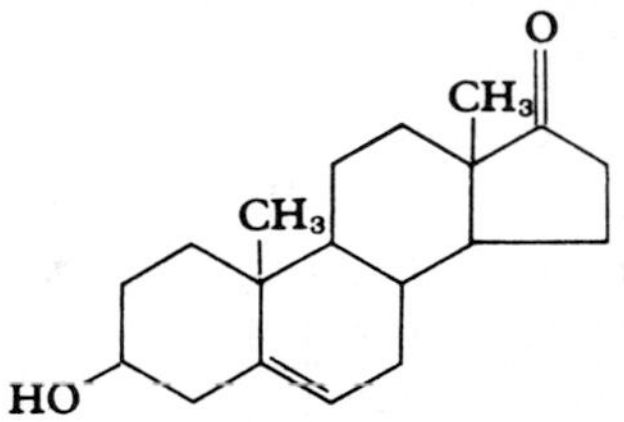

In 1935, several research groups determined that cholesterol (see Section 39) as the acetate dibromide could be degraded by vigorous chromic acid oxidation to give androstenolone as a minor product.

androstenolone

This substance, which had originally been isolated from male urine the year before, then served as a starting material for syntheses of the various sex hormones. The yield of androstenolone, isolated as the semicarbazone acetate, was originally less than 1%, but by various improvements in the method it was raised over the years to about 8%.

A far more satisfactory alternative for hormone syntheses was provided in 1939 by Russell Marker's discovery that certain substances of vegetable origin could be easily degraded in high yield to give suitable starting materials. For example, having established the spiroketal structure of the side chain of diosgenin, Marker showed that it could be converted in three steps to 16-dehydropregnenolone acetate in over 60% yield.

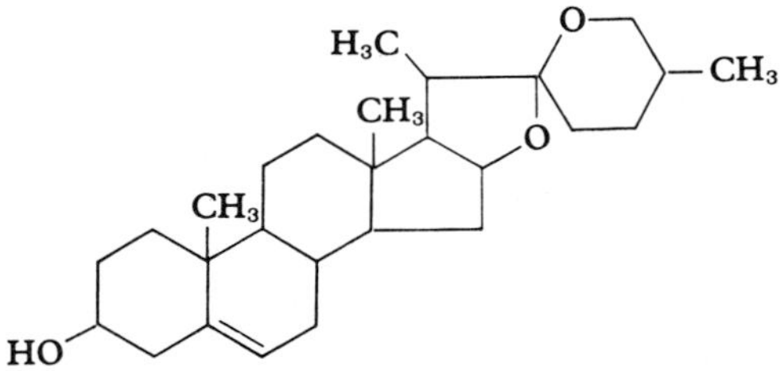

diosgenin

The double bond in the five-membered ring could then be selectively hydrogenated to give pregnenolone acetate.

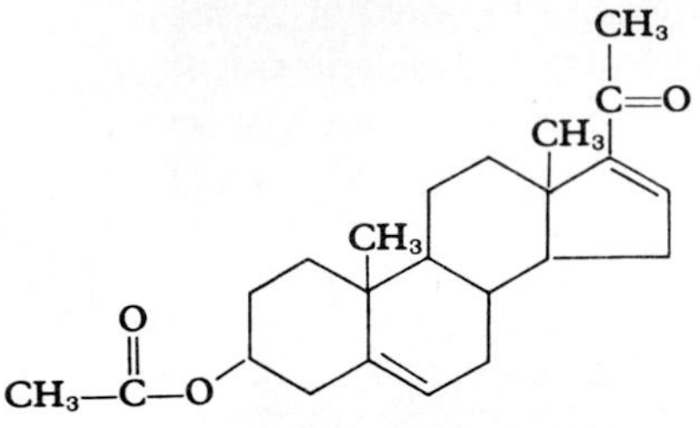

16-dehydropregnenolone acetate

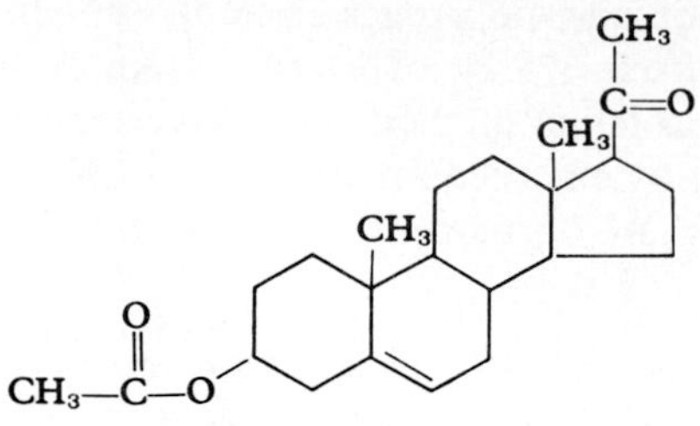

pregnenolone acetate

In 1943, Marker demonstrated the potential of his discovery by preparing 3 kg of progesterone from diosgenin. At that time, this was worth a quarter of a million dollars and was equivalent to the yield from the corpora lutea of 30 million sows. The source of the diosgenin was the roots of the wild Mexican yam. Marker worked directly in Mexico, achieving his feat under incredibly primitive conditions with the help of untrained assistants. Marker's unusual and generally unappreciated professional life is reviewed in an article in the *Journal of Chemical Education* (1).The article ends with the following quotation:

> When I retired from Chemistry in 1949, after 5 years of production and research in Mexico, I felt I had accomplished what I had set out to do. I had found sources for the production of steroidal hormones in quantity at low prices, developed the process of manufacture, and put them into production. I assisted in establishing many competitive companies in order to insure a fair price to the public and without patent protection or royalties from the producers.
>
> Since retiring from the laboratory 20 years ago, I have never returned to chemistry or consulting, and have no shares of stocks in any hormone or related companies. My only appearance in public was recently on April 23, 1969 to accept an award by the Mexican Chemical Society showing their appreciation for the work I had accomplished.
>
> *Russell E. Marker, May 15, 1969**

The book *Steroids* by Fieser and Fieser gives many interesting details about the isolation, structure, determination, preparation, and total synthesis of sex hormones and other steroids (2).

Synthesis of Sex Hormones

Androstenolone, obtained from cholesterol, was used for the preparation of testosterone (Figure 79.1), and pregnenolone, obtained from diosgenin via pregnenolone acetate, was an important starting material for the synthesis of progesterone (Figure 79.2). Both these syntheses require the transformation of a β,γ-unsaturated alcohol into the corresponding α,β-unsaturated ketone.

* Quoted with permission from the *Journal of Chemical Education* **50**, 199 (1973).

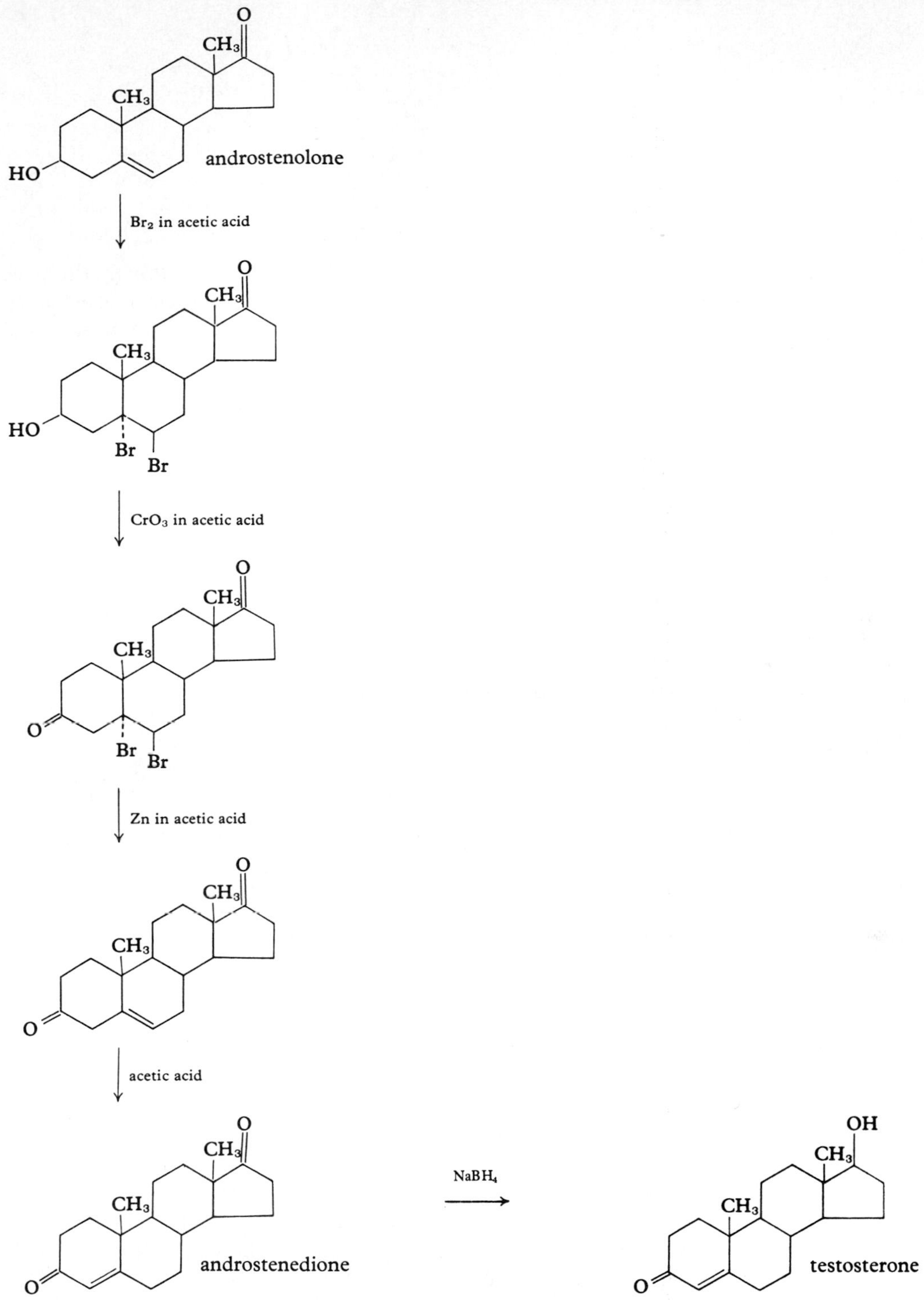

androstenolone

Br_2 in acetic acid

CrO_3 in acetic acid

Zn in acetic acid

acetic acid

androstenedione

$NaBH_4$

testosterone

Figure 79.1. Conversion of androstenolone to androstenedione and testosterone.

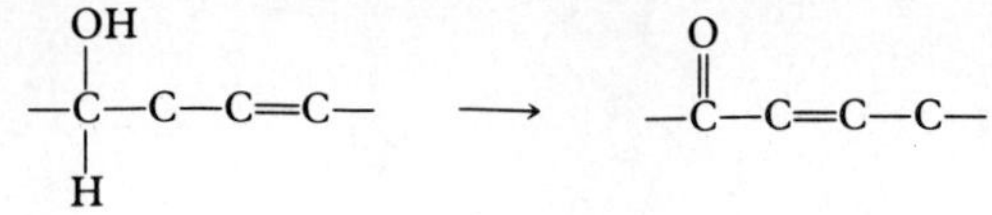

β,γ-unsaturated alcohol → α,β-unsaturated ketone

Steroid Transformations

In the sequence of reactions described in this section, exactly the same transformations are carried out, except that cholesterol is used as an inexpensive model compound. The reactions, which follow, consist of

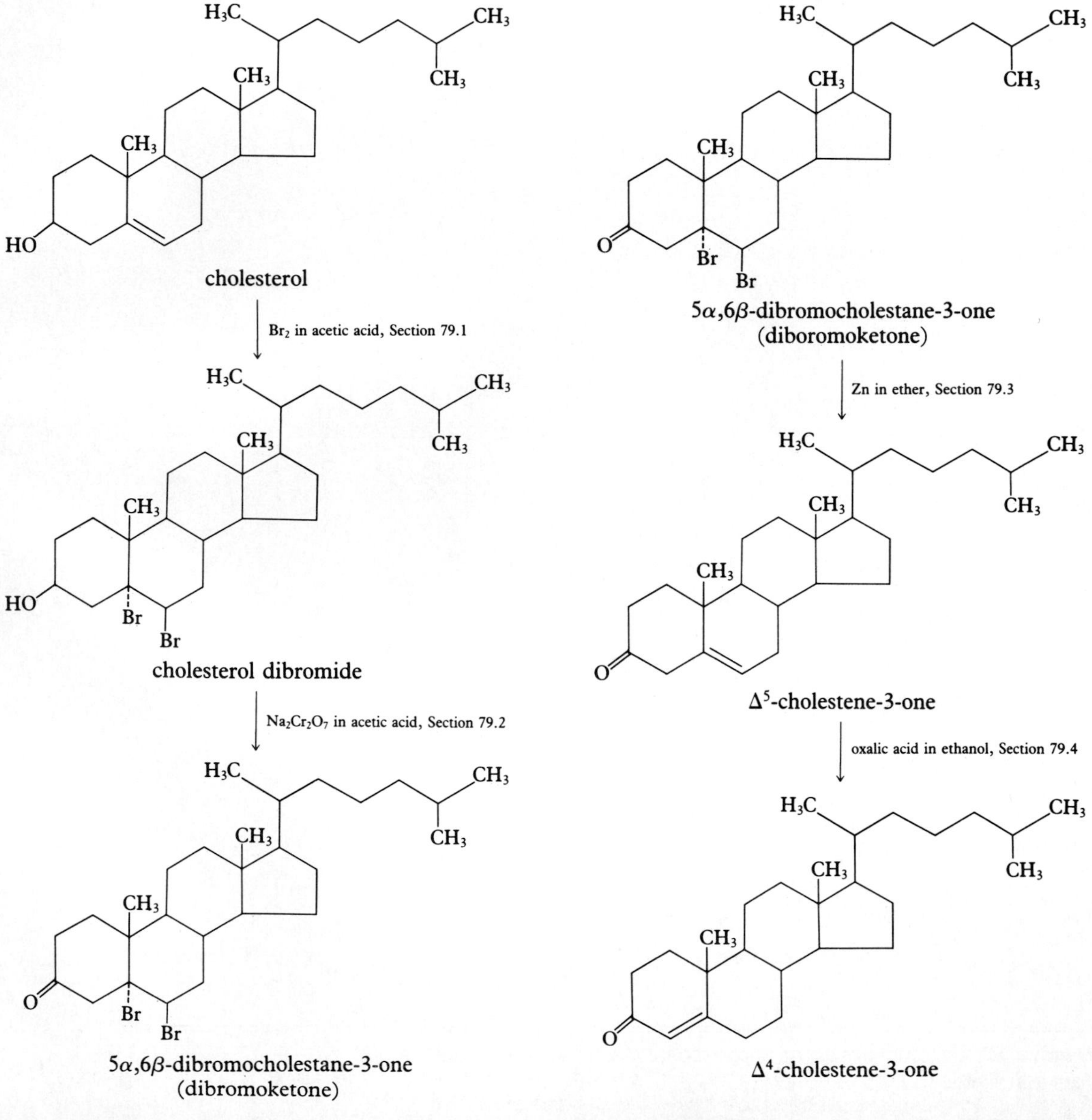

cholesterol

↓ Br_2 in acetic acid, Section 79.1

cholesterol dibromide

↓ $Na_2Cr_2O_7$ in acetic acid, Section 79.2

$5\alpha,6\beta$-dibromocholestane-3-one
(dibromoketone)

$5\alpha,6\beta$-dibromocholestane-3-one
(diboromoketone)

↓ Zn in ether, Section 79.3

Δ^5-cholestene-3-one

↓ oxalic acid in ethanol, Section 79.4

Δ^4-cholestene-3-one

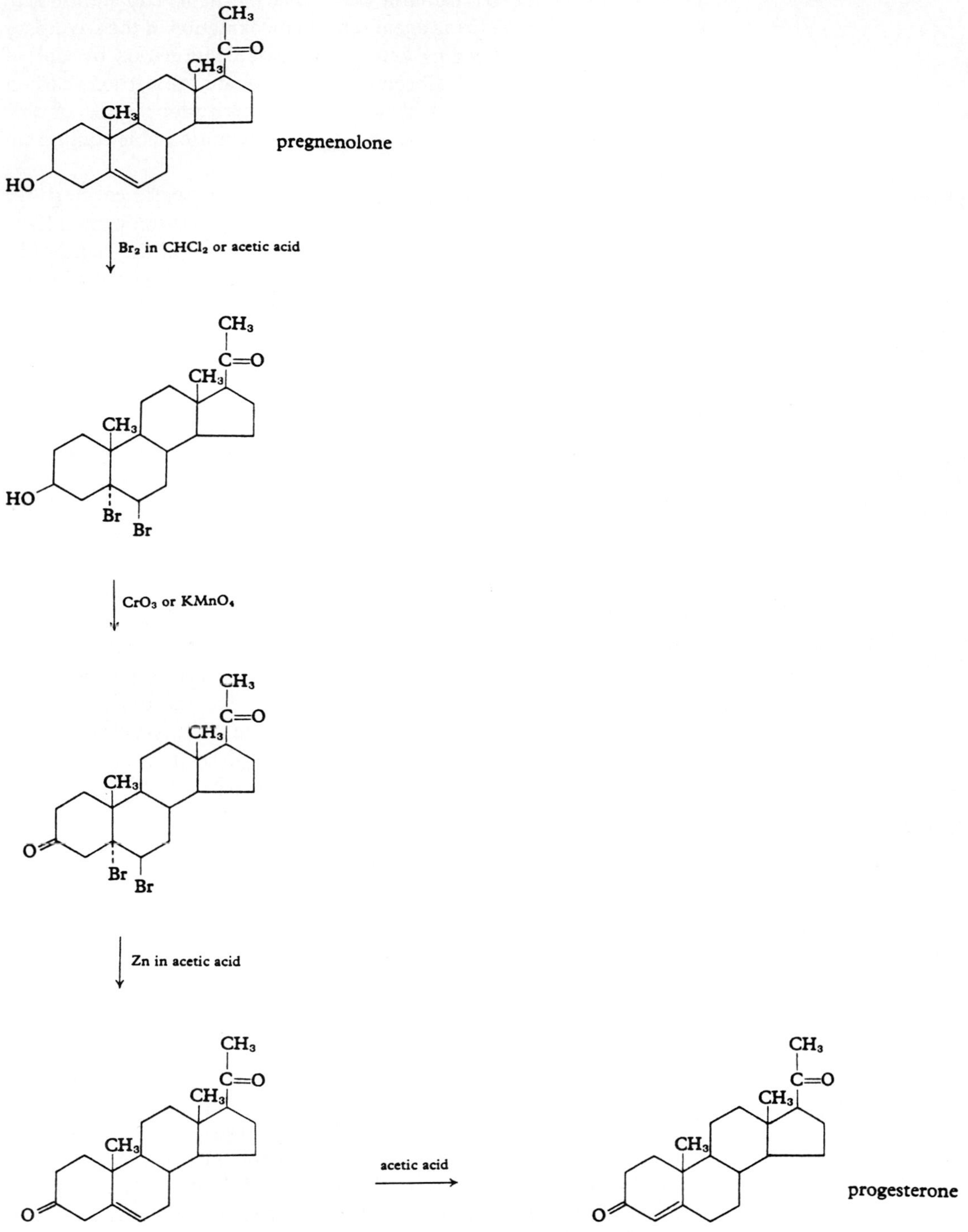

pregnenolone

Br_2 in $CHCl_3$ or acetic acid

CrO_3 or $KMnO_4$

Zn in acetic acid

acetic acid

progesterone

Figure 79.2. Conversion of pregnenolone to progesterone.

addition of bromine to the double bond (*anti* mechanism of addition) to protect it from the oxidizing agent, chromate oxidation of the secondary alcohol to the corresponding ketone, reductive elimination by zinc of bromine from the vicinal dibromide to regenerate the carbon–carbon double bond, and finally acid-catalyzed isomerization of the double bond from the unconjugated β,γ-position to the more stable conjugated α,β-position.

There are no good stopping points (■) in this experiment until the Δ^5-cholestene-3-one solution has been set aside for crystallization (Section 79.3). If the bromine solution (Section 79.1) and the sodium dichromate solution (Section 79.2) are prepared ahead of time, it will take an inexperienced person about four hours to reach this point.

79.1 CHOLESTEROL DIBROMIDE FROM CHOLESTEROL

Procedure. Place 3.0 grams (7.8 mmole) of cholesterol in a 50-mL Erlenmeyer flask and add 20 mL of anhydrous ether. Dissolve the cholesterol by swirling the flask in a beaker of warm water until the ether just starts to boil. Cool the solution to about room temperature by swirling the flask in a beaker of cool water. Add 12 mL of a solution of bromine and sodium acetate in glacial acetic acid (Note 1; 8.5 mmole of bromine) and swirl to mix. The contents of the flask should promptly solidify to a stiff paste of cholesterol dibromide. Now place the flask in a beaker of cold water to cool. In the meantime, prepare a mixture of 9 mL of anhydrous ether and 21 mL of glacial acetic acid. Cool this mixture in an ice bath while you collect the cholesterol dibromide by suction filtration. Use the cooled mixture of ether and acetic acid to rinse the product from the Erlenmeyer flask into the funnel and to wash the product in the funnel free of the yellow mother liquor. Continue to apply suction to the product until the wash liquid has almost stopped dripping from the stem of the funnel. The damp dibromide is suitable for use in the next reaction.

Note

1. The solution may be prepared by dissolving 1.0 gram of anhydrous sodium acetate (0.012 mole) in 120 mL of glacial acetic acid and then adding 4.10 mL (13.6 grams; 0.085 mole) of bromine. This is enough reagent for ten 3-gram runs.

Time: about 1 hour.

79.2 5α,6β-DIBROMOCHOLESTANE-3-ONE FROM CHOLESTEROL DIBROMIDE

Procedure. Add the moist dibromide obtained from the procedure of Section 79.1 to 40 mL of glacial acetic acid in a 125-mL Erlenmeyer flask. To this, add, all at once, 40 mL of a solution of sodium dichromate in acetic acid that has been preheated to 105°C. (Note 1; sufficient dichromate to oxidize 16 mmoles of secondary alcohol to the corresponding ketone). After the mixture has been swirled briefly, the temperature should rise to between 55 and 60°C. The temperature of the mixture must be maintained between 55 and 58°C for as long as it takes the solids to dissolve (from 3–5 minutes) and then for 2 minutes more (Note 2). After allowing the solution to stand at room temperature for about 20 minutes, add 8 mL of cold water, swirl to mix, and then cool the suspension of dibromoketone to about 15°C by means of a cold-water or ice bath. Collect the product by suction filtration, using 10–12 mL of cold methanol for rinsing and washing. Transfer the crude product to a 100-mL beaker containing 25 mL of ice-cold methanol. Stir the mixture so as to thoroughly wash the crystals and then collect the product again by suction filtration. The damp dibromoketone is suitable for use in the next reaction.

Notes

1. The solution can be prepared by dissolving 16 grams of sodium dichromate dihydrate (0.054 mole) in 400 mL of glacial acetic acid. This is enough reagent for ten runs of the scale described.
2. If the temperature fails to reach 55°C or shows signs of falling below 55°C, the mixture should be heated briefly in order to reach or maintain the specified temperature.

Time: 1 hour.

79.3 Δ^5-CHOLESTENE-3-ONE FROM 5α,6β-DIBROMOCHOLESTANE-3-ONE

If the reaction described in the procedure that follows is successful, the product will have an infrared spectrum like that of Figure 79.3. If the reaction does not go as expected, the dibromoketone will be recovered unchanged; see Note 4.

Procedure. Place the damp product obtained from the preceding oxidation in a 125-mL Erlenmeyer flask and add to the flask 40 mL of anhydrous ether and 0.5 mL of glacial acetic acid. Cool the contents of

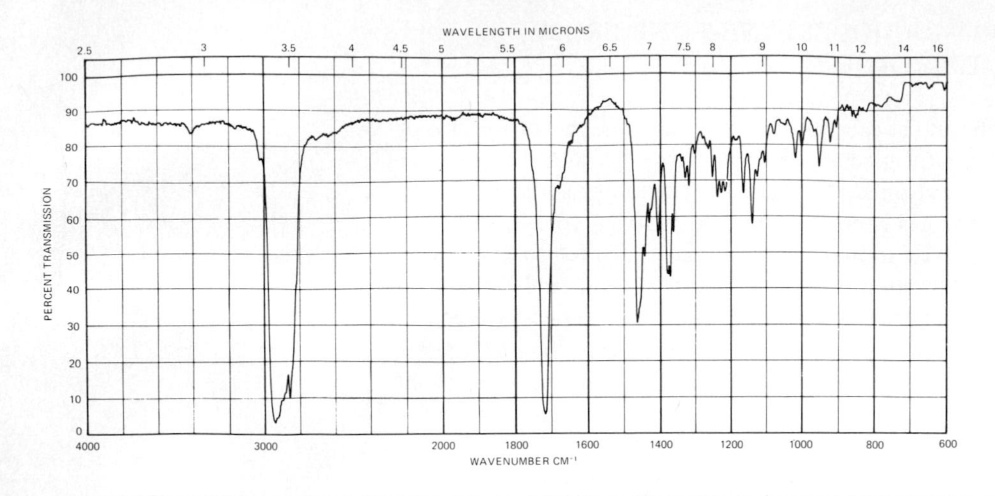

Figure 79.3. Infrared spectrum of Δ^5-cholestene-3-one; CCl_4 solution.

the flask to between 15 and 20°C by swirling it in a beaker of cold water. Then add, in several portions over about 3–4 minutes, 0.8 gram (12 mmole) of zinc dust. During this time, the flask must be swirled vigorously to suspend the zinc; the temperature of the contents of the flask should show a definite tendency to rise but should be held below 20°C by very brief swirling of the flask in the beaker of cold water. Most but not all of the zinc will dissolve. After the zinc has been added and the mildly exothermic reaction is over, allow the mixture to stand at room temperature for 10 minutes. Then add 1.4 mL (17 mmole) of pyridine (to precipitate ionic zinc by complex formation), swirl well to mix, and remove the white precipitate by suction filtration (Note 1). Wash the filter cake well with several small portions of ether, collecting these washings along with the original filtrate. Transfer the filtrate to a separatory funnel and wash it with three 15-mL portions of water and one 15-mL portion of 5% sodium bicarbonate solution (Note 2). Dry the ethereal solution over anhydrous magnesium sulfate and then filter it by gravity into a 50-mL Erlenmeyer flask that has been marked to show levels at which it contains approximately 20 and 15 mL (Note 3). Add a boiling stone and evaporate or distill the ether until the volume has been reduced to 20 mL. The flask should be warmed either on the steam bath or in a beaker of warm water; no burners should be used because of the risk of setting ether vapors on fire. Add 10 mL of methanol to the flask and continue to concentrate the solution until the volume has been reduced to 15 mL. Stopper the flask and set it aside for crystallization. ■

After crystallization has proceeded at room temperature for at least one-half hour, cool the flask in an ice bath and then collect the unsatu-

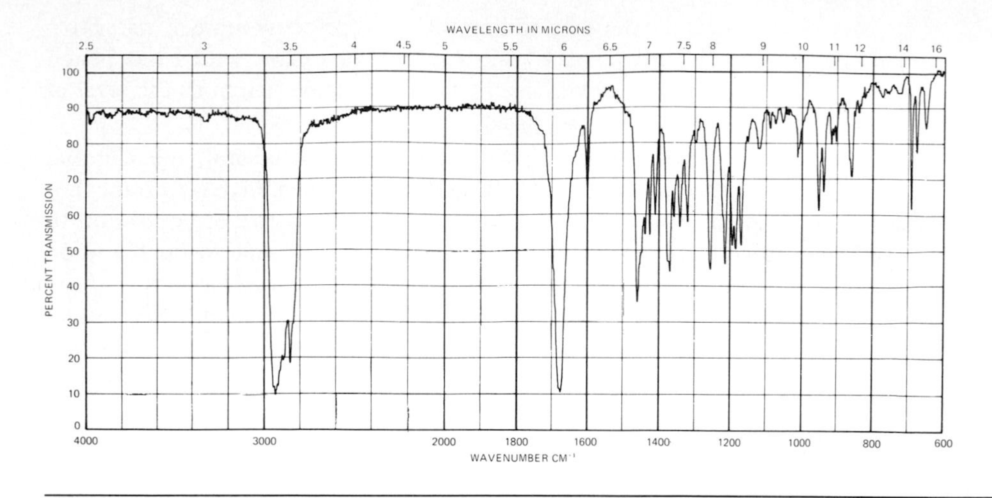

Figure 79.4. Infrared spectrum of 6β-bromocholest-4-ene-3-one; CCl_4 solution; see Note 4.

rated ketone by suction filtration (Note 4). A small amount of cold methanol may be used for rinsing and washing. Δ^5-Cholestene-3-one is obtained as white granular crystals in a yield of about 1.5 grams (about 50% from cholesterol) with a melting point of 126–128°C. Reported literature values (Reference 1) include: m.p., 126–129°C; $[\alpha]_D$, −2.5° (2.03 grams/100 mL of chloroform).

Δ^5-cholestene-3-one

6β-bromocholest-4-ene-3-one

Notes

1. It is the *filtrate* you want; be sure your filter flask is clean.
2. The purpose of the sodium bicarbonate wash is to remove any trace of acid that could prematurely catalyze the isomerization of the double bond. Dip a piece of moist blue litmus paper into the ethereal solution to make sure it is not acidic.

3. This can be done by adding the appropriate volume of water to a second, similar, flask and then making a mark with a wax pencil on the flask you will use at about the same height as the level of liquid in the second flask.
4. If the debromination with zinc was unsuccessful, the material obtained at this point will be the unchanged $5\alpha,6\beta$-dibromocholestane-3-one. It will melt at about 70°C with decomposition to give a brown-orange liquid and will turn pink to purple upon storage. If the debromination with zinc was unsuccessful *and* the solution was set aside for crystallization for a week, the product obtained may be 6β-bromocholest-4-ene-3-one. This substance is reported to have a melting point of 122–124°C, and can be prepared as follows. Add 1 gram of $5\alpha,6\beta$-dibromocholestane-3-one to 10 mL of pyridine at 5°C, stir the mixture at room temperature for 2 hours, pour the resulting solution into water, collect the precipitate by suction filtration, and recrystallize using 8 mL of hexane [see *Chemical Abstracts* **60,** 590d (1964)]. The infrared spectrum of 6β-bromocholest-4-ene-3-one is shown in Figure 79.4. This compound is recovered unchanged when treated according to the procedure of Section 79.4.

Time: 2 hours.

79.4 Δ^4-CHOLESTENE-3-ONE FROM Δ^5-CHOLESTENE-3-ONE

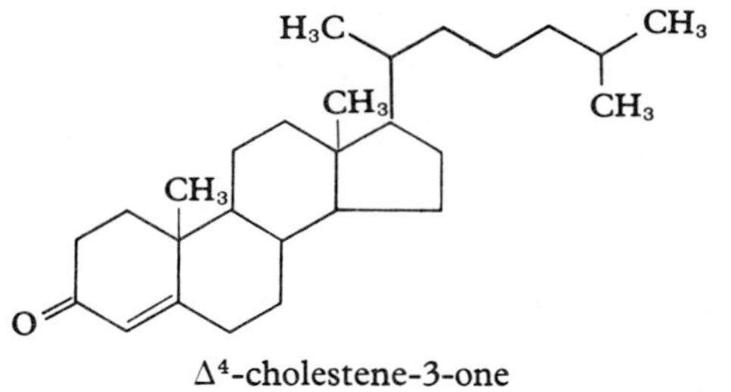

H₃C CH₃ CH₃ CH₃ CH₃ O

Δ^4-cholestene-3-one

The infrared spectrum of Δ^4-cholestene-3-one is shown in Figure 79.5.

Procedure. Place 1.0 gram (2.6 mmole) of Δ^5-cholestene-3-one and 100 mg of anhydrous oxalic acid in a 25-mL Erlenmeyer flask. Add 8 mL of 95% ethanol, fit the flask with a reflux condenser, and boil the mixture gently until all the solids have gone into solution; about 15 minutes of boiling is required. Continue to heat the solution for 10 more minutes and then allow it to cool to room temperature. If necessary, induce crystallization by scratching the inner surface of the flask

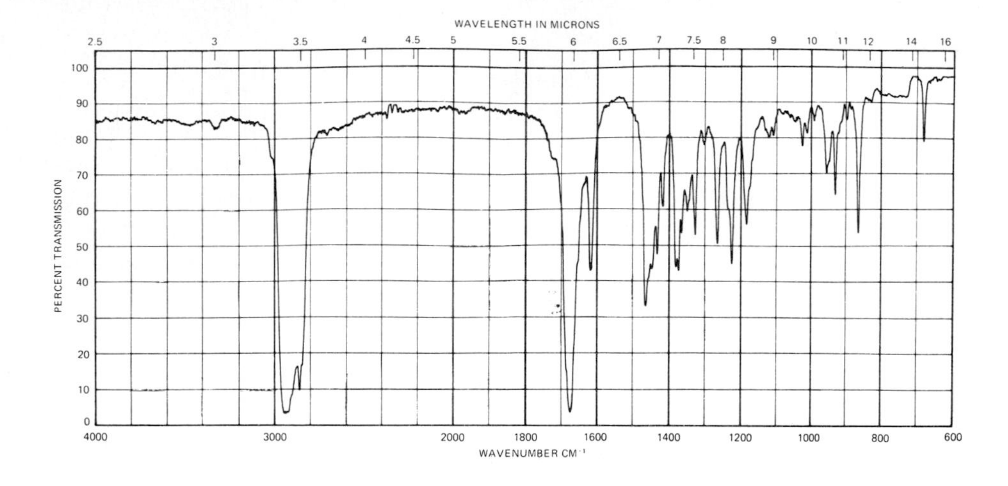

Figure 79.5. Infrared spectrum of Δ^4-cholestene-3-one; CCl_4 solution.

at the liquid level with a stirring rod or by adding seed crystals. After crystallization appears to be complete at room temperature, cool the flask in an ice bath for about 15 minutes and then collect the large colorless needles of Δ^4-cholestene-3-one by suction filtration. A small amount of cold methanol may be used for rinsing and washing. Yield: about 0.9 gram (about 90%) of crystals with a melting point of 79–81°C. Literature values (Reference 1) include: m.p., 81–82°C; $[\alpha]_D$, +92.0° (2.01 grams/100 mL of chloroform); λ_{max}, 242 nm; ϵ, 17,000 (ethanol).

Time: 2 hours.

Questions

1. How many stereoisomeric 5,6-dibromocholestane-3-ones can there be? How many of these stereoisomers could be formed by a *trans* mechanism of addition of bromine to cholesterol? Suggest an explanation for the fact that the $5\alpha,6\beta$-isomer is the only one formed. Which isomer would you expect to be the most stable?
2. Write a reasonable mechanism for the debromination of $5\alpha,6\beta$-dibromocholestane-3-one by zinc.
3. Comment on the mechanism and circumstances of formation of 6β-bromocholest-4-ene-3-one from $5\alpha,6\beta$-dibromocholestane-3-one as described in Note 4 of section 79.3.
4. The acid-catalyzed isomerization of the β,γ-unsaturated ketone to the α,β-unsaturated isomer probably goes via the intermediate formation of the conjugated enol. Write out the details of the mechanism, being sure to indicate the possibilities for resonance stabilization of each intermediate cation.

References

1. L. F. Fieser, *J. Am. Chem. Soc.* **75**, 5421 (1953).
2. L. F. Fieser, *Organic Syntheses*, Coll. Vol. 4, Wiley, New York, 1963, p. 195.

80 Tetraphenylcyclopentadienone from Benzaldehyde and Phenylacetic Acid

Tetraphenylcyclopentadienone can be prepared from benzaldehyde and phenylacetic acid in four steps according to the following scheme.

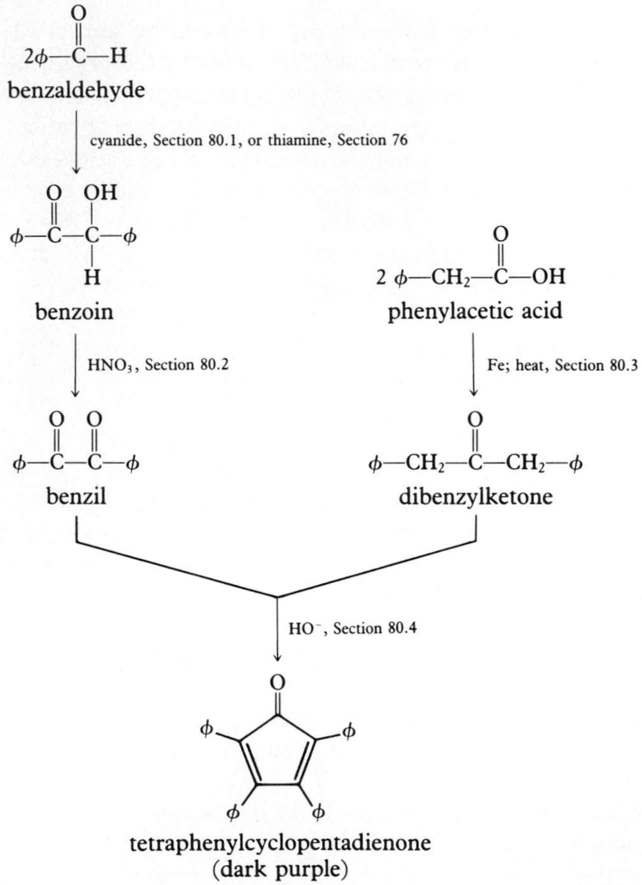

$2\phi\text{—C(=O)—H}$
benzaldehyde

↓ cyanide, Section 80.1, or thiamine, Section 76

$\phi\text{—C(=O)—CH(OH)—}\phi$
benzoin

↓ HNO_3, Section 80.2

$\phi\text{—C(=O)—C(=O)—}\phi$
benzil

$2\ \phi\text{—}CH_2\text{—C(=O)—OH}$
phenylacetic acid

↓ Fe; heat, Section 80.3

$\phi\text{—}CH_2\text{—C(=O)—}CH_2\text{—}\phi$
dibenzylketone

benzil + dibenzylketone ↓ HO^-, Section 80.4

tetraphenylcyclopentadienone
(dark purple)

80.1 BENZOIN FROM BENZALDEHYDE

In this procedure, benzoin is prepared from benzaldehyde by use of cyanide as the catalyst. It can also be prepared by catalysis with thiamine (Section 76). The mechanisms of catalysis are outlined in Section 76.

Caution: cyanide

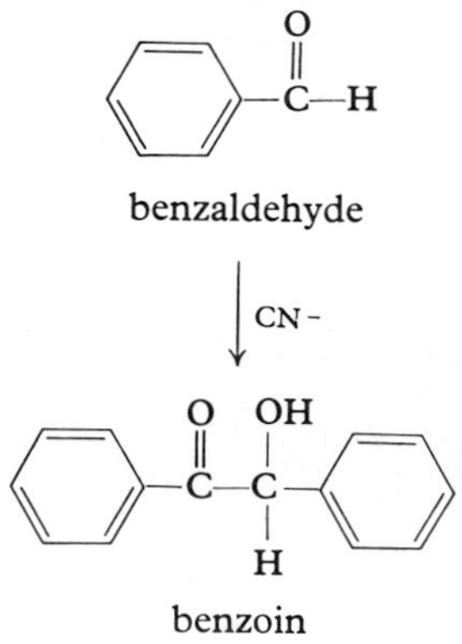

Procedure. Dissolve 1.5 g (0.023 mole) of potassium cyanide (*poison*, see Section 1.7) in 15 mL of water in a 125-mL Erlenmeyer flask. Add 30 mL of 95% ethanol followed by 15 mL (15.6 g; 0.147 mole) of benzaldehyde (Note 1). Fit the flask with a condenser and boil the mixture on the steam bath for 30 minutes. Allow the mixture to cool to room temperature, swirling occasionally to speed crystallization. Cool the mixture in an ice bath, collect the benzoin by suction filtration, and wash it, first with two 15-mL portions of cold 50% ethanol and finally with water. Benzoin may be recrystallized from methanol using 12 mL per gram.

O
C—H
benzaldehyde

CN-

O OH
C—C
H
benzoin

Note

1. The benzaldehyde should be free of benzoic acid to avoid liberating hydrogen cyanide gas.

Time: less than 2 hours.

80.2 BENZIL FROM BENZOIN

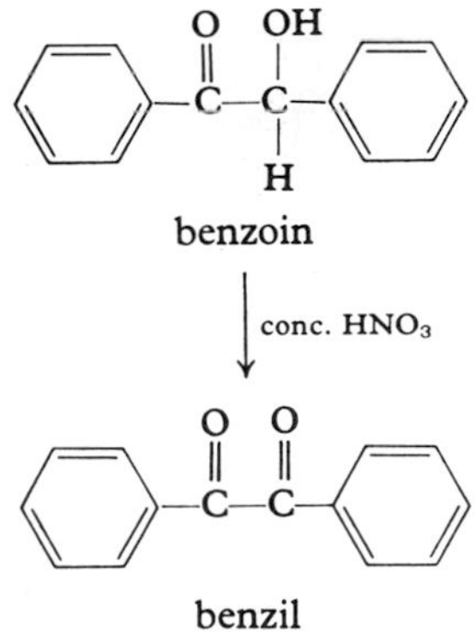

Procedure. Place 4.2 g (0.02 mole) of benzoin in a 125-mL ground-glass-stoppered Erlenmeyer flask and then add 14 mL (20 g; 0.22 mole) of conc. nitric acid. Heat this mixture in the hood (Note 1) on a steam bath for 10 to 12 minutes. During the heating period, swirl the mixture occasionally (Note 2). At the end of the heating period, add to the flask 75 mL of cold water and swirl the flask to mix the contents. Then add a seed crystal of benzil, stopper the flask, and shake it so as to cause the oily product to solidify in small lumps. Collect the bright yellow solidified benzil by suction filtration, washing it thoroughly with water to remove the nitric acid. Yield: about 4 g (about 95%). Benzil may be recrystallized from 95% ethanol, using 5–7 mL per gram (compare Exercise 5, Section 8.9).

O OH
C—C
H
benzoin

conc. HNO_3

O O
C—C
benzil

Notes

1. Nitrous oxide fumes are evolved.

2. The solid will gradually dissolve, and an oil (molten benzil) will form.

Time: 1 hour.

80.3 DIBENZYLKETONE FROM PHENYLACETIC ACID

$$Fe + 2\ \phi\text{—}CH_2\text{—}\overset{\overset{\displaystyle O}{\|}}{C}\text{—}OH \longrightarrow \phi\text{—}CH_2\text{—}\overset{\overset{\displaystyle O}{\|}}{C}\text{—}CH_2\text{—}\phi + H_2 + FeO + CO_2$$

This transformation takes place in two stages. The first is the reaction of the acid with elemental iron to form the ferrous salt of the acid:

$$2\ \phi\text{—}CH_2\text{—}\overset{\overset{\displaystyle O}{\|}}{C}\text{—}OH + Fe \longrightarrow (\phi\text{—}CH_2\text{—}\overset{\overset{\displaystyle O}{\|}}{C}\text{—}O^-)_2Fe^{2+} + H_2$$

The second stage is the loss of the elements of carbon dioxide and ferrous oxide by pyrolysis of the ferrous salt:

$$(\phi\text{—}CH_2\text{—}\overset{\overset{\displaystyle O}{\|}}{C}\text{—}O^-)_2Fe^{2+} \longrightarrow \phi\text{—}CH_2\text{—}\overset{\overset{\displaystyle O}{\|}}{C}\text{—}CH_2\text{—}\phi + CO_2 + FeO$$

The dibenzylketone produced in this experiment should have an infrared spectrum like that shown in Figure 80.1.

Figure 80.1. Infrared spectrum of dibenzylketone; thin film.

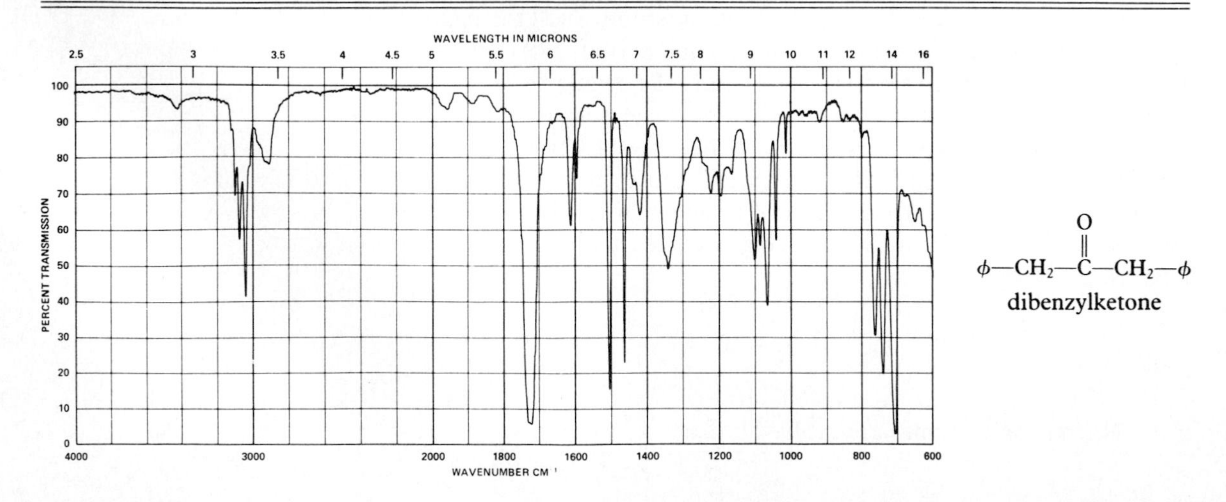

Procedure (Reference 1). Place 13.6 g (0.10 mole) phenylacetic acid and 3.07 g (0.055 mole) iron powder in a 50-mL distilling flask (Note 1). Fit the flask with a side-arm test tube as a receiver and connect a length of rubber tubing to the side arm; lead the tubing to an area free of flames (Note 2). Arrange a thermometer in the top of the distilling flask so that the bulb almost touches the bottom of the flask. Heat the mixture of solids until it melts and gradually raise the temperature to 320°C (Note 3). When the temperature of the mixture reaches 320°, stop heating and raise the thermometer into position for a normal distillation. Then distill the contents of the flask, collecting the product at about 324°C (Note 4). Yield: about 60%.

Notes

1. Use an inexpensive flask as this reaction mixture etches the glass badly.
2. Hydrogen gas (flammable, explosive) is evolved.
3. During the heating period, which should be about 15 minutes, the contents of the flask will liquefy, gas will be evolved (H_2), and the contents of the flask will darken and solidify and then will remelt with further gas evolution (CO_2). Do not heat so quickly that anything distills over.
4. The product will be colored yellow to brown. Do not attempt to distill to dryness. To clean the distilling flask, use acetone and soapy water until they seem to have no more effect. Then cover the residue in the flask with conc. hydrochloric acid and allow it to stand in the hood.

Time: 3 hours.

Reference

1. R. Davis and H. P. Schultz, *J. Org. Chem.* **27**, 854 (1962).

80.4 TETRAPHENYLCYCLOPENTADIENONE FROM BENZIL AND DIBENZYLKETONE

$$\phi\text{—}\overset{\overset{\displaystyle O}{\|}}{C}\text{—}\overset{\overset{\displaystyle O}{\|}}{C}\text{—}\phi + \phi\text{—}CH_2\text{—}\overset{\overset{\displaystyle O}{\|}}{C}\text{—}CH_2\text{—}\phi \xrightarrow{HO^-} \text{(tetraphenylcyclopentadienone)} + 2H_2O$$

tetraphenylcyclopentadienone
(dark purple)

Procedure (Reference 1). Dissolve 2.1 g (0.010 mole) of benzil (Section 80.2) and 2.1 g (0.010 mole) of dibenzylketone (Section 76.3) in 10 mL of triethylene glycol in a 25 mm × 150 mm test tube. While supporting the test tube in a clamp and stirring with a thermometer, heat the mixture with a small flame until the benzil is dissolved. Adjust the temperature to exactly 100°C, add 1 mL of 40% benzyltrimethylammonium hydroxide in methanol, and stir the mixture well with the thermometer.

When the temperature has dropped to 80°C, cool the mixture almost to room temperature, thin it with 10 mL of methanol, and collect the product by suction filtration. Wash it with methanol until the filtrate is purple-pink rather than brown. The very dark purple tetraphenylcyclopentadienone may be recrystallized from triethylene glycol, using 10 mL per gram.

Time: less than 2 hours.

Reference

1. L. F. Fieser, *Organic Experiments*, Heath, Boston, 1964, p. 303.

81 Sulfanilamide

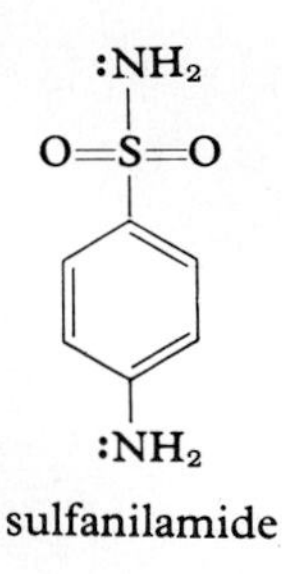

sulfanilamide

Sulfanilamide was the first substance to be used systematically and effectively as a chemotherapeutic agent for the prevention and cure of bacterial infections in humans. Although it was first prepared in 1908, more than 25 years had to pass before its therapeutic value was discovered. The discovery was, typically, indirect. In 1909, German dye chemists of the I. G. Farbenindustrie found that azo dyes containing sulfonamide groups were superior in colorfastness, especially toward the proteins of wool and silk. This led others to an investigation of the use of these dyes as selective staining agents for bacterial protoplasm. In the course of this work, it was gradually recognized that certain dyes had the ability to kill some bacteria *in vitro* (in a bacterial culture outside a living host). This, reasonably enough, inspired the hope that a similar antibacterial action could occur *in vivo* (within a living host), but research along these lines was not pursued very vigorously. Then in the early 1930s, further work at the I. G. Farbenindustrie with a new sulfonamide-containing dye, Prontosil, showed that this new substance could protect mice with streptococcal and other bacterial infections, despite the fact that it had no in vitro antibacterial action. This discovery was recognized by the awarding of the Nobel prize in medicine for 1939 to Gerhard Domagk, one of the directors of research of the I. G. Farbenindustrie. In 1935, research workers at the Pasteur Institute in

Paris who had been intrigued by the fact that Prontosil was effective in vivo but not in vitro reported a fascinating and far-reaching explanation: the reason was that, in a reaction that could take place only in vivo, Prontosil could be reduced by the host to sulfanilamide, which was the actual effective agent. Further research in England and the United States confirmed and extended this finding, with the result that within the next ten years over 5400 related substances had been prepared and tested for antibacterial activity. Fewer than 20 ever attained therapeutic importance.

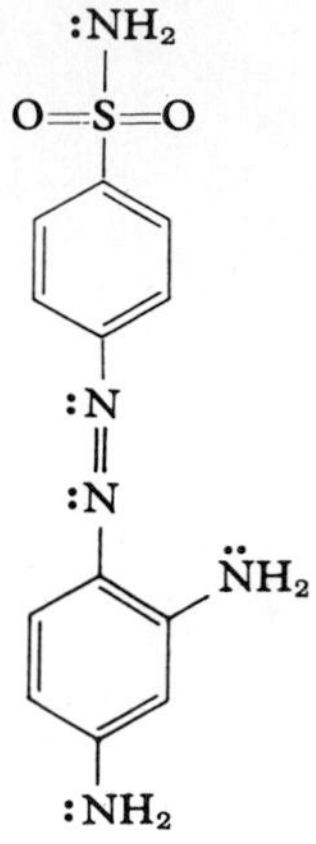

Prontosil

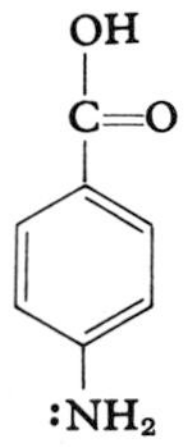

p-aminobenzoic acid

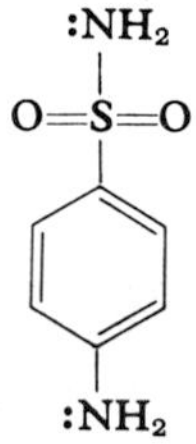

sulfanilamide

As with all drugs, the next question was "How does it work?" How is it that sulfanilamide attacks bacteria specifically, with no apparent effect on the host? First, sulfanilamide does not usually kill the bacteria, but it greatly reduces their rate of growth and reproduction. On the molecular level, sulfanilamide competes with *p*-aminobenzoic acid in the bacterial synthesis of folic acid, a substance required for bacterial growth. Sulfanilamide rather than *p*-aminobenzoic acid enters into the reaction that normally would produce folic acid, and the product, containing a O=S=O group in place of the C=O group of the amide function, is unable to carry out the function of folic acid. Since the bacteria are able under normal conditions to synthesize their own folic acid, they have no mechanism for obtaining folic acid from their environment at the low concentrations at which it is present. Inhibition or diversion of folic acid synthesis, then, accounts for the bacteriostatic action of sulfanilamide. On the other hand, the host, a "higher organism," does not have the ability to synthesize folic acid but has instead the capacity to make use of the folic acid present at low concentrations in its diet. Obviously, the presence or absence of sulfanilamide is irrelevant to this capacity to acquire folic acid from the environment, and this accounts for the lack of effect of sulfanilamide on the host organism.

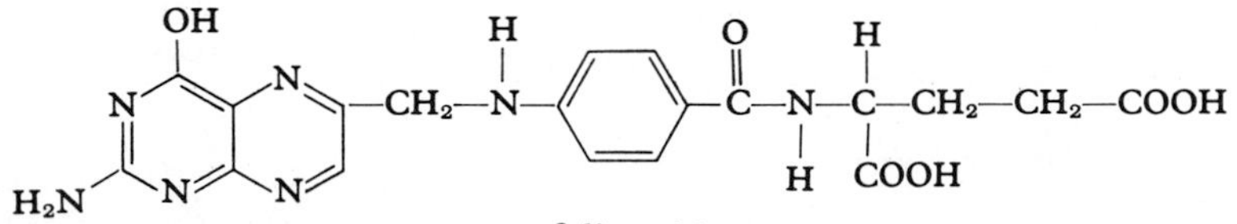

folic acid

Sulfanilamide can be prepared from benzene in six steps, as indicated by the scheme shown below. According to this approach, the aromatic amino group is first acetylated and then later deacetylated. There are at least two reasons for this. First, if one were to try to chlorosulfonate aniline itself, chlorosulfonic acid, an acid chloride, could be expected to react with the amino group of aniline. Then, even if it were possible to prepare *p*-aminobenzenesulfonyl chloride, the amino group of one molecule might be expected to react with the sulfonyl chloride group of another.

The last three steps of this synthesis are presented in this section; the procedures for the second and third steps can be found in earlier sections of the book. The reaction by which benzene is converted to nitrobenzene has been omitted from this edition of this book; many chemists believe that the toxicity of benzene is such that it should not be used in the introductory organic laboratory.

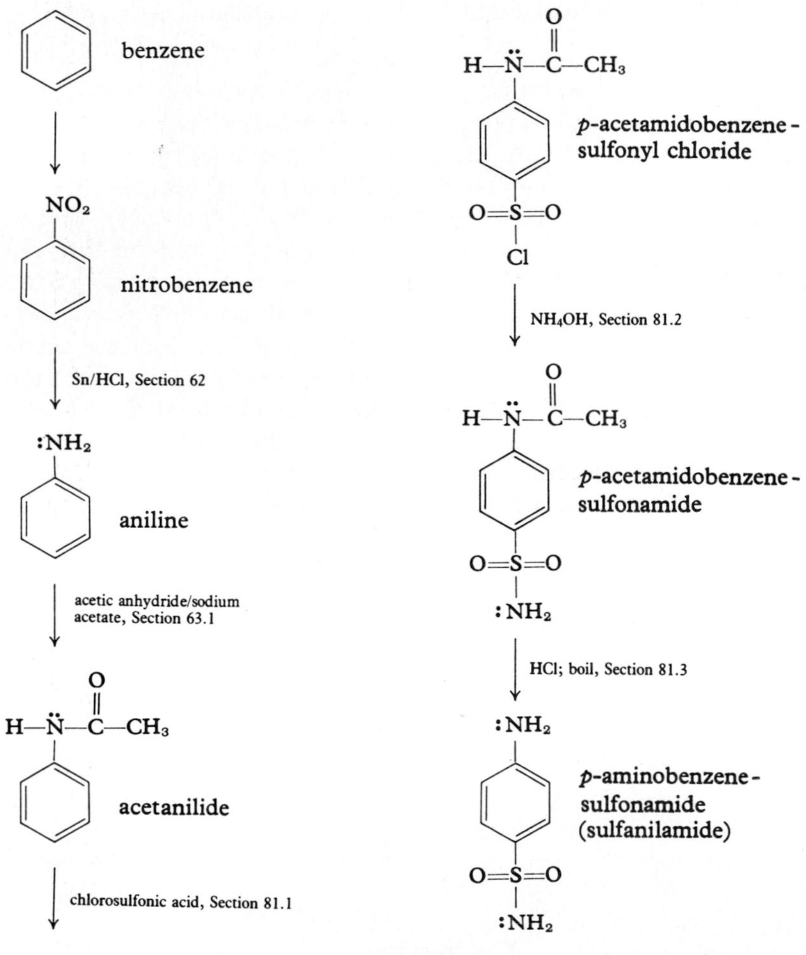

benzene

↓

NO_2

nitrobenzene

↓ Sn/HCl, Section 62

$:NH_2$

aniline

↓ acetic anhydride/sodium acetate, Section 63.1

O
||
H—N̈—C—CH_3

acetanilide

↓ chlorosulfonic acid, Section 81.1

O
||
H—N̈—C—CH_3

p-acetamidobenzene-sulfonyl chloride

O=S=O
|
Cl

↓ NH_4OH, Section 81.2

O
||
H—N̈—C—CH_3

p-acetamidobenzene-sulfonamide

O=S=O
|
$:NH_2$

↓ HCl; boil, Section 81.3

$:NH_2$

p-aminobenzene-sulfonamide (sulfanilamide)

O=S=O
|
$:NH_2$

81.1 *p*-ACETAMIDOBENZENESULFONYL CHLORIDE FROM ACETANILIDE

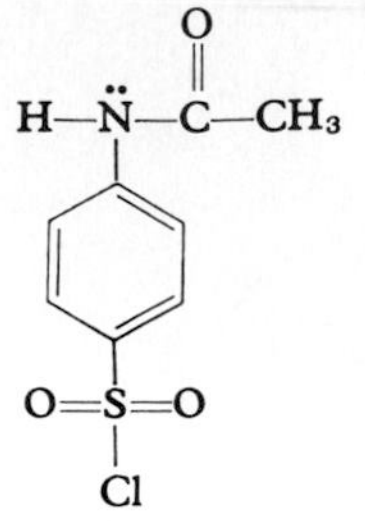

O
‖
H—N̈—C—CH_3
|
(benzene ring)
|
O=S=O
|
Cl

p-acetamidobenzene-sulfonyl chloride

Procedure. Place 6.75 g (0.050 mole) of completely dry (Note 1) acetanilide in a 125-mL Erlenmeyer flask. Gently heat the flask with a flame to melt the acetanilide, and swirl the flask containing the molten acetanilide as it cools, so that it will solidify in a thin layer on the bottom and lower walls of the flask. Prepare a trap to absorb hydrogen chloride and arrange for it to be connected to the Erlenmeyer flask by a length of rubber tubing (Section 35). Make certain that there is no possibility for water to be sucked back into the flask (*caution:* Note 1). Cool the flask in an ice bath. Then remove it and add all at once 17 mL (30 g; 0.26 mole) of chlorosulfonic acid (*caution:* Note 2) and connect the flask to the gas trap. Swirl the mixture until part of the solid has dissolved and hydrogen chloride is being rapidly evolved. Cool the flask in the ice bath if necessary to moderate the reaction. When all but a few lumps of acetanilide have dissolved (5–10 minutes), heat the mixture on the steam bath for 10 minutes and then cool the flask to room temperature. Slowly and cautiously (Note 1) pour the mixture onto 100 grams of ice in a beaker in the ice bath in the hood. Stir the precipitated product for a few minutes to obtain an even, granular suspension, and then collect it by suction filtration. The crude product should be used immediately.

Caution: chlorosulfonic acid

Notes

1. Chlorosulfonic acid reacts violently with water.
2. Measure the chlorosulfonic acid in a *dry* graduate, as it will react violently with water.

Time: less than 2 hours.

Questions

1. Why is an excess of chlorosulfonic acid used?
2. Write out the reaction that takes place between chlorosulfonic acid and water.
3. What might be the product of reaction of chlorosulfonic acid and aniline?
4. What might be the product of reaction of *p*-aminobenzenesulfonyl chloride with another molecule of the same compound?
5. Why would we not expect a similar reaction between two molecules of *p*-acetamidobenzenesulfonyl chloride?
6. Although *p*-aminobenzenesulfonic acid is readily available, it can be acetylated only in the form of its sodium salt. Why should this be so? As a matter of fact, the sodium salt of *p*-aminobenzenesulfonic acid is not very soluble in

acetic anhydride and the reaction proceeds so poorly that this is not a practical way to make *p*-acetamidobenzenesulfonic acid, which could then be converted to *p*-acetamidobenzenesulfonyl chloride.

81.2 *p*-ACETAMIDOBENZENESULFONAMIDE FROM *p*-ACETAMIDOBENZENESULFONYL CHLORIDE

p-acetamidobenzene-sulfonamide

Procedure. Transfer the crude product of the procedure of Section 81.1 to a 125-mL Erlenmeyer flask, and add 20 mL (19 g; 0.32 mole) of conc. ammonium hydroxide and 20 mL of water. By means of a flame and in the hood, heat the mixture almost to boiling; in about 5 minutes, with occasional swirling, the granular sulfonyl chloride will be converted to the more pasty sulfonamide. Cool the mixture well in an ice bath, collect the product by suction filtration, and suck it as dry as possible. ■

Time: 1 hour.

Question

1. Estimate the relative nucleophilicity of water, ammonia, and hydroxide ion toward the sulfonyl chloride group.

81.3 SULFANILAMIDE FROM *p*-ACETAMIDOBENZENESULFONAMIDE

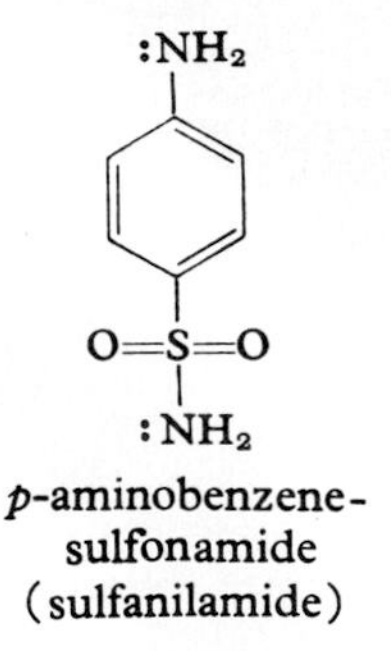

p-aminobenzene-sulfonamide (sulfanilamide)

Procedure. Transfer the moist product of the procedure of Section 81.2 to a 100-mL boiling flask, and add 7 mL (8.3 g; 0.09 mole) of conc. hydrochloric acid and 14 mL of water. Fit the flask with a condenser, and gently boil the mixture under reflux until all the solid has dissolved (about 10 minutes) and then for 10 minutes more. Cool the solution to room temperature (Note 1), add a little decolorizing carbon, and filter with suction. Transfer the filtrate to a beaker and add sodium bicarbonate as a saturated aqueous solution until the mixture is no longer acid to litmus (Note 2). Cool the suspension of precipitated sulfanilamide in the ice bath and then collect it by suction filtration. ■ Recrystallize the sulfanilamide from water, using 12 mL per gram (add decolorizing carbon, if necessary).

Notes

1. If solid separates upon cooling, reheat and boil the mixture for a while longer.
2. About 7.5 g (0.09 mole) will be needed.

Time: less than 2 hours.

Question

1. Estimate the relative rates of hydrolysis of carboxylic acid amides and sulfonic acid amides.

82 A Bootstrap Synthesis: *p*-Phenetidine from *p*-Phenetidine

This synthesis, originally presented as a project in the first edition of this book, was reported in the German patent literature as a method by which *p*-phenetidine could be prepared for the synthesis of phenacetin, the analgesic and antipyretic described in Sections 63.2 and 73.

The feature of greatest interest is the fact that the product from one molecule of the starting material is *two* molecules of that same compound! Briefly, *p*-ethoxyaniline is diazotized and coupled at the *para* position of phenol, the phenolic —OH group is ethylated, and the resulting symmetrical azo compound is reductively cleaved to give two molecules of *p*-ethoxyaniline.

Two moles from one!

$$\text{Ar—NH}_2 \xrightarrow[\text{Section 82.1}]{\text{HNO}_2;\ \phi\text{—OH}} \xrightarrow[\text{Section 82.2}]{\text{Et—I}} \text{Ar—N=N—Ar} \xrightarrow[\text{Section 82.3}]{\text{H}_2} 2\ \text{Ar—NH}_2$$

The procedures described here are essentially those of the patent (Reference 1), except that we carry out the reduction of diethyldioxyazobenzene to 2 moles of phenetidine with hydrazine in the presence of Raney nickel (Reference 2) rather than with zinc metal and hydrochloric acid.

82.1 ETHYLDIOXYAZOBENZENE FROM *p*-PHENETIDINE

$$CH_3CH_2\text{—O—}C_6H_4\text{—NH}_2 \xrightarrow[\text{diazotize}]{\text{HNO}_2} CH_3CH_2\text{—O—}C_6H_4\text{—N}_2^+\ Cl^-$$

p-phenetidine diazonium salt

$$CH_3CH_2\text{—O—}C_6H_4\text{—N≡N}^+ + C_6H_5\text{—O—H} \xrightarrow[\text{coupling}]{}$$

diazonium salt

$$CH_3CH_2\text{—O—}C_6H_4\text{—N=N—}C_6H_4\text{—O—H} + H^+$$

ethyldioxyazobenzene

Procedure. Place 6.4 mL (6.9 grams; 0.050 mole) of *p*-phenetidine in a 250-mL Erlenmeyer flask. To this add 10 mL of water and then 8.2 mL (0.098 mole) of conc. HCl. Swirl this mixture until all the amine has dissolved (Note 1). Next, prepare a solution of 3.45 grams (0.050 mole) of sodium nitrite in 25 mL of water. Finally, in a 500-mL Erlenmeyer flask, prepare a solution of 4.7 grams (0.050 mole) of phenol and 10.6 grams (0.10 mole) of sodium carbonate in 175 mL of water.

Why does the amine dissolve?

Now, add the solution of sodium nitrite to the solution of the amine; swirl to mix (Note 2). After allowing this mixture to stand for about 60 seconds, pour it slowly with good mixing into the solution of phenol and sodium carbonate. A voluminous yellow precipitate of ethyldioxyazobenzene forms immediately. After allowing the resulting suspension to stand for 10 minutes or so, collect the solid by suction filtration, using some cold water to rinse the material into the suction funnel and to wash the product on the funnel. The filter cake should be sucked free of as much water as possible so that it can be used in the next reaction without further drying. Yield of dried material: 11.9 grams (98%).

Suck it dry!

Notes

1. If the solution is highly colored, it can be treated with decolorizing carbon, which can be removed by gravity filtration; most of the color can be removed in this way.
2. The diazotization of this amine can be done at room temperature.

82.2 DIETHYLDIOXYAZOBENZENE FROM ETHYLDIOXYAZOBENZENE

$$CH_3CH_2{-}O{-}C_6H_4{-}N{=}N{-}C_6H_4{-}O^- \; Na^+ + CH_3CH_2{-}Br \xrightarrow[\text{1-propanol}]{\text{NaOH}}$$

ethyldioxyazobenzene

$$CH_3CH_2{-}O{-}C_6H_4{-}N{=}N{-}C_6H_4{-}O{-}CH_2CH_3 + Na^+ \; Br^-$$

diethyldioxyazobenzene
(yellow-orange)

Procedure. Place 6.0 grams (0.025 mole) of dry ethyldioxyazobenzene in a 100-mL ⊤ boiling flask, and then add to the flask 25 mL of 1-propanol, followed by 1.5 mL (0.029 mole) of 50% aqueous sodium hydroxide solution. Fit the flask with a reflux condenser, and support the apparatus so that the contents can be heated at about 80°C for 4

Why does the starting material dissolve?

hours (Note 1). Now add 8 mL (11.7 grams; 0.10 mole) (Note 2) of ethyl bromide down the condenser, and heat the deep-red solution under very gentle reflux for about 4 hours. During the heating period, a quantity of solid will crystallize from the solution.

Why is the product not soluble in base?

At the end of the heating period, cool the flask by immersing it in an ice bath for a few minutes, and then collect the crystalline diethyldioxyazobenzene by suction filtration, using about 25 mL of methanol to assist in transferring the solid to the suction funnel and to wash the product. The product should be sucked as free from solvent as possible. Yield of dried material: 6.6 grams (98%).

Notes

1. An electrically heated oil bath is ideal for this purpose since it can be set and left unattended. A steam bath can also be used, but with it maintaining a constant rate of heating is more difficult.
2. Only 1 equivalent of ethyl bromide is required. We use 4 equivalents so as to decrease the time needed for the reaction.

82.3 *p*-PHENETIDINE FROM DIETHYLDIOXYAZOBENZENE

In this experiment, *p*-phenetidine is formed in the reduction of diethyldioxyazobenzene by hydrazine in the presence of Raney nickel.

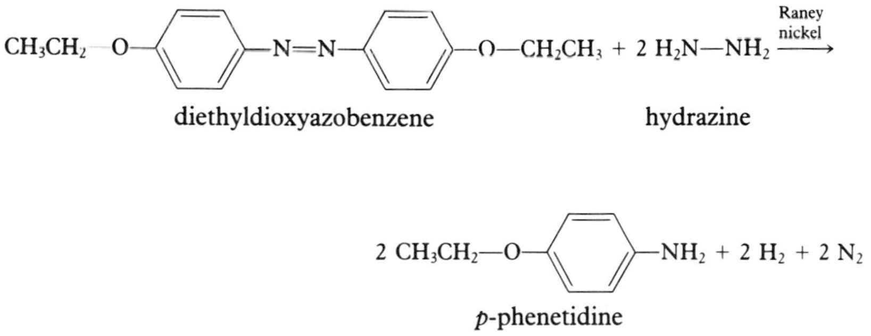

$$CH_3CH_2-O-C_6H_4-N{=}N-C_6H_4-O-CH_2CH_3 + 2\ H_2N{-}NH_2 \xrightarrow{\text{Raney nickel}}$$

diethyldioxyazobenzene hydrazine

$$2\ CH_3CH_2-O-C_6H_4-NH_2 + 2\ H_2 + 2\ N_2$$

p-phenetidine

Raney nickel, a finely divided form of nickel metal, catalyzes the decomposition of hydrazine into hydrogen gas and diimide, the actual reducing agent.

$$H_2N{-}NH_2 \xrightarrow{\text{Raney nickel}} H_2 + H{-}N{=}N{-}H$$

hydrazine diimide

Then, diimide first reduces the azo compound to the hydrazo compound

$$\underset{\text{azo compound}}{Ar{-}N{=}N{-}Ar} + \underset{\text{diimide}}{H{-}N{=}N{-}H} \longrightarrow \underset{\text{hydrazo compound}}{Ar{-}\underset{H}{N}{-}\underset{H}{N}{-}Ar} + N_2$$

and finally reduces the hydrazo compound to 2 moles of the aromatic amine.

$$\underset{\text{hydrazo compound}}{Ar{-}\underset{H}{N}{-}\underset{H}{N}{-}Ar} + \underset{\text{diimide}}{H{-}N{=}N{-}H} \longrightarrow \underset{\text{aromatic amine}}{Ar{-}NH_2 + H_2N{-}Ar} + N_2$$

Since in this experiment, the aromatic amine is a liquid, it is more convenient to isolate it as the solid acetyl derivative, phenacetin, a substance that is used as a mild analgesic and antipyretic (agent for the relief of pain and the reduction of fever; Section 63.2).

$$\underset{p\text{-phenetidine}}{CH_3CH_2{-}O{-}C_6H_4{-}\underset{H}{N}{-}H} + CH_3{-}\overset{O}{\overset{\|}{C}}{-}O{-}\overset{O}{\overset{\|}{C}}{-}CH_3 \xrightarrow{\text{sodium acetate}}$$

$$\underset{\text{phenacetin}}{CH_3CH_2{-}O{-}C_6H_4{-}\underset{H}{N}{-}\overset{O}{\overset{\|}{C}}{-}CH_3} + H{-}O{-}\overset{O}{\overset{\|}{C}}{-}CH_3$$

Procedure. Place 2.7 grams (0.01 mole) of diethyldioxyazobenzene in a 125-mL Erlenmeyer flask. Add to the flask 20 mL of methanol, 1.5 mL of hydrazine (1.5 grams; 0.05 mole) and about 100 milligrams of Raney nickel (Note 1). Fit the flask with a reflux condenser, and support the flask so that it can be heated on the steam bath. Cautiously warm the flask on the steam bath until the contents start to foam. Heat the flask as long as the contents foam and until both the yellow solid and the yellow color, which indicate the presence of unchanged diethyldioxyazobenzene, are gone; about 1 hour of heating will be required (Note 2).

Make the yellow go away!

After the starting material and excess hydrazine are both gone, remove the flask from the steam bath and disconnect the condenser from the flask. Add to the flask 20 mL of cold water and then 1.7 mL (0.02 mole) of conc. HCl. Add a modest portion of decolorizing carbon to the solution, and, after swirling the flask and then allowing it to stand for a few minutes, remove the carbon by gravity filtration. Collect the filtrate in a 125-mL Erlenmeyer flask.

While you are waiting for the solution to flow through the filter, prepare a solution of 3.0 grams (0.022 mole) of sodium acetate trihydrate in 10 mL of water. After filtration is complete, add 2.1 mL (0.022 mole) of acetic anhydride to the filtrate, and swirl the flask to mix the contents. Then add, all at once with swirling, the solution of sodium acetate; phenacetin will immediately separate from solution. Warm the suspension of phenacetin on the steam bath until all of the solid has dissolved. Then add 50 mL of hot water, swirl to mix, and set the solution aside to cool slowly for crystallization (Note 3).

After crystal formation is well underway, the flask can be placed in a beaker of cold water and then in an ice bath to speed up the process. When crystallization is judged to be complete, collect the product by suction filtration, using water to assist the transfer of the solid to the filter funnel and to wash the crystals. Yield: about 80% (Note 4).

Notes

1. Raney nickel can be purchased, or it can be prepared from the Raney alloy of nickel and aluminum. To prepare Raney nickel from the alloy, add *in several portions* about a gram of the alloy to a mixture of 5 mL of 50% sodium hydroxide and 20 mL of water. The mixture will foam strongly after each addition. When the foaming has subsided, heat the mixture on the steam bath until the bubbling ceases. Decant the basic aqueous supernatant liquid, add about 10 mL of water, swirl, and decant. Repeat this washing operation with a second 10-mL portion of water and then with a 10-mL portion of methanol. Transfer portions of the moist suspension to the reaction flask by means of a medicine dropper.

Caution: foaming!

2. The flask must be heated until the starting material is gone (as indicated by the disappearance of the yellow solid and the yellow color) *and* until the excess hydrazine has been decomposed (as indicated by the cessation of foaming). *If the foaming stops before the color has been discharged,* add a few more drops of hydrazine, down the condenser, to the reaction mixture. The reaction mixture must be heated until *both* the starting material and the excess hydrazine are gone. The time required for heating will depend on the activity of the Raney nickel, the amount of this catalyst that is used, and the vigor of heating.
3. If the solution is allowed to cool slowly, the product will separate in the form of large spars.
4. If a solution of 0.020 mole of phenetidine is dissolved in 20 mL of methanol and this solution is carried through the process of acetylation, phenacetin is isolated in a yield of about 80%. This implies that the reduction of the azo compound to the amine gives the amine in a yield of about 100%.

References

1. J. D. Riedel, *D. R. P.* 48543 (1889).
2. D. Balcom and A. Furst, *J. Am. Chem. Soc.* **75**, 4334 (1953).

83 1-Bromo-3-chloro-5-iodobenzene

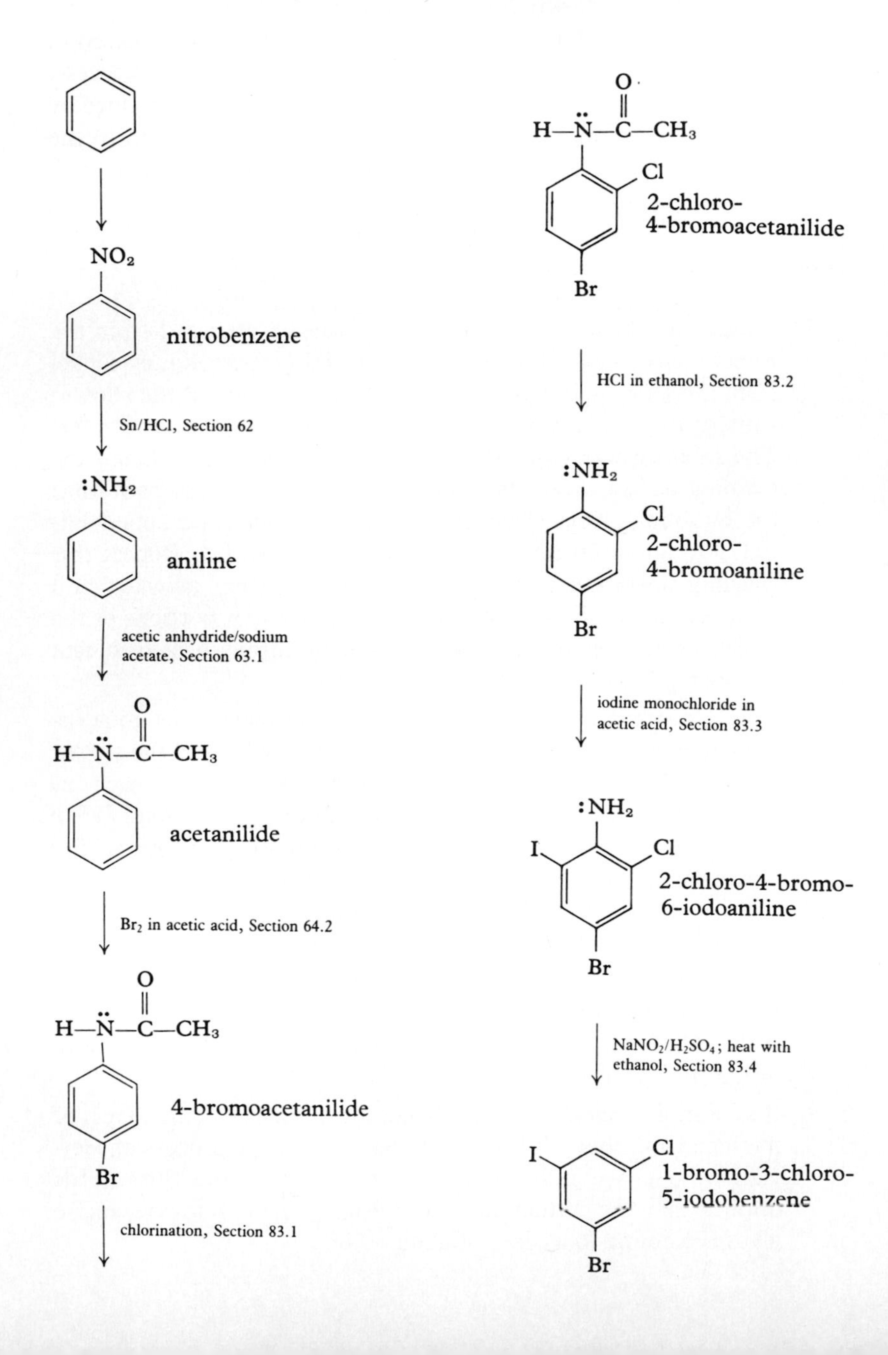

1-Bromo-3-chloro-5-iodobenzene can be prepared from benzene in eight steps, as indicated in the preceding scheme. The last four steps of the synthesis are presented in this section; procedures for steps two, three, and four can be found in other sections of the book. The reaction by which benzene is converted to nitrobenzene has been omitted from this edition of this book; many chemists believe that the toxicity of benzene is such that it should not be used in the introductory organic laboratory.

Look, Ma! NO BENZENE!

83.1 2-CHLORO-4-BROMOACETANILIDE FROM 4-BROMOACETANILIDE

Procedure a. Suspend 10.7 g (0.050 mole) 4-bromoacetanilide and 4.1 g (0.050 mole) of sodium acetate in 40 mL of glacial acetic acid in a 250-mL flask. The crude 4-bromoacetanilide may be wet with as much as an equal weight of water without affecting the reaction. While stirring the solution rapidly, pass in 0.055 mole chlorine gas (3.9 g; equivalent to 2.5 mL liquid chlorine) through a 7-mm tube which leads below the surface of the liquid in the flask (Note 1).

After the addition of chlorine is complete (in about 20 minutes), continue to stir the mixture for 2 or 3 minutes, and then slowly add 160 mL of cold water. During the addition of water, a homogeneous solution will be obtained, and then the product will begin to crystallize. Treat the resulting suspension with enough concentrated sodium bisulfite solution to remove the yellow-green color, and then collect the crude 2-chloro-4-bromoacetanilide by suction filtration, washing it with water and sucking it as dry as possible; yield: 11.8 g (95%). 2-Chloro-4-bromoacetanilide may be recrystallized from methanol, using 7 mL per gram, with 73% recovery; m.p., 154–156°C (literature value, 151°C).

Note

1. The chlorine is conveniently handled in the following way. A graduated receiver is fitted with a Claisen adapter, and a glass tube is positioned in the center neck of the Claisen adapter by means of a Teflon tubing adapter in such a way that the end of the tube extends to the top of the graduations of the receiver. A drying tube is placed in the outer neck of the Claisen adapter, and the glass tube is connected to a lecture bottle of chlorine gas by a length of Tygon tubing. While the receiver is cooled in a Dry Ice/acetone bath, chlorine gas is passed into it until the required volume of chlorine has been condensed. The end of the Tygon tubing attached to the lecture bottle is then connected to the 7-mm inlet tube of the reaction flask, the drying tube is replaced by a stopper, and the Dry Ice/acetone bath is

removed. As the chlorine evaporates, it will pass into the reaction mixture at a convenient rate. If the chlorine starts to boil, the graduated receiver may be cooled momentarily with the cold bath.

Time: 2 hours.

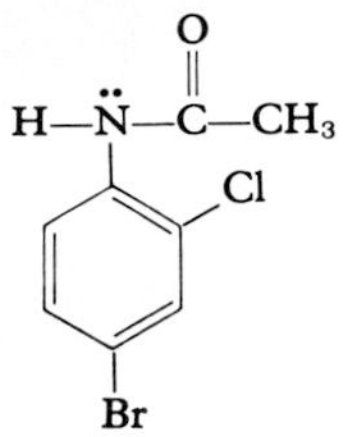

2-chloro-4-bromoacetanilide

Procedure b (Reference 2). Suspend 10.7 g (0.05 mole) of 4-bromoacetanilide in a mixture of 23 mL of conc. hydrochloric acid and 28 mL of glacial acetic acid in a 250-mL flask. Heat the mixture on the steam bath until all the solid has gone into solution. Cool the solution to 0°C in an ice bath. To the cold mixture gradually add a solution of 2.77 g (0.026 mole) of sodium chlorate dissolved in 7 mL of water. During the addition of the sodium chlorate solution some chlorine gas will be evolved (Note 1). As the addition is made, a yellow precipitate will form and the solution will turn yellow. After the addition is complete, allow the reaction mixture to stand at room temperature for an hour and then collect the solid by suction filtration.

Note

1. You must either work in the hood or make some other provision for removal of the chlorine gas produced (see Section 35).

Time: 2 hours.

Question

1. An attempt was made to prepare 2-chloro-4-bromoacetanilide from acetanilide without isolation of 4-bromoacetanilide, by treating 5 g of acetanilide and two equivalents of sodium acetate dissolved in 40 mL of glacial acetic acid with one equivalent of bromine followed by one equivalent of chlorine. Instead of the anticipated product, 2,4-dibromoacetanilide was isolated in 92% yield. Consideration of this result led us to believe that it might be possible to prepare 2-chloro-4-bromoacetanilide by treating acetanilide, two equivalents of sodium acetate, and one equivalent of sodium bromide in 80% acetic acid with two equivalents of chlorine. The product, however, appears to be a mixture of 2,4-dihalogenated acetanilides. Interpret the observations.

83.2 2-CHLORO-4-BROMOANILINE FROM 2-CHLORO-4-BROMOACETANILIDE

The infrared and proton NMR spectra of 2-chloro-4-bromoaniline are shown in Figures 83.1 and 83.2.

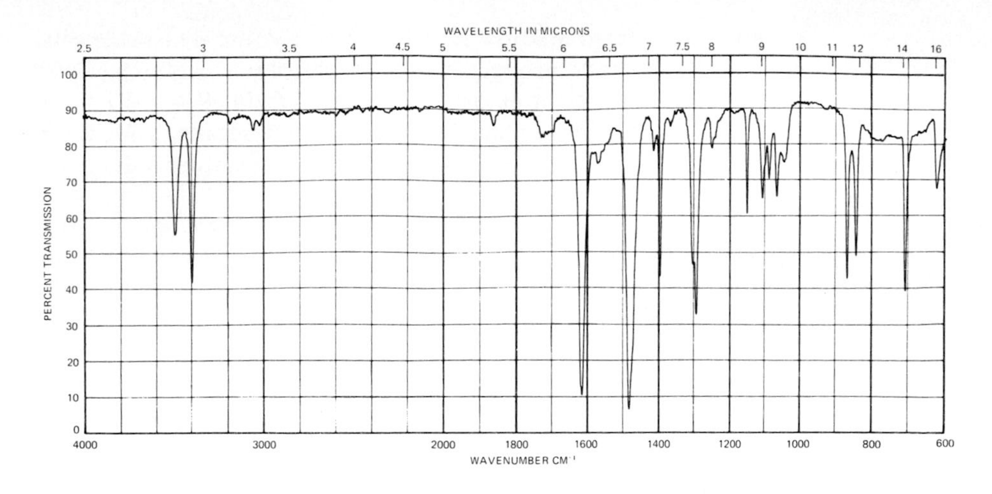

Figure 83.1. Infrared spectrum of 2-chloro-4-bromoaniline; CCl_4 solution.

Procedure. Place 12.4 g (0.050 mole) 2-chloro-4-bromoacetanilide in a 250-mL flask, and add 20 mL of 95% ethanol and 13 mL of conc. hydrochloric acid. Fit the flask with a reflux condenser and heat the mixture on the steam bath for 30 minutes. During this time, a clear solution will be obtained, and then a white precipitate will form. At the end of the heating period, add 80 mL of hot water, and swirl the flask to dissolve the solid. Pour the solution onto about 150 g of ice, and then

Figure 83.2. NMR spectrum of 2-chloro-4-bromoaniline; $CDCl_3$ solution.

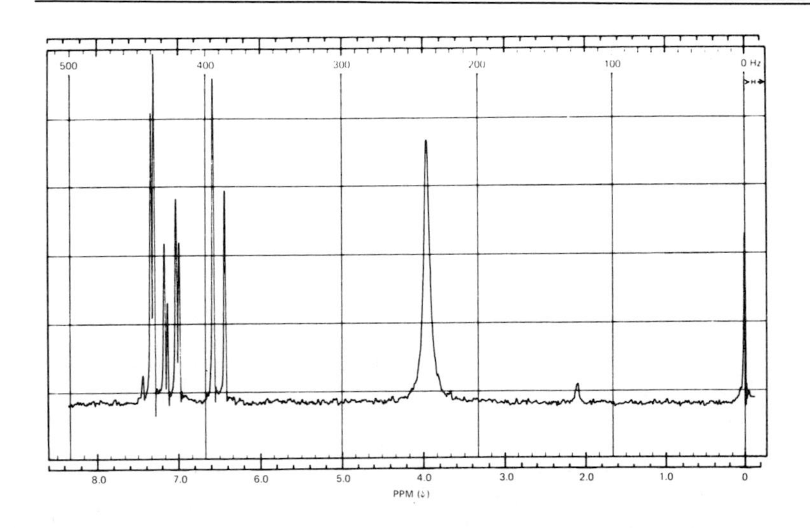

NH_2, Cl, Br

2-chloro-4-bromoaniline

add to the well-stirred mixture 12 mL of 50% sodium hydroxide solution. Collect the crude product by suction filtration, washing it well with water and sucking it as dry as possible. Yield: 9.4 g (91%). 2-Chloro-4-bromoaniline may be recrystallized from hexane, using 2 mL per gram, with 91% recovery; m.p., 69–71°C (literature value, 70–71°C).

Time: less than 2 hours.

83.3 2-CHLORO-4-BROMO-6-IODOANILINE FROM 2-CHLORO-4-BROMOANILINE

Procedure. Dissolve 5.0 g (0.024 mole) 2-chloro-4-bromoaniline in 80 mL of glacial acetic acid, and add to the solution 20 mL of water. (If the crude 2-chloro-4-bromoaniline is wet with water, reduce accordingly the amount of water added to the reaction mixture.) Then add, over a period of 5 minutes, a solution of 5.0 g (1.6 mL; 0.03 mole) of technical iodine monochloride in 20 mL of glacial acetic acid. Heat the resulting mixture to 90°C on the steam bath, and add enough concentrated aqueous sodium bisulfite solution to lighten the color of the solution to bright yellow; about 20 mL will be needed. Then add an amount of water such that the volume of sodium bisulfite solution plus water added will equal 25 mL; about 5 mL will be required. Allow the solution to cool slowly to room temperature, ■ and finally cool it in an ice bath. Collect the 2-chloro-4-bromo-6-iodoaniline, which will have

Figure 83.3. Infrared spectrum of 2-chloro-4-bromo-6-iodoaniline; CCl_4 solution.

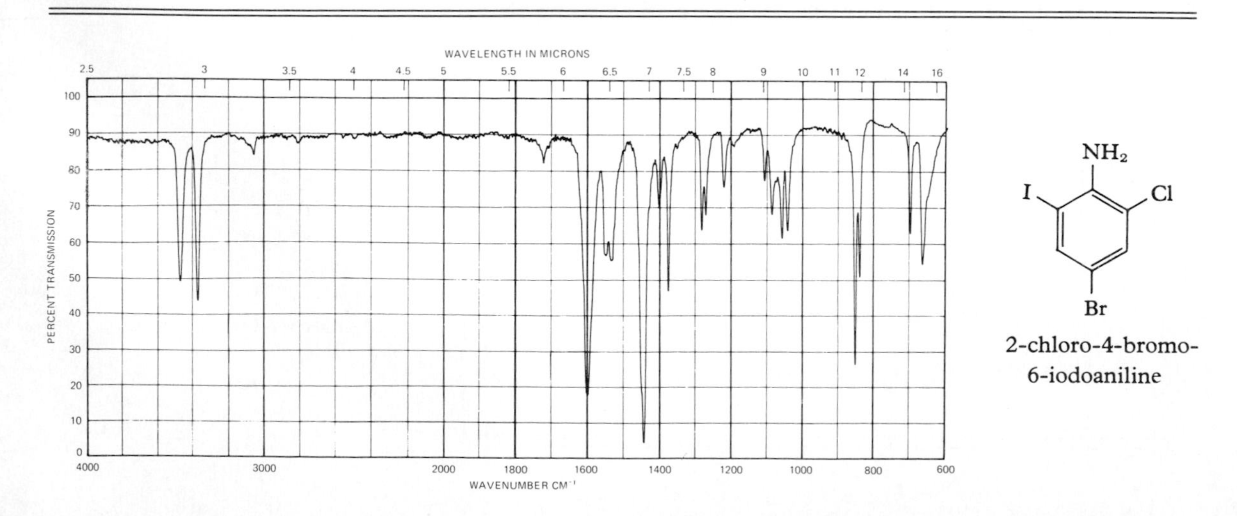

separated in a mass of long, almost colorless crystals, by suction filtration. Wash them with a little 33% acetic acid solution and then with water. 2-Chloro-4-bromo-6-iodoaniline may be recrystallized with 80% recovery by dissolving 1 g in 20 mL of glacial acetic acid, slowly adding 5 mL of water to the solution as it is heated on the steam bath, and allowing the resulting solution to cool slowly; m.p., 98–99°C (literature value, 97–97.5°C).

Time: less than 2 hours, plus the time required for the solution to cool.

Figure 83.3 shows the infrared spectrum of 2-bromo-4-chloro-6-iodoaniline.

83.4 1-BROMO-3-CHLORO-5-IODOBENZENE FROM 2-CHLORO-4-BROMO-6-IODOANILINE

Procedure. Suspend 2.0 g (0.006 mole) 2-chloro-4-bromo-6-iodoaniline in 10 mL absolute ethanol in a 250-mL boiling flask, and add 4.0 mL of conc. sulfuric acid dropwise while stirring the mixture. Fit the flask with a reflux condenser. Add 0.70 g (0.01 mole) of powdered sodium nitrite down the condenser, in several portions, to the stirred solution. After the addition of sodium nitrite is complete, heat the mixture on the steam bath for 10 minutes, and then add 50 mL of hot water down the condenser. Steam distill the mixture, and collect about 100 mL of distillate. The product, which will have solidified in the

Figure 83.4. Infrared spectrum of 1-bromo-3-chloro-5-iodobenzene; CCl_4 solution.

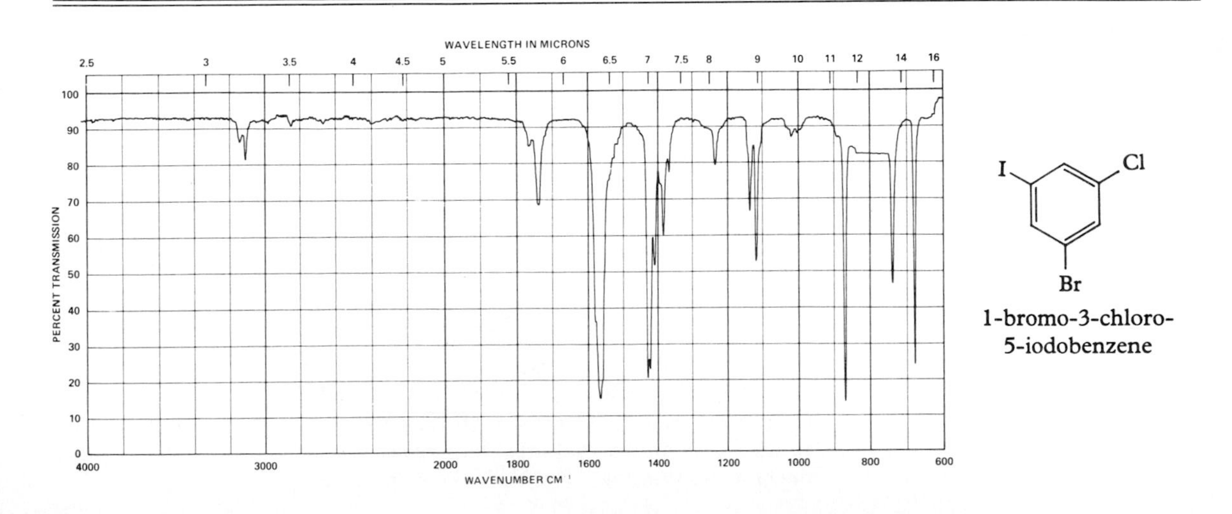

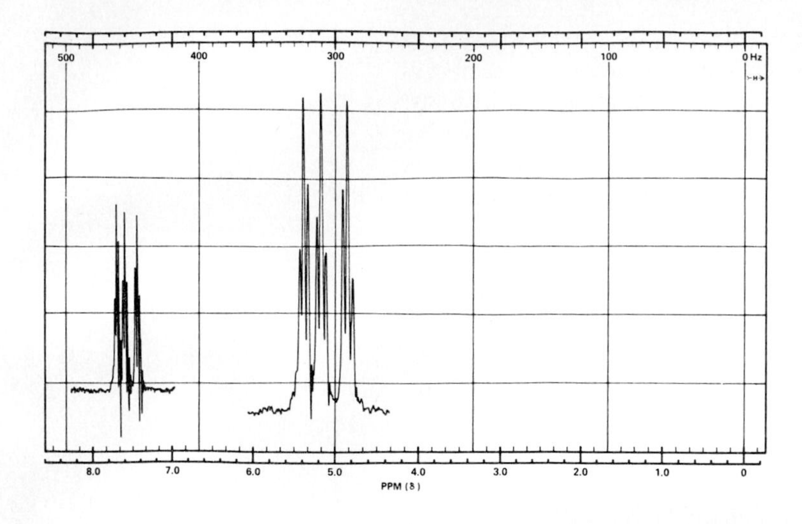

I Cl Br

1-bromo-3-chloro-5-iodobenzene

Figure 83.5. NMR spectrum of 1-bromo-3-chloro-5-iodobenzene; CCl_4 solution.

condenser and receiver, is most conveniently isolated by dissolution and extraction with ether. Dry the extract with anhydrous magnesium sulfate, filter it, and distill off the ether. A residue of about 1.5 g (80%) crude 1-bromo-3-chloro-5-iodobenzene will remain. Recrystallization of the crude product from 10 mL of methanol will give about 0.9 g (47%) 1-bromo-3-chloro-5-iodobenzene in the form of long, almost colorless needles; m.p., 87.5–89°C (literature value, 85.5–86°C).

Time: about 3 hours.

The IR and NMR spectra of the product of this procedure are shown in Figures 83.4 and 83.5.

Question

1. What do you suppose might happen if the "mixture of 2,4-dihalogenated acetanilides" described in the question in Section 79.1 were carried on through the rest of the procedures?

References

1. A. Ault and R. Kraig, *J. Chem. Educ.* **43,** 213 (1966).
2. R. M. Roberts, J. C. Gilbert, L. B. Rodewald, and A. S. Wingrove, *An Introduction to Modern Experimental Organic Chemistry,* 2nd edition, Holt, Rinehart and Winston, New York, 1974, p. 350.

84 MOED: A Merocyanine Dye

This experiment involves the preparation of a substance whose color depends upon the solvent in which it is dissolved. As the polarity of the solvent is decreased, the wavelength of maximum light absorption becomes longer, and the color changes from yellow through orange to red and finally to violet. The colors are quite intense, and the changes are remarkable.

The first step in the synthesis is the preparation of the quaternary methiodide of 4-methylpyridine from the amine and methyl iodide.

$$I{-}CH_3 + :N\langle\ \rangle{-}CH_3 \longrightarrow I^- \; CH_3{-}\overset{+}{N}\langle\ \rangle{-}CH_3$$

methyl iodide 4-methylpyridine methiodide

In the presence of a weak base, the conjugate base of the methiodide attacks the carbonyl group of *p*-hydroxybenzaldehyde to give an addition product, which then suffers elimination of water to give the bright red 4-(*p*-hydroxystyryl)-1-methylpyridinium iodide:

$$\left\{ CH_3{-}\overset{+}{N}\langle\ \rangle{-}\ddot{C}H_2^- \longleftrightarrow CH_3{-}\ddot{N}\langle\ \rangle{=}CH_2 \right\}$$

conjugate base of the methiodide
present in low concentration

$$I^- \; CH_3{-}\overset{+}{N}\langle\ \rangle{-}CH_2{-}\underset{}{CH(OH)}{-}\langle\ \rangle{-}OH$$

addition product

$$I^- \; CH_3{-}\overset{+}{N}\langle\ \rangle{-}CH{=}CH{-}\langle\ \rangle{-}OH$$

4-(*p*-hydroxystyryl)-1-methylpyridinium iodide
(red)

The maroon-colored merocyanine dye MOED, the conjugate base of the red substance, is then produced by treatment of the red substance with aqueous potassium hydroxide.

$$\left\{ CH_3{-}\overset{+}{N}\langle\ \rangle{-}CH{=}CH{-}\langle\ \rangle{-}\ddot{O}:^- \longleftrightarrow CH_3{-}N:\langle\ \rangle{=}CH{-}CH{=}\langle\ \rangle{=}\ddot{O}: \right\}$$

1-methyl-4-[(oxocyclohexadienylidene)ethylidene]-1,4-dihydropyridine; MOED
(a maroon solid)

84.1 1,4-DIMETHYLPYRIDINIUM IODIDE FROM 4-METHYLPYRIDINE AND METHYL IODIDE

$$CH_3-C_5H_4N: \; + \; CH_3I \longrightarrow CH_3-C_5H_4\overset{+}{N}-CH_3 \;\; I^-$$

4-methylpyridine methyl iodide 1,4-dimethylpyridinium iodide

Procedure. Place 20 mL of isopropyl alcohol and 10 mL (9.3 g; 0.10 mole) of 4-methylpyridine (4-picoline) in a 50-mL Erlenmeyer flask. Fit the flask with a reflux condenser and add to the flask, down the condenser, 6.2 mL (14.1 g; 0.10 mole) of methyl iodide. Swirl the flask to mix the contents. Lower a thermometer down the condenser so that the bulb rests on the bottom of the Erlenmeyer flask and indicates the temperature of the liquid. Heat the flask on the steam bath until the temperature of the liquid has reached 50°C. At this time, remove the flask from the heat and allow the exothermic reaction to take place spontaneously. During the course of the reaction, the temperature will rise to a maximum of about 70–75°C and the solid product will crystallize from the reaction mixture. Allow the mixture to cool spontaneously for about 15 minutes after the maximum temperature has been reached, and then collect the product by suction filtration. Portions of isopropyl alcohol can be used to rinse the reaction flask and to wash the product. Yield: about 20 grams (about 85%).

84.2 PREPARATION OF 4-(*p*-HYDROXYSTYRYL)-1-METHYLPYRIDINIUM IODIDE FROM 1,4-DIMETHYLPYRIDINIUM IODIDE AND *p*-HYDROXYBENZALDEHYDE

$$CH_3-\overset{+}{N}C_5H_4-CH_3 \; I^- + H-\overset{O}{\overset{\|}{C}}-C_6H_4-OH \xrightarrow[\text{1-propanol}]{\text{piperidine}} CH_3-\overset{+}{N}C_5H_4-CH{=}CH-C_6H_4-OH \; I^- + H_2O$$

(red)

Procedure. Add to a 50-mL Erlenmeyer flask 2.35 g (0.01 mole) of 1,4-dimethylpyridinium iodide, 1.22 g (0.01 mole) of *p*-hydroxybenzaldehyde, 15 mL of 1-propanol, and 1 mL of piperidine. Fit the flask with a reflux condenser, swirl the flask to mix the contents, and then heat the flask on the steam bath for 45 minutes. During the heating time, the reaction mixture will turn red and the product will separate from solution as red crystals. At the end of the heating period, cool the flask by immersing it in cold water for a few minutes and then collect the product by suction filtration. Methyl alcohol can be used to rinse the flask and to wash the crystals. Yield: about 3 grams (about 90%). The product can be recrystallized from methanol, using 75 mL per gram. Recrystallization occurs slowly and therefore the solution should be allowed to stand for several hours before the product is collected by suction filtration.

84.3 1-METHYL-4-[(OXOCYCLOHEXADIENYLIDENE)-ETHYLIDENE]-1,4-DIHYDROPYRIDINE (MOED) FROM 4-(*p*-HYDROXYSTYRYL)-1-METHYLPYRIDINIUM IODIDE

$$CH_3-\overset{+}{N}C_5H_4-CH{=}CH-C_6H_4-OH \; (I^-) + HO^- \longrightarrow CH_3-\ddot{N}C_5H_4{=}CH-CH{=}C_6H_4{=}O + H_2O + I^-$$

MOED
(maroon)

Procedure. Add 1.0 g (3 mmole) of 4-(*p*-hydroxystyryl)-1-methylpyridinium iodide to a solution of 0.25 mL of 45% aqueous potassium hydroxide (3 mmole KOH) in 50 mL of water in a 125-mL Erlenmeyer flask. Heat the mixture on the steam bath until all the solid has dissolved. Allow the solution to cool to room temperature (Note 1) and then collect the maroon crystals of MOED by suction filtration. Water can be used to rinse the flask and to wash the crystals. Yield: about 0.75 gram (about 75%).

Note

1. The product crystallizes rather slowly and so the mixture should be allowed to stand overnight for the best yield.

Demonstration of the Dependence of the Color of MOED on Solvent

Portions of the solution can be used as described below to demonstrate the dependence of the color of MOED on the solvent.

Procedure. Add 20 mg of MOED to 1 mL of water in a small test tube and heat the test tube in the steam bath until the solid has dissolved. Place two drops of the resulting red solution in a 25-mL Erlenmeyer flask and add 15 mL of water, methanol, isopropanol, or acetone. The resulting color of the solution will be yellow, orange, red, or violet, respectively.

Reference

1. M. J. Minch and S. Sadiq Shah, *J. Chem. Educ.* **54** 709 (1977).

Appendix

All boiling points and melting points are given in °C.

Table A.1. Derivatives of alcohols

			Derivative				
Alcohol	*b.p.*	*m.p.*	*p*NB	DNB	HNP	PU	NU
Methyl alcohol	65		96	108	153	47	124
Ethyl alcohol	78		57	93	158	52	79
2-Propanol	82		111	123	152	76	106
tert-Butyl alcohol	83	26	116	142		136	101
3-Buten-2-ol	96			44			
Allyl alcohol	97		28	50	124	70	108
1-Propanol	97		35	74	146	57	80
2-Butanol	99		26	76	131	65	97
2-Methyl-2-butanol	102		85	116		42	72
2-Methyl-1-propanol	108		69	87	181	86	104
3-Methyl-2-butanol	114			76	127	68	109
3-Pentanol	116		17	101	121	49	95
1-Butanol	118		70	64	147	61	71
3,3-Dimethyl-2-butanol	121			111		66	101
3-Methyl-3-pentanol	123			97		44	84
2-Methyl-2-pentanol	123			72			
2-Methyl-1-butanol	129			70	158	31	82
4-Methyl-2-pentanol	132		26	65		143	88
3-Methyl-1-butanol	132		21	61	166	57	68
2,2-Dimethyl-1-butanol	137				51	66	80
1-Pentanol	138		11	46	136	46	68
2-Hexanol	139		40	39			61
2,4-Dimethyl-3-pentanol	140		155		151	95	99
Cyclopentanol	141			115		133	118
4-Heptanol	156		35	64			80
1-Hexanol	158		5	58	124	42	62
Cyclohexanol	161	25	50	113	160	82	129
1-Heptanol	177		10	47	127	65	62
2-Octanol	179			32		114	63
1-Octanol	195		12	62	128	74	67
Benzyl alcohol	205		85	113	176	77	134
2-Phenylethyl alcohol	220		62	108	123	80	119

*p*NB = *p*-Nitrobenzoate ester; Section 29.1
DNB = 3,5-Dinitrobenzoate ester; Section 29.1
HNP = Hydrogen 3-nitrophthalate ester; Section 29.2
PU = Phenylurethane; Section 29.3
NU = α-Naphthylurethan; Section 29.3

Table A.2. Derivatives of aldehydes

			Derivative			
Aldehyde	*b.p.*	*m.p.*	Semi	DNP	Ox	Meth
Acetaldehyde	20		162	168	47	139
Propionaldehyde	49		154	150	40	156
Isobutyraldehyde	64		126	187		154
Butyraldehyde	75		104	123		142
Trimethylacetaldehyde	75		190	210	41	
3-Methylbutanal	93		107	123	49	155
2-Methylbutanal	93		105	120		
Pentanal	103		108	107	52	105
2,2-Dimethylpentanal	103			147		
2-Butenal	104		199	190	119	183
Hexanal	131		106	104	51	109
Heptanal	155		109	108	57	135
Octanal	171		101	106	60	90
Benzaldehyde	179		233	237	35	193
Phenylacetaldehyde	194	33	153	121	103	165
Cinnamaldehyde	252		216	255	139	213

Semi = Semicarbazone; Section 29.6
DNP = 2,4-Dinitrophenylhydroazone; Section 29.5
Ox = Oxime; Section 29.7
Meth = Methone derivative; Section 29.4

Table A.3. Derivatives of amides

		Components		*Derivative*
Amide	*m.p.*	*Acid; m.p.*	*Amine; m.p.*	Xan
Acetanilide	114	Acetic	Aniline	
Phenylacetanilide	117	Phenylacetic; 77	Aniline	
Benzamide	128	Benzoic; 122	Ammonia	124
Phenacetin	134	Acetic	*p*-Ethoxyaniline	
Phenylacetamide	154	Phenylacetic; 77	Ammonia	196
p-Toluamide	160	*p*-Toluic; 182	Ammonia	225
Benzanilide	161	Benzoic; 122	Aniline	
p-Bromoacetanilide	167	Acetic	*p*-Bromoaniline; 66	
p-Chloroacetanilide	179	Acetic	*p*-Chloroaniline; 72	

Xan = 9-Acylamidoxanthene; Section 29.9

Table A.4. Derivatives of primary and secondary amines

			Derivative				
Amine	*b.p.*	*m.p.*	Ac	Benz	*p*Tos	PTU	NTU
tert-Butylamine	46		102	134		120	143
n-Propylamine	49			84	52	63	
Diethylamine	56			42	60	34	108
Allylamine	58				64	98	
Isobutylamine	69			57	78	82	
n-Butylamine	77			42		65	109
Isoamylamine	96					102	
n-Amylamine	104				65	69	114
Piperidine	106			48	96	101	
Di-*n*-Propylamine	110					69	
n-Hexylamine	130			40		77	
Morpholine	130			75	147	136	
Cyclohexylamine	134		101	149		148	
n-Heptylamine	155					75	121
Di-*n*-Butylamine	159					86	123
Aniline	184		114	160	103	154	
Benzylamine	185		65	105	116	157	
α-Phenylethylamine	187		57	120			
N-Methylaniline	196		102	63	94	87	
β-Phenylethylamine	198		51	116		135	
2-Methylaniline	200		110	146	186	136	
3-Methylaniline	203		65	125	172	94	
2-Chloroaniline	209		87	99	193	156	
3-Chloroaniline	236		78	120	210	124	
n-Butylaniline	241			56	56		
p-Ethoxyaniline	248	3	137	173	106	136	
N-Benzylaniline	298	37	58	107	149	103	
4-Methylaniline	200	45	147	158	118	141	
4-Bromoaniline	245	66	167	204		148	
4-Chloroaniline	232	72	179	192	119	152	

Ac = N-Substituted acetamide; Section 29.11
Benz = N-Substituted benzamide; Section 29.12
*p*Tos = N-Substituted *p*-toluenesulfonamide; Section 29.13
PTU = Phenylthiourea; Section 29.14
NTU = Naphthylthiourea; Section 29.14

Table A.5. Derivatives of tertiary amines

Amine	*b.p.*	*m.p.*	n_D	den	*Derivative*		
					Pic	MeI	Tos
Triethylamine	89		1.400	0.726	173		
Pyridine	116		1.509	0.978	167	117	139
4-Methylpyridine	143		1.506	0.957	167		
N,N-Dimethylaniline	193	3	1.558	0.956	163	228	161
Tri-*n*-butylamine	211			0.778	106	180	
N,N-Diethylaniline	218			0.935	142	102	
Quinoline	239		1.627	1.023	203	133	126
Isoquinoline	243	26	1.615	1.097	222	159	163
Tribenzylamine	380	91			190	184	

n_D = Index of refraction; Section 19
den = Density; Section 18
Pic = Picrate; Section 29.15
MeI = Methiodide; Section 29.16
Tos = Methyl *p*-toluenesulfonate; Section 29.16

Table A.6. Derivatives of carboxylic acids

			Derivative				
Carboxylic Acid	*b.p.*	*m.p.*	Am	An	*p*Tol	*p*Br	*p*NB
Acetic	118	17	82	114	149	86	78
Propionic	141		81	106	126	63	31
Isobutyric	155		129	105	109	77	
Butyric	164		116	97	75	63	35
3-Methylbutanoic	176		135	110	107	68	
Pentanoic	186		106	63	74	75	
Hexanoic	205		101	95	75	72	
Heptanoic	223		97	71	81	72	
Cyclohexanecarboxylic	233	31	186	146			
Dodecanoic	299	44	100	78	87	76	
3-Phenylpropionic	280	49	105	98	135	104	36
Chloroacetic	189	61	121	137	162	104	
Trichloroacetic	198	58	141	97	113		80
Stearic		70	109	96	102	92	
Phenylacetic	256	76	156	118	136	89	65
Phenoxyacetic	285	100	102	99		149	
2-Methylbenzoic	257	107	143	125	144	57	91
3-Methylbenzoic	263	113	97	126	118	108	87
Benzoic		122	130	160	158	119	89
trans-Cinnamic		133	148	153	168	146	117
2-Chlorobenzoic		142	141	118	131	106	106
Diphenylacetic		148	168	180	173		
3-Chlorobenzoic		158	134	125		116	107
4-Methylbenzoic		180	160	145	165	153	105
4-Methoxybenzoic		186	167	171	186	152	132
3,5-Dinitrobenzoic		205	183	234		159	157
4-Nitrobenzoic		241	201	211	204	137	168
4-Chlorobenzoic		243	179	194		126	130

Am = Amide; Section 29.17
An = Anilide; Section 29.18
*p*Tol = *p*-Toluidide; Section 29.18
*p*Br = *p*-Bromophenacyl ester; Section 29.19
*p*NB = *p*-Nitrobenzyl ester; Section 29.20

Table A.7. Derivatives of esters

Ester	*b.p.*	*m.p.*	n_D	den	*Derivative* NBz	DNB	Acid
Methyl formate	32		1.346	0.974	60	108	
Ethyl formate	54		1.360	0.922	60	93	
Methyl acetate	57		1.362	0.927	61	108	
Ethyl acetate	77		1.372	0.901	61	93	
Methyl propionate	80		1.378	0.915	47	108	
Isopropyl acetate	88		1.374	0.842	61	123	
Methyl isobutyrate	93		1.384	0.891	88	108	
Ethyl propionate	99		1.385	0.889	47	93	
n-Propyl acetate	102		1.385	0.883	61	74	
Methyl *n*-butyrate	102		1.388	0.892	38	108	
Ethyl isobutyrate	110		1.390	0.869	88	93	
sec-Butyl acetate	112		1.386	0.872	61	76	
Isobutyl acetate	118		1.390	0.875	61	87	
Ethyl *n*-butyrate	122		1.400	0.879	38	93	
n-Propyl propionate	123		1.392	0.874	47	74	
n-Butyl acetate	126		1.396	0.881	61	64	
Methyl chloroacetate	132		1.422	1.238		108	61
Isoamyl acetate	142		1.400	0.867	61	61	
Ethyl chloroacetate	145		1.423	1.156		93	61
n-Amyl acetate	149		1.403	0.876	61	46	
n-Hexyl acetate	169		1.411	0.873	61	58	
n-Heptyl acetate	192		1.417	0.871	61	47	
Phenyl acetate	197		1.503	1.078	61	146	
Methyl benzoate	199		1.516	1.089	106	108	122
2-Tolyl acetate	208			1.048	61	135	
3-Tolyl acetate	212	12	1.498	1.049	61	165	
Ethyl benzoate	213		1.506	1.047	106	93	122
4-Tolyl acetate	213		1.500	1.051	61	189	
Methyl 2-toluate	215			1.068		108	108
Methyl 3-toluate	215			1.061	76	108	113
Benzyl acetate	217		1.520	1.055	61	113	
Methyl phenylacetate	220		1.507	1.068	122	108	77
Methyl salicylate	224		1.537	1.184		108	158
Ethyl phenylacetate	228		1.499	1.033	122	93	77
n-Propyl benzoate	231		1.500	1.023	106	74	122
Ethyl 4-toluate	234		1.509	1.027	133	93	180
Methyl phenoxyacetate	245			1.150	86	108	99
n-Butyl benzoate	250		1.497	1.000	106	64	122
Ethyl phenoxyacetate	251			1.104	86	93	99
Ethyl cinnamate	271	7	1.560	1.049	226	93	133
Methyl anthranilate	300	24				108	146
Methyl 4-toluate	223	33			133	108	180
Methyl cinnamate	261	36			226	108	133

n_D = Index of refraction; Section 19
den = Density; Section 18
NBz = N-Benzylamide of acidic component; Section 29.21
DNB = 3,5-Dinitrobenzyl ester of alcohol component; Section 29.22
Acid = Carboxylic acid produced upon hydrolysis; Section 29.23

Table A.8. Derivatives of aromatic ethers

					Derivative	
Aromatic Ether	*b.p.*	*m.p.*	n_D	den	Pic	Br
Methoxybenzene	154		1.5221	0.993	81	61
Benzyl methyl ether	171		1.5008	0.965	116	
2-Methoxytoluene	171		1.505	0.985	116	64
Phenyl ethyl ether	172		1.508	0.967	92	
4-Methoxytoluene	176		1.512	0.970	89	
3-Methoxytoluene	177		1.513	0.972	114	55
Benzyl ethyl ether	186		1.496	0.948		
n-Butyl phenyl ether	206		1.505	0.950	112	
1,3-Dimethoxybenzene	217		1.423	1.055	58	
n-Butyl benzyl ether	221		1.483	0.923		
1-Methoxynaphthalene	271		1.694	1.092	130	
Dibenzyl ether	290	4		1.043	78	108
1,2-Dimethoxybenzene	205	23	1.529	1.080	56	93
Diphenyl ether	259	28	1.583	1.073	110	55
1,4-Dimethoxybenzene	213	56			48	142
2-Methoxynaphthalene	273	73			117	84

n_D = Index of refraction; Section 19
den = Density; Section 18
Pic = Picrate; Section 15
Br = Bromo derivative; Section 24

Table A.9. Derivatives of aliphatic halides

Aliphatic Halide	*b.p.*	*m.p.*	n_D	den	Derivative SRP	An	Nap
2-Chloropropane	37		1.378	0.859	196	103	
Ethyl bromide	38		1.425	1.460	188	104	126
Methyl iodide	43		1.532	2.282	224	104	160
1-Chloropropane	47		1.388	0.899	181	92	121
tert-Butyl chloride	51		1.386	0.846		128	147
2-Bromopropane	60		1.425	1.314	196	103	
sec-Butyl chloride	68		1.397	0.804	190	108	129
Isobutyl chloride	69		1.398	0.881	174	109	125
1-Bromopropane	71		1.434	1.353	181	92	121
Ethyl iodide	72		1.514	1.940	188	104	126
tert-Butyl bromide	73			1.211		128	147
n-Butyl chloride	78		1.402	0.886	180	63	112
2-Iodopropane	90		1.499	1.703	196	103	
Isobutyl bromide	91		1.435	1.253	174	109	125
sec-Butyl bromide	91		1.437	1.256	190	108	129
n-Butyl bromide	101		1.440	1.274	180	63	112
1-Iodopropane	102		1.505	1.743	181	92	121
tert-Butyl iodide	103					128	147
1-Chloropentane	106		1.412	0.882	154	96	112
Chlorocyclopentane	115		1.451	1.005			
sec-Butyl iodide	120		1.499	1.592	190	108	129
Isobutyl iodide	120		1.496	1.602	174	109	125
1-Bromopentane	129		1.445	1.219	154	96	112
n-Butyl iodide	130		1.499	1.616	180	63	112
1-Chlorohexane	133		1.420	0.878	157	69	106
Bromocyclopentane	137		1.489	1.387			
Chlorocyclohexane	143		1.462	0.989		146	188
1-Iodopentane	155		1.496	1.512	154	96	112
1-Bromohexane	157		1.448	1.175	157	69	106
Bromocyclohexane	165		1.495	1.336		146	188
Iodocyclopentane	167		1.545	1.710			
Iodocyclohexane	179			1.626		146	188

n_D = Index of refraction; Section 19
den = Density; Section 18
SRP = S-Alkylthiuronium Picrate; Section 29.25
An = Anilide; Section 29.28
Nap = α-Naphthalide; Section 29.28

Table A.10. Derivatives of aromatic halides

Aromatic Halide	*b.p.*	*m.p.*	n_D	den	*Derivative* Acid
Chlorobenzene	132		1.525	1.107	
Bromobenzene	156		1.560	1.494	
2-Chlorotoluene	159		1.524	1.082	141
3-Chlorotoluene	162		1.521	1.082	158
4-Chlorotoluene	162	7	1.521	1.071	240
1,3-Dichlorobenzene	173		1.546	1.288	
1,2-Dichlorobenzene	179		1.522	1.305	
1,4-Dichlorobenzene	173	53			
2-Bromotoluene	182		1.425		150
3-Bromotoluene	184		1.410		155
Iodobenzene	188		1.620	1.831	
2,4-Dichlorotoluene	199		1.249	1.545	164
2,6-Dichlorotoluene	199		1.269	1.551	139
3-Iodotoluene	204			1.698	187
2-Iodotoluene	211			1.698	162
1,3-Dibromobenzene	219		1.606	1.952	
1,2-Dibromobenzene	224	7	1.609	1.956	
4-Bromotoluene	184	28			251
4-Iodotoluene	211	35			270
1-Bromo-4-chlorobenzene	195	67			
1,4-Dibromobenzene	219	89			

n_D = Index of refraction; Section 19
den = Density; Section 18
Acid = Acid formed upon oxidation with permanganate; Section 29.27

Table A.11. Derivatives of aromatic hydrocarbons

					Derivative			
Aromatic Hydrocarbon	*b.p.*	*m.p.*	n_D	den	*o*Ar	Acid	Pic	TNF
Toluene	110		1.496	0.867	137	122	88	
Ethylbenzene	136		1.496	0.867	128	122	97	
p-Xylene	138	13	1.496	0.861	148	subl	90	
m-Xylene	139		1.497	0.864	142	348	91	
o-Xylene	144		1.505	0.880	178	208	88	
Isopropylbenzene	152		1.491	0.861	133	122		
1,3,5-Trimethylbenzene	165		1.499	0.865	212	380	97	
tert-Butylbenzene	169		1.493	0.867		122		
p-Isopropyltoluene	177		1.491	0.834	124	subl		
n-Butylbenzene	183		1.490	0.860	97	122		
Bibenzyl	284	53						
Biphenyl	255	71			225			
Naphthalene	218	80			172		149	154
Fluorene	295	114			228		87	179
trans-Stilbene	305	124						
Anthracene	340	216					138	194

n_D = Index of refraction; Section 19
den = Density; Section 18
*o*Ar = *o*-Aroylbenzoic acid; Section 29.26
Acid = Acid formed upon oxidation with permanganate; Section 29.27
Pic = Picrate; Section 29.15
TNF = 2,4,7-Trinitrofluorenone adduct; Section 29.29

Table A.12. Derivatives of ketones

Ketone	*b.p.*	*m.p.*	*Derivative*		
			Semi	DNP	Ox
Acetone	56		190	128	59
2-Butanone	82		136	117	129
2-Methyl-3-butanone	94		114	120	
3-Pentanone	102		139	156	69
2-Pentanone	102		112	144	58
4-Methyl-2-pentanone	117		135	95	58
3-Methyl-2-pentanone	118		95	71	
2,4-Dimethyl-3-pentanone	124		160	98	
3-Hexanone	125		113	130	
2-Hexanone	128		125	110	49
Cyclopentanone	131		216	146	57
2-Heptanone	151		127	89	
Cyclohexanone	156		167	162	91
2,6-Dimethyl-4-heptanone	168		126	92	210
2-Octanone	173		123	58	
Cyclohexyl methyl ketone	180		177		60
Acetophenone	205	20	199	250	60
Propiophenone	220	20	174	191	54
4-Methylacetophenone	226	28	205	258	88
Butyrophenone	230		191	190	50
4-Chloroacetophenone	232		204	231	95
Valerophenone	249		160	166	52

Semi = Semicarbazone; Section 29.6
DNP = 2,4-Dinitrophenylhydrazone; Section 29.5
Ox = Oxime; Section 29.7

Table A.13. Derivatives of nitriles

Nitrile	*b.p.*	*m.p.*	*Derivative*		
			n_D	den	Acid
Benzonitrile	190		1.529	1.010	122
2-Toluonitrile	205		1.527	0.991	104
3-Toluonitrile	212			1.032	113
Phenylacetonitrile	234		1.521	1.021	77
Phenoxyacetonitrile	240			1.09	99
Cinnamonitrile	256	20			133
4-Toluonitrile	217	29			180
3-Chlorobenzonitrile		41			158
2-Chlorobenzonitrile	232	43			141
4-Chlorobenzonitrile	223	96			240

n_D = Index of refraction; Section 19
den = Density; Section 18
Acid = Acid formed upon hydrolysis; Section 29.8

Table A.14. Derivatives of phenols

Phenol	*b.p.*	*m.p.*	Derivative NU	Br	ArO
2-Chlorophenol	176	7	120	76	145
2-Bromophenol	195	5	129	95	
3-Methylphenol	203	12	128	84	103
2-Methylphenol	192	31	142	56	152
2-Methoxyphenol	205	32	118	116	116
3-Bromophenol	236	33	108		108
3-Chlorophenol	214	33	158		110
4-Methylphenol	202	16	146	199	
4-Chlorophenol	217	43	166	90	156
Phenol	183	42	133	95	99
4-Ethylphenol	219	47	128		97
Thymol	234	52	160	55	149
3,4-Dimethylphenol	225	63	142	171	163
4-Bromophenol	238	66	169	171	157
3,5-Dimethylphenol	220	68	109	166	111
2,4,6-Trimethylphenol	220	70		158	142
2,5-Dimethylphenol	212	75	173	178	118
1-Naphthol	280	94	152	105	194
3-Nitrophenol		97	167	91	156
4-*tert*-Butylphenol	237	100	110	67	87
4-Nitrophenol		114	151	142	187
2-Naphthol	286	123	157	84	95

NU = α-Naphthylurethan; Section 29.3
Br = Bromination product; Section 29.30
ArO = Aryloxyacetic acid; Section 29.31

Chemical Substance Index

p-Acetamidobenzenesulfonamide:
from *p*-acetamidobenzene-sulfonyl chloride, 495
conversion to *p*-aminobenzene-sulfonamide, 496
conversion to sulfanilamide, 496
p-Acetamidobenzenesulfonyl chloride:
from acetanilide, 495
conversion to *p*-acetamidobenzenesulfonamide, 495
p-Acetamidophenol:
conversion to phenacetin, 457
from *p*-aminophenol, 456
Acetanilide:
from aniline, 416
conversion to *p*-acetamidobenzenesulfonyl chloride, 490
conversion to *p*-bromoacetanilide, 423
IR spectrum, 417
recrystallization, 61
Acetic acid, 8, 38
dimerization of, 160
properties of, 49
reaction with aniline, 416
reaction with isoamyl alcohol, 385
Acetic anhydride, 8
reaction with *p*-aminophenol, 456
reaction with aniline, 416
reaction with glucose, 397, 399, 400
reaction with *p*-phenetidine, 417
reaction with salicylic acid, 395
Acetone, properties of, 49
Acetonitrile, 8
properties of, 49
Acetyl chloride, 8
Acetylsalicylic acid:
IR spectrum, 396
isolation from APC tablets, 117
isolation from aspirin tablets, 325
NMR spectrum, 397
from salicylic acid, 395
Adamantane:
inclusion compound with thiourea, 346
IR spectrum, 347
from *endo*-tetrahydrodicyclopentadiene, 346
Adipic acid, solubility of, 114
Aluminum chloride, 9, 492
American flag red, preparation of, 436
p-Aminobenzenesulfonamide:
from *p*-acetamidobenzene-sulfonamide, 496
synthesis from benzene, 492
p-Aminobenzoic acid, 493
p-Aminophenol, acetylation of, 456
Ammonia, 9
tert-Amyl alcohol, conversion to *tert*-amyl chloride, 375
tert-Amyl chloride, from *tert*-amyl alcohol, 375
Androstenedione:
from androstenolone, 479
conversion to testosterone, 479
formula, 208, 479
IR spectrum, 209
Androstenolone:
conversion to androstenedione, 479
formula, 477
Aniline, 9, 38
conversion to acetanilide, 416
conversion to benzenediazonium chloride, 430
diazotization of, 430
IR spectrum, 415
from nitrobenzene, 414
reaction with acyl halides, 282
reaction with anhydrides, 282
reaction with 2, 4-dinitrobromobenzene, 422
p-Anisidine, reaction with 2,4-dinitrobromobenzene, 429
Anthracene, purification by chromatography, 126
Anthrone, bromination of, 454
APC tablets, 117
Aspirin (*see* Acetylsalicylic acid)

Benzaldehyde:
conversion to benzoin by cyanide, 489
conversion to benzoin by thiamine, 462
conversion to *trans,trans*-1,4-diphenylbutadiene, 451
conversion to stilbene, 450
Benzene, 8
conversion to 1-bromo-3-chloro-5-iodobenzene, 502
conversion to sulfanilamide, 492
NMR spectrum, 219
properties, 49

Benzenediazonium chloride:
from aniline, 430
conversion to chlorobenzene, 432
reaction with β-naphthol, 435
Benzenesulfonyl chloride, 254
Benzidine, 8
Benzil:
from benzoin, 489
conversion to tetraphenylcyclopentadienone, 491
reaction with dibenzylketone, 491
recrystallization, 61
Benzoic acid:
conversion to methyl ester, 392
solubility, 60
sublimation, 105
Benzoin:
from benzaldehyde by cyanide catalysis, 457, 462, 489
from benzaldehyde by thiamine catalysis, 464
oxidation to benzil, 489
Benzophenone, reaction with phenylmagnesium bromide, 413
Benzoyl chloride, 8
benzoylation by means of, 271, 278
reaction with cholesterol, 390
Benzoyl peroxide, 8
Benzylamine, 285
Benzyl chloride, 8
reaction with nicotinamide, 490
reaction with triethylphosphite, 450
1-Benzyldihydronicotinamide:
IR spectrum, 471
preparation from 1-benzylnicotinamide chloride, 470
reduction to Malachite green by, 470, 472
synthesis from nicotinamide, 472
1-Benzylnicotinamide chloride:
preparation from nicotinamide, 470
reduction to 1-benzyldihydronicotinamide, 472
2-Benzylpyridine, dinitration of, 458
Black pepper, 328
Bromine, 9, 38
in carbon tetrachloride, 255
reaction with aniline, 423
reaction with aromatic ethers, 287
Bromoacetanilide:
from acetanilide, 423
conversion to 2-chloro-4-bromoacetanilide, 503
IR spectrum, 424
p-Bromoaniline, reaction with 2,4-dinitrobromobenzene, 429
9-Bromoanthrone:
from anthrone, 454
conversion to dianthraquinone, 455
Bromobenzene:
conversion to Grignard reagent, 411
dinitration of, 424
1-Bromo-4-chlorobenzene, NMR spectrum, 232
1-Bromo-2-chloroethane, NMR spectrum, 233
1-Bromo-3-chloro-5-iodobenzene:
from 2-chloro-4-bromo-6-iodoaniline, 507
IR spectrum, 507
NMR spectrum, 508
synthesis from benzene, 502
6β-Bromocholest-4-ene-3-one:
from 5α,6β-dibromocholestane-3-one, 486
IR spectrum, 486
p-Bromophenacyl bromide, 8, 283
Butadiene, reaction with maleic anhydride, 436
n-Butyl alcohol, conversion to *n*-butyl bromide, 372
tert-Butyl alcohol, conversion to *tert*-butyl chloride, 373
n-Butyl bromide:
from *n*-butyl alcohol, 372
IR spectrum, 372
tert-Butyl chloride:
from *tert*-butyl alcohol, 373
IR spectrum, 374
kinetics of hydrolysis of, 375
NMR spectrum, 375
4-*tert*-Butylcyclohexanol, *cis* and *trans*, from 4-*tert*-Butylcyclohexanone, 370
4-*tert*-Butylcyclohexanone, borohydride reduction of, 369

Caffeine:
from APC tablets, 117
IR spectrum, 326
NMR spectrum, 327
from NoDoz, 325
sublimation of, 105
from tea, 325
Calcium chloride, anhydrous, as drying agent, 145–149
Calcium hydride, 11
Calcium sulfate, anhydrous, as drying agent, 145–149
Camphor, sublimation of, 105
Carbon, decolorizing, 50
Carbon disulfide, 96
IR spectrum, 205
properties, 49
Carbon tetrachloride, 9
IR spectrum, 205
mass spectrum, 243
properties, 49
(R)-(−)-Carvone:
IR spectrum, 335
from oil of spearmint, 335
(S)-(+)-Carvone:
IR spectrum, 336
from oil of caraway, 335
Chlorine, 9
reaction with 4-bromoacetanilide, 503
reaction with 2, 4-dimethylpentane, 139
Chloroacetic acid, 291
Chlorobenzene:
from benzenediazonium chloride, 452
IR spectrum, 432
2-Chloro-4-bromoacetanilide:
from 4-bromoacetanilide, 503, 504
conversion to 2-chloro-4-bromoaniline, 504
2-Chloro-4-bromoacetanilide:
from 2-chloro-4-bromoacetanilide, 504
conversion to 2-chloro-4-bromo-6-iodoaniline, 506
IR spectrum, 505
NMR spectrum, 505
1-Chloro-4-bromobenzene, NMR spectrum, 232
2-Chloro-4-bromo-6-iodoaniline:
from 2-chloro-4-bromoaniline, 506
conversion to 1-bromo-3-chloro-5-iodobenzene, 507
IR spectrum, 506
Chloroethane, NMR spectrum, 226

Chloroform, 9
 IR spectrum, 245
 properties, 49
p-Chlorophenacyl bromide, 8, 283
Chlorosulfonic acid, 9
 reaction with acetanilide, 495
o-Chlorotoluene:
 IR spectrum, 431
 from *o*-toluidine, 431
p-Chlorotoluene:
 IR spectrum, 433
 from *p*-toluenediazonium chloride, 433
Δ^4-Cholestene-3-one:
 from Δ^5-cholestene-3-one, 486
 IR spectrum, 487
 synthesis from cholesterol, 476, 480
Δ^5-Cholestene-3-one:
 conversion to Δ^4-cholestene-3-one, 486
 from $5\alpha,6\beta$-dibromocholestane-3-one, 483
 IR spectrum, 484
Cholesterol:
 conversion to cholesterol dibromide, 482
 conversion to cholesteryl benzoate, 390
 IR spectrum, 322
 isolation from gallstones, 321
Cholesterol dibromide:
 from cholesterol, 482
 conversion to $5\alpha,6\beta$-dibromocholestane-3-one, 475
Cholesteryl benzoate:
 from cholesterol, 390
 IR spectrum, 391
 as liquid crystal, 389, 391
 as solvent for asymmetric synthesis, 389
Cinnamaldehyde, conversion to *trans,trans*-1,4-diphenylbutadiene, 451
Chromic anhydride, 256
Clove oil:
 isolation from cloves, 328
 isolation of eugenol from, 328
Cloves, isolation of oil of cloves from, 328
Coconut aldehyde (*see* γ-Nonanolactone)
Cumene, 12
Cyclohexane, properties, 49
Cyclohexanol:
 conversion to cyclohexene, 352
 conversion to cyclohexyl bromide, 358
 from cyclohexanone, 369
 IR spectrum, 361
 oxidation to cyclohexanone, 366
Cyclohexanone:
 from cyclohexanol, 366
 IR spectrum, 367
 reduction to cyclohexanol, 369
Cyclohexene:
 conversion to cyclohexanol, 360
 from cyclohexanol, 352
 IR spectrum, 354
 reaction with dichlorocarbene, 363
Cyclohexyl bromide:
 from cyclohexanol, 358
 IR spectrum, 359
1,5-Cyclooctadiene, reaction with dichlorocarbene, 363
Cyclopentadiene:
 from dicyclopentadiene, 440
 reaction with maleic anhydride, 440

Decalin, 12
Deuterocholoroform, as NMR solvent, 234
Dianthraquinone:
 from 9-bromoanthrone, 455
 thermochromism of, 452, 455
cis-1,2-Dibenzoylethylene:
 from *trans*-1,2-dibenzoylethylene, 350
 IR spectrum, 350
 NMR spectrum, 351
trans-1,2-Dibenzoylethylene:
 IR spectrum, 349
 isomerization of, 348, 350
 NMR spectrum, 349
 recrystallization of, 348
Dibenzylketone:
 IR spectrum, 490
 from phenylacetic acid, 490
 reaction with benzil, 491
$5\alpha,6\beta$-Dibromocholestane-3-one:
 from cholesterol dibromide, 483
 conversion to 6β-bromocholest-4-ene-3-one, 486
 conversion to Δ^5-cholestene-3-one, 483
o-Dichlorobenzene, NMR spectrum, 232
2,2-Dichlorobicyclo[4.1.0]heptane (*see* Dichloronorcarane)
Dichlorocarbene:
 addition to cyclohexene, 272
 addition to 1,5-cyclooctadiene, 365
 addition to styrene, 365
1,1-Dichloroethane, NMR spectrum, 225
1,2-Dichloroethane, NMR spectrum, 226
Dichloromethane, 10
 properties, 49
Dichloronorcarane:
 from cyclohexene, 362
 IR spectrum, 364
Dicyclohexylcarbodiimide, 8
Dicyclopentadiene, conversion to cyclopentadiene, 440
1,2-Diethoxyethane, NMR spectrum, 234
Diethylamine, reaction with *m*-toluyl chloride, 420
Diethyl benzylphosphonate, 450
Diethyl ether (*see* Ether)
Diethylethoxyazobenzene:
 conversion to *p*-phenetidine, 499
 from ethyldioxyazobenzene, 498
Diethyl malonate, reaction with mesityl oxide, 446
N,N-Diethyl-*m*-toluamide:
 IR spectrum, 419
 preparation from *m*-toluic acid, 419
Diimide, 499
Diisopropyl ketone, NMR spectrum, 181
Dimedon, 272
 from diethyl malonate and mesityl oxide, 446
Dimethylallene, NMR spectrum, 236
N,N-Dimethylaniline, reaction with nitrobenzenediazonium sulfate, 436
3,3-Dimethyl-2-butanol, dehydration of, 355
5,5-Dimethyl-1,3-cyclohexanedione (*see* Dimedon)
Dimethyldihydroresorcinol (*see* Dimedon)
Dimethylformamide:
 properties, 49
 as reaction solvent, 387, 450

2,4-Dimethylpentane, chlorination of, 139
2,4-Dimethyl-3-pentanone, NMR spectrum, 236
1,4-Dimethylpyridinium iodide:
conversion to 4-(*p*-hydroxystyryl)-1-methylpyridinium iodide, 510
from 4-methylpyridine, 510
Dimethyl sulfate, 6, 10
Dimethylsulfoxide, 10
properties, 49
as reaction solvent, 461, 470
2,4-Dinitroaniline, from 2,4-dinitrobromobenzene, 422
3,5-Dinitrobenzoic acid, 285
3,5-Dinitrobenzoyl chloride, 271
2-(2,4-Dinitrobenzyl)pyridine:
from 2-benzylpyridine, 458
photochromism of, 458
2,4-Dinitrobromobenzene:
from bromobenzene, 424
conversion to 2,4-dinitroaniline, 426
conversion to 2,4-dinitrodiphenylamine, 427
conversion to 2,4-dinitrophenylhydrazine, 427
conversion to 2,4-dinitrophenylpiperidine, 428
conversion to 4′-substituted 2,4-dinitrophenylanilines, 428
IR spectrum, 426
NMR spectrum, 426
nucleophilic substitution reactions of, 426
2,4-Dinitrodiphenylamine, from 2,4-dinitrobromobenzene, 426
2,4-Dinitrophenylhydrazine, 8, 256, 273
from 2,4-dinitrobromobenzene, 427
2,4-Dinitrophenylpiperidine, from 2,4-dinitrobromobenzene, 428
Diosgenin, 477
Dioxane, 10
properties, 49
trans,trans-1,4-Diphenylbutadiene:
IR spectrum, 452
preparation by Wittig reaction, 451
(7R,8S)-(+)-Disparlure, as pheromone, 408
Dixanthylene:
thermochromism of, 452, 454
from xanthone, 453
DMF (*see* Dimethylformamide)
DMSO (*see* Dimethylsulfoxide)
Drierite, as drying agent, 145–149

Estradiol, 476
Ethanol, properties, 49
Ether, 4, 10
peroxides in, 10
properties, 49
removal of peroxides from, 10
solubility in, 191
Ether, diethyl (*see* Ether)
p-Ethoxyacetanilide (phenacetin):
from *p*-acetamidophenol, 457
from *p*-aminophenol, 455
from APC tablets, 119
IR spectrum, 419
NMR spectrum, 418
from *p*-phenetidine, 417
p-Ethoxyaniline (*see* *p*-Phenetidine)
Ethyl acetate, properties, 49
Ethyl bromide:
reaction with *p*-acetamidophenol, 457
reaction with ethyldioxyazobenzene, 498
Ethyl chloride, NMR spectrum, 226
Ethyldioxyazobenzene:
conversion to diethyldioxyazobenzene, 498
from *p*-phenetidine, 497
Ethyl ether (*see* Ether)
Ethyl iodide, NMR spectrum, 221, 228
Eugenol:
from clove oil, 331
IR spectrum, 331

Ferric chloride, 257
Folic acid, 493
Furan, reaction with maleic anhydride, 441

Gallstones, 321
D-Glucose, conversion to α- or β-D glucose pentaacetate, 397
α-D-Glucose pentaacetate, from D-glucose, 399, 400
β-D-Glucose pentaacetate, from D-glucose, 399

Heptaldehyde, conversion to 3-nonenoic acid, 466
Hexachlorobutadiene, IR spectrum, 207
Hexachloroethane, sublimation of, 105
10E, 12Z-Hexadecadien-1-ol, as pheromone, 406
Hexane, properties, 49
Hydrazine, 11
reaction with diethyldioxyazobenzene, 499
reaction with 2,4-dinitrobromobenzene, 429
reaction with 3-nitrophthalic acid, 460
Hydriodic acid, 11
Hydrobromic acid, 11
Hydrochloric acid, 11
solubility in, 192
Hydrogen bromide gas, 11
Hydrogen chloride gas, 11
Hydrogen peroxide, 11
Hydrogen sulfide gas, 11
p-Hydroxybenzaldehyde, conversion to 4-(*p*-hydroxystyryl)-1-methylpyridinium iodide, 510
Hydroxylamine, 274
4-(*p*-Hydroxystyryl)-1-methylpyridinium iodide:
conversion to MOED, 513
preparation of, 510

Iodine monochloride, reaction with 2-chloro-4-bromoaniline, 506
Iodoethane, NMR spectrum, 221, 228
Iodomethane (*see* Methyl iodide)
Isoamyl acetate:
IR spectrum, 385
NMR spectrum, 386
as pheromone, 384
preparation from isoamyl alcohol, 385
preparation from isoamyl bromide, 387
Isoamyl alcohol:
conversion to acetate ester, 385
conversion to isoamyl bromide, 373

Isoamyl bromide:
  conversion to isoamyl acetate, 387
  preparation from isoamyl alcohol, 373
Isocyanates, 8
Isopentyl (*see* Isoamyl)
p-Isopropylbenzaldehyde, NMR spectrum, 237
Isopropyl ether, 10

Lactose, from powdered milk, 323
Ligroin, properties, 49
(R)-(+)-Limonene:
  IR spectrum, 332
  isolation from grapefruit peel, 332
  isolation from orange peel, 332
  as pheromone, 406
Lithium aluminum hydride, 11
Lithium metal, 11
Luciferin, 459
Luminol:
  chemiluminescence of, 459, 461
  from 3-nitrophthalhydrazide, 460

Magnesium sulfate, anhydrous, as drying agent, 147, 148
Malachite green, reduction by 1-benzyldihydronicotinamide, 470, 472
Maleic anhydride:
  reaction with butadiene, 437
  reaction with cyclopentadiene, 438
  reaction with furan, 441
Malonic acid, conversion to 3-nonenoic acid, 466
Merocyanine dye (*see* MOED)
Mesityl oxide, reaction with diethyl malonate, 446
Methanol, properties, 49
Methone (*see* Dimedon)
p-Methoxybenzene,
  reaction with 2,4-dinitrobromobenzene, 429
Methyl benzoate:
  from benzoic acid, 392
  conversion to triphenylmethanol, 412
  IR spectrum, 203, 393
  nitration of, 422
  NMR spectrum, 393
(R),(S)-α-Methylbenzylamine:
  IR spectrum, 341
  resolution of, 339
2-Methylcyclohexanol, dehydration of, 354
Methylene chloride (*see* Dichloromethane)
Methyl heptanoate, IR spectrum, 202
4-Methyl-3-heptanol:
  IR spectrum, 409
  oxidation to 4-methyl-3-heptanol, 408
  as pheromone, 408
  preparation of, 408
4-Methyl-3-heptanone:
  IR spectrum, 409
  as pheromone, 408
  preparation of, 408
Methyl *p*-hydroxybenzoate, mass spectrum, 243
N-Methylimidazole, 400
Methyl iodide:
  NMR spectrum, 221
  reaction with 4-methylpyridine, 510
  reaction with tertiary amines, 281
Methyl *m*-nitrobenzoate:
  IR spectrum, 423
  from methyl benzoate, 422
2-Methyl-2-pentenal:
  IR spectrum, 443
  NMR spectrum, 444
  from propionaldehyde, 443
  UV spectra, 444, 445
4-Methylpyridine, reaction with methyl iodide, 510
Methyl salicylate:
  IR spectrum, 394
  NMR spectrum, 395
  from salicylic acid, 395
Methyl *p*-toluenesulfonate, 281
Methyl trimethylacetate, NMR spectrum, 234
Milk, powdered, 323
Mineral oil:
  IR spectrum, 207
  mull, 206
MOED (merocyanine dye):
  preparation, 511
  variable color, 509, 511

NAD$^+$, 469
NADH, 469
β-Naphthol:
  reaction with benzenediazonium chloride, 435
  reaction with *p*-nitrobenzenediazonium sulfate, 436
α-Naphthylamine, 8
β-Naphthylamine, 8
α-Naphthylisocyanate, 272, 290
α-Naphthylisothiocyanate, 279
Nicotinamide, reaction with benzyl chloride, 470
Nitric acid, 12
  fuming, 11
m-Nitroaniline, recrystallization of, 61
p-Nitroaniline:
  conversion to *p*-nitrobenzenediazonium sulfate, 431
  recrystallization, 61
Nitrobenzene, 6, 12
  reduction to aniline, 414
p-(4-Nitrobenzeneazo)-dimethylaniline, 433
p-(4-Nitrobenzeneazo)-phenol, 436
p-Nitrobenzenediazonium sulfate:
  from *p*-nitroaniline, 431
  reaction with N,N-dimethylaniline, 436
  reaction with β-naphthol, 436
  reaction with phenol, 435
p-Nitrobenzyl bromide, 284
p-Nitrobenzyl chloride, 271, 284
1-(Nitrophenylazo)-2-naphthol, 436
3-Nitrophthalhydrazide:
  conversion to luminol, 460
  from 3-nitrophthalic acid, 460
3-Nitrophthalic acid, conversion to 3-nitrophthalhydrazide, 460
3-Nitrophthalic anhydride, 271
3-Nonenoic acid:
  conversion to coconut aldehyde, 467
  from heptaldehyde and malonic acid, 466
γ-Nonanolactone:
  from 3-nonenoic acid, 467
  IR spectrum, 466
  synthesis of, 465
endo-Norbornene-2,3-dicarboxylic acid, recrystallization of, 61
endo-Norbornene-2,3-dicarboxylic acid anhydride:

endo-Norbornene-2,3-dicarboxylic acid anhydride (*cont.*)
hydrolysis and crystallization, 61
preparation of, 440
Nujol (mineral oil):
IR spectrum, 207
mull, 206

"Off" (*see* N,N-Diethyl-*m*-toluamide)
Oleum (fuming sulfuric acid), 12

Para red, preparation of, 436
1,1,2,3,3-Pentachloropropane, NMR spectrum, 225
Pentane, 8
properties, 49
Pepper, 328
Peracids, 12
Perchloric acid, 12
Peroxides, 10, 12
removal of, 10
test for, 10
Petroleum ether, 8
properties, 49
(S)-(+)-α-Phellandrene, 335
(S)-(+)-β-Phellandrene, 335
Phenacetin (*p*-ethoxyacetanilide):
from *p*-acetamidophenol, 457
from *p*-aminophenol, 455
from APC tablets, 119
IR spectrum, 419
NMR spectrum, 418
from *p*-phenetidine, 417
Phenacyl bromide, 8, 238
p-Phenetidine:
acetylation of, 417
conversion to ethyldioxyazobenzene, 498
conversion to *p*-phenetidine, 497
diazotization of, 497
from *p*-phenetidine, 497
reaction with 2,4-dinitrobromobenzene, 429
Phenol, 6, 12
ionization constant of, 114
reaction with *p*-ethoxybenzenediazonium chloride, 498
reaction with *p*-nitrobenzenediazonium sulfate, 436
Phenylacetic acid:
conversion to dibenzylketone, 490
as pheromone, 406
1-Phenylazo-2-naphthol, preparation of, 435
(R),(S)-α-Phenylethylamine:
IR spectrum, 341
resolution of, 340
Phenyl ethyl malonic acid, asymmetric decarboxylation of, 389
Phenylhydrazine, 6, 12
Phenylisocyanate, 272, 290
Phenylisothiocyanate, 279
Phenylmagnesium bromide:
preparation of, 411
reaction with benzophenone, 413
reaction with diethyl carbonate, 413
reaction with dimethyl carbonate, 413
reaction with methyl benzoate, 412
synthesis of triphenylmethanol from, 411
p-Phenylphenacyl bromide, 8, 238
Phosphoric acid, 12
solubility in, 193
Phthalic anhydride, 288
Picric acid, 8, 280, 287
ionization constant of, 114
Piperidine, reaction with 2,4-dinitrobenzene, 428
Piperine, from pepper, 328
Potassium carbonate, anhydrous as drying agent, 147, 148
Potassium cyanide, 12
as catalyst in benzoin condensation, 457, 489
Potassium hydroxide, 12
Potassium metal, 12
Potassium permanganate, 289
Powdered milk, isolation of lactose from, 323
Pregnenolone:
conversion to progesterone, 481
formula, 481
Pregnenolone acetate:
formula, 209, 478
IR spectrum, 209
Progesterone:
formula, 209, 476
IR spectrum, 209
from pregnenolone, 478
Prontosil, 493
Propionaldehyde, self-condensation of, 443
Pyridine, 38

Raney nickel, 499, 501

Salicylic acid:
conversion to acetylsalicylic acid, 395
conversion to aspirin, 395
conversion to methyl salicylate, 394
Semicarbazide, 273
Sodium amide, 12
Sodium bicarbonate, solubility in, 192
Sodium borohydride, 11
reduction of ketones by, 369
Sodium cyanide, 12
Sodium dichromate, 367
Sodium dithionite, reduction of 1-benzylnicotinamide chloride by, 460
Sodium hydride, 11, 12
Sodium hydroxide, 13
solubility in, 191
Sodium hypochlorite, 368
Sodium metal, 13
Sodium sulfate, anhydrous, as drying agent, 147
trans-Stilbene:
IR spectrum, 449
preparation by Wittig reaction, 450
4′-Substituted 2,4-dinitrophenylanilines, 428
Sudan I, preparation of, 435
Sulfanilamide:
from *p*-acetamidobenzenesulfonamide, 496
as antibiotic, 492
synthesis from benzene, 492
3-Sulfolene, reaction with maleic anhydride, 437, 438
Sulfuric acid, 13
fuming, 11
solubility in, 193

Testosterone:
from androstenedione, 479
formula, 208, 476
IR spectrum, 209
properties, 416
Tetra-*n*-butylammonium bromide, as phase-transfer catalyst, 362

Tetrachloroethane, 8, 13
properties, 49
(E)-11-Tetradecenyl acetate, as pheromone, 406
(Z)-11-Tetradecenyl acetate, as pheromone, 406
endo-Tetrahydrodicyclopentadiene, 346
Tetrahydrofuran (THF), 10
properties, 49
Tetralin, 12
Tetramethylsilane (TMS), 220, 234
Tetraphenylcyclopentadienone, from benzaldehyde and phenylacetic acid, 488
THF (*see* Tetrahydrofuran)
Thiamine pyrophosphate, as catalyst in benzoin condensation, 462
Thionyl chloride, 13, 38
reaction with *m*-toluic acid, 419
use in preparation of acid chlorides, 282
Thiourea, 287
inclusion complex with adamantane, 346
in preparation of S-alkylthiuronium picrates, 287
TMS (*see* Tetramethylsilane)
Toluene, properties, 49
p-Toluenediazonium chloride:
conversion to *p*-chlorotoluene, 433
from *p*-toluidine, 430
p-Toluenesulfonyl chloride, 278
m-Toluic acid, conversion to N,N-diethyl-*m*-toluamide ("Off"), 419
o-Toluidine, conversion to *o*-toluenediazonium chloride, 434
p-Toluidine:
conversion to *p*-toluenediazonium chloride, 430
reaction with acyl halides, 282
reaction with anhydrides, 282
reaction with 2,4-dinitrobromobenzene, 429
p-Tolylisocyanate, 290
TPP (*see* Thiamine pyrophosphate)
Trichloroacetic acid, ionization constant of, 114
1,2,3-Trichlorobenzene, NMR spectrum, 231
1,1,1-Trichloroethane, NMR spectrum, 226
1,1,2-Trichloroethane, NMR spectrum, 221
Triethylphosphite:
reaction with benzyl chloride, 450
use in Wittig reaction, 450
1,3,5-Trinitrobenzene, 8
2,4,7-Trinitrofluorenone, 290
2,4,6-Trinitrophenol (*see* Picric acid)
2,4,6-Trinitrotoluene, 8
Triphenylmethanol:
preparation from benzophenone, 413
preparation from diethyl carbonate, 413
preparation from dimethyl carbonate, 413
preparation by Grignard synthesis, 410
preparation from methyl benzoate, 411

Undecanal, as pheromone, 406

Valeric acid, as pheromone, 406
Vanillin, reduction to vanillyl alcohol, 371
Vanillyl alcohol, from vanillin, 371

Water:
density (table), 174
in IR samples, 204, 206
properties, 49
removal of, 142

Xanthone, reductive dimerization to dixanthylene, 418
Xanthydrol, 276
p-Xylene, NMR spectrum, 220

Zinc dust, debromination with, 483

General Subject Index

Abbe refractometer, 176
Absorbance, 197
Acetamides, substituted (derivatives):
  melting points, 515
  preparation, 277
Acetylation:
  of *p*-aminophenol, 456
  of aniline, 416
  of D-glucose, 394, 400
  of *p*-phenetidine, 417
  of salicylic acid, 395
Acid halides, 9
Acids, 9
  isolation of, 314
  molecular weights, 37
  solutions of, 37
9-Acylamidoxanthenes (derivatives):
  melting point, 514
  preparation, 276
Adapters, 17
Addition of reagents, 304
  gases, 306
  liquids, 306
  solids, 304
  solutions, 306
Alcohol derivatives:
  benzoates, 271
  3,5-dinitrobenzoates, 271
  hydrogen 3-nitrophthalates, 271
  melting points, 513
  α-naphthylurethans, 292
  p-nitrobenzoates, 271
  phenylurethans, 272
Alcohols:
  conversion to alkenes, 351
  conversion to alkyl halides, 358
Aldehyde derivatives:
  2,4-dinitrophenylhydrazones, 273
  melting points, 514
  methone derivative, 272
  oximes, 274
  semicarbazones, 273
Aldol condensations, 441
  benzil with dibenzylketone, 491
  mesityl oxide with diethylmalonate, 445
  propionaldehyde with propionaldehyde, 443
Aldrich Catalog Handbook of Fine Chemicals, 14
Alkenes from alcohols, 351
Alkyl halides from alcohols, 372
  tert-amyl chloride, 375
  n-butyl bromide, 372
  tert-butyl chloride, 373
  competitive substitution, 375
  cyclohexyl bromide, 358
  isoamyl bromide, 373
S-Alkylthiuronium picrates (derivative):
  melting points, 520
  preparation, 287
American flag red, 436
Amide derivatives:
  9-acylamidoxanthenes, 276
  hydrolysis, 274, 276
  melting points, 514
Amides, 415
  p-acetamidophenol from *p*-aminophenol, 456
  acetanilide from aniline, 416
  N,N-diethyl-*m*-toluamide ("Off"), 419
  p-ethoxyacetanilide from *p*-phenetidine, 305
Amine derivatives, primary and secondary amines:
  α-naphthylthioureas, 279
  melting points, 515
  phenylthioureas, 279
  N-substituted acetamides, 277
  N-substituted benzamides, 278
  N-substituted *p*-toluenesulfonamides, 278
Amine derivatives, tertiary amines:
  methiodide, 281
  melting points, 516
  picrate, 280
  p-toluenesulfonate, 281
Ammonia (test), 253
Anilides (derivatives):
  from alkyl halides, 290
  from carboxylic acid anhydrides, 282
  from carboxylic acid halides, 282
  from carboxylic acids, 282
  melting points, 517
APC tablets, 117
Aqueous workup, 311
  flow diagrams, 312, 314, 315
Aromatic acids by oxidation, 289

o-Aroylbenzoic acids (derivatives):
  melting points, 522
  preparation, 288
Aryloxyacetic acids (derivatives):
  melting points, 524
  preparation, 291
Aspirator, water, 91
Aspirin tablets, 325
Asymmetric synthesis, 286, 473
Atmosphere, inert, 310
Azeotropic mixtures, 74
  maximum boiling point (table), 79
  minimum boiling point (table), 80

Baeyer test (unsaturation), 260
Bases:
  isolation of, 315
  molecular weights, 37
  solutions of, 37
Beaker, 16
Beilsteins Handbuch der Organischen Chemie, 30
Beilstein test (halogen), 247
Benzamides (derivatives):
  melting points, 515
  preparation, 278
Benzenesulfonyl chloride test (amines), 254
Benzoates (derivatives), 271
N-Benzylamides (derivatives):
  melting points, 518
  preparation, 285
Bioluminescence, 459
Bleeding, control of, 7
Boiling flask, 15
Boiling point, 66, 153
  and degree of branching, 156
  by distillation, 152
  experimental determination, 152
  and molecular structure, 154
  and molecular weight, 154
  and nature of functional group, 154
  nomograph, 87
  normal, 66
  small-scale determination, 153
Boiling point diagram:
  benzene/methanol, 77
  benzene/toluene, 69
  chloroform/acetone, 78
  water/methanol, 75
Boiling points of solvents (table), 49
Boiling stones, 50, 81, 93
"Bootstrap synthesis," 497
Borohydride reduction:
  of cyclohexanone, 369
  of 4-*tert*-butylcyclohexanone, 370
  of vanillin, 371
Borohydrides, 11
Bromide ion (test), 251
Bromination:
  of acetanilide, 423
  of anthrone, 454
  of aromatic ethers, 287
  of cholesterol, 482
  of phenols, 291
Bromine in carbon tetrachloride test (unsaturation), 255
p-Bromoanilides (derivatives):
  from carboxylic acid anhydrides, 282
  from carboxylic acid halides, 282
  from carboxylic acids, 282
p-Bromophenacyl esters (derivatives):
  melting points, 517
  preparation, 217
Bumping, 81, 93
Burns, 5

Calibration of thermometer, 164
Carbene reactions, 363, 365
Carboxylic acid anhydride derivatives:
  anilides, 282
  p-bromoanilides, 282
  melting points, 517
  p-toluidides, 282
Carboxylic acid derivatives:
  amide, 281
  anilide, 282
  p-bromoanilide, 282
  melting points, 517
  p-nitrobenzyl ester, 283
  phenacyl ester, 283
  p-toluidides, 282
Carboxylic acid ester derivatives:
  N-benzylamide, 285
  3,5-dinitrobenzoate, 285
  hydrolysis, 274
  melting points, 518
Carboxylic acid halide derivatives:
  anilides, 282
  p-bromoanilides, 282
  melting points, 517
  p-toluidides, 282
Carcinogens, 8
Chemical Abstracts, 27
Chemical literature, 26
Chemical shift, 218
  and molecular structure (table), 229
Chemiluminescence, 459
Chloride ion (test), 251
Chlorination:
  of 4-bromoacetanilide, 503, 504
  of 2,4-dimethylpentane, 139
p-Chlorophenacyl esters (derivatives), 283
Chromatography, 120
  adsorbents (table), 121
  column (*see* Column chromatography)
  eluting solvents (table), 122
  high pressure liquid, 141
  paper (*see* Paper chromatography)
  thin-layer (*see* Thin-layer chromatography)
  vapor-phase (*see* Vapor-phase chromatography)
Chromic acid oxidation of cyclohexanol, 367
Chromic anhydride test (alcohols), 256
Classification tests (table), 253
Cleaning up, 19
Clove oil from cloves, 328
Cloves, 328
Cold trap, 91, 92
Collections of spectra, 35
Color and molecular structure, 214
Column chromatography:
  adsorbents (table), 121
  eluting solvents (table), 121
  exercises, 126
  of oil of caraway, 338
  of oil of spearmint, 338
  of technical anthracene, 126
  technique, 122
  theory, 120
Competitive nucleophilic substitution, 375
Concentration, 307
  by distillation, 309
  by evaporation, 309
  by use of rotary evaporator, 309, 310

Concentrations:
- of acids, 37
- of bases, 38

Condenser, 15

Cooling, 301
- with a Dry Ice bath, 302
- efficient, 302
- with an ice bath, 301

Cooling curves, 163, 178

Coupling constants, 221
- and molecular structure (table), 230

Coupling reactions of diazonium salts, 434
- benzenediazonium chloride and β-naphthol, 435
- *p*-nitrobenzenediazonium sulfate and dimethylaniline, 436
- *p*-nitrobenzenediazonium sulfate and β-naphthol, 435
- *p*-nitrobenzenediazonium sulfate and phenol, 432

Cow, 89

Crystallization (*see also* Recrystallization):
- of acetylsalicylic acid, 325
- of caffeine, 326
- of cholesterol, 321
- of lactose, 323

Cuts, 6

Cyanide ion (test), 248

Cylinder, graduated, 19

Dangerous compounds (list), 8

Dangerous Properties of Industrial Materials, 13

Decarboxylation of β-keto-acid, 445, 466

Decolorizing, 50, 141

Decolorizing carbon, 50

Dehydration:
- of cyclohexanol, 350
- of 3,3-dimethyl-2-butanol, 357
- of 2-methylcyclohexanol, 357

Density, 173
- experimental determination of, 173
- and molecular structure, 175
- of selected reagents (table), 38
- of solvents (table), 49

Derivatives:
- as a means of identification of substances, 264
- melting points (Table A.1 through Table A.14), 513–524
- procedures, summary and index (table), 269

Desiccants for desiccators (table), 145

Desiccator, vacuum, 144

Detector:
- flame ionization, 134
- thermal conductivity, 134

Diazonium salts, preparation:
- from aniline, 430
- from 2-chloro-4-bromo-6-iodoaniline, 507
- from *p*-nitroaniline, 431
- from *p*-phenetidine, 497
- from *o*-toluidine, 434
- from *p*-toluidine, 430

Dichlorocarbene, 361

Diels-Alder reaction, 437
- butadiene and maleic anhydride, cyclopentadiene and maleic anhydride, 438
- furan and maleic anhydride, 441
- 3-sulfolene and maleic anhydride, 437

Diimide reduction of diethyldioxyazobenzene, 499

Dimedon derivatives:
- melting points, 514
- procedures, 272

3,5-Dinitrobenzoates (derivatives):
- from alcohols, 271
- from esters, 285
- melting points, 513, 518

2,4-Dinitrophenylhydrazine test (aldehydes and ketones), 256

2,4-Dinitrophenylhydrazones (derivatives):
- melting points, 514, 523
- preparation, 273

Disposal of waste, 21

Distillation, 62
- apparatus, 65, 82, 83
- of azeotropic mixtures, 85
- boiling stones, 81
- bumping, 81
- exercises, 85
- flask, 19
- fractional, 71
- fractionating columns, 73
- holdup, 72
- of methanol, 85
- of miscible pair of liquids, 66
- of oil of caraway, 303
- of oil of spearmint, 303
- of a pure liquid, 66
- rate, 81
- at reduced pressure (*see* Reduced pressure distillation)
- reflux ratio, 73
- steam (*see* Steam distillation)
- superheating, 81
- technique, 80
- theoretical plates, 72
- of water, 85

Dry Ice bath, 302

Drying, 142
- by azeotropic distillation, 148
- of crystals, 55
- of gases, 149
- of glassware, 21
- of liquids, 148
- of small samples, 143
- of solids, 142
- of solutions, 146
- of solvents, 148
- tube, 19

Drying agents:
- for gases (table), 149
- for solutions in organic solvents (table), 147
- for solvents (table), 148

Drying oven, 143

Drying pistol, 145

Electromagnetic spectrum (table), 196

Electrophilic aromatic substitution, 421, 431
- *p*-bromoacetanilide from acetanilide, 423
- chlorosulfonation of acetanilide, 495
- by diazonium ions, 434, 497
- 2-(2,4-dinitrobenzyl)pyridine from 2-benzylpyridine, 458
- 2,4-dinitrobromobenzene from bromobenzene, 424
- methyl *m*-nitrobenzoate from methyl benzoate, 422

Elimination reactions:
- butadiene from 3-sulfolene, 437
- cyclohexene from cyclohexanol, 352
- cyclopentadiene from dicyclopentadiene, 438

Elimination reactions: (*cont.*)
  debromination of 5α,6β-dibromocholestane-3-one, 483
  dehydration of 3,3-dimethyl-2-butanol, 355
  dehydration of 2-methylcyclohexanol, 356
  dehydrobromination of 5α,6β-dibromocholestane-3-one, 486
Eluting solvents (table), 122
Emulsions, 112
Enthalpy:
  of fusion, 170
  of mixing, 184
  of vaporization, 64, 157
Entropy:
  of fusion, 170
  of mixing, 184
  of vaporization, 69, 159
Erlenmeyer flask, 16
Esters, 384, 388
  acetylsalicylic acid from salicylic acid, 395
  cholesterol benzoate from cholesterol, 389
  "coconut aldehyde," 465
  derivatives (*see* Carboxylic acid ester derivatives)
  α-D-glucose pentaacetate from D-glucose, 397, 400
  β-D-glucose pentaacetate from D-glucose, 400
  isoamyl acetate from isoamyl alcohol, 385
  isoamyl acetate from isoamyl bromide, 387
  methyl benzoate from benzoic acid, 392
  methyl salicylate from salicylic acid, 394
  γ-nonanolactone, 465
Ethers, 10
Ethers, aromatic (derivatives):
  bromination product, 287
  melting points, 518
  picrate, 280
Evaporation (*see* Concentration)
Exothermic reactions, control of, 296
Explosion hazards, 5, 8
Extinction coefficient, 197
Extraction, 105
  of acids and bases, 115, 311–315
  of caffeine from NoDoz, 325
  of caffeine from tea, 325
  of eugenol from clove oil, 331
  of mixture of aspirin, phenacetin, and caffeine, 117
Extraction of solvents, 105
  emulsions, 112
  exercises, 115
  technique, 103
  theory, 105
Eyes:
  chemicals in, 7
  protection of, 7

Ferric chloride test (phenols), 257
Ferric hydroxamate tests, 257, 258
Filter paper, how to fold, 43
Filtration, 43
  cold, 54
  flask, 18
  gravity, 44
  hot, 51
  suction, 49
  vacuum, 44
Fire extinguisher, 4
Fire hazards, 4
First-order spectrum (NMR)(*see* N + 1 Rule)
Fisher-Davidson gravitometer, 174
Flame ionization detector, 134
Flasks:
  boiling, 15
  distillation, 17
  Erlenmeyer, 16
  suction filtration, 18, 45
Flavors and fragrances:
  banana oil, 385
  (R)-(−)-carvone, 334
  (S)-(+)-carvone, 334
  "coconut aldehyde," 465
  eugenol, 331
  isoamyl acetate, 385
  (R)-(+)-limonene, 332
  methyl salicylate, 394
  γ-nonanolactone, 465
  oil of caraway, 334
  oil of cloves, 331
  oil of spearmint, 334
  oil of wintergreen, 394
  pear oil, 385
Flow diagrams:
  recrystallization, 46
  isolation of an acidic substance, 314
  isolation of a basic substance, 315
  isolation of a neutral substance, 313
Fractional distillation, 71
Fractionating column, 72, 73
Fraction collectors, 89
Fragrances (*see* Flavors and fragrances)
Free energy:
  of activation, 294
  of fusion, 170
  of mixing, 184
Functional group tests:
  as a means of identification of substances, 251
  summary and index (table), 253
Funnel:
  Buchner, 18
  Hirsch, 18
  separatory, 17, 109
  stemless, 44, 51

Gallstones, 321
Gases:
  absorption of, 307
  addition to reaction mixture, 306
  control of noxious, 307
  drying of, 149
  measurement of, 306
Gas trap, 307
Glassware:
  cleaning, 19
  drying, 21
Gravitometer, 174
Gravity filtration, 30
Grease, 294
Grignard reaction, 401
  benzophenone and phenylmagnesium bromide, 413
  diethyl carbonate and phenylmagnesium bromide, 413
  dimethyl carbonate and phenylmagnesium bromide, 413
  methyl benzoate and phenylmagnesium bromide, 411
  preparation of aliphatic alcohols, 402
  preparation of 4-methyl-3-heptanol, 406
  preparation of triphenylmethanol, 411
  propionaldehyde and 2-pentylmagnesium bromide, 406

Ground-glass-jointed glassware, 14

Halide ion (test), 250
Halides, aliphatic, derivatives:
  S-alkylthiuronium picrate, 287
  anilide, 290
  melting points, 520
  α-naphthalide, 290
  p-toluidide, 290
Halides, aromatic, derivatives:
  anilide, 290
  o-aroylbenzoic acid, 288
  melting points, 521
  α-naphthalide, 290
  oxidation, 289
  p-toluidide, 290
Halogenated solvents, 11
Halogens (test), 247
Handbook of Laboratory Safety, 19
Handbooks, 29
Hazardous materials, 8
  properties of (table), 8–13
Heat:
  of fusion, 170
  of mixing, 184
  of vaporization, 64, 157
Heating, 297
  with a flame, 297
  with a heating mantle, 298
  with a heat lamp, 143
  with a hotplate, 299
  with a hot water bath, 300
  with an oil bath, 299
  with a steam bath, 300
Heating baths for melting-point determination, 166
Heating mantle, 229
High-pressure liquid chromatography, 141
Hinsberg's test (amines), 254
Hood, 307
Hotplate, 299
Hot water bath, 300
Hydrocarbons, aromatic, derivatives for:
  o-aroylbenzoic acid, 288
  melting points, 522
  oxidation, 289
  picrate, 280
  2,4,7-trinitrofluorenone adduct, 290
Hydrocarbons, preparation of:
  adamantane, 346
  trans,trans-1,4-diphenylbutadiene, 451
  trans-stilbene, 448
Hydrochloric acid/zinc chloride test (alcohols), 259
Hydrogen 3-nitrophthalates (derivatives):
  melting points, 513
  preparation, 271
Hydrolysis:
  of *p*-acetamidobenzenesulfonamide, 488
  of amides, 274, 276
  of 2-chloro-4-bromoacetanilide, 504
  of esters, 286
  of nitriles, 274, 276
  of *endo*-5-norbornene-2, 3-dicarboxylic anhydride, 61

Ice bath, 301
Ignition test, 246
Immersion heater, 294
Index of refraction, 175
  experimental determination, 175
  and molecular structure, 176
Inert atmosphere, 310
Infrared spectrometry (*see* IR spectra)
Insect attractants (*see* Pheromones)
Insolubility, 184
Integral (NMR), 225
Iodide ion (test), 250
Iodination of 2-chloro-4-bromoaniline, 506
Iodoform test (ketones), 260
IR spectra, 195 (*see also specific substances*)
  absorption-spectra correlations (table), 200–201
  interpretation, 198
  and molecular structure, 197
  sample preparation, 203
Isolation of the product, 311
  by aqueous workup, 311
  by distillation, 313
  by extraction, 313–315
  flow diagrams, 312, 314, 315
  by suction filtration, 311
Isomerizations:
  of Δ^5-cholestene-3-one to Δ^4-cholestene-3-one, 486
  of *trans*-1,2-dibenzoylethylene, 348
  of β-D-glucose pentaacetate to α-D-glucose pentaacetate, 399
  of *endo*-tetrahydrodicyclopentadiene to adamantane, 346

Ketone derivatives:
  2,4-dinitrophenylhydrazones, 273
  melting points, 523
  oximes, 274
  semicarbazones, 273

Laboratory notebook, 22
Lachrymators, 8
"Like dissolves like," 185
Liquid crystals, 389
Liquids, addition to reaction mixture, 306
Literature, chemical, 26
Lucas' test (alcohols), 259

Magnetic equivalence (NMR), 224
Magnetic nonequivalence (NMR), 233
Magnetic stirring, 303
Manometer, 92
Marker degradation, 477
Marker, Russell, 477
Mass spectrometry, 238
  determination of molecular weight, 240
  high resolution, 242
  interpretation of spectra, 240
  and molecular formula, 241
  and molecular structure, 242
  and molecular weight, 240
  parent peak, 240
  theory, 238
McLeod gauge, 91
Mechanical stirring, 302, 311
Melting point, 162
  capillary, 165
  from cooling curve, 167
  as criterion of purity, 167
  experimental determination, 163
  as means of identification, 169
  micro hot-stage, 167
  mixture, 169
  and molecular structure, 170
Melting points of derivatives (Table A.1 through Table A.14), 513-524

Melting point bath, 166
Melting range, 165 (*see also* Melting point)
Merck Index, 13
Merocyanine dye, 509
Metals (test), 246
Methiodides of tertiary amines (derivative):
  melting points, 516
  preparation, 281
Methones (derivatives):
  melting points, 514
  preparation, 272
Michael condensation, 445
Milk, powdered, 323
Mixture melting points, 169
Mixtures, azeotropic, 74
Molar absorptivity (*see* Extinction coefficient)
Molecular refraction, 177
Molecular refractivity, 177
Molecular weight, 182
  by freezing point depression, 183
  by mass spectrometry, 182, 240
  by titration, 183
Molecular weights:
  of acids, 37
  of bases, 37
  of selected reagents, 38
Multiplet (NMR), 222

N + 1 Rule (NMR), 223, 224
α-Naphthalides from alkyl halides (derivatives):
  melting points, 520
  preparation, 290
β-Naphthol:
  reaction with benzenediazonium chloride, 435
  reaction with *p*-nitrobenzenediazonium sulfate, 436
α-Naphthylthioureas (derivatives):
  melting points, 515
  preparation, 279
α-Naphthylurethans (derivatives):
  melting points, 513, 524
  preparation, 272
Natural products:
  black pepper, 328
  caraway oil, 334
  clove oil, 331
  cloves, 329
  gallstones, 321
  grapefruit peel, 332
  orange peel, 332
  pepper, 328
  powdered milk, 323
  spearmint oil, 334
  tea, 325
Nitration:
  of 2-benzylpyridine, 458
  of bromobenzene, 424
  of methyl benzoate, 422
Nitriles, hydrolysis of (derivative):
  melting points of acids, 523
  preparation, 274, 275
p-Nitrobenzoates (derivatives):
  melting points, 513
  preparation, 271
p-Nitrobenzyl esters (derivatives):
  melting points, 513
  preparation, 284
NMR spectra, 218 (*see also specific substances*)
  chemical shift, 218
  chemical shift correlations (table), 229
  coupling constant, 222
  coupling constant correlations (table), 230
  integral, 228
  interpretation, 230
  magnetic equivalence, 224
  magnetic nonequivalence, 233
  mechanism of energy absorption, 281
  and molecular structure, 228
  N + 1 Rule, 223, 234
  "pointing," 227
  sample preparation, 234
  shielding, 218
  splitting, 221
  tube, 234
Nomograph, for estimation of boiling point, 87
Notebook, laboratory, 22
  format of, 23
  sample page, 25
Nuclear magnetic resonance spectrometry (*see* NMR spectra)
Nucleophilic substitution:
  aliphatic, competitive, 375
  aromatic, 422
  of 2,4-dinitrobromobenzene, 422–425
Oil:
  of caraway, 334
  of cloves, 331
  of Niobe (methyl benzoate), 392
  of spearmint, 334
  of wintergreen, 394
Oil bath, 299
Oiling out, 56
Oil pump, 91
Optical activity, 177
  experimental determination, 178
  and molecular structure, 180
Optical rotation, 178
Optical rotatory dispersion, 181
Oxidation:
  of benzoin to benzil, 489
  of 1-benzyldihydronicotinamide, 492
  of cholesterol dibromide, 483
  of cyclohexanol to cyclohexanone, 366
  by permanganate, 289
Oximes (derivatives):
  melting points, 514, 523
  preparation, 274

Paper chromatography:
  analysis of carrots, 132
  analysis of food coloring, 132
  analysis of spinach leaves, 132
  exercises, 132
  separation of ink pigments, 132
  technique, 131
  theory, 130
Parent peak (MS), 240
Partial pressure, 64
Partial solubility, 185
Percent transmittance, 196
Periodic table, 39
Permanganate test (unsaturation), 260
Peroxides:
  hazards, 10
  removal, 10
  test for, 10
Phase transfer catalysis, 363
Phenacyl esters (derivatives):
  melting points, 517
  preparation, 283
Phenacyl halides, 8
Phenol derivatives:
  aryloxyacetic acid, 291
  bromination product, 291

Phenol derivatives: (*cont.*)
melting points, 524
α-naphthylurethan, 272
p-Phenylphenacyl esters (derivatives), 283
Phenylthioureas (derivatives):
melting points, 515
preparation, 279
Phenylurethans (derivatives):
melting points, 513
preparation, 272
Pheromones:
10E,12Z-hexadecadien-1-ol, 407
isoamyl acetate, 380, 385, 407
(R)-(+)-limonene, 332, 406
4-methyl-3-heptanol, 406
4-methyl-3-heptanone, 407
phenylacetic acid, 406
(E)-11-tetradecenylacetate, 407
(Z)-11-tetradecenylacetate, 407
undecanal, 406
valeric acid, 406
Photochemical reactions:
chlorination of 2,4-dimethylpentane, 139
isomerization of *trans*-1,2-dibenzoylethylene, 350
Photochromic compound, 458
Picrates (derivatives):
melting points, 516, 519, 522
preparation, 280
Poisoning hazards, 6
Polarimeter, 178
Polynuclear hydrocarbons, 8
Potassium permaganate test (unsaturation), 260

Qualitative characterization tests:
as means of identification of substances, 251
summary and index (table), 253
Qualitative organic analysis, 245
Qualitative test for elements, 246
Beilstein test, 247
for bromide, 251
for chloride, 251
for cyanide, 248
for halogens, 247, 250
ignition test, 246
for iodide, 250
for metals, 246
for nitrogen, 248
sodium fusion, 247
for sulfur, 249
Quaternary ammonium salts (derivatives):
melting points, 516
preparation, 281

R_f, 127, 131
Raney nickel reduction, 499
Raoult's law, 67
Reagents for Organic Synthesis, 13, 35
Recrystallization, 44
of acetanilide, 61
of acetylsalicylic acid, 325
of benzil, 61
of cholesterol, 321
cold filtration, 54
cooling the solution, 52
decolorizing the solution, 50
of *trans*-1,2-dibenzoylethylene, 346
dissolving the sample, 50
drying the crystals, 55
exercises, 61
failure to crystallize, 57
flow chart, 46
folding the filter paper, 43
hot filtration, 51
low-melting compounds, 58
of *m*-nitroaniline, 61
of *p*-nitroaniline, 61
of *endo*-norbornene-2,3-dicarboxylic acid, 61
oiling out, 56
second crops, 59
seeding, 48
small samples, 58
solvent pairs, 55
solvents for, 49
washing of crystals, 54
wet samples, 58
Reduced-pressure distillation, 86
apparatus, 87, 88
estimation of boiling point, 86
exercises, 94
of oil of caraway, 336
of oil of spearmint, 336
pressure measurement, 91
source of vacuum, 91
technique, 92
Reduction:
of 1-benzylnicotinamide chloride, 472
of 4-*tert*-butylcyclohexanone, 370
of 2-chloro-4-bromo-6-iodoaniline, 507
of cyclohexanone, 369
of diethyldioxyazobenzene, 499
of ketones to alcohols, 369
of Malachite green, 472
of nitrobenzene to aniline, 414
of vanillin, 371
Reference liquids for boiling point determination (table), 155
Reference solids for thermometer calibration (table), 164
Refractometer, 176
Registry of Toxic Effects of Chemical Substances, 14
Resolution of (R),(S)-α-phenylethylamine, 328
Rotary evaporator, 309, 310

Safety, 3
chemicals in the eye, 6
cuts, 5
explosions, 5
eye protection, 7
fire, 4
hazardous materials (list), 8-11
poisoning, 6
spills, 7
Safety shield, 5
Salting out, 112, 187
Sample preparation:
IR, 203-208
NMR, 234
UV, 215
Scratching, to induce crystallization, 48
Second crops, 59
Seed crystals, 48
Semicarbazones (derivatives):
melting points, 514, 523
preparation, 273
Separations, 321
Separatory funnel, 17, 109
Sex hormones, 476
Shielding (NMR), 218
Silver nitrate test (alkyl halides), 261
Snyder column, 73
Sodium borohydride reduction:
of ketones, 369
of vanillin, 371
Sodium dichromate oxidation of cyclohexanol, 367
Sodium fusion test (sulfur, nitrogen, and halogen), 247
Sodium hydroxide test, 262
Sodium hypochlorite oxidation of cyclohexanol, 368

Sodium iodide test (alkyl halides), 263
Solids, addition to reaction mixture, 305
Solubility, 183
  and boiling point, 186
  classification by, 189
  in concentrated sulfuric acid, 192
  in diethyl ether, 191
  in 5% hydrochloric acid, 192
  of liquids in liquids, 184
  and melting point, 189
  and molecular structure, 189
  partial, 185
  in 85% phosphoric acid, 193
  in 5% sodium bicarbonate, 192
  in 5% sodium hydroxide, 191
  of solids in liquids, 188
  techniques for determination, 193
  temperature dependence, 188
  in water, 190
Solutions:
  of acids, 37
  of bases, 38
Solvent pairs, 55
Solvents:
  eluting (table), 122
  flammable, 4, 8
  halogenated, 11
  for IR spectrometry, 204
  for NMR spectrometry, 234
  properties of (table), 49
  for UV-visible spectrometry, 216
Specific gravity, 173
Specific gravity bottle, 174
Specific rotation, 180
Spectra, collections of, 35
Spectrometry:
  infrared (IR), 195
  mass (MS), 238
  nuclear magnetic resonance (NMR), 218
  ultraviolet-visible (UV), 211
Spills, 7
Splitting (NMR), 221
Steam bath, 300
Steam distillation, 95
  of aniline, 411
  apparatus, 98, 99, 332
  of 1-bromo-3-chloro-5-iodobenzene, 507
  of cloves, 331
  exercises, 101
  of grapefruit peel, 332
  of orange peel, 332
  technique, 98
  theory, 95
Steroids, 476
  transformations, 480
Stirring, 302
  with magnetic stirrer, 303
  with mechanical stirrer, 303
  by swirling, 302
Sublimation, 101
  of benzoic acid, 105
  of caffeine, 105
  of camphor, 105
  exercises, 105
  of hexachloroethane, 105
  technique, 104
  theory, 101
Substituted acetamides (derivatives):
  melting points, 515
  preparation, 277
4′-Substituted 2,4-dinitrodiphenylanilines, 428
Suction filtration, 54
Sulfide ion (test), 249
Sulfur (test), 249
Supercooling, 164
Superheating, 81
Supersaturation, 47
Swirling, 302

Table, periodic, 39
Temperature control, 294
  by boiling point of solvent, 296
  by rate of addition of reagent, 296
  by scale of the reaction, 297
Thermal conductivity detector, 134
Thermochromic compounds, 452
  dianthraquinone, 454
  dixanthylene, 453
Thermometer calibration, 164
Thermometer correction, 153, 167
Thin-layer chromatography:
  of aspirin, phenacetin, and caffeine, 130
  exercises, 130
  of leaf pigments, 130
  technique, 128
  theory, 127
Tollens test (aldehydes), 264
p-Toluenesulfonamides (derivatives):
  melting points, 515
  preparation, 278
p-Toluenesulfonates of tertiary amines (derivatives):
  melting points, 516
  preparation, 281
p-Toluidides (derivatives):
  from alkyl halides, 290
  from carboxylic acid anhydrides, 282
  from carboxylic acid halides, 282
  from carboxylic acids, 282
  melting points, 517
Transmittance, 196
2,4,7-Trinitrofluorenone adducts (derivatives):
  melting points, 522
  preparation, 290
Tube, drying, 19
Tube, NMR, 234

Ultraviolet-visible spectrometry (*see* UV-visible spectra)
Unsaturation, tests for:
  bromine in carbon tetrachloride, 255
  permanganate, 260
UV-visible spectra, 211 (*see also specific substances*)
  color and molecular structure, 214
  correlations for dienes (table), 212
  correlations for α,β-unsaturated ketones (table), 213
  interpretation, 213
  mechanism of energy absorption, 211
  and molecular structure, 211
  sample preparation, 215

Vacuum desiccator, 144
Vacuum distillation (*see* Reduced-pressure distillation)
Vacuum filtration, 44
Vapor-phase chromatography (VPC):
  exercises, 139
  preparative, 136
  qualitative analysis, 136
  quantitative analysis, 135
  stationary phases (table), 133
  technique, 137
  theory, 133
Vapor pressure, 63
Vapor pressure-composition diagram:
  benzene/methanol, 76

Vapor pressure-composition diagram: (*cont.*)
  benzene/toluene, 68
  chloroform/acetone, 77
  ideal system *A*/*B*, 67
  water/methanol, 75
Variations:
  addition of dichlorocarbene to 1,5-cyclooctadiene, 365
  addition of dichlorocarbene to styrene, 365
  conversion of *tert*-amyl alcohol to *tert*-amyl chloride, 375
  conversion of isoamyl alcohol to isoamyl bromide, 373
  conversion of *o*-toluidine to *o*-chlorotoluene, 434
  conversion of *p*-toluidine to *p*-chlorotoluene, 433
  dehydration of 2-methylcyclohexanol, 356
  preparation of *trans,trans*-1, 4-diphenylbutadiene, 451
  preparation of 4-methyl-3-heptanol, 406
  preparation of 4-methyl-3-heptanone, 406
  preparation of triphenylmethanol from benzophenone, 413
  preparation of triphenylmethanol from diethyl carbonate, 414
  preparation of triphenylmethanol from dimethyl carbonate, 414
Vesicants, 8
Vigreux column, 73
VPC (*see* Vapor-phase chromatography)

Washing:
  crystals, 54
  glassware, 20
  solutions, 108, 313-315
Waste disposal, 21
Water:
  density (table), 174
  in IR samples, 204
  removal of, 142
Widmer column, 73
Wittig reactions, 447
  preparation of *trans,trans*-1, 4-diphenylbutadiene, 451
  preparation of *trans*-stilbene, 448
Working up the reaction, 311–315

Solutions of Acids

Solution	Density (grams/ml)	Concentration (moles/liter)*	To make a liter of solution
95–98% H_2SO_4	1.84	18.1	Concentrated sulfuric acid
6 *M* H_2SO_4	1.34	6.0	332 ml conc. H_2SO_4 + 729 ml H_2O
3 *M* H_2SO_4	1.18	3.0	166 ml conc. H_2SO_4 + 875 ml H_2O
10% H_2SO_4	1.07	1.09	60 ml conc. H_2SO_4 + 957 ml H_2O
1 *M* H_2SO_4	1.06	1.0	55 ml conc. H_2SO_4 + 962 ml H_2O
69–71% HNO_3	1.42	15.7	Concentrated nitric acid
6*M* HNO_3	1.19	6.0	382 ml conc. HNO_3 + 648 ml H_2O
3 *M* HNO_3	1.10	3.0	191 ml conc. HNO_3 + 831 ml H_2O
10% HNO_3	1.06	1.67	106 ml conc. HNO_3 + 905 ml H_2O
1 *M* HNO_3	1.03	1.0	64 ml conc. HNO_3 + 943 ml H_2O
36.6–38% HCl	1.18	12.0	Concentrated hydrochloric acid
6 *M* HCl	1.10	6.0	500 ml conc. HCl + 510 ml H_2O
3 *M* HCl	1.05	3.0	250 ml conc. HCl + 756 ml H_2O
10% HCl	1.05	2.9	242 ml conc. HCl + 763 ml H_2O
1 *M* HCl	1.02	1.0	83 ml conc. HCl + 920 ml H_2O
99.7% CH_3COOH	1.05	17.5	Glacial acetic acid
6 *M* CH_3COOH	1.04	6.0	343 ml glacial acetic acid + 682 ml H_2O
3 *M* CH_3COOH	1.03	3.0	171 ml glacial acetic acid + 845 ml H_2O
10% CH_3COOH	1.01	1.69	97 ml glacial acetic acid + 912 ml H_2O
1 *M* CH_3COOH	1.01	1.0	57 ml glacial acetic acid + 949 ml H_2O
48% HBr	1.50	8.9	Constant-boiling HBr
57% HI	1.70	7.6	Constant-boiling HI
85% H_3PO_4	1.70	14.7	Syrupy phosphoric acid
70% $HClO_4$	1.67	11.7	Concentrated perchloric acid

* Also mmole/ml.